Lecture Notes in Mathematics 1574

Editors:
A. Dold, Heidelberg
B. Eckmann, Zürich
F. Takens, Groningen

V. P. Havin   N. K. Nikolski (Eds.)

Linear and Complex Analysis Problem Book 3

Part II

Springer-Verlag

Berlin Heidelberg New York
London Paris Tokyo
Hong Kong Barcelona
Budapest

Editors

Victor P. Havin
Department of Mathematics and Mechanics
St. Petersburg State University
Staryi Peterhof
St. Petersburg, 198904, Russia

Nikolai K. Nikolski
UFR de Mathématiques
Université Bordeaux-I
351, cours de la Libération
33405 Talence CEDEX, France

Mathematics Subject Classification (1991): 30B, 30C, 30D, 30E, 30H, 31, 32, 46B, 46D, 46H, 46J, 46K, 46L, 42A, 42B, 45, 47A, 47B, 93B

ISBN 3-540-57871-4 Springer-Verlag Berlin Heidelberg New York
ISBN 0-387-57871-4 Springer-Verlag New York Berlin Heidelberg

CIP-data applied for

© Springer-Verlag Berlin Heidelberg 1994
Printed in Germany

SPIN: 10078835   46/3140-543210 - Printed on acid-free paper

CONTENTS

VOLUME 1

Chapter 1. Banach spaces (ed. by S. Kisliakov) 1

Chapter 2. Banach algebras (ed. by H. Dales and A. Helemskii) 51

CONTENTS

Chapter 10. Singular integrals, BMO, H^p

VOLUME 2

Chapter 11. Spectral analysis and synthesis (ed. by N. Nikolski) 1

Chapter 12. Approximations and capacities
(ed. by J. Brennan, A. Volberg, and V. Havin) 73

CONTENTS

CONTENTS

CONTENTS

CONTENTS

PREFACE

In 1978 we published a book entitled "99 unsolved problems of Linear and Complex Analysis" (Volume 81 of "Zapiski nauchnyh seminarcv LOMI"; English translation in Journal of Soviet Mathematics, **26** (1984), No. 5). It consisted of short problem articles sent by mathematicians of many countries in response to our invitation headed by the following lines:

'Which problems of Linear and Complex Analysis would you propose to your numerous colleagues if you had a possibility to address them all simultaneously?
The editorial board of "Investigations in Linear Operators and Function Theory" edited by the Leningrad Branch of the V. A. Steklov Mathematical Institute of the Academy of Sciences of the USSR (LOMI) has decided to put this question to a hundred specialists joined in an invisible collective working on the common circle of problems and to publish their answers as "Collection of unsolved Problems of Linear and Complex Analysis". Such "Collection ..." may be useful not only to its authors but to their colleagues including the beginner analysts.'

It seems we were right. In 1984 the second edition appeared. Instead of 99, its title mentioned 199 problems.* Both editions have interested many colleagues. A big part of problems is now solved, but time has brought with it more new problems and questions. That is why one more (yet again enlarged) publication of the Collection seemed desirable. Its third version reproduces a large part of the second with addition of new problems and of information concerning the old ones. Our motives, the style and the general direction of the book were described in detail in the preface to the 2nd edition. Excerpts from that preface are reproduced below. There is not much to add. We only make several remarks on some *new* moments.

The first is the increase of the size. Instead of 13, the number of chapters is now 20; the total number of problems is 341 (they were 199 in 1983). The book consists of two volumes (both preceding editions were one-volume books). This growth can be explained by the abundance of new results and ideas in Spectral Operator-and-Function Theory. Our purely operator-theoretic chapters are now six (namely, Chapters 4–9); they were only two in the second edition. This fact bears witness to the intense activity of operator theorists gaining new areas and discovering new connections. So much for Operator Theory, a key subject underlying and unifying the whole book; the word "Linear" in the title refers mainly to this theme. As to the word "Complex", this part of the book is also enriched by the inclusion of new chapters 13, 18, 19, not to mention new problems gathered under the "old" titles.

The second moment is the new technique of preparation of the text. Both preceding versions of the book (as a whole) were prepared by its editors (though assisted by a collective of collaborators). This time every chapter had its own editor (or editors). The

---

Linear and complex Analysis Problem Book. 199 Research Problems. Lect. Notes Math. **1043**, Springer–Verlag, 1984

initiative and organization and coordination problems were ours, a difficult task, to say the least (see also the explanations in *Acknowledgements* below). Almost all chapters are provided with introductions by the chapter editors. In these introductions they try to help the reader to grasp the general direction of the chapter, to record additional bibliography, and sometimes also to explain their point of view on the subject or to make historical comments.

Chapters are divided into sections. They total 341 (in 1984 and 1978 there were 199 and 99 respectively). We treat the words "section" and "problem" as synonymous for the purposes of classification (though a section may contain more than one problem). "Problem 1.25" means the 25-th section of the first chapter; "Problem 1.26 old" ("Problem 1.26 v. old") mean that Problem 1.26 is reproduced from the 1984 edition (1978 edition, respectively) and has not been completely solved (as far as we know); "Problem S.1.27" means the 27-th section of Chapter 1 representing a solution of a problem from the previous edition. Some notation (used sometimes without further explanations) is indicated at the end of the book. A subject index and an author index are provided. We took the liberty to modify the section titles in "Contents" to make it shorter.

And **the third moment** in which this edition differs from its predecessors is the unfavorable situation in former Soviet mathematics caused by the well-known events that interfered brutally with our project just when it was started and could not be stopped. As we already mentioned, both preceding versions were prepared by "an informal editorial board" consisting for the most part of the members of the (then) Leningrad Seminar of the Spectral Function-and-Operator Theory. It was a numerous and energetic group of enthusiasts whose participation ensured the success of the undertaking. In 1990, proposing the project of the 3rd edition to our colleagues throughout the world, we hoped that we still could rely upon the same group. We also reckoned with the technical group of LOMI (now POMI), the Leningrad Branch of the Steklov Institute, remembering our experience of 1978 and 1983. But when our project was really started the situation changed dramatically. Our group melted away and soon became unable to achieve a joint effort, and POMI couldn't support us anymore (such things as, say, keyboarding, paper and so on, are now a big problem in Russia). In fact, the project turned out to be a purely private enterprise of the editors.

But now, after all, thanks to generous help of our friends and colleagues (see *Acknowledgements* below) this book lies before its reader. We hope that it will serve "the invisible community" of analysts working in Linear and Complex Analysis and will help them in solving and discovering many new and exciting problems.

FROM THE PREFACE TO THE PREVIOUS EDITION

This volume offers a collection of problems concerning analytic functions, linear function spaces and linear operators.

The most exciting challenge to a mathematician is usually not what he understands, but what still eludes him. This book reports what eluded a rather large group of analysts in 1983 whose interests have a large overlap with those of our Seminar.* Consequently,

---

*i.e., the Seminar on Spectral Theory and Complex Analysis consisting principally of mathematicians working in the Leningrad Branch of the V. A. Steklov Mathematical Institute (LOMI) and in Leningrad University.

therefore, the materials contained herein are chosen for some sort of mild homogeneity, and are not at all encyclopaedic. Thus, this volume differs markedly from some well-known publications which aim at universality. We confine ourselves to the (not very wide) area of Analysis in which we work, and try—within this framework—to make our collection as representative as possible. However, we confess to obeying the Bradford law (the exponential increase of difficulties in obtaining complete information). One of our purposes is to publish these problems promptly, before they lose the flavour of topicality or are solved by their proposers or other colleagues.

This Problem Book evolved from the earlier version published as volume 81 of "Zapiski Nauchnyh Seminarov LOMI" in 1978 (by the way, much of the work arising from the above mentioned Seminar is regularly published in this journal). It is now twice the size, reflecting the current interests of a far wider circle of mathematicians. For five years now the field of interests of the "invisible community" of analysts we belong to has enlarged and these interests have drifted towards a more intense mixing of Spectral Theory with Function Theory. And the volume as a whole is rather accurate reflection of this process.

We are pleased that almost a half of the problems recorded in the first edition, 50 of 99, have been solved, partly or completely. The problems of 1978 (we call them "old" problems) are sometimes accompanied with commentary reporting what progress towards their solution has come to our attention. Moreover, those "old" problems which have been almost completely solved are assembled under the title "SOLUTIONS" at the end of each chapter (including information as to how and by whom they have been solved).

When we decided to prepare this new edition, we solicited the cooperation of many colleagues throughout the world. Some two hundred responded with ample and helpful materials, doubling the number of collaborators of the first edition. Their contributions ranged from carefully composed articles (not always short) to brief remarks. This flow it was our task to organize and to compress into the confines of a single volume. To effectuate this we saw no alternative to making extensive revisions (more exactly, abbreviations) in the texts supplied. We hope that we have succeeded in preserving the essential features of all contributions and have done no injustice to any.

At first sight the problems may appear very heterogeneous. But they display a certain intrinsic unity, and their approximate classification (i.e. division into chapters) did not give us much trouble. We say "approximate" because every real manifestation of life resists systematization. Some problems did not fit into our initial outline and so some very interesting ones are collected under the title "Miscellaneous Problems" ...

EDITORS

PREFACE

Acknowledgements

The publication of these volumes would have been impossible without the generous and self-denying help of our colleagues. To explain this, we start by describing some of the obstacles we had to overcome.

Our instructions, sent to all chapter editors, were very thorough and detailed. They contained a lot of technical explanations, and TEX-macros* (prepared by A. V. SU-DAKOV). Unfortunately, they were largely ignored or neglected (except by those editors who worked in LOMI). We got a huge collection of texts in disorder; they required enormous work to coordinate and unify them (which could have been dispensed with almost completely if our instructions had been followed). We had to typeset hundreds of pages anew; many solved problems had to be detected and separated from the unsolved ones. Innumerable instances "to appear" from the preceding edition had to be replaced by correct bibliographical data (actually, some 1000 new references have been added!). Dozens of new commentaries had to be written (some chapter editors practically didn't revise "old" problems). And we had no technical staff necessary to turn a motley set of chapters into a book.

Meanwhile, as a result of the deteriorating situation and the decay of all structures in the former USSR (a malignant process whose rapidity we underestimated starting the project), the collective we could rely upon had disappeared and its members dispersed throughout the world. The e-mail became the only way of communication between them (including the authors of these lines). The situation looked desperate and the project could not have been rescued without assistance of our colleagues. This assistance was really invaluable to us. These skilled mathematicians in their most active years put their research aside and did a huge amount of purely technical work, making it possible for this book, to see the light of day.

The job has been done by three consecutive "technical teams". The first was headed by A. A. BORICHEV, the second by V. V. KAPUSTIN, the third by V. I. VASYUNIN, whose contribution to the project was especially great. The teams spent much time and energy retyping the text, tidying it up, hunting out inconsistencies and omissions to make the book a well-organized and handy tool for the user. We are not sure the debugging process has been completed and all defects have been discovered and removed: we apologize to the reader for remaining flaws. Our possible excuse is the fact that collectives capable of doing gratuitous work of such proportions existed only in the USSR, the country where the project was conceived, but which disappeared just at the final (and the hardest) stage of editing.

We are happy to thank the Mathematical Department of the University Bordeaux-I for its financial support. Its Graduate School (then headed by J.-L. JOLY and P. FAB-RIE) made possible the work of V. Vasyunin, putting at his disposal all necessary technical facilities.

We hope that this introduction is sufficient to explain why our gratitude to all who contributed to the book is especially deep and sincere. It is our duty and pleasure to name the following colleagues.

---

*This book was typeset using $\mathcal{AMS}$-TEX macro package.

General Technical Directors:

> Aleksander BORICHEV
> Vladimir KAPUSTIN
> Vasily VASYUNIN

Proof-reading and English Editing:

> Serguei KISLIAKOV

Proof-reading:

> Maria GAMAL

Keyboarding Advisor:

> Andrei SUDAKOV

Keyboarding:

> Cathy ANTONOVSKAYA
> Yuri YAKUBOVICH

Checking References and Addresses:

> Evgueni ABAKUMOV

Indices Compilers:

> Maria GAMAL
> Andrei GROMOV

Technical Advice and Various Help:

> Ludmila DOVBYSH
> Dmitri YAKUBOVICH

Episodical Advice:

> William BADE
> Philip CURTIS
> Anton SERGEYEV
> Rouslan SIBILEV
> Elisabeth STROUSE
> Yuri VYMENETS

We are indebted very much to all of them as well as to the Mathematical Editorial Board of Springer–Verlag for patience and support of our project.

EDITORS

LIST OF PARTICIPANTS

Adams D. R. 12.31
Adamyan V. M. 4.2
Ahern P. 10.19
Aizenberg L. A. ch.17, 1.18, 17.4
Aleksandrov A. B. 10.10, S.10.25, 11.15
Alexander H. 17.12
Anderson J. 3.7, S.10.24
Arov D. Z. 4.2, 4.3, 4.7
van Assche W. 13.6
Atzmon A. 5.4
Axler S. S.7.22
Azarin V. 16.8, 16.9
Azizov T. Ya. 5.9, 20.3
Bade W. 2.8
Baernstein A. 10.15, 18.10, 18.16
Bagby T. 12.14
Belitskiĭ G. 16.13
Belyi V. I. 12.7, 18.9
Ben-Artzi M. 6.7, 6.8
Berenstein C. A. 11.8, 11.9, 17.15, 17.16
Berg Ch. 12.5
Bielefeld B. ch.19, 19.1
Birman M. Sh. ch.6, 5.15, 6.5, 6.6
Bishop C. J. 18.1
Boivin A. 12.8
Bollobás B. S.2.33
Böttcher A. 7.18
Bourgain J. 1.2
de Branges L. 2.31, 6.1, 14.10
Brennan J. E. ch.12, ch.14, 12.9, 12.10
Brown G. 2.22
Brudnyi Yu. A. 15.8
Bruna J. 11.26, 15.5, 15.7
Carleson L. 19.5
Casazza P. G. 10.21
Chang S.-Y. A. 10.17, 10.18
Clark D. N. 5.11
Coburn L. 7.14
Conway J. B. ch.8, 8.7
Curtis P. C. Jr. 2.3
Dales H. G. ch.2, 2.6, 2.7
Davidson K. R. 5.3
Davis Ch. S.5.18
Devaney R. 19.9
Devinatz A. 6.7, 6.8, S.14.20
Dijksma A. 4.8
Djrbashyan M. M. 14.1
Domar Y. 11.17

Douglas R. 7.8
Duren P. L. ch.18, 18.11, 18.15
Dyakonov K. M. 10.23
Dym H. 12.4
Dyn'kin E. M. ch.10, 11.19, 14.6
Ecalle J. 14.9
Eiermann M. 13.1
Eremenko A. E. 16.8, 16.15, S.16.19, S.16.20, 19.10
Esterle J. 17.18
Faddeev L. D. 4.5, S.6.11
Fan Q. 7.6
Fel'dman I. A. 2.20, 5.14
Forelli F. 10.22, S.11.27, 17.13
Frankfurt R. 11.12
Fritzsche B. 4.7
Fuglede B. V. 16.15
Gamelin T. W. 2.14
Gapcshkin V. F. 3.4
Garnett J. 10.14
Gauthier P. 12.8
Gay R. 17.15, 17.16, 17.17
Ginzburg Yu. 4.4
Gohberg I. ch.4, 4.9
Goldberg A. A. ch.16, 16.2, 16.6, 16.7, S.16.19, S.16.20
Gonchar A. A. 12.17
Goodman A. W. 18.5
Gorin E. A. 2.9, 2.12, 20.4
Gorkin P. 2.17
Grimmett G. 14.14
Grishin A. 16.8
Grinshpan A. Z. 18.7
Gulisashvili A. B. 20.7
Gubreev G. 15.4
Guivarc'h Y. 3.8
Gurariĭ V. P. 11.16, 11.18
Haslinger F. 1.17
Hasumi M. 10.20
Havin V. P. ch.12, ch.14, S.10.25, 14.3, 14.4
Havinson S. Ya. 16.14
Hayman W. 12.24, 16.16
Hedenmalm H. 12.12, 12.13, 17.11
Hedberg L. I. S.12.33
Heinonen J. 18.14
Helemskiĭ A. Ya. ch.2, 2.1
Helson H. 11.23
Henkin G. 12.22, 17.1

LIST OF PARTICIPANTS

Herrero D. A. 5.6
Holbrook J. A. R. 5.1
Hruščëv S. V. 3.3, 14.3, 14.12
Ibragimov I. A. 3.2
Igari S. 2.21
Iohvidov I. S. 5.9
Iserles A. 13.2
Ivanov L. D. 12.26
Ivanov O. V. 2.24
Janas J. 7.15
Janson S. 7.3
Jarnicki M. 17.3
Jones P. W. 1.6, 10.2, 10.9, 12.2, 12.32
Jöricke B. 14.4, 17.6
Kaashoek M. A. ch.4, 4.9
Kadec M. I. 16.11
Kahane J.-P. ch.3, 3.9, 11.20
Kapustin V. V. 9.3
Kargaev P. P. 10.10
Karlovich Yu. 7.12
Kaufman R. 14.7, 16.18
Kérchy L. 9.4
Khavinson D. 12.20
Khurumov Yu. V. 17.5
Kirstein B. 4.7
Kisliakov S. V. ch.1, ch.10, 10.12
Kitover A. K. 2.9, 5.13
Komarchev I. A. 1.10
Koosis P. 14.5
Korenblum B. I. 11.13, 18.2
Král J. 10.3, 12.27, 16.17
Krasichkov-Ternovskiĭ I. F. 11.4, 11.6
Krein M. G. 4.2, 7.16, 15.11
Krein S. G. 20.2
Kriete T. 12.3, 14.2
Krupnik N. Ya. 2.20, 7.11, 10.6
Krushkal S. L. 17.14
Krzyż J. G. 10.4, 18.6
Kurina G. A. 20.2
Langer H. 4.8
Langley J. 16.12
Latushkin Yu. D. 7.20
Leiterer J. 4.6
Leont'ev A. F. 15.2
Levin B. ch.16, 11.22, 16.1, S.16.21, S.16.22
Lewis J. 14.16, 14.17
Lin V. Ya. 20.5, 20.6
Littlejohn L. L. 13.2
Litvinchuk G. S. 7.20
Lubinsky D. S. 13.3
Lykova Z. A. 2.2
Lyubarskiĭ Yu. I. 1.8
Lyubich M. Yu. ch.19, 19.7, 19.10
Lyubich Yu. I. S.5.19
MacCluer B. 14.2

Magnus A. P. 13.4
Makarov B. M. 1.10, 1.11
Makarov N. G. S.6.12, 8.11, 14.4
Manfredi J. 14.15
Markus A. 2.20, 5.14
Marshall D. 12.23
Matsaev V. I. 14.8
Maz'ya V. 12.30
McGuire P. J. 8.1
McKean H. P. 3.1, 6.3
McMullen C. 19.2
Meise R. 11.7
Mel'nikov M. S. 12.25
Méril A. 17.17
Milin I. M. 18.13
Milman V. 1.1
Milnor J. 19.3, 19.6
Momm S. 11.7
Moran W. 2.22
Mortini R. 2.26
Muckenhoupt B. 10.5
Müller P. 1.4, 10.11
Murai T. 12.28
Naboko S. N. 9.2
Napalkov V. 14.13
Nevai P. ch.13
Nikolski N. K. ch.9, ch.11, ch.15, 8.11, 9.1, S.9.5,
        11.11, 15.3
Novikov R. G. 17.1
O'Farrell A. 12.15
Olin R. F. 3.2
Ostrovskii I. V. ch.16, 16.2–16.7, S.16.20, S.16.21
Ovcharenko I. 15.12
Palamodov V. 11.2
Pavlov B. S. S.6.11, S.9.5
Pedersen H. 12.5
Peetre J. ch.7
Pełczyński A. 1.9
Peller V. V. 3.3, 5.12, 7.2, 7.7
Peng L. 7.6
Perez-Gonzalez F. 12.11
Pflug P. 17.3
Power S. C. 7.1
Prössdorf S. 7.19
Przytycki F. 19.8
Pták V. 2.28
Putinar M. 4.10, 8.5, 8.6
Putnam C. R. 8.3, 8.9, 8.10
Radjavi H. 5.2
Ransford T. J. 12.21
Rees M. 19.4
Reshetihin N. Yu. 4.5
Rochberg R. 2.19, 7.4, 10.8
Rodman L. 4.9, 5.5
Ronkin L. I. 17.9

LIST OF PARTICIPANTS

Rosenthal P. ch.5
Rovnyak J. 18.8
Rubel L. A. 12.19, 16.12, 17.2
Sakhnovich L. A. 6.9, 7.13, 15.13
Saitoh S. 1.7, 18.4
Samokhin M. 2.25
Sarason D. 2.16, 7.9, 10.16
Semënov E. M. 1.5, 1.12, 1.13
Semënov-Tian-Shansky M. A. 7.21
Semiguk O. S. 1.15
Semmes S. 10.7
Sergeev A. G. 17.7, 17.8
Shamoyan F. A. 11.24
Shields A. L. 11.14
Shirokov N. A. 15.6
Shishkin S. 20.1
Shul'man V. S. 2.13
Shwartsman P. A. 15.8
Siddiqi J. A. 14.11
Silbermann B. 7.18, S.7.23
Simon B. 5.17
Sjögren P. 10.13
Skiba N. I. 1.15
de Snoo H. S. V. 4.8
Sodin M. L. 16.4, 16.5, 16.10, 16.15
Solev V. N. 3.2
Solomyak M. Z. 5.15, 5.16
Spitkovskiĭ I. M. 5.5, 7.12, 7.16
Stahl H. B. 13.1
Stephenson K. 18.12
Stray A. 12.6
Struppa D. C. 11.8
Sudakov V. N. 3.6
Sundberg C. 7.10
Sutherland S. 19.11
Szökefalvi-Nagy B. S.4.11
Tarkhanov N. 12.1
Taylor B. 15.1
Teodorescu R. S.9.6
Thomas M. P. 11.21
Tkachenko V. A. 11.5, 16.13

Tolokonnikov V. A. 2.27
Totik V. 13.5
Trutnev V. M. 1.19, 11.3
Tumarkin G. C. 10.1
Vasyunin V. I. ch.9, S.9.5, S.9.6
Verbitskiĭ I. È. 7.11, 10.6
Verdera J. 12.16
Vershik A. M. 3.5, 5.10
Villamor E. 14.15
Vinogradov S. A. 15.9
Vitushkin A. G. 12.25
Vladimirov V. 7.17
Voiculescu D. 2.18, 6.2, 8.8
Volberg A. L. ch.12, ch.14, S.14.19
Volovich I. V. 7.17
Waelbroeck L. 11.1
Wallin H. 12.18
Wermer J. 2.15, 2.32
Widom H. 6.4
Williams D. 11.25
Willis G. 2.5
Wodzicki M. 1.14, 2.4
Wojtaszczyk P. 1.3
Wolff T. 11.10, 14.18
Wolniewicz T. M. 1.4
Wu J.-M. 18.14
Wu P. Y. 5.7
Xia D. 8.4
Yafaev D. R. 6.10
Yakubovich D. V. 5.8
Yakubovich V. A. 20.1
Yger A. 11.9, 17.15–17.17
Young N. J. 2.29, 4.1
Zafran M. 15.10
Zaidenberg M. G. 20.6
Zakhariuta V. P. 1.15, 1.16
Żelazko W. 2.11, 2.23
Zemánek J. 2.10, 2.30, 12.29
Zhu K. 7.5, 18.3
Znamenskiĭ S. V. 17.10

Chapter 11

SPECTRAL ANALYSIS AND SYNTHESIS

Edited by

N. K. Nikolski

Université Bordeaux-I
UFR Mathématiques
351, cours de la Libération
33405 Talence CEDEX
France

and

Steklov Mathematical Institute
St. Petersburg Branch
Fontanka 27
St. Petersburg, 191011
Russia

INTRODUCTION

Problems of Spectral Analysis–Synthesis arose long before they were stated in a precise form. They stimulated, e. g. , the development of Linear Algebra ("The Fundamental Theorem of Algebra", Jordan Theory) and of basic ideas of Fourier Analysis. The success and the universal character of the last are the reasons why the present theme was confined for a very long time to the sphere of influence of Harmonic Analysis. The well developed theory of trigonometric series and integrals, group representations, Abstract Harmonic Analysis — all these disciplines are directed to the same twofold problem: what are "the elementary harmonics" of an object (= a function, an operator, ...) which is undergoing the action of a semigroup of transformations; what are the ways of reconstructing the object, once its spectrum, i. e. the intensity of every harmonic, is known? Another apparently different, but essentially identical aspect stimulating the development of the theme has roots in Differential Equations. The ritual of writing down the general solution of $p(\frac{d}{dx})f = 0$ using the zeros of the symbol p generated numerous investigations of differential-difference and more general convolution operators. The results always reflect the same routine: the general solution is the limit of linear combinations of elementary solutions $z^k e^{\lambda z}$ corresponding to the zeros λ of the symbol (Ritt, Valiron, Delsarte). It was L. Schwartz who formulated the circle of ideas in its real meaning and appropriate generality (in his classical paper in *Ann. Math.*, 48(1947), No. 4, 857–927). Now the Problem of Spectral Analysis — Synthesis can be stated as follows: given a linear topological space X and a semi-group τ of its endomorphisms, — describe τ-invariant closed subspaces, containing non-trivial τ-invariant finite-dimensional parts ("Analysis"), and then
— describe subspaces spanned topologically by the above parts ("Synthesis").

If τ has a single generator then our problem actually deals with eigen- and root-subspaces of the generator and with the subsequent recovery of all its invariant subspaces via the "elementary" ones. Systems of differential and general convolution equations lead to finitely-generated τ-invariant subspaces, τ being the corresponding group (or semi-group) of translations (in $\mathbb{R}^n, \mathbb{C}^n, \mathbb{T}^n$ etc). Annihilators of such subspaces become (via Fourier transform) modules over the ring of trigonometrical (resp. "analytic" trigonometrical) polynomials; the Analysis–Synthesis Problem converts into the well-known problem of "localization of ideals". Roughly speaking the principal role is played in this context by the concept of the divisor of an analytic function, and the Problem reduces to the description of divisorial ideals (or submodules). After this reduction is accomplished, we may forget the origin of our problem and confine ourselves exclusively to Function Theory. Namely, we are led to one of its key questions, the interplay of local and global properties of analytic functions. Thus, starting with Analysis–Synthesis, we come to the multiplicative structure of analytic functions (Weierstrass products and their generalizations), the factorization theory of Nevanlinna–Smirnov, uniqueness theorems characterizing non-trivial divisors, free interpolation problems and to many other accouterments of Complex Analysis.

The problems of this chapter treat the above ideas in various ways. Localization of ideals (submodules) in spaces of analytic functions determined by growth conditions

2

is discussed in Problems 11.1–11.9, and in more special spaces in 11.10–11.15. These Problems overlap essentially. We add to the references given in the text of Problems books of L. Schwartz (*Théorie des distributions*, Paris Hermann, 1966), L. Ehrenpreis (*Fourier analysis in several complex variables*, N. Y., 1970) and J. -P. Ferrier (*Spectral Theory and Complex Analysis*, N. Y., 1973) (see also the bibliography survey [1] cited in Problem 11.11). Analyzing spaces of holomorphic functions defined by a family of majorants requires a study of the intrinsic properties of majorants (see, e.g., Problem 16.14, §7.3 of Ferrier's book and B. Ya. Levin, *Completeness of systems of functions, quasi-analyticity and subharmonic majorants*, Zap. Nauchn. Semin. LOMI, 170(1989), 102–156 (Russian); English transl. in J. Soviet Math. 63(1993), no. 2).

Problem 11.10 deals with an interesting question concerning finitely generated (algebraically) ideals in H^∞, a generalization of the Corona Problem. Problem 11.9 treats the same subject for "Hörmander algebras" in multidimensional settings. Item 11.8 contains a polynomial version of the problem.

The more "rigid" is the topology of a space, the more profound is the concept of divisor (and, as a rule, the more difficult it is to prove that z-invariant subspaces are divisorial). The series of Problems 11.2–11.6 (and then 11.24–11.26 for functions smooth up to the boundary) is very instructive in this respect. Another feature they have in common is that they aim at the well-known "secondary" approximation problem of Analysis–Synthesis: to prove or to disprove that any subspace with a trivial divisor is dense (cf. Wiener's Tauberian Theorem). This problem s implicit in arguments of items 11.11–11.23 concerning weakly invertible (cyclic) functions in corresponding spaces.

Free interpolation and unconditional bases methods became important tools in studies of closed ideals and convolution equations (in particular, partial differential equation). Problems 11.3, 11.7–11.9 touch this direction; for more information see references in the problems mentioned.

Classical Harmonic Analysis has led to very delicate and difficult theorems in Spectral Synthesis and to a vast variety of problems — from numerous generalizations of periodicity (which corresponds to the simplest convolution equation $\delta_0 - \mathcal{E}_a * f = 0$) to the theory of resolvent sets of Malliavin–Varopoulos. This direction is represented by Problems 11.16–11.23.

Other problems related to Analysis–Synthesis are 9.1, 10.16, S.10.24, 11.23, 12.9, S.12.33, 14.1, 14.3, 14.13, 15.1, 15.2, 15.9, 15.10. The references in 11.11 contain several items concerning the localization of ideals (for $n = 1$ in the spirit of 11.2. Many problems in 11.12 are discussed in the book [3] cited in 11.11.

ABOUT HOLOMORPHIC FUNCTIONS
WITH LIMITED GROWTH

L. Waelbroeck

Can one develop a theory of holomorphic functions satisfying growth conditions analogous to the theory of holomorphic functions on Stein manifolds?

Let δ be a continuous non-negative function on $\mathbb{C}^n$ which tends to zero at infinity; $\mathcal{O}(\delta)$ will be the set of all holomorphic functions u on the set $S_\delta : \delta > 0$ such that $\delta^N u$ is bounded for N large enough.

Research about the holomorphic functional calculus [1] led the author to the consideration of the algebras $\mathcal{O}(\delta)$. The only relevant algebras however were the algebras $\mathcal{O}(\delta)$ where δ is Lipschitz and $|s|\delta(s)$ is bounded.

L. Hörmander [2–4] has obtained results concerning algebras that he called $A(\varphi)$, but $A(\varphi) = \mathcal{O}(e^{-\varphi})$. His proofs used assumptions about φ which imply (up to equivalence) that $e^{-\varphi}$ is Lipschitz and $|s|e^{-\varphi(s)}$ is a bounded function of s.

He also assumed that φ, i.e. $-\log\delta$, is a plurisubharmonic function. This is an expected hypothesis, it means that $\mathcal{O}(\delta)$ behaves like the algebra of holomorphic functions on a domain of holomorphy. From the point of view of the holomorphic functional calculus, the condition "$-\log\delta$ is p.s.h." is also significant, as I. Cnop [5] showed (using Hörmander's results).

The reason why L. Hörmander and the author looked more specially at the algebras $\mathcal{O}(\delta)$, δ Lipschitz, were quite different. For Hörmander it appears that better estimates can be obtained when δ is Lipschitz. For the author, the only algebras relevant to a significant application of the theory were the algebras $\mathcal{O}(\delta)$, with δ Lipschitz. This coincidence suggests that the Lipschitz property is an important property δ has to possess if we want $\mathcal{O}(\delta)$ to behave somewhat like holomorphic functions on an open set.

Unfortunately, it is not clear what should take place of this Lipschitz property when we investigate holomorphic functions on manifolds. The Lipschitz property is expressed in global coordinates. Manifolds only have local coordinate systems. An auxiliary Riemann or Kähler metric could be defined on the manifold. Or one may notice that $\mathcal{O}(\delta)$ is nuclear when δ is Lipschitz.

The plurisubharmonicity of $-\log\delta$ involves the structure of the complex manifold only. It generalizes the holomorphic convexity of Stein manifolds.

L. Hörmander has proved an analogue of Cartan's theorem B for holomorphic functions satisfying growth conditions. The full force of the Oka–Cartan theory of ideals and modules of holomorphic functions does not follow, until an analogue of Cartan's theorem on invertible matrices has been proved, with bounds, and bounds have been inserted in Oka's theorem on the coherence of the sheaf of relations.

We shall call $B(s, \varepsilon\delta(s))$ the open ball with center s and radius $\varepsilon\delta(s)$, and shall assume that ε is small. This ensures that $B(s, \varepsilon\delta(s)) \subset S_\delta$, also that $\delta(z)/\delta(s)$ is bounded from above and bounded away from below when $z \in B(s, \varepsilon\delta(s))$, and that

4

$\delta(s)/\delta(t)$ is bounded from above and bounded away from below when $B\big(s,\varepsilon\delta(s)\big)$ and $B\big(t,\varepsilon\delta(t)\big)$ have a non-empty intersection (because δ is lipschitzian).

The following results should be a part of the theory.

CONJECTURE. Let $f_1,\ldots,f_k$ be elements of $\mathcal{O}(\delta)^q$. Let $g \in \mathcal{O}(\delta)^q$ be such that $u_{1,s},\ldots,u_{k,s}$ can be found for each s, holomorphic on $B\big(s,\varepsilon\delta(s)\big)$, with

$$g = \sum_{1\leqslant i\leqslant k} u_{i,s}f_i$$

on $B\big(s,\varepsilon\delta(s)\big)$, and $|u_{i,s}(z)| < M\delta(s)^{-N}$ for some M, $M \in \mathbb{R}_+$, and N, $N \in \mathbb{N}$. Then g is in the submodule of $\mathcal{O}(\delta)^q$ generated by $f_1,\ldots,f_k$.

CONJECTURE. With the same conventions, assume that g_s is given for each s, $s \in S_\delta$, such that $\|g_s(z)\| < M\delta(s)^{-N}$, when $z \in B\big(s,\varepsilon\delta(s)\big)$, if M and N are large enough. Assume also that

$$g_s - g_t = \sum u_{i,st}f_i$$

on $B\big(s,\varepsilon\delta(s)\big)\bigcap B\big(t,\varepsilon\delta(t)\big)$ with $g_s - g_t$ holomorphic on this open set, and less than $M\delta(s)^{-N}$. Then it is possible to find g, M', N', ε' such that $g \in \mathcal{O}(\delta)$, and

$$g - g_s = \sum V_{i,s}f_i$$

on $B\big(s,\varepsilon'\delta(s)\big)$, with $V_{i,s}$ holomorphic on $B\big(s,\varepsilon'\delta(s)\big)$ and $|V_{i,s}(z)| < M'\delta(s)^{-N'}$ when $z \in B\big(s,\varepsilon'\delta(s)\big)$.

A local description of the submodules of $\mathcal{O}(\delta)^q$ would also be welcome. Let M be a submodule of $\mathcal{O}(\delta)^q$. Then, for each s, M generates a submodule M_s of $\mathcal{O}\big(B(s,\varepsilon\delta(s))\big)^q$. When $B\big(s,\varepsilon\delta(s)\big)$ and $B\big(t,\varepsilon\delta(t)\big)$ intersect, M_s and M_t generate the same submodule of $\mathcal{O}\big(B(s,\varepsilon\delta(s))\big)\bigcap B(t,\varepsilon\delta(t))\big)^q$. Is it possible to find conditions from functional analysis which ensure that a family of modules M_s, which agree in the manner described, would be generated by a submodule M of $\mathcal{O}(\delta)^q$? J.-P. Ferrier [6], [7] considers Runge's theorem in the above context. Assuming $\delta \geqslant \delta'$ to be two Lipschitz functions he shows that the set of limits in $\mathcal{O}(\delta')$ of restrictions of elements of $\mathcal{O}(\delta)$ is—or can be identified with—some $\mathcal{O}(\delta_1)$, and δ_1 has some analogy with an "$\mathcal{O}(\delta)$-convex hull" of δ'.

However the limits that Ferrier handles are bornological, not topological. Ferrier cannot show that $\mathcal{O}(\delta_1)$ is a closed subspace of $\mathcal{C}(\delta)$. It might very well be that the limits of elements of $\mathcal{O}(\delta_1)$ would be elements of $\mathcal{O}(\delta_2)$ with $\delta_1 \geqslant \delta_2 \geqslant \delta'$, etc. This specific problem is therefore open. So is the generalization of Ferrier's results to algebras of holomorphic functions satisfying growth conditions on a Stein manifold ... once we know what is a good analogue to the condition "δ is Lipschitz".

The general problem described in this note is vaguer than the editors of the series wish. It intrigued the author eighteen years ago,* when [1] was published, but the author could not make any headway and went on to other things. Hörmander's breakthrough came later. The author has not taken the time to investigate all of the consequences of Hörmander's results. Results have been obtained by several authors, after Hörmander. They do not solve the problem as it is put. But they indicate that significant progress at the boundary of complex and functional analysis would follow from a good understanding of the question.

---

*Now, thirty-three years ago

REFERENCES

1. Waelbroeck L., *Étude spectrale des algèbres complètes*, Acad. Royale Belg. Mém. Cl. Sci. **2** (1960), 31.
2. Hörmander L., L^2-*estimates and existence theorems for the $\bar{\partial}$-operator*, Acta Math. **113** (1965), 85–152.
3. Hörmander L., *An Introduction to Complex Analysis in Several Variables*, Van Nostrand, New York, 1966.
4. Hörmander L., *Generators for some rings of analytic functions*, Bull.Amer.Math.Soc. **73** (1967), 943–949.
5. Cnop I., *Spectral study of holomorphic functions with bounded growth*, Ann.Inst.Fourier **22** (1972), 293–309.
6. Ferrier J.-P., *Approximation des fonctions holomorphes de plusiers variables avec croissance*, Ann. Inst.Fourier **22** (1972), 67–87.
7. Ferrier J.-P., *Spectral Theory and Complex Analysis*, North Holland Math. Stud., vol. 4, North Holland, Amsterdam, 1973.

UNIV. LIBRE DE BRUXELLES
DÉP. DE MATH.
CAMPUS PLAINE C.P.214
BRUXELLES
BELGIQUE

LOCALIZATION OF POLYNOMIAL SUBMODULES
IN SOME SPACES OF HOLOMORPHIC FUNCTIONS
AND SOLVABILITY OF THE $\bar{\partial}$-EQUATION

V. P. Palamodov

Let K be a compact in $\mathbb{R}^n$. Consider its support function

$$m_K(y) = \max\{ (y, \xi) \colon \xi \in K \}, \qquad y \in (\mathbb{R}^n)'.$$

For every positive integer q define a norm $\| \cdot \|_{q,K}$ on the space of complex-valued functions in $\mathbb{C}^n$ by

$$\|f\|_{q,K} = \sup\{ |f(z)|(|z| + 1)^q \exp(-m_K(y)) \colon z = x + iy \in \mathbb{C}^n \}.$$

Let S_K be the space of all entire functions f in $\mathbb{C}^n$ with $\|f\|_{q,K} < \infty$ for every q. This space can be considered as a module over the algebra $\mathbb{C}[z]$ of polynomials in $\mathbb{C}^n$ with respect to the pointwise multiplication. Therefore each ideal I of $\mathbb{C}[z]$ generates a submodule $I \cdot S_K$ of S_K.

DEFINITION. A submodule $I \cdot S_K$ is called *local* if it contains all functions $f \in S_K$ satisfying the following condition: for every $w \in \mathbb{C}^n$ the Taylor series of f in w

$$\sum \frac{f^{(j)}(w)}{j!}(z - w)^j, \qquad j = (j_1, \ldots, j_n), \ j! = j_1! \ldots j_n$$

belongs to the submodule $I \cdot T_w$, where T_w is the $\mathbb{C}[z]$-module of all formal power series in $z - w = (z_1 - w_1, \ldots, z_n - w_n)$.

CONJECTURE 1. *For any compact set K and for any ideal I in $\mathbb{C}[z]$ the submodule $I \cdot S_K$ is local.*

The Conjecture can be generalized to the case where the ideal of $\mathbb{C}[z]$ is replaced by an arbitrary submodule I of $\overset{l}{\oplus}\mathbb{C}[z]$ (the direct sum of l copies of $\mathbb{C}[z]$). This more general Conjecture is easily reduced to the case of the ideal I.

Since the support function of a compact set coincides with that of its convex hull, we can suppose K to be a convex compact set. In this case the space of the Fourier transforms of S_K coincides with the space $\mathcal{D}_K$ of infinitely differentiable functions in $\mathbb{R}^n$ supported on K. The validity of Conjecture 1 would lead, in view of this connection, to some interesting consequences in the theory of differential equations with constant coefficients. Let us mention one of them.

COROLLARY. *Let P be a $(t \times s)$ matrix of differential operators in $\mathbb{R}^n$ with constant coefficients. Then the system of equations $Pu = f$, $u = (u_1, \ldots, u_s)$ has a solution in the class $\overset{s}{\oplus}\mathcal{D}'_K$ of distributions on K for any $f \in \overset{t}{\oplus}\mathcal{D}'_K$ satisfying the formal compatibility condition (i.e. $Qf = 0$ for any matrix Q of operators with constant coefficients such that $QP = 0$).*

Conjecture 1 is induced by the following result.

THEOREM OF MALGRANGE AND PALAMODOV ([1], [2]). *Let Ω be a convex domain in $\mathbb{R}^n$, S_Ω be the union of S_K over all compact subsets K of Ω. Then for any ideal I of $\mathbb{C}[z]$ the submodule $I \cdot S_\Omega$ is local.*

The proof of this Theorem depends on the triviality of the Čech cohomologies for holomorphic cochains in $\mathbb{C}^n$ with an estimation of the growth at infinity or on the equivalent theorem on the solvability of the $\bar{\partial}$-equation in $\mathbb{C}^n$ with the estimation at infinity as well. To use this way for the proof of Conjecture 1 one needs the following assertion.

Let S_K^* be the space of $\bar{\partial}$-differential forms

$$f = \sum_{j_1, \ldots, j_s} f_{j_1, \ldots, j_s} \, d\bar{z}_{j_1} \wedge \cdots \wedge d\bar{z}_{j_s}$$

such that all derivatives $f_{j_1, \ldots, j_s}^{(l)}$ have finite norms $\| \cdot \|_{q,K}$ for every q.

CONJECTURE 2. *For every $K \subset \mathbb{R}^n$ and every α in S_K^* such that $\bar{\partial}\alpha = 0$ there exists β in S_K^* satisfying $\bar{\partial}\beta = \alpha$.*

In this Conjecture the essential point is not the local properties of the coefficients but their growth at infinity. We can assume them to be locally square summable or even to be distributions. The operator $\bar{\partial}$ being elliptic the complexes corresponding to the different local conditions are homotopic and therefore can satisfy Conjecture 2 only simultaneously.

The following result obtained for another purpose can be considered as an approach to Conjecture 2.

LEMMA ([3]). *Let G be a ball in $\mathbb{R}^n$ centered at the origin, G^+ be the intersection of G and a half-space of $\mathbb{R}^n$. Then for every q and for every $\bar{\partial}$-closed form $\alpha \in S_K^*$ there exists a $\bar{\partial}$-form β such that $\bar{\partial}\beta = \alpha$ and $\|\beta^{(l)}\|_{q, K+G^+} < \infty$ with $l < q$.*

The following result was obtained recently.

THEOREM (Dufresnoy [4]). *Conjecture 2 is valid for any convex compact set with C^2 boundary.*

The proof is based on a well-known Hörmander's theorem on solvability of the $\bar{\partial}$-equation. A non-trivial point is the choice of an appropriate weight e^α with plurisubharmonic α. It is here where the smoothness of boundary is used.

PROBLEM 11.2

REFERENCES

1. Malgrange B., *Sur les systèmes differentiel à coefficients constants* Coll. Int. CNRS, Paris, 1963.
2. Palamodov V. P., *Linear differential operators with constant coefficients*, "Nauka", Moscow, 1967. (Russian)
3. Palamodov V. P., *A complex of holomorphic waves*, Trudy Sem. Petrovsk. (1975), no. 1, 177–210. (Russian)
4. Dufresnoy A., *Un résultat de d''-cohomologie; applications aux systèmes differentiel à coefficients constants*, Ann. Inst. Fourier **27** (1977), no. 2, 125–143.
5. Hörmander L., *Linear Partial Differential Operators*, Springer-Verlag, Berlin–Göttingen–Heidelberg, 1963.

UL. 26 BAKINSKIH KOMISSAROV 3-1, 422
MOSCOW 117526
RUSSIA

INVARIANT SUBSPACES AND THE SOLVABILITY
OF DIFFERENTIAL EQUATIONS

V. M. Trutnev

1. Let Ω be a convex domain in $\mathbb{C}^n$ and let $H(\Omega)$ be the space of all functions analytic in Ω supplied with the natural topology. L. Schwartz posed and solved (for $\Omega = \mathbb{C}$) [1] the following

PROBLEM. *Does any closed subspace $W \subset H(\Omega)$ invariant under the operator of differentiation contain exponential monomials, and if it does then do such monomials span W?*

This problem is completely explored in [2] for $n = 1$. In case $n > 1$ the problem has not been solved so far even for principal subspaces such as

$$W = \{ f \in H(\Omega) \colon \int f(\xi + z)\, d\mu(\xi) = 0, \quad \mathrm{supp}(\mu) \subset \Omega \}$$

for example.

The positive answer to the question of L. Schwartz has been obtained only for special domains in $\mathbb{C}^n$, namely for $\Omega = \mathbb{C}^n$ [3], [4], for half-spaces in $\mathbb{C}^n$ [5]; for tube domains [6] and for domains in $\mathbb{C}^n$ satisfying $\Omega + \Omega \subset \Omega$ [7]. The proof in all listed cases, besides the tube domains, exploits essentially the fact that W is invariant under the translations. The condition $\Omega + \Omega \subset \Omega$ embraces a general class of domains with required invariance property. As in the one-dimensional case the proof of the following conjecture could be the key to the solution of the whole problem. Let $T \to \ <e^{(z,\xi)}, T_\xi>$ be the generalized Laplace transform and let E_Ω be the space of entire functions coinciding with the Laplace transform of continuous linear functionals on $H'(\Omega)$. The space E_Ω is endowed with the natural topology borrowed from $H'(\Omega)$.

CONJECTURE. *Given $\varphi, \psi \in E_\Omega$ such that ψ/φ is an entire function there exists a sequence $\{P_n\}_{n \geqslant 0}$ of polynomials satisfying $E_\Omega - \lim_n \varphi \cdot P_n = \psi$.*

The proof of this statement in case $n = 1$ hinges on the employment of canonical products and therefore cannot be directly transferred to the case of several variables.

2. It is well known in the theory of differential equations that $PC^\infty(\Omega) = C^\infty(\Omega)$ for every differential operator P with constant coefficients if and only if Ω is a convex domain. A natural complex analog of this statement can be formulated as follows.

CONJECTURE. *Let Ω be a pseudo-convex domain in $\mathbb{C}^n$. Then $PH(\Omega) = H(\Omega)$ for every differential operator $P = P\left(\dfrac{\partial}{\partial z_1}, \ldots, \dfrac{\partial}{\partial z_n}\right)$ with constant coefficients if and only if Ω is strongly linearly convex (see 1.18 of this volume for the definition).*

The following facts are in favour of the conjecture. The property of strong linear convexity is a sufficient condition [8]. Conversely if Ω is a pseudo-convex domain and

$PH(\Omega) = H(\Omega)$ then all slices of Ω by one-dimensional complex planes are simply connected (the proof follows the lines of [9]). It is known (references [10], [11] in Problem 1.18) that this implies that Ω is strongly linearly convex provided all slices of Ω are connected.

References

1. Schwartz L., *Théorie générale des fonctions moyenne-périodiques*, Ann. Math. **48** (1947), no. 4, 857–925.
2. Krasichkov-Ternovskii I. F., *Invariant subspaces of analytic functions. I. Spectral synthesis on convex domains*, Mat. Sb. **87** (1972), no. 4, 459–487 (Russian); English transl. in Math. USSR Sb. **16** (1972), 471–500; II., Mat. Sb. **88** (1972), no. 1, 3–30 (Russian); English transl. in Math. USSR Sb. **17** (1972), no. 1, 1–29.
3. Malgrange B., *Existence at approximation des solution des équations aux dernées partielles et des équations de convolution*, Ann. Inst. Fourier 6 (1955), 271–354.
4. Ehrenpreis L., *Mean periodic functions*, Amer. J. Math. **77** (1955), no. 2, 293–328.
5. Napalkov V. V., *On subspaces of analytic functions invariant relative to a shift*, Izv. Akad. Nauk SSSR, ser. matem. **36** (1972), 1269–1281 (Russian); English transl. in Math. USSR Izvestija **6** (1972), no. 6, 1251–1264.
6. Napalkov V. V., *On equation of convolution type in tube domains of* $\mathbb{C}^2$, Izv. Akad. Nauk SSSR, ser. matem. **38** (1974), 446–456 (Russian); English transl. in Math. USSR Izvestija **8** (1974), no. 2, 452–464.
7. Trutnev V. M., *On convolution equations in convex domains of* $\mathbb{C}^n$, Topics in mathematics (Voprosy matematiki), vol. 510, Tashkent State University, 1976, pp. 148–150. (Russian)
8. Martineau A., *Sur la notion d'ensemble fortement linéellement convexe*, Ann. Acad. Brasil., Ciens. **40** (1968), no. 4, 427–435.
9. Pinčuk S. I., *On the existence of holomorphic primitives*, Dokl. Akad. Nauk SSSR **204** (1972), no. 2, 292–294 (Russian); English transl. in Soviet Math. Doklady **13** (1972), no. 3, 654–657.

Krasnoyarsk State University
ul. Maerchaka 6
Krasnoyarsk 660075
Russia

Commentary

D. I. Gurevich proved in [10] that in the space $H(\mathbb{C}^2)$ there exist closed non-trivial translation invariant subspaces without exponential polynomials. The same holds in $C(\mathbb{R}^3)$, $C^\infty(\mathbb{R}^3)$, $\mathcal{D}'(\mathbb{R}^3)$ too.

See also a series of papers by R. Meise, S. Momm and others on right inverses of $P(D)$'s; [11] and references therein.

References

10. Gurevič D. I., *Counterexample to a problem of L. Schwartz*, Functs. Anal. Prilozhen. **9** (1975), no. 2, 29–35 (Russian); English transl. in Funct. Anal. Appl. **9** (1975), no. 2, 116–120.
11. Momm S., *Convex univalent functions and continuous linear inverses*, Preprint 20pp., 1992, Univ. of Düsseldorf (Germany).

LOCAL DESCRIPTION OF CLOSED SUBMODULES
AND THE PROBLEM OF OVER-SATURATION

I. F. KRASICHKOV-TERNOVSKII

The space $\mathcal{H}^q$ of all $\mathbb{C}^q$-valued functions analytic in a domain G of the complex plane $\mathbb{C}$ becomes a module over the ring of all polynomials $\mathbb{C}[z]$ under pointwise algebraic operations. Consider a submodule P of $\mathcal{H}^q$ endowed with the structure of a Hausdorff locally convex space such that the multiplication operators by polynomials are continuous. A great many problems in Analysis, such as the problem of polynomial approximation [1], convolution equations [2], mean periodic functions [3], the problem of spectral synthesis [4], [5] etc., is connected with the problem of local description of closed submodules $I \subset P$. Such a submodule I defines a *divisor* div(I). The *divisor* is a mapping which transforms any point $\lambda \in G$ into a submodule I_λ of the module $\mathcal{O}_\lambda^q$ of all germs at λ of $\mathbb{C}^q$-valued analytic functions. The mapping $\varphi_\lambda \colon \mathcal{H}^q \to \mathcal{O}_\lambda^q$ transforms every function in $\mathcal{H}^q$ into its germ at λ. The module I_λ is the smallest $\mathcal{O}_\lambda$-submodule of $\mathcal{O}_\lambda^q$ containing $\varphi_\lambda(I)$.

A submodule I is called a *divisorial submodule* if

$$I = I(\mathrm{div}) \overset{\mathrm{def}}{=} \{\, f \in P \colon \varphi_\lambda(f) \in I_\lambda, \quad \forall \lambda \in G \,\}.$$

The module $P = \mathcal{H}^q$ equipped with the topology of uniform convergence on compact subsets of G provides an example of a module whose all closed submodules are divisorial [6]. Many antipodal examples can be found in [1], [4], [7].

THE PROBLEM OF LOCALIZATION consists in *the characterization of those conditions which ensure that every submodule of a given module is divisorial.*

The following concepts are useful for the solution of the problem of localization. Namely, these are the concepts of stability and saturation, which separate the algebraic and analytic difficulties of the problem. Define $I_\lambda = \mathcal{H}^q$ if $\lambda \notin G$.

DEFINITION 1. A submodule I is called *stable* if for every $\lambda \in \mathbb{C}$

$$f \in I, \quad f/(z - \lambda) \in I_\lambda \implies f/(z - \lambda) \in I.$$

It is natural to consider stable submodules for modules P possessing the property of the uniform stability. This property ensures a certain kind of "softness" of the topology in P.

DEFINITION 2. A module P is called *uniformly stable* if for every neighbourhood $V \subset P$ of zero there exists a neighbourhood $U \subset P$ of zero satisfying

$$f \in U, \quad f/(z - \lambda) \in \mathcal{H}^q \implies f/(z - \lambda) \in V.$$

The following theorem explains the importance of the concept of saturation which will be defined later.

THEOREM 1. *Let P be a uniformly stable module. Then the submodule $I \subset P$ is divisorial iff it is stable and saturated.*

The saturated submodules for $q = 1$ can be described as follows. Let V be a neighbourhood in P and let $f \in I(\text{div})$. Set

$$C_{f,V}(z) = 1 + \inf\left\{\left|\frac{f(z)}{\varphi(z)}\right| : \varphi \in I \cap V\right\}.$$

Suppose that for each $f \in I(\text{div})$ and each $\Phi \in \mathcal{H}^1$

(1) $$|\Phi(z)| \leqslant C_{f,V}(z), \quad z \in G \implies |\Phi| \leqslant \text{const}$$

Then I is called *saturated*. Note that (1) automatically holds for $f \in I$.

In general we proceed as follows. The dimension $\dim I_\lambda$ of I_λ over $\mathcal{O}_\lambda$ ($\lambda \in G$) is clearly not greater than q. Put $k \overset{\text{def}}{=} \max_{\lambda \in G} \dim I_\lambda$. Then it is easy to show, using standard arguments with determinants, that $\dim I_\lambda \equiv k$ in G. Moreover there exists a family $u^{(1)}, \ldots, u^{(k)}$ in I such that $\varphi_\lambda(u^{(1)}), \ldots, \varphi_\lambda(u^{(k)})$ form a basis of I_λ for every $\lambda \in G$. Set $\dim I \overset{\text{def}}{=} k$ (the local rank of I). If $\varphi_\lambda(f) \in I_\lambda$ then

(2) $$f = c_1 u^{(1)} + \cdots + c_k u^{(k)}, \qquad e_j \in \mathcal{O}_\lambda$$

and the germs c_j can be found as follows. Consider in $\mathbb{C}^q$ the orthogonal projection $P_{\mathcal{J}}$ onto the subspace spanned by $e_{j_1}, \ldots, e_{j_k}$. Here $\mathcal{J} = (j_1, \ldots, j_k)$ and $\{e_j\}_{j=1}^q$ is the standard basis in $\mathbb{C}^q$. The system of linear equations (with a $(k \times k)$-matrix) $P_{\mathcal{J}}f = c_1 P_{\mathcal{J}}u^{(1)} + \cdots + c_k P_{\mathcal{J}}u^{(k)}$ can be solved which leads to the formula

$$c_s = \frac{\det_s(f, \mathcal{J})}{\det(u, \mathcal{J})}, \qquad s = 1, \ldots, k,$$

where the determinants are defined in accordance with Kramer's rule.

DEFINITION 3. *A submodule I, $\dim I = k$ is called saturated with respect to $f \in P$ if for every neighbourhood of zero $V \subset P$ the following holds*

$$|\Phi| \leqslant 1 + \inf\left\{\sum_{s=1}^k \left|\frac{\det_s(f, \mathcal{J})}{\det(u, \mathcal{J})}\right| : u^{(1)}, \ldots, u^{(k)} \in I \cap V, \ \forall \mathcal{J}\right\}$$

$$\implies |\Phi| \leqslant \text{const}.$$

A submodule I is called *saturated* if it is saturated with respect to every $f \in I(\text{div})$ and I is called *over-saturated* provided it is saturated with respect to every $f \in P$.

The existence of suitable estimates for holomorphic ratios f/g (see [1], [8], [9]) in many cases permits to prove that a given submodule I is saturated. In particular the local description of ideals in algebras can be obtained in this way [10–12]. If P is an algebra then every ideal of P is stable (as a rule). But if P is only a module, as for example in [4], then the role of stability may turn out to be dominant.

THEOREM 2. *Suppose that for every collection $f^{(1)}, \ldots, f^{(n)}$ of elements of P the set*

$$(3) \qquad \left\{ f \in \mathcal{H}^q : |f_j(z)| \leqslant |f_i^{(1)}(z)| + \cdots + |f_i^{(n)}(z)|, \ z \in G, \ i = 1, \ldots, q \right\}$$

is contained in P and bounded. Then every divisorial submodule $I \subset P$, $\dim I = 1$ is over-saturated.

For the proofs of Theorems 1 and 2 see [13].

It follows from Theorems 1 and 2 that for a uniformly stable module P satisfying (3) every submodule of local rank 1, containing a submodule with the same properties, is divisorial. This shows how important is it to extend Theorem 2 to submodules of an arbitrary local rank.

THE PROBLEM OF SATURATION. *Let P be a uniformly stable submodule satisfying (3). Is it true that every divisorial submodule $I \subset P$ is over-saturated? If not, what are general conditions ensuring that I is over-saturated?*

The solution of the problem would clarify obscure points in the theory of the local description and in its own turn would lead to solutions of some problems of real and complex analysis.

REFERENCES

1. Nikol'skii N. K., *Selected problems of weighted approximation and spectral analysis*, Trudy Mat. Inst. Steklov **120** (1974) (Russian); English transl. in Proc. Steklov Inst. Math. **120** (1974).
2. Krasichkov-Ternovskii I. F., *A homogeneous equation of convolution type on convex domains*, Dokl. Acad. Nauk SSSR **197** (1971), 29–31 (Russian); English transl. in Soviet Math. Dokl. **12** (1971), no. 1, 396–398.
3. Schwartz L., *Théorie générale des fonctions moyenne-périodiques*, Ann. Math. **48** (1947), no. 4, 857–929.
4. Krasichkov-Ternovskii I. F., *Invariant subspaces of analytic functions. I. Spectral synthesis on convex domains*, Mat. Sb. **87** (1972), no. 4, 459–488 (Russian); English transl. in Math. USSR Sb. **16** (1972), 471–500.
5. Krasichkov-Ternovskii I. F., *Invariant subspaces of analytic functions. II. Spectral synthesis on convex domains*, Mat. Sb. **88** (1972), no. 1, 3–30 (Russian); English transl. in Math. USSR Sb. **17** (1972), no. 1, 1–29.
6. Cartan H., *Idéaux et modules de fonctions analytiques de variables complexes*, Bull. Soc. Math. France **78** (1950), no. 1, 29–64.
7. Kelleher J. J., Taylor B. A., *Closed ideals in locally convex algebras of analytic functions*, J. reine und angew. Math. **225** (1972), 190–209.
8. Krasichkov-Ternovskii I. F., *An estimate for subharmonic difference of subharmonic functions I*, Mat. Sb. **102** (1977), no. 2, 216–247 (Russian); English transl. in Math. USSR Sb. **31** (1977).
9. Krasichkov-Ternovskii I. F., *An estimate for subharmonic difference of subharmonic functions II*, Mat. Sb. **103** (1977), no. 1, 69–111 (Russian); English transl. in Math. USSR Sb. **32** (1977).
10. Raševskii P. K., *Closed ideals in a countably-normed algebra of entire analytic functions*, Dokl. Akad. Nauk SSSR **162** (1965), 513–515 (Russian); English transl. in Soviet Math. Dokl. **6** (1965), 717.
11. Krasičkov I. F., *Closed ideals in locallyconvex algebras of entire functions. I, II*, Izv. Akad. Nauk SSSR, ser. matem. **31** (1967), 37–60 (Russian); English transl. in Math. USSR Izvestija **31** (1967), 35–55; Izv. Akad. Nauk SSSR, ser. matem. **32** (1968), 1024–1032 (Russian); English transl. in Math. USSR Izvestija **32** (1968), 979–986.
12. Matsaev V. I., Mogul'skii E. Z., *Completeness of weak perturbations of self-adjoint operators*, Zap. Nauchn. Sem. LOMI **56** (1976), 90–103 (Russian); English transl. in J. Soviet Math. **14** (1980), no. 2, 1093–1103.

13. Krasichkov-Ternovskii I. F., *Local description of closed ideas and submodules of analytic functions of one complex variable. I and II*, Izvestiya AN SSSR, ser. matem. **43** (1979), no. 1, 44–66; *and* **43** (1979), no. 2, 309–341. (Russian)

MATHEMATICAL INSTITUTE UrO AN OF RUSSIA
CHERNYSHEVSKOGO 112
UFA 450025
RUSSIA

ON THE SPECTRAL SYNTHESIS IN SPACES
OF ENTIRE FUNCTIONS OF NORMAL TYPE

V. A. TKACHENKO

Let ρ be a positive real number and let H be a 2π-periodic lower semi-continuous trigonometrically ρ-convex function with values in $(-\infty, \infty]$. Denote by $\mathcal{M}(H)$ the set of all trigonometrically ρ-convex functions h satisfying $h(\theta) < H(\theta)$ for every $\theta \in [0, 2\pi]$. For $h \in \mathcal{M}(H)$ let $F(h)$ be the Banach space of all entire functions ψ with the norm $\|\psi\|_h = \sup\limits_{r,\theta} |\psi(re^{i\theta})| \exp(-h(\theta)r^\rho)$. The family $\{F(h)\}_{h \in \mathcal{M}(H)}$ is inductive with respect to natural imbeddings and its inductive limit $[\rho, H(\theta))$ is a space of (LN^*)-type in the sense of J. Sebastião-e-Silva.

Multiplication by the independent variable is a continuous mapping of $[\rho, H(\theta))$ into itself, so that $[\rho, H(\theta))$ is a topological module over the ring of polynomials and one may consider the lattice of the closed (invariant) submodules. The submodule $I_{a,k} = \{\psi \in [\rho, H(\theta)): \psi(a) = \cdots = \psi^{(k)}(a) = 0\}$ defined by $a \in \mathbb{C}$ and $k \in \mathbb{Z}^+$ is of the simplest structure.

By a commonly accepted definition (cf. [1]) a submodule $I \subset [\rho, H(\theta))$ *admits the spectral synthesis* (or is *localized*) if it coincides with the intersection of all submodules $I_{a,k}$ containing it.

PROBLEM. *Find necessary and sufficient conditions (on $H(\theta)$) for every closed submodule $I \subset [\rho, H(\theta))$ to admit spectral synthesis.*

In 1947 L. Schwartz [2] proved that every ideal is localized in the algebra of all entire functions of exponential type ($\rho = 1$, $H(\theta) \equiv +\infty$). The progress in the localization theory for spaces invariant with respect to multiplication by the independent variable in the weighted algebras and modules of entire functions is described in the survey by N. K. Nikol'skii [1]; there is an extensive list of references there.

For $\rho = 1$ the discussed problem was solved by I. F. Krasičkov-Ternovskii: every closed submodule $I \subset [1, H(\theta))$ is localized iff H is unbounded [3].

For an arbitrary $\rho > 0$ it was proved in [4] that if the length of every interval, where H is finite, does not exceed π/ρ then every closed submodule $I \subset [\rho, H(\theta))$ admits localization.

There is an indirect evidence that the last condition is not only sufficient but necessary. Unfortunately, all my attempts to prove its necessity failed.

REFERENCES

1. Nikol'skii N. K., *Invariant subspaces in operator theory and function theory*, Itogi Nauki i Tekhniki: Mat. Anal. **12** (1974), 199–412, VINITI, Moscow (Russian); English transl. in J. Soviet Math. **5** (1976), no. 2.
2. Schwartz L., *Théorie générale des fonctions moyenne-périodiques*, Ann. Math. **48** (1947), no. 4, 857–929.

3. Krasichkov-Ternovskii I. F., *Invariant subspaces of analytic functions. I. Spectral synthesis on convex domains*, Mat. Sb. **87** (1972), no. 4, 459–487 (Russian); English transl. in Math. USSR Sb. **16** (1972), 471–500; II., Mat. Sb. **88** (1972), no. 1, 3–30 (Russian); English transl. in Math. USSR Sb. **17** (1972), no. 1, 1–29.
4. Tkachenko V. A., *On spectral synthesis in spaces of analytic functionals*, Dokl. Akad. Nauk SSSR **223** (1975), no. 2, 307–309 (Russian); English transl. in Soviet Math. Doklady **16** (1975), no. 4, 895–898.

UL. 23 AVGUSTA 39, 31
KHARKOV 310103
UKRAINE

SPECTRAL SYNTHESIS
FOR A DIFFERENTIAL OPERATOR
WITH CONSTANT COEFFICIENTS

I. F. KRASICHKOV-TERNOVSKIĬ

Let G be a convex domain in $\mathbb{C}$, H the space of functions analytic in G, endowed with the topology of uniform convergence on compacta, $\pi(D) = D^q + a_1 D^{q-1} + \cdots + a_q D^0$ differential operator. The root elements of $\pi(D)$ corresponding to an eigenvalue λ are exponential polynomials of the form

$$P_0(z)e^{\lambda_0 z} + P_1(z)e^{\lambda_1 z} + \cdots + P_{q-1}(z)e^{\lambda_{q-1} z},$$

where $P_k(z)$ are polynomials and $\lambda_0, \ldots, \lambda_{q-1}$ are roots of the algebraic equation $\pi(z) = \lambda$, $\pi(z) := z^q + a_1 z^{q-1} + \cdots + a_q$. A closed subspace $W \subset H$ is said to be *invariant* if it is invariant under $\pi(D)$: $f \in W \implies \pi(D)f \in W$. An invariant subspace W admits *spectral synthesis* if it coincides with the closure of linear span of some root elements of $\pi(D)$. The spectral synthesis problem is to determine conditions under which a given invariant subspace admits spectral synthesis. The approach to this problem goes back to a device due to L. Ehrenpreis [1]. Let T be the mapping taking each continuous linear functional $S \in H^*$ to its characteristic function $\varphi(\lambda) = \langle S, e^{\lambda z} \rangle$. As S runs through H^*, φ runs through the class P of entire functions of exponential type satisfying the condition $h_\varphi(\theta) < H(\theta)$, where $H(\theta)$ is the support function of the domain $\overline{G} = \{ w = \bar{z} : z \in G \}$, $h_\varphi(\theta)$ the indicator of φ. P is endowed with the natural locally convex topology of an inductive limit; moreover, P is invariant under multiplication by polynomials. We consider P as a locally convex module over the ring $\mathbb{C}[\pi(z)]$ of polynomials in $\pi(z)$.

Suppose that W is an invariant subspace of H, $W^0 = \{ S \in H^* : \langle S, f \rangle = 0 \ \forall f \in W \}$ and $I = T(W^0)$. The set I has the structure of a closed submodule of the module P. The relation $I = T(W^0)$ establishes a one-to-one correspondence between the invariant subspaces $W \subset H$ and the closed submodules $I \subset P$. The submodule $I = T(W^0)$ is called the *annihilator submodule* of the subspace W. We need special notion of independence of elements of P. Its definition requires some preparation. The set $\tilde{\lambda} := \pi^{-1}(\lambda)$ is called a *π-fiber* over λ. A set is π-symmetric if it may be represented as a union of some collection of π-fibers. Let V be a π-symmetric set. A function $f \colon V \to \mathbb{C}$ is called *π-symmetric* if it takes equal values on π-fibers $\tilde{\lambda} \subset V$.

For a fixed point $\lambda \in \mathbb{C}$ denote $O(\tilde{\lambda})$ the set of germs of functions locally analytic in neighbourhoods of π-fiber $\tilde{\lambda}$ over λ. $O_\pi(\tilde{\lambda})$ is a subring of $O(\tilde{\lambda})$ consisting of π-germs of π-symmetric functions locally analytic in π-symmetric neighbourhoods of $\tilde{\lambda}$. $O(\tilde{\lambda})$ has the structure of a module over the ring $O_\pi(\tilde{\lambda})$. A system of elements $u^{(1)}, \ldots, u^{(k)} \in O(\tilde{\lambda})$ is *independent* over $O_\pi(\tilde{\lambda})$ if

$$\sum c_i u^{(i)} = 0, \ c_i \in O_\pi(\tilde{\lambda}) \implies c_i = 0, \ i = 1, \ldots, k.$$

The rank of a submodule $I \subset O(\tilde{\lambda})$ is the maximal number of elements in independent systems $u^{(1)}, \ldots, u^{(k)} \in I$. For any fixed $\lambda \in \mathbb{C}$ a submodule I of $\mathbb{C}[\pi(z)]$-module P generates a *local submodule* $I(\tilde{\lambda})$ of $O(\tilde{\lambda})$ consisting of all possible linear combinations $c_1 u^{(1)} + \cdots + c_m u^{(m)}$, where $u^{(1)}, \ldots, u^{(m)} \in I$ and $c_1, \ldots, c_m \in O(\tilde{\lambda})$. The rank of $I(\tilde{\lambda})$ does not depend on λ; therefore it makes sense to call it the *rank* of the submodule I. A domain G in $\mathbb{C}$ is said to be starlike in the direction θ, $0 \leqslant \theta < 2\pi$, if

$$z \in G \implies z + te^{i\theta} \in G \text{ for all } t > 0.$$

The set of directions in which G is starlike is denoted by $\Theta(G)$. The following theorem has been proved by the author and announced in [2].

THEOREM. *Suppose that G is an unbounded convex domain and that*

$$\bigcap_{k=0}^{m} \Theta\left(e^{ik\frac{2\pi}{q}} G\right) \neq \emptyset$$

for an integer m with $0 \leqslant m \leqslant q - 1$. Let $\pi(D) = L^q$. Then an arbitrary invariant subspace $W \subset H$ with annihilator submodule of rank not exceeding $m + 1$ admits spectral synthesis.

This result contains earlier results by the author for the case $q = 1$ [3] and by Merzlyakov for the cases $q = 2$, G is a strip [4]; q is arbitrary, $G = \mathbb{C}$ [5].

PROBLEM. *Prove the same statement for arbitrary differential operator*

$$\pi(D) = D^q + a_{q-1}D^{q-1} + \cdots + a_q D^0.$$

REFERENCES

1. Ehrenpreis L., *Mean periodic functions. I. Varieties whose annihilator ideals are principal*, Amer. J. Math. **XX** (1955), 293–328.
2. Krasichkov-Ternovskiĭ I. F., Shishkin A. B., *Spectral synthesis for a multiple differentiation operator*, Soviet Math. Dokl. **40** (1990), no. 1, 16–19.
3. Krasichkov-Ternovskiĭ I. F., *Invariant subspaces of analytic functions. I. Spectral synthesis on convex domains*, Mat. Sb. **87** (1972), no. 4, 459–487 (Russian); English transl. in Math. USSR Sb. **16** (1972), 471–500; II., Mat. Sb. **88** (1972), no. 1, 3–30 (Russian); English transl. in Math. USSR Sb. **17** (1972), no. 1, 1–29.
4. Merzlyakov S. G., *Invariant subspaces of a multiple differentiation operator*, Mat. Zametki **33** (1983), 701–713. (Russian)
5. Merzlyakov S. G., *On subspaces of analytical functions invariant under multiple differentiation operator*, Mat. Zametki **40** (1986), 635–639. (Russian)

MATHEMATICAL INSTITUTE UrO AN RUSSIA
CHERNYSHEVSKOGO 112
UFA 450025
RUSSIA

THE LOCALLY CONVEX STRUCTURE
OF THE SPACE OF ALL ZERO SOLUTIONS
OF CERTAIN CONVOLUTION OPERATORS

R. Meise,   S. Momm

Denote by $\mathcal{E}_\omega$ the Fréchet space of all ω-ultradifferentiable functions on $\mathbb{R}$ in the sense of Beurling–Björck (see [2]). Examples are $\omega(t) = \log^+ t$ and $\omega(t) = t^\alpha$, $0 < \alpha < 1$, $t \geqslant 0$. Then every $\mu \in \mathcal{E}'_\omega$ induces a convolution operator

$$T_\mu \colon \mathcal{E}_\omega \to \mathcal{E}_\omega, \quad T_\mu f \colon x \mapsto \langle \mu, f(x - \cdot) \rangle, \quad f \in \mathcal{E}_\omega,\, x \in \mathbb{R}.$$

PROBLEM. *Describe the locally convex structure of* $\ker T_\mu$ *for* $\mu \in \mathcal{E}'_\omega$.

To reformulate the problem, define $p \colon \mathbb{C} \to [0, \infty[$ by $p(z) := |\operatorname{Im} z| + \omega(|z|)$ and denote by A_p the (DFN)-space of all entire functions on $\mathbb{C}$ which satisfy for some $A, B > 0$ the estimate $|f(z)| \leqslant A \exp(B p(z))$ for all $z \in \mathbb{C}$. Then the Fourier–Laplace transform $\widehat{} \colon \mathcal{E}'_\omega \to A_p$, $\widehat{\mu}(z) := \langle \mu, \exp(-iz\cdot) \rangle$, is an isomorphism. Identifying $\mathcal{E}'_\omega$ with A_p, the adjoint of T_μ identifies with the operator of multiplication by $\tilde{\mu} \colon z \mapsto \widehat{\mu}(-z)$. Since $\tilde{\mu} A_p = I_{\mathrm{loc}}(\tilde{\mu}) = \{ f \in A_p : f/\tilde{\mu}$ is entire $\}$, duality theory of locally convex spaces implies

$$(\ker T_\mu)' = \mathcal{E}'_\omega / \overline{\operatorname{im} T_\mu^t} \cong A_p / \tilde{\mu} A_p = A_p / I_{\mathrm{loc}}(\tilde{\mu}).$$

Hence we have the following

REFORMULATED PROBLEM. *Describe the locally convex structure of* $A_p / I_{\mathrm{loc}}(\tilde{\mu})$.

If $\tilde{\mu}$ is slowly decreasing in the sense of Berenstein and Taylor (see [3], § 3) then the reformulated problem can be solved, as [3] and [5] have shown. Using this and a characterization of the slowly decreasing condition (see [9] and [10]), the following partial solution of the problem was given in [8].

THEOREM. *Let* $T_\mu \colon \mathcal{E}_\omega \to \mathcal{E}_\omega$ *be a surjective convolution operator and assume that* $\tilde{\mu}$ *has the zeros* $(a_j)_{j \in \mathbb{N}}$, *counted with multiplicities. Then* $\ker T_\mu$ *has an absolute basis consisting of exponential solutions of* T_μ *and* $\ker T_\mu$ *is isomorphic to the sequence space*

$$\left\{ x \in \mathbb{C}^{\mathbb{N}} : \|x\|_k := \sum_{j=1}^{\infty} |x_j| \exp\bigl(k(p(a_j))\bigr) < \infty \text{ for all } k \in \mathbb{N} \right\}.$$

Remark. L. Ehrenpreis has indicated that the method used in [3] and [5] cannot be applied to solve the problem for arbitrary $\mu \in \mathcal{E}'_\omega$ (see [7], 3.9).

Variations of the problem. The problem above can obviously be stated also for convolution operators on other spaces of ultradifferentiable functions, ultradistributions and holomorphic functions on convex domains. In all these cases partial solutions of the problem have been given in the spirit of the theorem above and using similar methods (see [6], [1], [4], [12], [11]).

PROBLEM 11.7

REFERENCES

1. Braun R. W., Meise R., *Generalized Fourier expansions for zero-solutions of surjective convolution operators on $\mathcal{D}_{\{\omega\}}(\mathbb{R})'$*, Arch. Math. **55** (1990), 55–63.
2. Braun R. W., Meise R., Taylor B. A., *Ultradifferentiable functions and Fourier analysis*, Result. Math. **17** (1990), 206–237.
3. Berenstein C. A., Taylor B. A., *A new look at interpolation theory for entire functions of one variable*, Adv. Math. **33** (1979), 109–143.
4. Franken U., Meise R., *Generalized Fourier expansions for zero solutions of surjective convolution operators on $\mathcal{D}'(\mathbb{R})$ and $\mathcal{D}'_{\{\omega\}}(\mathbb{R})$*, Note di Mat. (to appear).
5. Meise R., *Sequence space representations for (DFN)-algebras of entire functions modulo closed ideals*, J. Reine Angew. Math. **363** (1985), 59–95.
6. Meise R., *Sequence space representations for zero-solutions of convolution equations on ultradifferentiable functions of Roumieu type*, Studia Math. **92** (1989), 211–230.
7. Meise R., Momm S., Taylor B. A., *Splitting of slowly decreasing ideals in weighted algebras of entire functions*, in Complex Analysis II LNM (C. A. Berenstein, ed.), vol. 1276, Springer, Berlin, 1987, pp. 229–252.
8. Meise R., Schwerdtfeger K., Taylor B. A., *Kernels of slowly decreasing convolution operators*, Doga Tr. J. Math. **10** (1986), 176–197.
9. Meise R., Taylor B. A., Vogt D., *Equivalence of slowly decreasing conditions and local Fourier expansions*, Indiana Univ. Math. J. **36** (1987), 729–756.
10. Momm S., *Closed principal ideals in nonradial Hörmander algebras*, Arch. Math (to appear).
11. Momm S., *Convolution equations on the analytic functions on convex domains in the plane*, preprint (1991).
12. Napalkov V. V., *A basis in the space of solutions of a convolution equation*, Math. Notes Acad. Sci. USSR **43** (1988), 27–33.

MATHEMATISCHES INSTITUT
HEINRICH-HEINE-UNIVERSITÄT DÜSSELDORF
UNIVERSITÄTSSTRASSE 1
4000 DÜSSELDORF 1
GERMANY

MATHEMATISCHES INSTITUT
HEINRICH-HEINE-UNIVERSITÄT DÜSSELDORF
UNIVERSITÄTSSTRASSE 1
4000 DÜSSELDORF 1
GERMANY

INTERPOLATING VARIETIES AND COMPLEX ANALYSIS

C. A. BERENSTEIN,   D. C. STRUPPA

0.   The problem which we describe here have a common background in the study of solutions of systems of convolution equations in Analytically Uniform spaces [10]. Such a study naturally leads to the study of interpolation on the analytic varieties defined by holomorphic functions with growth conditions. A fairly complete reference is [6].

1. The Bezout equation with rational coefficients.   Many questions both in pure and in applied mathematics lead the problem of solving the polynomial Bezout equation in an explicit way: given polynomials $p_1, \ldots, p_r$ with no common zeros in $\mathbb{C}^n$, one wants to construct polynomials $q_1, \ldots, q_r$ in $\mathbb{C}^n$ such that

$$(1.1) \qquad\qquad p_1 q_1 + \cdots + p_r q_r \equiv 1.$$

While the existence of q_j's is assured by the classical Hilbert's Nullstellensatz, their explicit construction can be done by using some integral formulas from the theory of several complex variables [7] (this result has already been conjectured in [1]). There are however, some concrete situation, see e.g. [3], in which the polynomials $p_1, \ldots, p_r$ have a non-empty variety of common zeros

$$V = \{ z \in \mathbb{C}^n : p_1(z) = \cdots = p_r(z) = 0 \};$$

in this case (1.1) can be solved by rational functions q_j and one would like to control the variety of poles of the q_j. More specifically, if V is contained in some open set U, one would like to know that all the singularities of $q_1, \ldots, q_r$ are contained, as well, in U. In [3], we have provided an explicit result in this direction at least in a simple case. More precisely we have proved the following result:

THEOREM 1.1.   *Let $U \subseteq \mathbb{C}^n$ be an open set such that its complement $U^c = \mathbb{C}^n \setminus U$ is the closure of a complete Reinhardt domain. Then, if $p_1, \ldots, p_r$ are polynomials without common zeros in U^c, it is possible to construct rational functions $q_1, \ldots, q_r$ such that*

$$p_1 q_1 + \cdots + p_r q_r \equiv 1,$$

and such that all the poles of the q_j are contained in U.

The condition we are requesting of U is, of course, quite strong, and we are therefore led to formulating the following general problem:

PROBLEM 1.1. *Characterize those domains D such that the conclusion of Theorem 1.1 holds for $U = \mathbb{C}^n \setminus D$.*

Such a problem is asking to a result of Lech [13]; in [13] it is shown that if an algebraic variety V in $\mathbb{C}^n$ satisfies the condition

$$(1.2) \qquad \liminf \left[\operatorname{dist}(z, \mathbb{R}^n)(1 + |z|)^N \right] > 0, \qquad z \in V,$$

then there exists an algebraic hypersurface S which separates V from $\mathbb{R}^n$ and that, moreover, (1.2) holds if V is replaced by S. In view of this result, we may restrict our open problem to the case in which convex cones are concerned. To be precise we ask:

PROBLEM 1.2. *Let K be a closed cone in $\mathbb{C}^n$. Prove (or disprove) that, given an algebraic variety V for which*

$$\liminf \left[\operatorname{dist}(z, K)(1 + |z|)^N \right] > 0, \qquad z \in V,$$

then

a) *there exists an algebraic hypersurface S which separates V from K,*
b) *$\liminf \left[\operatorname{dist}(z, K)(1 + |z|)^N \right] > 0, \ z \in S$.*

Note that a similar problem can be posed for the case of polydisks. In this case, one may suppose that V does not intersect the polydisk but may intersect its closure, and a weakened form of the problem would consist in looking for an *analytic separating hypersurface*.

Finally, let us conclude this section by pointing out that our proof of Theorem 1.1, as given in [3], is analytic, while one might expect a purely algebraic proof:

PROBLEM 1.3. *Give an algebraic proof of Theorem 1.1.*

2. A refined version of Noether's normalization.

The following question arose in the preparation of a recent work with Kawai [1], when trying to construct examples of slowly decreasing systems of infinite order differential operators:

PROBLEM 2.1. *Let $U \subseteq \mathbb{C}^n$ be an open set, and let A be an algebra of functions holomorphic on U; we will assume A to have identity. Let us assume we have exactly m polynomials in m variables, with coefficients in A. We suppose that the content of each polynomial is 1 (i.e. its coefficients generate, as an ideal, all of A) and we also assume that, for any $\zeta \in U$ fixed, the variety of common zeros in $\mathbb{C}^n$ of the polynomials is discrete or empty. We want to know whether it is possible (maybe by shrinking U, or even by representing it as the union of a finite collection of subsets) to perform the Noether normalization in such a way that the coefficients in each one of the new polynomials also generate A.*

The interest of solving this question lies in the fact that if this were true, we could easily construct a large class of examples of functions to which the interpolation theory developed in [1] for the algebra of entire functions of infraexponential type could be applied.

23

3. Interpolation and Dirichlet series. In an important paper [12], Kawai has shown how to use some microhyperbolicity properties of infinite order differential operators to generalize and put into a different perspective some classical overconvergence theorems such as the Fabry gap theorem. Kawai's results, which we have somehow generalized in [4], and which originated by some remarks of Ehrenpreis [10], enable to study questions of analytic continuation for functions, which are analytic in a half-plane, and have a representation of the form

$$(3.1) \qquad\qquad f(z) = \sum c_j \exp(ia_j z),$$

for $\{a_j\}$ the sequence of zeros of an entire function of infinite order. When, however, one intends to apply these ideas to the study of Dirichlet series, one needs to consider the case of frequencies which are zeros of functions of faster growth. This was first pointed out in the works of V. Bernstein [8] and of Mandelbrojt [14] where originally, appears the idea of studying interpolation of the variety $\{a_j\}$. In our paper [5], and in its outgrowth [11], we develop an interpolation theory for varieties defined by functions in the space of Laplace transform of $A_{p,0}(\Gamma)'$ (see [5] for notations) and many new results are obtained on analytic continuation of mean-periodic functions in $A_{p,0}(\Gamma)$; all of these results, however, require that the variety of frequencies do not intersect the real axis (except, possibly, for a finite number of points); we give here an example for $\Gamma = \Pi_+ = \{\, z \in \mathbb{C} : \operatorname{Re} z > 0 \,\}$ (we refer the reader to [5] for the specific terminology):

THEOREM 3.1. *Every generalized Dirichlet series such as*

$$\sum_{n=1}^{+\infty} \sum_{j=1}^{N_n} c_{n,j}(z) \exp\left(-\lambda_{n,j} z\right)$$

convergent in $A_{p,0}(\Pi_+)$ does actually converge in $A_{p,0}(\mathbb{C})$, provided that $V = \{(\lambda_n, N_n)\}$ is the zero variety of a slowly decreasing convolutor $\mu \in A_{p,0}(\Pi_+)'$, and that V intersects, for some $\vartheta \in (0, \pi/2)$, in at most a finite number of points the set $\{\, w \in \mathbb{C} : -\vartheta < \arg w < \vartheta \,\}$.

The unpleasant condition on the location of the frequencies λ_n is a direct consequence of the methods employed, but, cuts off from our field of applicability a certain number of interesting situations. It is therefore reasonable to pose the following question:

PROBLEM 3.1. *Remove from Theorem 3.1 the hypothesis on the intersection of V with $\{-\vartheta < \arg w < \vartheta\}$.*

REFERENCES

1. Berenstein C. A., Kawai T., Struppa D. C., in preparation.
2. Berenstein C. A., Struppa D. C., *On explicit solutions to the Bezout equation*, Systems and Control Lett. **4** (1984), 33–39.
3. Berenstein C. A., Struppa D. C., *1-inverses of polynomial matrices of non-constant rank*, Systems and Control Lett. **6** (1986), 309–314.
4. Berenstein C. A., Struppa D. C., *On the Fabry–Ehrenpreis–Kawai gap theorem*, Publ. RIMS Kyoto Univ. **23** (1987), 565–574.
5. Berenstein C. A., Struppa D. C., *Dirichlet series and convolution equations*, Publ. RIMS Kyoto Univ. **24** (1988), 783–810.

6. Berenstein C. A., Struppa D. C., *Convolution equations and complex analysis*, Contemporary Problems in Mathematics, Fundamental Directions, vol. 5, pp. 5–110. (Russian)
7. Berenstein C. A., Yger A., *Effective Bezout identities in* $\mathbb{Q}[z_1, \ldots, z_n]$, Acta Math. **166** (1991), 69–120.
8. Bernstein V., *Series de Dirichlet*, Paris, 1933.
9. Brownawell D. W., *Distance to common zeros and lower bounds for polynomials*, preprint.
10. Ehrenpreis L., *Fourier Analysis in Several Complex Variables*, New York, 1970.
11. Neyman R., *Interpolation of entire functions of infinite order*, Ph.D. Thesis, Univ. of Maryland, 1990.
12. Kawai T., *The Fabry-Ehrenpreis gap theorem and linear differential equations of infinite order*, Am. J. Math. **109** (1987), 57–64.
13. Lech C., *A metric result about the zeros of a complex polynomial ideal*, Ark. Math. **52** (1958), 543–554.
14. Mandelbrojt S., *Dirichlet Series*, Dordrecht, 1972.

DEPARTMENT OF MATHEMATICS
UNIVERSITY OF MARYLAND
COLLEGE PARK, MD 20742
USA

VIALE UMBRIA 107
20135, MILAN,
ITALY

SOME PROBLEMS ABOUT IDEALS
GENERATED BY EXPONENTIAL–POLYNOMIALS

C. A. BERENSTEIN AND A. YGER

Given a collection of m exponential polynomials in n variables of the form

$$f(Z) = \sum_{\gamma \in \Gamma} p_\gamma(Z) \exp(< \gamma, Z >)$$

($p_\gamma \in \mathbb{C}[z_1, \ldots, z_n]$) with frequencies in a finitely generated subgroup

$$\Gamma = \mathbb{Z}\gamma_1 \oplus \cdots \oplus \mathbb{Z}\gamma_N \subset \mathbb{C}^n$$

let I be the ideal they generate in the algebra $\mathcal{A}_\Gamma$ of entire functions F with growth conditions of the form

$$|F(Z)| \leqslant C(1 + \|Z\|)^C \exp \left(C \max_{1 \leqslant j \leqslant N} |\operatorname{Re}(< Z, \gamma_j >)| \right)$$

The analytic methods used in the resolution of division problems (based either on $\bar{\partial}$-cohomology with bounds (as in [1]) or on an extensive use of the analytic continuation of distributions (as in [2] or [3]) suggest the three following questions:

PROBLEM 1. *Is any global version of the Nullstellensatz valid for the ideal I, considered as an ideal in $\mathcal{A}_\Gamma$, i.e. does there exist some positive integer $M = M(I)$ such that, for any element F in $\mathcal{A}_\Gamma$ vanishing on the zero set of the ideal I the following inclusion $F^M \in I$ holds?*

PROBLEM 2. *Is any global version of Briançon – Skoda theorem ([7]) valid for the ideal I, considered as an ideal in $\mathcal{A}_\Gamma$, i.e. for any F in $\mathcal{A}_\Gamma$ such that, for any $x \in \mathbb{C}^n$, the germ of F at x is in the integral closure of the ideal I_x in $\mathcal{O}_x$, does one have*

$$F^{\inf(n,m)} \in I$$

(or at least $F^\kappa \in I$, with some exponent κ depending only on n)?

PROBLEM 3. *When $m \leqslant n$ and the zero set of I is $n - m$ dimensional, is the ideal I closed in $\mathcal{A}_\Gamma$?*

A positive answer to these three questions (with $\kappa = 2\inf(m + 1, n)$) is given in [4] in the very particular case where $\operatorname{rank} \Gamma = 1$. When $\operatorname{rank}(\Gamma) > 1$, the example of the ideal generated in $\mathbb{C}^3$ by $(\cos(z_1), \cos(z_2), z_2 - az_1)$, $a \in \mathbb{R} \backslash \mathbb{Q}$ (the variable z_3 does not appear, the zero set is empty) shows that in order to get a positive answer to any of the above questions, additional conditions (of arithmetic nature) have certainly to be made (for example in the above situation, the fact that a is Liouville or non Liouville is crucial). Some natural conjecture one could formulate in one of the most interesting (and difficult!) case, that is when $\Gamma \subset i\mathbb{R}^n$ (i.e. the algebra $\mathcal{A}_\Gamma$ coincide with the Paley – Wiener algebra) would be that the answer to these three questions is positive when $\Gamma \subset \overline{\mathbb{Q}}$ and the coefficients of all polynomials involved in the generators of I are algebraic numbers. For some partial results and remarks in this direction, we refer to [5] and [6]. A positive answer to the first question under such hypothesis would solve the following conjecture due to L. Ehrenpreis in [8]: if f is an exponential polynomial in one variable with algebraic coefficients and algebraic purely imaginaries frequencies, then there exist a positive integer q and a strictly positive constant c such that

$$f(z_1) = f(z_2) = 0, \quad z_1 \neq z_2 \quad \text{implies} \quad |z_1 - z_2| \geq c(1 + |z_1|)^{-q}.$$

These questions are of deep interest in signal theory.

References

1. Berenstein C.A., Taylor B.A, *Interpolation problems in $\mathbb{C}^n$ with applications to harmonic analysis*, J. Analyse Math. **38** (1980), 188–254.
2. Berenstein C. A., Gay R., Yger A., *Analytic continuation of currents and division problems*, Forum Math. **1** (1989), 15–51.
3. Berenstein C.A., Yger A., *Formules de représentation intégrale et problèmes de division*, Diophantine Approximations and Transcendental Numbers (P. Philippon, ed.), Luminy 1990, Walter de Gruyter, Berlin, 1992, pp. 15–37.
4. Berenstein C. A., Yger A., *Exponential polynomials and D-modules* (to appear).
5. Berenstein C. A., Yger A., *Ideals generated by exponential polynomials*, Advances in Math. **60** (1986), 1–80.
6. Berenstein C. A., Yger A., *On Lojasiewicz inequalities for exponential polynomials*, J. Math. Anal. Applications **129** (1988), 166–195.
7. Briançon J., Skoda H. *Sur la clôture intégrale d'un idéal de germes de fonctions holomorphes en un point de $\mathbb{C}^n$*, Comptes Rendus Acad. Sci. Paris, ser. A **278** (1974), 949–951.
8. Ehrenpreis L., *Fourier Analysis in several Complex variables*, Wiley Interscience, New York, 1970.

DEPARTMENT OF MATHEMATICS
AND SYSTEM RESEARCH CENTER
UNIVERSITY OF MARYLAND
MD 20742, USA

CEREMAB, UA ASSOCIÉE 226
UNIVERSITÉ BORDEAUX 1
33405 TALENCE, CEDEX
FRANCE

A REFINEMENT OF THE CORONA THEOREM

T. WOLFF

The usual methods for proving corona type theorems [1,2] use existence of solutions with bounded radial limits for equations $\dfrac{\partial u}{\partial \bar{z}} = f$ when $|f|\, dx\, dy$, or something similar to $|f|\, dx\, dy$, is a Carleson measure. Our problem is a variant of the corona theorem for which apparently no Carleson measure is in sight.

PROBLEM. *Suppose* $f, f_1, f_2 \in H^\infty$ *with* $|f(z)| \leqslant |f_1(z)| + |f_2(z)|$ *for all* $z \in \mathbb{D}$. *Must there be* $g_1, g_2 \in H^\infty$ *with* $g_1 f_1 + g_2 f_2 = f^2$?

The answer is known to be yes if the exponent 2 is replaced by 3 (or $2 + \varepsilon$ if f has no zeroes) and no if 2 is replaced by 1 (or $2 - \varepsilon$). See [2]. The answer is also yes if g_1, g_2 are only required to be in H^1. This is by a $\bar{\partial}$ argument using the estimate

$$\iint\limits_{z\in\mathbb{D}:\, 4^{-1}\leqslant \left|\frac{f_1(z)}{f_2(z)}\right|\leqslant 4} \left|\left(\frac{f_1}{f_2}\right)'(z)\right|^2 (1 - |z|)\, dx\, dy < \infty,$$

which comes from the Ahlfors–Shimizu form of the first fundamental theorem. If

$$\mu \overset{\text{def}}{=} \chi(z)\left|\left(\frac{f_1}{f_2}\right)'(z)\right|^2 (1 - |z|)\, dx\, dy$$

(χ = characteristic function of $\{\, z \in \mathbb{D} : 4^{-1} \leqslant \left|\frac{f_1(z)}{f_2(z)}\right| \leqslant 4\,\}$) were always a Carleson measure our problem could be answered affirmatively by arguments in [2]; but in general μ is not Carleson.

REFERENCES

1. Carleson L., *The corona theorem*, Proc. 15th Scandinavian Congress, Lect. Notes in Math., vol. 118, Springer–Verlag, 1970, pp. 121–132.
2. Garnett J., *Bounded Analytic Functions*, Academic Press, New York, 1981.

DEPARTMENT OF MATHEMATICS
UNIVERSITY OF CALIFORNIA
BERKELEY, CA 94720
USA

PROBLEM 11.10

Editors' Note

The results similar to the ones mentioned in the Problem were obtained by V. A. Tolokonnikov [3] in the following slightly different setting: for which increasing functions α does the inequality

$$|f| \leqslant \alpha \left(\left(\sum_{k \geqslant 1} |f_k|^2 \right)^{1/2} \right); \qquad f, f_k \in H^\infty,$$

imply $f = \sum_{k \geqslant 1} f_k g_k$ with $g_k \in H^\infty$, $\sup_{\mathbb{D}} \sum_{k \geqslant 1} |g_k|^2 < \infty$? See also [4],[5].

References

3. Tolokonnikov V. A., *Estimates in the Carleson corona theorem, ideals of the algebra H^∞, a problem of Sz.-Nagy*, Zap. Nauchn. Sem. LOMI **113** (1981), 178–198 (Russian); English transl. in J. Soviet Math. **22** (1983), 1814–1828.
4. Tolokonnikov V. A., *Interpolating Blaschke products and the ideals of the algebra H^∞*, Zap. Nauchn. Sem. LOMI **126** (1983), 196–201 (Russian); English transl. in J. Soviet Math. **27** (1984), 2549–2553.
5. Tolokonnikov V. A., *The corona theorem in algebras of bounded analytic functions* VINITI (1984), no. 251–84DEP, 1–61 (Russian); English transl. in Amer. Math. Soc. Transl. **149** (1991), no. 2, 61–95.

11.11
v.old

TWO PROBLEMS ON THE SPECTRAL SYNTHESIS

N. K. Nikolski

1. Synthesis is impossible.

We are concerned with the synthesis of (closed) invariant subspaces of $\mathbf{z}^*$, the adjoint of the operator $\mathbf{z}$ of multiplication by the independent variable z on some space of analytic functions. More precisely, let X be a Banach space of functions defined in the unit disc $\mathbb{D}$ and analytic there, and suppose that $\mathbf{z}X \subset X$ and the natural imbedding $X \to \mathrm{Hol}(\mathbb{D})$ is continuous, $\mathrm{Hol}(\mathbb{D})$ being the space of all functions holomorphic in $\mathbb{D}$. If $f \in X$ then $k_f(\zeta)$ denotes the multiplicity of zero of f at a point ζ in $\mathbb{D}$, and for any function k from $\mathbb{D}$ to nonnegative integers let

$$X_k \stackrel{\text{def}}{=} \{\, f \colon f \in X, \quad k_f \geqslant k \,\}.$$

A closed $\mathbf{z}$-invariant subspace E of X is said to be *divisorial* (or to have *the d-property*) if $E = X_k$ for some k (necessarily $k(\zeta) = k_E(\zeta) \stackrel{\text{def}}{=} \min_{f \in E} k_f(\zeta)$, $\zeta \in \mathbb{D}$).

CONJECTURE 1. *In every space X as above there exist non-divisorial $\mathbf{z}$-invariant subspaces.*

The dualized d-property means that the spectral synthesis is possible. To be more precise, let Y be the space dual (or predual) to X equipped with the weak topology $\sigma(Y, X)$ (the duality of X and Y is determined by the Cauchy pairing, i.e. $<f, g> = \sum_{n \geqslant 0} \hat{f}(n) \hat{g}(n)$ for polynomials f, g). A $\mathbf{z}^*$-invariant subspace E of Y is said to be *synthesable* (or simply *s-space*) if

$$(1) \qquad E = \mathrm{span}\big((1 - \lambda z)^{-n} z^{n-1} \colon 1 \leqslant n \leqslant k(\lambda)\big)$$

with $k = k_{E^\perp}$. In other words E is an s-space if it can be recovered by the root vectors of $\mathbf{z}^*$ it contains.

All known results on $\mathbf{z}$-invariant subspaces (cf. [1]) support Conjecture 1. The main hypothesis on X here is that X should be a *Banach space*. The problem becomes non-trivial if, e.g. the set of polynomials $\mathcal{P}_A$ is contained and dense in X and $\{\zeta \colon \zeta \in \mathbb{C} \colon \sup_{p \in \mathcal{P}_A} |p(\zeta)| \, \|p\|_X^{-1} < \infty\} = \mathbb{D}$. The existence of a single norm defining the topology should lead to some limit stable peculiarities of the boundary behaviour of elements of X, and it is these peculiarities that should be responsible for the presence of non-divisorial $\mathbf{z}$-invariant subspaces. Spaces topologically contained in the Nevanlinna class provide leading examples. The aforementioned boundary effect consists here in the presence of a non-trivial inner factor (i.e. other than a Blaschke product) in the canonical factorization. Analogues of inner functions are discovered in classes of functions defined by growth restrictions ([2], [3], [4]); these classes are even not necessarily Banach spaces

30

but their topology is still "sufficiently rigid" (i.e. the seminorms defining the topology are of "comparable strength"). On the contrary, in spaces X with a "soft" topology the invariant subspaces are usually divisorial. Sometimes the "softness' of the topology can be expressed in purely quantitative terms (for example, under some regularity restrictions on λ, all ideals in the algebra $\{ f : f \in \mathrm{Hol}(\mathbb{D}), \ |f(\zeta)| = O(\lambda^c(\zeta)), \ c = c_f \}$ are divisorial if and only if $\int_0^1 \left(\frac{\log \lambda(r)}{1-r} \right)^{1/2} dr = +\infty$, [5], [10]). This viewpoint can be given a metric character; it can be connected with the multiplicative structure of analytic functions, with some problems of weighted polynomial approximation, with generalizations of the corona theorem, etc. (cf. [1,3,6])

2. Approximative synthesis is possible. Let us read formula (1) in the following manner: there is an increasing sequence $\{E_n\}$ of $\mathbf{z}^*$-invariant subspaces of finite dimension that approximates E:

$$E = \lim_n E_n \overset{\text{def}}{=} \{ f : f \in X, \lim_n \mathrm{dist}(f, E_n) = 0 \}.$$

Removing one word from this sentence seems to lead to a universal description of $\mathbf{z}^*$-invariant subspaces.

CONJECTURE 2. *Let Y be a space from section 1 and E be a $\mathbf{z}^*$-invariant subspace of Y. Then there exist subspaces E_n with $\mathbf{z}^* E_n \subset E_n$, $\dim E_n < \infty (n \in \mathbb{N})$ so that $E = \lim_n E_n$.*

There is a further extension of this Conjecture that still could look probable. Namely, let T be a continuous linear operator on a linear space Y and suppose that the system of root vectors of T is complete in Y. Is it true that $TE \subset E \implies E = \lim_n E_n$ for some sequence E_n with $TE_n \subset E_n$, $\dim E_n < \infty$ $n \in \mathbb{N}$)? But it is easy to see that without additional restrictions on T the answer to the last question is "no". A counterexample is provided by the left shift (i.e. still $\mathbf{z}^*$) $(a_0, a_1, \dots) \mapsto (a_1, a_2, \dots)$ on $\ell^p(w_n)$ with an appropriate weight $\{w_n\}_{n \geqslant 0}$ (decreasing rapidly and irregularly). This operator possesses invariant subspaces that can not be approximated by root subspaces, [3]. In examples of such kind it is essential that the spectrum of the operator reduces to the single point 0.

A plenty of classical theorems on $\mathbf{z}$-invariant subspaces (as e.g., Beurling's theorem) not only support Conjecture 2, but also allow to describe $\mathbf{z}^*$-cyclic vectors (that is, functions f with the property span $(\mathbf{z}^{*n} f : n \geqslant 0) = Y$) in terms of the approximation by rational functions with bounded "X-capacities". If r is a rational function with poles in $\mathbb{C} \setminus \mathrm{clos}\, \mathbb{D}$, $r(\infty) = 0$ then $\mathrm{cap}_X r \overset{\text{def}}{=} \inf \{ \|g\|_X : <z^n g, \cdot> = 0, \ n \geqslant 1; \ g(0) = 1 \}$. The capacity of an arbitrary $\mathbf{z}^*$-invariant subspace is defined similarly. If $f = (Y) - \lim_n r_n$ and $\sup \mathrm{cap}_X r_n < \infty$ then f is not cyclic for $\mathbf{z}^*$; analogously, $\sup_n \mathrm{cap}_X E_n < \infty \implies \lim_n E_n \neq Y$. The last assertion can be converted, after a slight modification of the notion of "capacity" [7,8]. Probably techniques of rational approximation should allow to prove Conjecture 2 avoiding estimates of "X-capacities" of rational functions (that appears to be a more difficult question; it is worth mentioning that this question is a quantitative form of the uniqueness theorem for X). The results on this matter known up to now use, on the contrary, not only classical uniqueness theorems but also the explicit description of z-invariant subspaces in terms of the inner-outer factorization.

REFERENCES

1. Nikol'skii N. K., *Invariant subspaces in operator theory and function theory*, Itogi Nauki i Tekhniki: Mat. Anal. **12** (1974), 199–412, VINITI, Moscow (Russian); English transl. in J. Soviet Math. **5** (1976), no. 2.
2. Krasichkov-Ternovskii I. F., *Invariant subspaces of analytic functions. II. Spectral synthesis on convex domains*, Mat. Sb. **88** (1972), no. 1, 3–30 (Russian); English transl. in Math. USSR Sb. **17** (1972), no. 1, 1–29.
3. Nikol'skii N. K., *Selected problems of weighted approximation and spectral analysis*, Trudy Mat. Inst. Steklov **120** (1974) (Russian); English transl. in Proc. Steklov Inst. Math. **120** (1974).
4. Korenbljum B., *A Beurling-type theorem*, Acta Math. **135** (1975), 187–219.
5. Apresyan S. A., *A description of the algebra of analytic functions admitting localization of ideals*, Zap. Nauchn. Sem. LOMI **70** (1977), 267–269 (Russian); English transl. in J. Soviet Math. **23** (1983), no. 1, 2091–2093.
6. Nikol'skii N. K., *The technique of using a quotient operator for the localization of z-invariant subspaces*, Dokl. Acad. Nauk SSSR **240** (1978), no. 1, 24–27 (Russian); English transl. in Soviet Math. Dokl. **19** (1978), no. 3, 545–549.
7. Gribov M. B., Nikolskii N. K., *Invariant subspaces and rational approximation*, Zap. Nauchn. Sem. LOMI **92** (1979), 103–114. (Russian)
8. Nikolskii N. K., *Lectures on the shift operator. I*, Zap. Nauchn. Sem. LOMI **39** (1974), 59–93 (Russian); English transl. in J. Soviet Math. **8** (1977), no. 1.
9. Hilden H. M., Wallen L. J., *Some cyclic and non-cyclic vectors of certain operators*, Indiana Univ. Math. J. **23** (1974), no. 7, 557–565.
10. Shamoyan F. A., *Division theorems and closed ideals in algebras of analytic functions with a majorant of finite growth*, Izv. Akad. Nauk Arm.SSR, Matematika **15** (1980), no. 4, 323–331. (Russian)

STEKLOV MATHEMATICAL INSTITUTE
ST. PETERSBURG BRANCH
FONTANKA 27   AND
ST. PETERSBURG, 191011
RUSSIA

UNIVERSITÉ BORDEAUX-I
UFR DE MATHÉMATIQUES
351, COURS DE LA LIBÉRATION
33405 TALENCE CEDEX
FRANCE

WEAK INVERTIBILITY AND FACTORIZATION
IN CERTAIN SPACES OF ANALYTIC FUNCTIONS

R. Frankfurt

A measure μ on $\mathbb{D}$ is called a *symmetric measure* if μ has the form $d\mu(r, \theta) = (2\pi)^{-1} d\nu(r) d\theta$, where ν is a finite, positive Borel measure on $[0, 1]$, having no mass at 0, and such that $\nu([r, 1]) > 0$ for all $0 \leqslant r < 1$. For any function f analytic in $\mathbb{D}$ and any p, $0 < p < \infty$, we define the generalized mean

$$(1) \qquad M_p(r; f; \mu) = \left(\int_{\text{clos}\,\mathbb{D}} |f(rw)|^p \, d\mu(w) \right)^{\frac{1}{p}},$$

$0 \leqslant r < 1$. The class $E^p(\mu)$ consists of all functions f analytic in $\mathbb{D}$ such that

$$(2) \qquad \|f\|_{p,\mu} = \sup_{r<1} M_p(r; f; \mu) < \infty.$$

In the special case where ν is a single unit point mass at 1, the means (1) reduce to the classical H^p means, and the $E^p(\mu)$ classes to the standard Hardy classes on $\mathbb{D}$. In all cases, $E^p(\mu)$ is isometrically isomorphic to the $L^p(\mu)$-closure of the polynomials. General properties of these classes are outlined in [1,2,3]. Numerous investigations of special cases (e.g., the Bergman classes, $\mu =$area measure) are scattered throughout the literature. A complete bibliography would be quite extensive, and so references here are restricted to those which have had the most direct influence upon the author's work.

A function $f, f \in E^p(\mu)$ is said to be *weakly invertible* if there is a sequence of polynomials $\{p_n\}$ such that $p_n f \to 1$ in the metric of $E^p(\mu)$. From an operator-theoretic point of view, such functions are significant in that an element of $E^p(\mu)$ is weakly invertible if and only if it is a *cyclic vector* for the operator of multiplication by z on $E^p(\mu)$. (When $p = 2$, this operator is unitarily equivalent to a subnormal weighted shift.) In the special case of the Hardy classes, Beurling [4] showed that a function is weakly invertible if and only if it is outer. In the more general context of the $E^p(\mu)$ classes, a complete characterization of the weakly invertible functions awaits discovery. At this juncture, however, it is not even clear what general shape such a characterization might take. We know of only a handful of scattered results which are applicable to these special classes. The earliest of these can be found in three papers by Shapiro [5,6,7] and in the survey article by Mergelyan [8]. More recent contributions have been made by the author [1,2], Aharonov, Shapiro and Shields [9]; and Hedberg (see Shields [10, p.112]).

Many of the known results on weakly invertible functions in the $E^p(\mu)$ classes are essentially either multiplication or factorization theorems. It is well known that the product of two outer functions is outer, and that any factor of an outer function is outer.

Do these properties carry over to weakly invertible functions in the $E^p(\mu)$ classes? We list a number of specific questions along these lines.

(a) Suppose $f, g, h \in E^p(\mu)$ and $f = gh$. If g and h are weakly invertible, is f weakly invertible? Conversely, if f is weakly invertible, are g and h weakly invertible?

(b) If $f \in E^p(\mu)$ is nonvanishing and $\frac{1}{f} \in E^q(\mu)$ for some $q, q > 0$, is f weakly invertible?

(c) If $f \in E^p(\mu)$ is weakly invertible and $\alpha > 0$, is f^α weakly invertible in $E^{p/\alpha}(\mu)$?

(d) If $f \in E^p(\mu)$ and f is weakly invertible in $E^q(\mu)$ for some $q, q < p$, is f weakly invertible in $E^p(\mu)$?

(e) Let $f \in E^p(\mu)$, $g \in E^q(\mu)$ and $h \in E^s(\mu)$, and let $f = gh$. If g and h are weakly invertible in $E^q(\mu)$ and $E^s(\mu)$, respectively, is f weakly invertible in $E^p(\mu)$? What about the converse?

Of course, all these things are trivially true in the special case of the Hardy classes. Question (e) is the most general of the list. The reader can easily convince himself that affirmative answers to (e) would imply affirmative answers to all the others. Conversely, affirmative answers to (a) and (d) together would yield affirmative answers to (e).

The answers to question (a) are known to be affirmative if $h \in H^\infty$ or if $g \in E^{p'}(\mu)$ and $h \in E^{q'}(\mu)$ with $\frac{1}{p'} + \frac{1}{q'} = \frac{1}{p}$ (see [2])*. These results are inspired by an earlier result of Shapiro [5, Lemma 2]. The question remains unanswered for unrestricted g and h.

Question (b) has a long history, and versions of it appear in numerous sources. An affirmative answer may be obtained by imposing the additional condition $f \in E^{p+\delta}(\mu)$ for some $\delta, \delta > 0$. The legacy of results of this type seems to begin with the paper of Shapiro [6], and has been carried forth into a variety of different settings in the separate researches of Brennan [11], Hedberg [12], and the author [2]. A similar result with a different kind of side condition is to be found in the work of Aharonov, Shapiro and Shields [9]. In its full generality, however, the question remains unanswered.

Question (d) seems in some sense to be the crucial question, certainly in moving from the setting of question (a) to that of question (e), but perhaps also in removing the side conditions from the results cited above. Presently, however, there seems to be little evidence either for or against an affirmative answer, nor can we offer any tangible ideas on how to attack the problem. The key to its solution in the special case of the Hardy classes rests upon the fact that, there. weak invertibility can be accounted for in terms of behavior within the larger Nevanlinna class. Unfortunately, in the more general setting of the $E^p(\mu)$ classes, none of the several different generalization of the Nevanlinna theory discovered to date seems to shed any light upon the matter. It may very well be that the answer to the question is negative. Clearly, a negative answer would introduce complications which have no parallel in the Hardy classes. However, in view of the negative results of Horowitz [13] concerning the zero sets of functions in the Bergman classes, such complications would not be too surprising, and perhaps not altogether unwelcome.

---

*See also [3] of 11.11 — Ed.

PROBLEM 11.12

REFERENCES

1. Frankfurt R., *Subnormal weighted shifts and related function spaces*, J. Math. Anal. Appl. **52** (1975), 471–489.
2. Frankfurt R., *Subnormal weighted shifts and related function spaces. II*, J. Math. Anal. Appl. **55** (1976), 1–17.
3. Frankfurt R., *Function spaces associated with radially symmetric measures*, J. Math. Anal. Appl. **60** (1977), 502–541.
4. Beurling A., *On two problems concerning linear transformations in Hilbert space*, Acta Math. **81** (1949), 239–255.
5. Shapiro H. S., *Weakly invertible elements in certain function spaces, and generators of ℓ_1*, Mich. Math. J. **11** (1964), 161–165.
6. Shapiro H. S., *Weighted polynomial approximation and boundary behaviour of analytic functions*, Modern problems of the theory of analytic functions, Nauka, Moscow, 1966, pp. 326–335.
7. Shapiro G., *Some observations concerning weighted polynomial approximation of holomorphic functions*, Matem. Sbornik **73** (1967), no. 3, 320–330. (Russian)
8. Mergeljan S. N., *On completeness of a system of analytic functions*, Uspekhi Matem. Nauk **8** (1953), no. 4, 3–63. (Russian)
9. Aharonov D., Shapiro H. S., Shields A. L., *Weakly invertible elements in the space of square-summable holomorphic functions*, J. London Math. Soc. **9** (1974), 183–192.
10. Shields A. L., *Weighted shift operators and analytic function theory*, Topics in Operator Theory, Amer. Math. Soc., Providence, R. I., 1974, pp. 49–128.
11. Brennan J., *Invariant subspaces and weighted polynomial approximation*, Ark. Mat. **11** (1973), 167–189.
12. Hedberg L. I., *Weighted mean approximation in Caratheodory regions*, Math. Scand. **23** (1968), 113–122.
13. Horowitz C., *Zeros of functions in the Bergman spaces*, Duke Math. J. **41** (1974), 693–710.

DEPT. OF MATH.,
COLLEGE OF ARTS AND SCIENCES
UNIVERSITY OF KENTUCKY
LEXINGTON 40506
USA

EDITORS' NOTE

See also 11.13, 11.14 and Commentary to 11.14.

WEAKLY INVERTIBLE ELEMENTS
IN BERGMAN SPACES

Boris Korenblum

DEFINITION 1. $\mathcal{H}^2$ is the Hilbert space of analytic functions f in $\mathbb{D}$ with the norm

$$(1) \qquad \|f\| = \left(\frac{1}{\pi} \int_{\mathbb{D}} |f(x+iy)|^2 \, dx \, dy\right)^{\frac{1}{2}}.$$

DEFINITION 2 [1]. Let $\mathcal{K}$ be the set of all open, closed and half-closed arcs $I, I \subset \mathbb{T}$, including all single points, $\mathbb{T}$ and $\emptyset$. A function $\mu \colon \mathcal{K} \to \mathbb{R}$ is called a *premeasure* iff

   (i) $\mu(I_1 \cup I_2) = \mu(I_1) + \mu(I_2)$, where $I_1, I_2 \in \mathcal{K}$, $I_1 \cup I_2 \in \mathcal{K}$, $I_1 \cap I_2 = \emptyset$;
   (ii) $\lim_{n} \mu(I_n) = 0$, where $I_n \in \mathcal{K}$, $I_1 \supset I_2 \supset \ldots$ and $\cap_n I_n = \emptyset$.

DEFINITION 3. A closed set $F, F \subset \mathbb{T}$ is called *Beurling–Carleson* (B.-C. set) iff

   (i) $|F| = 0$;
   (ii) $\sum_{n} |I_n| \log \frac{2\pi}{|I_n|} < +\infty,$

where I_n are the components of $\mathbb{T} \setminus F$ and $|\cdot|$ denotes the linear Lebesgue measure.

PROPOSITION 1 [1;2]. *Let $f \in \mathcal{H}^2$ and $f(z) \neq 0$ ($z \in \mathbb{D}$). Then the following properties hold:*

   (i) *The limit*

$$(2) \qquad \mu^*(I) = \lim_{r \to 1-0} \int_I \log |\hat{f}(r\xi)| \, |d\xi|$$

   exists for any arc $I, I \subset \mathbb{T}$;
   (ii) *The limit*

$$(3) \qquad \mu(I) = \lim_{n} \mu^*(I_n)$$

   exists for any sequence of closed arcs (I_n) such that $I_1 \subset I_2 \subset \ldots$ and $\bigcup_n I_n = I$, I being any open arc;
   (iii) *$\mu(I)$, defined by (3) for open arcs $I, I \subset \mathbb{T}$, admits a unique extension to a premeasure;*
   (iv) *for any B.-C. set F, whose complementary arcs are I_n, the series $\sum_n \mu(I_n)$ is absolutely convergent;*
   (v) *if we define*

$$(4) \qquad \sigma_f(F) = 2\pi \log |f(0)| - \sum_n \mu(I_n),$$

   for B.-C. sets F, then σ_f admits a unique extension to a finite non-positive Borel measure on every B.-C. set.

DEFINITION 4. The measure σ_f (defined on the set of all Borel sets contained in a B.-C. set) is called *the $\mathcal{X}$-singular measure associated with f, $f \in \mathcal{H}^2$* (it is assumed that $f(z) \neq 0$ in $\mathbb{D}$).

Proposition 1 follows immediately from the results of [1,2], since $f \in \mathcal{H}^2$ implies

$$(5) \qquad \log|f(z)| \leqslant c_f + \log\frac{1}{1 - |z|}.$$

DEFINITION 5. An element f, $f \in \mathcal{H}^2$, is called *weakly invertible (or cyclic)* iff $\text{clos}\{fg: g \in H^\infty\} = \mathcal{H}^2$.

PROPOSITION 2. *The following conditions are necessary for an element f, $f \in \mathcal{H}^2$, to be weakly invertible:*

$$(6) \qquad f(z) \neq 0 \quad (s \in \mathbb{D});$$

$$(7) \qquad \sigma_f = 0.$$

This proposition follows easily from the main theorem in [2] which gives a description of closed ideals in the topological algebra $A^{-\infty}$ of analytic functions f satisfying

$$|f(z)| \leqslant c_f(1 - |z|)^{-n_f} \qquad (s \in \mathbb{D}).$$

CONJECTURE 1. *Conditions (6) and (7) are sufficient for an f, $f \in \mathcal{H}^2$, to be weakly invertible.*

CONJECTURE 2. *The same conditions also describe weakly invertible elements in any Bergman space $\mathcal{H}^p$ $(1 \leqslant p < \infty)$ of analytic functions f with the norm*

$$\|f\| = \left(\frac{1}{\pi} \int_{\mathbb{D}} |f(x + iy)|^p \, dx \, dy\right)^{\frac{1}{p}}.$$

REFERENCES

1. Korenblum B., *An extension of the Nevanlinna theory*, Acta Math. **135** (1975), 187–219.
2. Korenblum B., *A Beurling-type theorem*, Acta Math. **138** (1977), 265–293.

DEPT. OF MATHEMATICS
STATE UNIVERSITY OF NEW YORK AT ALBANY
1400 WASHINGTON AVENUE
ALBANY, NEW YORK, 12222
USA

COMMENTARY

Both Conjectures are supported by the results cited in Commentary to 11.14. See also [3], [4], [5].

REFERENCES

3. Hedenmalm H., Shields A., *Invariant subspaces in Banach spaces of analytic functions*, Michigan Math. J. **37** (1990), 91–104.
4. Shields A., *Cyclic vectors in Banach spaces of analytic functions*, Operators and Function Theory (S. Power, ed.), 1984, pp. 315–350.
5. Korenblum B., *Outer functions and cyclic elements in Bergman spaces*, J. Funct. Anal. **115** (1993), 104–118.

11.14

v.old

CYCLIC VECTORS IN SPACES OF ANALYTIC FUNCTIONS

ALLEN L. SHIELDS

Let X denote a Banach space of analytic functions in $\mathbb{D}$ satisfying the following two conditions: (i) for each ζ, $\zeta \in \mathbb{D}$, the map $f \to f(\zeta)$ is a bounded linear functional on X, (ii) $zX \subset X$. It follows from (ii), by means of the closed graph theorem, that multiplication by z is a bounded linear transformation (more briefly, an operator) on X. Finally, $f \in X$ is said to be a *cyclic* vector for the operator of multiplication by z if the finite linear combinations of the vectors $f, zf, z^2 f, \ldots$ are dense in X (when the constant function **1** is in X, one also says that f is *weakly invertible* in X; this terminology was first used in [1]).

QUESTION 1. *Does strong invertibility imply weak invertibility?* (*That is, if* **1**, $\frac{1}{f}$, f *are all in* X, *is* f *cyclic?*)

Consider the special case when X is the Bergman space, that is the set of square–integrable analytic functions: $\|f\|^2 = \int_{\mathbb{D}} |f|^2 < \infty$.

CONJECTURE 1. *If* f *is in the Bergman space and if* $|f(z)| > c(1 - |z|)^a$ *for some* c, $a > 0$, *then* f *is cyclic.*

If correct this would imply an affirmative answer to QUESTION 1 when X is the Bergman space. The Conjecture is known to be correct under mild additional assumptions (see [2], [3], [4]). In particular it is correct when f is a singular inner function. In this case the condition in the hypothesis of the Conjecture is equivalent to the condition that the singular measure associated with f has the modulus of continuity $O(\delta \log 1/\delta)$ (see [1]).

CONJECTURE 2. *A singular inner function is cyclic in the Bergman space if and only if its associated singular measure puts no mass on any Carleson set.* (*For the definition of Carleson set see* [5], *pp. 326–327, and* 11.13).

For more discussion of the cyclicity of inner functions see §6 of [6], pages 54–58, where the possibility of an "inner–outer" factorization for inner functions is considered.

QUESTION 2. *Does there exist a Banach space of analytic functions, satisfying* (i) *and* (ii), *in which a function* f *is cyclic if and only if it has no zeros in* $\mathbb{D}$?

N. K. Nikolskii has shown [7] that no weighted sup–norm space of a certain type has this property. If such a space X existed then the operator of multiplication by z on X would have the property that its set of cyclic vectors is non-empty, and is a closed subset of the space $X \setminus \{0\}$ (this follows since the limit of non-vanishing analytic functions is either non-vanishing or identically zero). No example of an operator with this property and non-trivial lattice of invariant subspaces is known. H. S. Shapiro has shown that for any operator the set of cyclic vectors is always a G_δ set (see [8], §11, Proposition 40,

p. 110). For a discussion of some of these questions from the point of view of weighted shift operators, see [8], §§11, 12.

QUESTION 3. *Let X be as before, and let $f, g \in X$ with g cyclic. If $\left|f(z)\right| \geqslant \left|g(z)\right|$ in $\mathbb{D}$, is f cyclic?*

This question has a trivial affirmative answer in spaces like the Bergman space, since bounded analytic function multiply the space into itself. It is unknown for the Dirichlet space (that is, the space of functions with $\int_{\mathbb{D}} |f'|^2 < \infty$); the special case $g = \text{constant}$ is established in [9].

REFERENCES

1. Shapiro Harold S., *Weakly invertible elements in certain function spaces, and generators in ℓ^1*, Mich. Math. J. **11** (1964), 161–165.
2. Shapiro Harold S., *Weighted polynomial approximation and boundary behaviour of holomorphic functions*, Contemporary problems of analytic function theory, Nauka, Moscow, 1966, pp. 326–335.
3. Shapiro H., *Some remarks on weighted polynomial approximation of holomorphic functions*, Mat. Sbornik **73** (1967), 320–330.
4. Aharonov D., Shapiro H. S., Shields A. L., *Weakly invertible elements in the space of square-summable holomorphic functions*, J. London Math. Soc. **9** (1974), 183–192.
5. Carleson L., *Sets of uniqueness for functions regular in the unit circle*, Acta Math. **87** (1952), 325–345.
6. Duren P. L., Romberg B. W., Shields A. L., *Linear functionals on H^p spaces with $0 < p < 1$*, J. für Reine und Angew. Math. **238** (1969), 32–60.
7. Nikolskii N. K., *Spectral synthesis and weighted approximation problem in spaces of analytic functions*, Izv. AN Arm. SSR. Ser. Matem. 6 (1971), no. 5, 345–367. (Russian)
8. Shields Allen L., *Weighted shift operators and analytic function theory*, Topics in operator theory, vol. 13, Amer. Math. Soc., Providence, 1974, pp. 49–128.
9. Shields Allen L., *Cyclic vectors in some spaces of analytic functions*, Proc. Royal Irish Acad. **74**, Section A (1974), 293–296.

COMMENTARY

QUESTION 1 has been answered in the negative by Shamoyan [10].

THEOREM ([10]). *Let*

$$\varphi(\zeta) = \left|\frac{1+\zeta}{1-\zeta}\right|^2 + \left|\frac{1+\zeta}{1-\zeta}\right|^\alpha, \qquad \zeta \in \mathbb{D}, \quad 0 < \alpha < 1$$

and let X_φ denote the space of all functions f analytic in the unit disc $\mathbb{D}$, continuous in the $\text{clos}\,\mathbb{D} \setminus \{1\}$ and such that $|f(\zeta)| = o\big(e^{\varphi(\zeta)}\big)$ for $|\zeta| < 1, \zeta \to 1$. Then polynomials are dense in X_φ and for $f \stackrel{\text{def}}{=} \exp\left(-\left(\frac{1+z}{1-z}\right)^2\right)$ we have: $f, f^{-1} \in X_\varphi$ but f is not weakly invertible in X_φ.

CONJECTURE 2. The "only if" part can be found in [2] of 11.13, the "if" part is proved in [11]. The same criterion of weak invertibility of inner functions holds in all Bergman spaces $\mathcal{H}^p$, $1 \leqslant p < \infty$, in spaces $A^p \stackrel{\text{def}}{=} \{ f : f$ is analytic in $\mathbb{D}$ and $|f(\zeta)| = o\big((1 - |\zeta|)^{-p}\big), |\zeta| \to 1 \}$ and in $\cup_{p>0}\mathcal{H}^p = \cup_{p>0} A_p$.

Note, by the way, that 11.11, 11.12 contain conjectures in the spirit of QUESTIONS 2–3, and that both QUESTION 2 and 3 (together with some others) are discussed in ref. [3] of 11.11. For more information see also [3], [4] of 11.13.

REFERENCES

10. Shamoyan F. A., *Weak invertibility in some spaces of analytic functions*, Dokl. AN Arm. SSR **74** (1982), no. 4, 157–161.
11. Korenblum B., *Cyclic elements in some spaces of analytic functions*, Bull. Amer. Math. Soc. **5** (1981), no. 3, 317–318.

INVARIANT SUBSPACES
OF THE BACKWARD SHIFT OPERATOR
IN THE SMIRNOV CLASS

A. B. ALEKSANDROV

Denote by N_* the Smirnov class i.e., the space of all functions f holomorphic in the unit disc $\mathbb{D}$ and such that $\{\log^+ |f_r|\}_{0<r<1}$ is uniformly integrable in $L^- = L^1(\mathbb{T}, m)$. Here $f_r \stackrel{\text{def}}{=} f(rz)$, m is the normalized Lebesgue measure on the unit circle $\mathbb{T}$. The space N_* can be identified with the closure of the set of polynomials (in z) in $\log L$, where $\log L$ is the space of all measurable functions f on $\mathbb{T}$ such that $\log(1 + |f|) \in L^1$. In $\log L$ the distance ρ is introduced by $\rho(f, g) \stackrel{\text{def}}{=} \int_{\mathbb{T}} \log(1 + |f - g|) \, dm$. Let S^* denote the backward shift operator, $S^* f = \frac{f - f(0)}{z}$.

1. Invariant subspaces and rational approximation.

PROBLEM. *Describe the invariant subspaces of* $S^*\colon N_* \to N_*$.

It should be noted that an analogous problem for the shift operator $S\colon N_* \to N_*$, $Sf = zf$, can be reduced easily to the famous Beurling theorem describing the invariant subspaces of $S\colon H^2 \to H^2$ (see [4]).

The Problem is connected with the description of the closure in N_* of the linear span of the Cauchy kernels $\left\{\frac{1}{1-\bar\zeta z}\right\}_{\zeta \in F}$, F being a closed subset of $\mathbb{T}$. In [1] and [2] analogous problems are solved for the case of the Hardy spaces H^p $(0 < p < 1)$. In this case the real variable characterization of H^p (see [6]) plays an important role.

CONJECTURE 1. *If* F *has no isolated points then the closure of the linear span of* $\left\{\frac{1}{1-\bar\zeta z}\right\}_{\zeta \in F}$ *is the set of all functions* $f \in N_*$ *such that* $z\bar f \in N_*$ *and* f *has an analytic continuation to* $\hat{\mathbb{C}} \setminus F$.

The case $F = \mathbb{T}$ is considered in [4].

2. Examples of S^*-invariant subspaces.

Let $X \subset N_*$ and let I be an inner function. Set $I^*(X) \stackrel{\text{def}}{=} \{ f \in X : zI\bar f \in N_* \}$. Denote by $\sigma(I)$ the spectrum of I (see [7]). Let F be a closed subset of $\mathbb{T}$, $F \supset \sigma(I) \cap \mathbb{T}$. We say that a function $k\colon F \to \mathbb{N} \cup \{\infty\}$ is I-admissible if $k(\zeta) = \infty$ for all $\zeta \in \sigma(I) \cap \mathbb{T}$ and for all non-isolated points $\zeta \in F$. Denote by $I^*(N_*, F, k)$ the set of all functions $f \in I^*(N_*)$ having a meromorphic continuation $\tilde f$ to $\hat{\mathbb{C}} \setminus F$ such that ζ is a pole of $\tilde f$ of order at most $k(\zeta)$ for all $\zeta \in F$ with $k(\zeta) \neq \infty$. It is easy to see that $I^*(N_*, F, k)$ is an invariant subspace of $S^*\colon N_* \to N_*$.

Let E be an invariant subspace of $S^*\colon N_* \to N_*$, $E \neq N_*$. Then $E \cap H^\infty$ is an invariant subspace of $S^*\colon H^\infty \to H^\infty$ and $E \cap H^\infty \neq H^\infty$. Hence $E \cap H^\infty = I^*(H^\infty)$

for some inner function $I = I_E$ (see [5]). Let us construct a closed subset $F = F_E$ of $\mathbb{T}$ and an I-admissible function $k = k_E$ as follows

$$F \overset{\text{def}}{=} \{\zeta \in \mathbb{T}: (1 - \bar{\zeta}z)^{-1} \in E\}, \qquad k(\zeta) \overset{\text{def}}{=} \sup\{n \in \mathbb{N}: (1 - \bar{\zeta}z)^{-n} \in E\}.$$

CONJECTURE 2. $E = I^*(N_*, F, k)$.

In case $F = \mathbb{T}$ the conjecture is true, see [4, Corollary 5.2.3].

Results of the following section imply the inclusion $E \subset I^*(N_*)$.

3. Cyclic vectors of S^*. Let P_+ denote the Riesz projection. The following proposition can be obtained from [5, Remark 1].

PROPOSITION. *Let (X, Y) be a Smirnov dual pair having Properties $1°$ and $2°$ (see [5]). Suppose that $X \supset H^\infty$, $Y \subset L^1/H^1_- \overset{\text{def}}{=} P_+(L^1)$ and $(f, P_+g) = \int f\bar{g}\,dm$ if $f \in H^\infty$, $g \in L^1$, $P_+g \in Y$. Let E be an invariant subspace of $S^*: X \to X$. Then if $E \cap H^\infty \subset I^*(H^\infty)$ for some inner function I, then $E \subset I^*(X)$.*

The proposition allows to generalize Theorem 5.2.4 in [4]. Let Ω be a function holomorphic in $\mathbb{D}$, $|\Omega| \geqslant 1$ everywhere in $\mathbb{D}$ and $\Omega \in N_*$. Using the Proposition for the Smirnov dual pair $(\Omega H^\infty, P_+(\Omega^{-1}L^1))$, $(f, P_+g) \overset{\text{def}}{=} \int f\bar{g}\,dm$, we can prove the following

THEOREM. *Let X be a Hausdorff topological vector space, $X \subset N_*$, $S^*X \subset X$. Suppose that X has the following property:*

$$f_n \in X,\ g \in X,\ f \in N_*,\ f_n \to f \quad \text{a.e.},\ |f_n| \leqslant |g| + 1 \quad \text{a.e.} \Longrightarrow f \in X,\ \lim_{n \to \infty} f_n = f$$

(in X). Let E be an invariant subspace of $S^*: X \to X$, $E \cap H^\infty = I^*(H^\infty)$ for some inner function I. Then $E \subset I^*(X)$.

COROLLARY 1. *If E is an invariant subspace of $S^*: N_* \to N_*$, $E \neq N_*$, then $E \subset I^*(N_*)$, where $I = I_E$.*

COROLLARY 2. *The closure of the linear span of the family $\{S^{*n}f\}_{n \geqslant 0}$ ($f \in N_*$) does not coincide with N_* iff f has a pseudocontinuation (see [7] for the definition).*

4. Distribution of values of functions in $1^*(N_*)$. Denote by $f^*: (0,1) \to \mathbb{R}$ the decreasing rearrangement of $|f|$, where f is a measurable function on $\mathbb{T}$. In [3] (see also [4]) it is proved that if $f \in 1^*(N_*)$ and $\varlimsup_{t \to 0} f^*(t) \cdot t = 0$, then $f \equiv 0$.

CONJECTURE 3. *If $f \in 1^*(N_*)$ and $\varliminf_{t \to 0} f^*(t) \cdot t = 0$, then $f \equiv 0$.*

Results of [3,7] imply that if $f \in 1^*(N_*)$, $\varlimsup_{t \to 0} f^*(t) \cdot t < +\infty$, then $\varliminf_{t \to 0} t f^*(t) = \varlimsup_{t \to 0} f^*(t) \cdot t$.

REFERENCES

1. Aleksandrov A. B., *Approximation by rational functions and an analog of the M. Riesz theorem on conjugate functions for L^p-spaces with $p \in (0, 1)$*, Math. USSR Sbornik **35** (1979), 301–316.
2. Aleksandrov A. B., *Invariant subspaces of the backward shift operator in the space H^p ($p \in (0, 1)$)*, Zapiski nauchn. semin. LOMI **92** (1979), 7–29. (Russian)
3. Aleksandrov A. B., *On the A-integrability of boundary values of harmonic functions*, Matem. zametki **30** (1981), no. 1, 59–72. (Russian)

4. Aleksandrov A. B., *Essays on non locally convex Hardy classes*, Lect. Notes Math. **864** (1981), 1–89.
5. Aleksandrov A. B., *Invariant subspaces of the shift operators. Axiomatic approach*, Zapiski nauchn. semin. LOMI **113** (1981), 7–26 (Russian); English transl in J. Soviet Math. **22** (1983), no. 6, 1695–1708.
6. Coifman R. R., *A real variable characterization of H^p*, Studia Math. **51** (1974), no. 3, 269–274.
7. Nikolskii N. K., *Treatise of the Shift Operator*, Springer-Verlag, 1986.

STEKLOV MATHEMATICAL INSTITUTE
ST. PETERSBURG BRANCH
FONTANKA 27
ST. PETERSBURG, 191011
RUSSIA

COMPLETENESS OF TRANSLATES
OF A GIVEN FUNCTION IN A WEIGHTED SPACE

V. P. GURARII

Let $L^1_\varphi(\mathbb{R})$ be the Banach space of measurable functions f with $f\varphi \in L^1(\mathbb{R})$, the norm in L^1_φ being defined by: $\|f\| = \int_\mathbb{R} |f|\varphi$. Here the weight φ is a measurable function satisfying

(1)
$$1 \leqslant \varphi(t+\tau) \leqslant \varphi(t)\varphi(\tau), \qquad \forall t, \tau \in \mathbb{R}.$$

The multiplication being defined by

$$(f_1 * f_2)(t) \stackrel{\text{def}}{=} \int_\mathbb{R} f_1(t-\tau) f_2(\tau) \, d\tau$$

the space becomes a Banach algebra without unit; that follows from (1).

Condition (1) ensures the existence of finite limits $\lim\limits_{t \to \pm\infty} \frac{\log \varphi(t)}{t} = \alpha_\pm$ and so for f in L^1_φ the Fourier transform $\mathcal{F}f = \frac{1}{2\pi} \int f(t) e^{itz} \, dt$ turns out to be continuous in the strip $\Pi = \{ z : \alpha_- \leqslant \operatorname{Im} z \leqslant \alpha_+ \}$ and analytic in int Π. The maximal ideal space for L^1_φ is homeomorphic to Π ([1]), and any maximal ideal $M(z_0)$ ($z_0 \in \Pi$) has the form:

$$M(z_0) = \{ f \in L^1_\varphi : \mathcal{F}f(z_0) = 0 \}.$$

Let us consider together with L^1_φ its closed subalgebra

$$L^1_\varphi(\mathbb{R}_+) = \{ f \in L^1_\varphi(\mathbb{R}) : f(t) = 0, \ \forall t \in \mathbb{R}_- \}.$$

The maximal ideal space of $L^1_\varphi(\mathbb{R}_+)$ is homeomorphic to the half-plane $\operatorname{Im} z \leqslant \alpha_+$, and the Fourier transform of function f, $f \in L^1_\varphi(\mathbb{R}_+)$ turns out to be continuous in this half-plane and analytic in its interior.

THE PROBLEM we treat here is the following: *let $\mathfrak{M}$ be a family of functions in $L^1_\varphi(\mathbb{R})$ ($\mathfrak{M}$ may consist of a single element) and let $I_\mathfrak{M}$ be the space spanned by all translates of functions from $\mathfrak{M}$. What are the conditions on $\mathfrak{M}$ for $I_\mathfrak{M}$ to coincide with $L^1_\varphi(\mathbb{R})$, i.e. when every function in $L^1_\varphi(\mathbb{R})$ can be approximated in $L^1_\varphi(\mathbb{R})$ by linear combinations of translates of functions in $\mathfrak{M}$?*

The problem for $\varphi \equiv 1$ ($L^1_\varphi(\mathbb{R}) = L^1(\mathbb{R})$) was stated by N. Wiener [2], he proved that $I_\mathfrak{M} = L^1(\mathbb{R})$ if and only if the Fourier transforms of $\mathfrak{M}$ have no common zero on $\mathbb{R}$. Since $I_\mathfrak{M}$ is the smallest closed ideal of L^1 containing $\mathfrak{M}$, Wiener's theorem means that a closed ideal of $L^1(\mathbb{R})$ is contained in no maximal ideal if and only if it is equal to $L^1(\mathbb{R})$.

A. Beurling [3] discovered the validity of Wiener's theorem for the space $L^1_\varphi(\mathbb{R})$ if the weight φ satisfies condition (1) and

$$(2) \qquad \int_\mathbb{R} \frac{\log \varphi(t)}{1+t^2}\, dt < \infty.$$

Later it turned out ([1]) that simple "Banach algebra" arguments prove both Wiener's and Beurling's theorem and work in a more general case of any regular Banach algebra. The regularity of L^1_φ is ensured by (2).

The Wiener-type theorem for non-regular Banach algebras were obtained by B. Nyman [4], who proved the following theorem for the case $\varphi(t) = e^{\alpha|t|}$, $\alpha > 0$: $I_{\mathfrak{M}} = L^1_\varphi(\mathbb{R})$ iff Fourier transforms of $\mathfrak{M}$ have no common zero in the strip $|\operatorname{Im} z| \leqslant \alpha$ and

$$(3) \qquad 0 = \sup_{f \in \mathfrak{M}} \varlimsup_{x \to \infty} \frac{\log |\mathcal{F}f(x)|}{\exp \frac{\pi x}{\alpha}} = \sup_{f \in \mathfrak{M}} \varlimsup_{x \to -\infty} \frac{\log |\mathcal{F}f(x)|}{\exp \frac{\pi x}{\alpha}}.$$

Scrutinizing conditions (3) we can see that in the algebra $L^1_\varphi(\mathbb{R})$ there are closed proper ideals contained in no maximal ideal. Such ideals will be called *prime ideals corresponding to infinity points of the strip* $|\operatorname{Im} z| \leqslant \alpha$.

Independently Nyman's result was rediscovered by B. Korenblum, who described completely the prime ideals corresponding to infinity points of the strip $|\operatorname{Im} z| \leqslant \alpha$. The question concerning Wiener-type theorems in algebras $L^1_\varphi(\mathbb{R})$, where the weight φ satisfies (1) and

$$(4) \qquad \int_\mathbb{R} \frac{\log \varphi(t)}{1+t^2}\, dt = \infty,$$

$\alpha_+ = \alpha_- = 0$, is still open.

The methods used earlier don't work in this case. The author does not know a necessary and sufficient condition for the validity of approximation theorem even under very restrictive conditions of regularity of the weight φ (for example $\varphi(t) = \exp(\frac{|t|}{\log(1+|t|)})$).

For a weight φ, satisfying (1), (4) one can find chains of prime ideals corresponding to infinity points (see, for example, [6]). The reason for existence of prime ideals of this sort is that by (4) there are the functions in $L^1_\varphi(\mathbb{R})$ with the greatest possible rate of decrease of Fourier transform (see [7], [8]). It remains unknown whether all prime ideals are of this sort.

There is a similar question for the algebra $L^1_\varphi(\mathbb{R}_+)$. We have the following theorem [9]: Let φ satisfy (1), (2), and let $I_{\mathfrak{M}}$ be the closure of the linear span of all right translates of $\mathfrak{M}$.

Then $I_{\mathfrak{M}} = L^1_\varphi(\mathbb{R}_+)$ if and only if the following two conditions are fulfilled:

1) there is no interval I adjacent to the origin and such that all functions in $\mathfrak{M}$ vanish a.e. on I,

2) Fourier transforms of functions in $\mathfrak{M}$ have no common zero in $\operatorname{Im} z \leqslant 0$.

It is worthwhile to note, that this case is simpler than the case of $L^1_\varphi(\mathbb{R})$ because there are no prime ideals of the above type. (The case when $\varphi = e^{\alpha t}$ doesn't differ from the case when $\varphi = 1$). We conjecture that the previous assertion remains true also for the case when φ satisfies condition (4). In particular the following conjecture can be stated:

CONJECTURE. *Let* φ *satisfy* (1), $f \in L^1_\varphi(\mathbb{R}_+)$ *and* $g \in L^\infty_\varphi(\mathbb{R}_+)$. *If the Fourier transform* $\mathcal{F}f$ *doesn't vanish in* $\operatorname{Im} z \leqslant 0$ *and if* $\varlimsup_{y \to -\infty} \dfrac{\log |\mathcal{F}f(iy)|}{|y|} = 0$, *then the equality* $\int_{\mathbb{R}_-} f(t)g(t+\tau)\, dt = 0 \;\forall \tau > 0$ *implies* $g(t) = 0$ *a.e. on* $\mathbb{R}_+$.

There are some reasons to believe the conjecture is plausible. We have no possibility to describe them in detail here, but we can note that condition (2) is much weaker than the condition

$$\int_{\mathbb{R}_+} \frac{\log \varphi(t)}{1 + t^{3/2}} < \infty$$

which is a well-known condition of "non-quasianalyticity" of $L^1_\varphi(\mathbb{R}_+)$. A more detailed motivation of the above problem and a list of related problems of harmonic analysis can be found in [10].

In conclusion we should like to call attention to a question on the density of right translates in $L^1(\mathbb{R})$. Let $f \in L^1(\mathbb{R})$, let I_f be the closure in $L^1(\mathbb{R})$ of linear span of all right translates of f. It is easy to prove that $I_f = L^1(\mathbb{R})$ implies that

$$\int_{\mathbb{R}} \frac{\log |\mathcal{F}f(x)|}{1 + x^2}\, dx = -\infty$$

and that $\mathcal{F}f$ is nowhere zero. However these conditions are not sufficient. There are some sufficient conditions but unfortunately they are far from being necessary. I think that it deserves attention to find necessary and sufficient conditions.

REFERENCES

1. Gelfand I. M., Raikov D. A., Shilov G. E., *Commutative Normed Rings*, Phys.-Math., Moscow, 1960. (Russian)
2. Wiener N., *Fourier Integral and some its Applications*, Phys.-Math., Moscow, 1963. (Russian)
3. Beurling A., *Sur les integrales de Fourier absolument convergentes et leur application a une transformation fonctionelle*, Congres des Math. Scand., Helsingfors, 1938.
4. Nyman B., *On the one-dimensional translations group and semi-group in certain function spaces*, Thesis, Uppsala, 1950.
5. Korenblum B. I., *Generalization of Tauber's Wiener's theorem and harmonic analysis of quickly growing functions*, Trudy Moskovskogo mat. ob-va **7** (1958), 121–148. (Russian)
6. Vretblad A., *Spectral analysis in weighted L^1-spaces on $\mathbb{R}$*, Ark. Math. **11** (1973), 109–138.
7. Djarbashyan M. M., *Uniqueness theorems for Fourier transforms and for infinitely differentiable functions*, Izv. AN Arm.SSR ser.f.-m. nauk **10** (1957), no. 6, 7–24. (Russian)
8. Babenko K. I., *About some classes of infinitely differentiable function's spaces*, Dokl.AN USSR **132** (1960), no. 6, 1231–1234. (Russian)
9. Gurarii V. P., Levin B. Ya., *About completeness of system of shifts in the space $L(0, \infty)$ with weight*, Zap. mekh.-mat. facult. KhGY and KhMO **30** (1964), no. 4, 178–185. (Russian)
10. Gurarii V. P., *Harmonic analysis in the spaces with weight*, Trudy Mosk. mat. ob-va **35** (1976), 21–76. (Russian)

CHEMISTRY AND PHYSICS INSTITUTE
OF ACADEMY OF SCIENCES OF RUSSIA
TCHERNOGOLOVKA, MOSCOW REGION, 142432
RUSSIA

COMMENTARY (1993)

The conjecture is answered in the affirmative. The first step is to deal with weighted Hilbert spaces $L^2_\varphi(\mathbb{R}_+)$; for $1 \leqslant p < \infty$ we say that f is in $L^p_\varphi(\mathbb{R}_+)$ if and only if $\int_{\mathbb{R}_+} |f(t)|^p \varphi(t)^p \, dt < \infty$.

THEOREM 1([11]). *Let φ be such that for some ε, $\varepsilon > 0$, $\log \varphi(x) - (\frac{1}{2} + \varepsilon) \log x$ is concave for big x. If $\mathfrak{M}$ is a family of elements in $L^2_\varphi(\mathbb{R}_+)$, then the set of right translates of elements $f \in \mathfrak{M}$ is complete in $L^2_\varphi(\mathbb{R}_+)$ if and only if conditions 1) and 2) mentioned before the conjecture are fulfilled.*

It is possible to extend this result to the general L^p situation.

THEOREM 2([12]). *Let $1 \leqslant p < \infty$, and φ be such that $\log \varphi(x) - 4 \log x$ is concave for big x. Then the conclusion of Theorem 1 holds with $L^2_\varphi(\mathbb{R}_+)$ replaced by $L^p_\varphi(\mathbb{R}_+)$.*

REFERENCES

11. Borichev A., Hedenmalm H., *Completeness of translates in weighted Hilbert spaces on the half-line*, Centre de recerca matemàtica, Barcelona, Preprint No. 196 (1993).
15. Borichev A., Hedenmalm H., *Completeness of translates in weighted L^p spaces on the half-line*, Uppsala University, Dept. of Math. Report (1993).

11.17
v.old

A CLOSURE PROBLEM FOR FUNCTIONS ON $\mathbb{R}_+$

YNGVE DOMAR

A *weight function* w is here a positive, bounded, decreasing function on $\mathbb{R}_+$, satisfying $x^{-1}\log w(x) \to -\infty$, as $x \to \infty$. L_w is the Banach space of functions f on $\mathbb{R}_+$ with $fw \in L^1(\mathbb{R}_+)$. For every $a, a \in \mathbb{R}_+ \cup \{0\}$ the translation T_a, defined by

$$T_a f(x) = \begin{cases} 0, & 0 < x \leqslant a \\ f(x-a), & x > a \end{cases},$$

is a contraction in L_w. A_w is the set of all f, $f \in L_w$ which do not vanish almost everywhere near 0, and B_w is the set of *cyclic* elements in L_w, i.e. elements f such that the translates $T_a f$, $a > 0$, span a dense subspace. Obviously $B_w \subset A_w$.

$$\text{Is} \quad B_w = A_w \, ?$$

Some light is thrown on this problem by the corresponding problem on $\mathbb{Z}_+ \cup \{0\}$. A *weight sequence* is a positive decreasing sequence $w = (w_n)_{n \geqslant 0}$, satisfying

$$n^{-1}\log w_n \to -\infty \quad \text{as} \quad n \to \infty.$$

ℓ_w is the Banach space of sequences $c = (c_n)_{n \geqslant 0}$ with $cw = (c_n w_n)_{n \geqslant 0} \in \ell^1(\mathbb{Z}_+ \cup \{0\})$ and the translations T_m are defined as above, giving contractions of ℓ_w. A_w is the set of c, $c \in \ell_w$, with $c_0 \neq 0$, B_w is the set of cyclic elements. $B_w \subset A_w$, and we ask whether $B_w = A_w$. This time results are easier to obtain. Let us say that w is of submultiplicative type if $w_{m+n} \leqslant C w_m w_n$, $m, n \in \mathbb{Z}_+$, for some constant C. In that case, ℓ_w is a unital Banach algebra under convolution, with $\ell_w \setminus A_w$ as its only maximal ideal, and $B_w = A_w$ follows from elementary Banach algebra theory. It should be observed that the submultiplicativity condition is an assumption on the regularity of w, not a restriction of its growth at ∞. If this condition is not fulfilled, there are cases, when $B_w = A_w$ [1], and other cases when $B_w \neq A_w$ [2], [1].

In an analogous way we say that a weight function w is of submultiplicative type, if $w(x-y) \leqslant C w(x) w(y)$, $x, y \in \mathbb{R}_+$, for some constant C. Using the results of Nikolskii and Styf it is easy to produce weight functions w, of non-submultiplicative type, for which $B_w = A_w$. But if we from now on restrict the attention to weight functions of submultiplicative type, we can in no single case answer the question whether $B_w = A_w$. It is tempting to conjecture, in analogy to the discrete case, that the answer is affirmative for every w. Now again we have a convolution Banach algebra, but the absence of a unit prevents us from carrying over the arguments from the discrete case. A vague indication that the answer perhaps is yes, at least if w tends to zero rapidly at infinity, is given by the circumstance that $B_w = A_w$ if the corresponding problem is formulated

48

in the limiting case when w is non-negative and vanishing for large x. (This follows from Titchmarsh's theorem).

It is a direct consequence of Hahn–Banach's theorem that $f \in B_w$ if and only if the convolution equation

$$\int_0^\infty \varphi(x+y)f(y)\,dy = 0, \qquad x \in \mathbb{R}_+$$

has the zero functions as only solution with $\varphi/w \in L^\infty(\mathbb{R}_+)$. Thus $B_w \neq A_w$ if and only if there exists $f \in A_w$ such that the equation has a non-zero solution. Maybe function theory, in particular the theory of special functions, can provide an example showing that $B_w = A_w$ for at least some w.

Here are some *sufficient conditions* for $f \in B_w$

1. $f \in B_w$ if $f \in A_w$ and $\int_0^\infty |f(x)|e^{bx}\,dx < \infty$ for some b, $b \in \mathbb{R}$.

   (This follows directly from the results in [3] or [4], and is valid also for w of non-submultiplicative type.)

2. Suppose $-\log w$ is convex and $x^{-2}\log w \to -\infty$, as $x \to \infty$. Let $f \in L_w$ and suppose that for some f_1, $f_1 \in L^1(\mathbb{R}_+)$, with compact support and coinciding with f near 0,

$$\left| \int_0^\infty f_1(x)e^{-yx}\,dx \right| \geq \exp\left\{ -Cy\left[w^{-1}\left(\int_0^\infty w(x)e^{yx}\,dx \right)^{-1} \right]^{-1} \right\}$$

   for large y, $y \in \mathbb{R}_+$, where C is a constant and w^{-1} denotes the inverse of w. Then $f \in B_w$. (In particular, $\log w(x) \sim -x^p$, $p > 2$ yields the right hand member $\sim \exp\{-Cy^\alpha\}$, for some C, where $\alpha = (p-2)/(p-1)$.)

3. Suppose $-\log w$ is convex and $(x\log x)^{-1}\log w \to -\infty$, as $x \to \infty$. Let $f \in L_w$ and suppose that f is of bounded variation near zero with $f(+0) \neq 0$. Then $f \in B_w$.

REFERENCES

1. Styf B., *Closed translation invariant subspaces in a Banach space of sequences, summable with weights*, Uppsala University, Dept. of Math., Report **3** (1977).

2. Nikolskii N. K., *About invariant subspaces of weighted shift operators*, Matem. Sbornik **74** (1967), no. 2, 171–190 (Russian); English transl. in Math. USSR Sbornik **74(116)** (1967), 172–190.

3. Nyman B., *On the one-dimensional translation group and semi-group in certain function spaces*, Uppsala, 1950.

4. Gurarii V. P., *Spectral synthesis of bounded functions on a half-axis*, Funktsional. Anal. i Prilozhen. **3** (1969), no. 4, 34–48 (Russian); English transl. in Funct. Anal. Appl. **3** (1969), no. 4, 282–294.

UPPSALA UNIVERSITET
MATEMATISKA INSTITUTIONEN
BOX 480
S–751 06 UPPSALA
SWEDEN

COMMENTARY

The proofs of Proposition 1–3 can be found in [5]. In an important paper [6] it is shown that $A_w = B_w$ if $\log w$ is eventually concave and $(\log x)^{1/2} = o(\log|x^{-1}\log w(x)|)$ for $x \to \infty$. An analogous theorem is proved under the same hypotheses for the spaces L_w^p, $1 \leqslant p < \infty$. These results are derived from a general theorem on convolution equations in L_w^p-spaces which is a strong form of the famous Titchmarsh theorem.

Concerning all these and many other problems on translation invariant subspaces and ideals in $L_w^p(\mathbb{R}_+)$, $L_w^p(\mathbb{R})$ see also [7], a very informative book.

REFERENCES

5. Domar Y., *Cyclic elements under translation in weighted L^1 spaces on $\mathbb{R}^+$*, Ark. mat. **19** (1981), no. 1, 137–144.
6. Domar Y., *Extensions of the Titchmarsh convolution theorem with applications in the theory of invariant subspaces*, Proc. London Math. Soc. **46** (1983), no. 3, 288–300.
7. Bachar J. M., Bade W. G., Curtis P. C. Jr., Dales H. G., Thomas M. P. (eds.), *Radical Banach algebras and automatic continuity*, Proceedings, Long Beach 1981, Lect. Notes in Math., vol. 975, 1983.

COMMENTARY (1993)

In [8] the results of [6] have been generalized to weighted l^2-spaces on the discrete real line $\mathbb{R}_d$. To extend to weighted l^p-spaces, $p \neq 2$, one could need a Titchmarsh theorem on $\mathbb{R}_d$ for convolution of l^p with l^q, q being conjugate of p.

For results of a similar nature as those in [6], see [9], [10]. In particular, the following result extending (A) is proved in [10]:

Let $p = -\log w$ be a convex function, and $p(x)x^{-c}$ decrease for some $c < \infty$. If $f \in A_w$ and $fw^a \in L^1(\mathbb{R}_+)$ for every $a > 0$, then $f \in B_w$.

REFERENCES

8. Domar Y., *Convolution theorems of Titchmarsh type on discrete $\mathbb{R}^n$*, Proc. Edinburg Math. Soc. **32** (1989), 449–457.
9. Borichev A. A., *A Titchmarsh-type convolution theorem on the group $\mathbb{Z}$*, Arkiv för matematik **27** (1989), 179–187.
10. Borichev A. A., *The generalized Fourier transformation, Titchmarsh's theorem and asymptotically holomorphic functions*, Leningrad Math. J. **1** (1990), 825–857.

TWO PROBLEMS OF HARMONIC
ANALYSIS IN WEIGHTED SPACES

V. P. Gurarii

We consider the space $L_\varphi^\infty(\mathbb{R})$ of measurable functions on $\mathbb{R}$ with the norm $\|g\| =$ ess sup $|g(t)|/\varphi(t)$. The weight φ is supposed to be measurable and to satisfy the conditions

(a) $1 \leqslant \varphi(t + \tau) \leqslant \varphi(t)\varphi(\tau)$, $\forall t, \tau \in \mathbb{R}$;

(b) $\lim\limits_{t \to \pm\infty} \dfrac{\log \varphi(t)}{t} = 0$.

Assign to each function $g \in L_\varphi^\infty(\mathbb{R})$ the smallest w^*-closed subspace of $L_\varphi^\infty(\mathbb{R})$ (denoted by B_g) invariant under all translations and containing g. The set

$$\Lambda_g = \{\, x \in \mathbb{R} : \exp(-itx) \in B_g \,\}$$

is called the spectrum of g.

Denote $L_\varphi^\infty(\mathbb{R}_+) = \{\, g \in L_\varphi^\infty(\mathbb{R}) : g(t) = 0 \text{ for } t < 0 \,\}$.

For each g_+ in $L_\varphi^\infty(\mathbb{R}_+)$ a spectrum $\Lambda_{g_+}^+$ is defined as follows:

$$\Lambda_{g_+}^+ = \{\, z : \operatorname{Im} z \leqslant 0, \exp(-itz) \in B_{g_+}^+ \,\}.$$

Here $B_{g_+}^+$ is the smallest w^*-closed subspace of $L_\varphi^\infty(\mathbb{R}_+)$ invariant under translations to the left and containing g_+.

The spectrum Λ_g (resp. $\Lambda_{g_+}^+$) is a closed subset of the real line (resp. of the lower half-plane).

The spectrum $\Lambda_{g_+}^+$ (or Λ_g) is said to be "simple" if the only functions in $B_{g_-}^+$ (resp. B_g) that have one-point spectrum, are exponentials times constants.

PROBLEM 1. *Describe the subsets* Λ *of* $\mathbb{R}$ *with the following property: every function* $g_+ \in L_\varphi^\infty(\mathbb{R}_+)$ *with a "simple" spectrum* $\Lambda_{g_+}^+ \subset \Lambda$ *admits an extension* g *to the whole of* $\mathbb{R}$ *so that* $g \in L_\varphi^\infty(\mathbb{R})$ *and*

$$(1) \qquad\qquad \Lambda_g = \Lambda_{g_+}^+.$$

If $\varphi(t) \equiv 1$ the set $\mathbb{Z}$ of all integers is an example of such set. Indeed, if $g_+ \in L^\infty(\mathbb{R}_+)$ and $\Lambda_{g_+}^+ \subset \mathbb{Z}$ then the theorem on spectral synthesis in $L^\infty(\mathbb{R}_+)$ proved in [1] implies that g_+ lies in the w^*-closure of the trigonometric polynomials with frequencies in $\mathbb{Z}$. Thus g_+ admits a 2π-periodic extension g to the whole of $\mathbb{R}$, and clearly (1) holds for this g. There also exist more refined examples.

When treating the spectral synthesis in $L_\varphi^\infty(\mathbb{R}_+)$, the following problem might be useful.

PROBLEM 2. *Let* $f \in L^1_\varphi(\mathbb{R}_+)$, $g \in L^\infty_\varphi(\mathbb{R}_+)$ *and suppose* $t^k \in L^\infty_\varphi(\mathbb{R}_+)$ *for* $0 \leqslant k \leqslant \omega$ (ω *is a positive integer or the symbol* ∞). *Suppose also that*

$$(2) \qquad\qquad \int_0^\infty t^k f(t)\, dt = 0, \qquad 0 \leqslant k < \omega$$

and

$$(3) \qquad\qquad \int_0^\infty f(t) g(t + \tau) = C, \qquad \forall \tau > 0$$

where C *is a constant. Describe weights* φ *such that* (2) *and* (3) *imply* $C = 0$.

For the weight $\varphi, \varphi(t) \equiv 1$ this implication has been proved in [1]. If $\varphi(t) = 1 + |t|$, a proof has been proposed by E. L. Suris.

Some considerations concerning Problems 1 and 2 are implicit in [2].

REFERENCES

1. Gurarii V. P., *Spectral synthesis of bounded functions on half-axe*, Funkts. Anal. i Prilozh. **3** (1969), no. 4, 34–48. (Russian)
2. Gurarii V. P., *Harmonic analysis in the spaces with weight*, Trudy Mosk. mat. ob-va **35** (1976), 21–76. (Russian)

CHEMISTRY AND PHYSICS INSTITUTE OF
ACADEMY OF SCIENCES OF RUSSIA
TCHERNOGOLOVKA, MOSCOW REGION, 142432
RUSSIA

HARMONIC SYNTHESIS AND COMPOSITIONS

E. M. DYN′KIN

Let $\mathcal{F}l^1$ be the algebra of all absolutely convergent Fourier series on the circle $\mathbb{T}$:

$$f(\xi) = \sum_{n \in \mathbb{Z}} \hat{f}(n)\xi^n, \xi \in \mathbb{T}; \qquad \|f\|_1 \overset{\text{def}}{=} \sum |\hat{f}(n)| < \infty.$$

Let $f \in \mathcal{F}l^1$ and suppose $f(\mathbb{T}) = [-1,1]$.

We say f admits *the harmonic synthesis* (= h.s.) if there is a sequence $\{\varphi_n\} \subset \mathcal{F}l^1$ such that $\|\varphi_n - f\|_1 \to 0$ and $f^{-1}(0) \subset \text{Int} \, \varphi_n^{-1}(0), n = 1, 2, \ldots$. The algebra $\mathcal{F}l^1$ contains functions not admitting h.s. though every sufficiently smooth function admits h.s.

QUESTION 1. *Let f admit h.s. Is it possible to choose functions φ_n in the above definition so that $\varphi_n = F_n \circ f$, F_n being some functions on $[-1,1]$?*

Denote by $[f]$ the set of all functions F on $[-1,1]$ such that $F \circ f \in \mathcal{F}l^1$. This set is a Banach algebra with the norm $\|F\|_{[f]} = \|F \circ f\|_1$. It contains the identity function $F(x) \equiv x$. Now we can reformulate our question.

QUESTION 2. *Let f admit h.s. Is it possible to approximate $F(x) \equiv x$ in the algebra $[f]$ by functions vanishing near the point $x = 0$?*

If $[f] \subset C^1[-1,1]$ (this embedding has to be continuous by the Banach theorem), then the functional $\delta' : F \to F'(0)$ separates x from functions in $[f]$ vanishing at a vicinity of zero. So the following question is a particular case of our problem.

QUESTION 3. *Let f admit h.s. Is it possible that $[f] \subset C^1[-1,1]$?*

Obviously $\|e^{2\pi i n x}\|_{C^1[-1,1]} \asymp n, n \to \pm\infty$, and so we have a further specialization.

QUESTION 4. *Is it possible to construct a function $f \in \mathcal{F}l^1$ admitting h.s. with*

$$\varlimsup_n \frac{1}{n} \|e^{inf}\|_1 > 0?$$

Now (1983) very little is known about the structure of the ring $[f]$. The theorems of Wiener–Levy type ([1] Ch.VI [2]) give some sufficient conditions for the inclusion $F \in [f]$, but these conditions are much stronger than C^1-smoothness of F. On the other hand, let the function $t \to f(e^{it})$ be even on $[-\pi, \pi]$ and strictly monotone on $[0, \pi]$. Thus any even function on $[-\pi, \pi]$ has a form $F \circ f$ and our Question 1 has the affirmative answer. Hence $[f] \not\subset C^1$ and all known theorems of Wiener–Levy type are a priori too rough for this f. Kahane [3] has constructed examples of functions f with $[f] \subset C^\infty[-1,1]$. Thus, the ring $[f]$ is quite mysterious.

A possible way to answer our questions is the following. If $[f] \subset C^1$ then the functional $\delta' : F \to F'(0)$ is well-defined on $[f]$ and generates a functional $\delta'(f)$ on the subalgebra $[[f]] = \{ F \circ f \colon F \in [f] \} \subset \mathcal{F}l^1$. If there is an extension of this functional to $\mathcal{F}l^1$ with $\langle \delta'(f), g \rangle = 0$ for $f^{-1}(0) \subset \text{Int}\, g^{-1}(0)$, then f cannot admit h.s. It is in this way that Malliavin's lemma on the absence of h.s. has been proved. Namely, if (in addition) $\int_0^\infty u\|e^{iuf}\|_{(\mathcal{F}l^1)^*}\, du < \infty$ and $\int_{\mathbb{T}} dm \int_{-\infty}^\infty e^{iuf}\, du = 1$, then Malliavin's functional

$$g \to \int_{\mathbb{T}} dm \int_{-\infty}^\infty iue^{iuf} g\, du, \qquad g \in \mathcal{F}l^1$$

gives the desired extension of $\delta'(f)$.

The author thanks professor Y. Domar for a helpful discussion in 1978 in Leningrad.

REFERENCES

1. Kahane J.-P., *Séries de Fourier absolument convergentes*, Springer, 1970.
2. Dyn'kin E. M., *Wiener–Levy type theorems and estimates for Wiener–Hopf operators*, Matem. issled. 8 (1973), no. 3, 14–25. (Russian)
3. Kahane J.-P., *Une nouvelle réciproque du théorème de Wiener–Lévy*, C.R. Acad. Sci. Paris **264** (1967), 104–106.

DEPT. OF MATHEMATICS
TECHNION
32000 HAIFA
ISRAEL

COMMENTARY BY A. ATZMON

Paper [4] contains results which are closely related to these problems.

In Section 3 of [4] (Theorem 3.1) it is shown that there exists a closed subset E of the unit circle $\mathbb{T}$ such that the restriction algebra $A(E)$ (of $A(\mathbb{T}) = \mathcal{F}l^1$ to E) is an algebra of spectral synthesis although it contains a function f such that $f(\mathbb{T}) \subset [-1; 1]$ and $[f] = \{F \in C[-1; 1], F \circ f \subset A(E)\} \subset C^\infty[-1; 1]$.

Hence the analog of Question 2 in Dyn'kin's note for the algebra $A(E)$ admits a negative solution. That is, f is of synthesis in this algebra but it is not possible to approximate the function $F(x) = x$ in the algebra $[f]$ by functions vanishing near the point $x = 0$.

If the above function f admits an extension $\tilde{f}$ in $A(\mathbb{T})$ which is of spectral synthesis, we would obtain a negative answer to Question 2, since clearly

$$[\tilde{f}] \overset{\text{def}}{=} \{F \in C[-1; 1], F \circ \tilde{f} \in A(\mathbb{T})\} \subset [f] \subset C^\infty[-1; 1].$$

We do not know whether f admits such an extension but it seems conceivable that it does.

REFERENCE

4. Atzmon A., *Spectral synthesis in regular Banach algebras*, Israel J. Math. 8 (1970), no. 3, 197–212.

DEUX PROBLÈMES CONCERNANT
LES SÉRIES TRIGONOMÉTRIQUES

J.-P. KAHANE

1. Soit $\sum_{n \in \mathbb{Z}} a_n e^{int}$ une série trigonométrique dont les coefficients tendent vers 0 et dont les sommes partielles tendent vers 0 sur un ensemble fermé $F \subset \mathbb{T}$:

$$\lim_{N \to \infty} \sum_{n=-N}^{N} a_n e^{int} = 0 \qquad \text{quand } t \in F,$$

$$\lim_{|n| \to \infty} a_n = 0.$$

Soit $\mu \in M^+(F)$ une mesure positive portée par F,

$$\mu \sim \sum_{n \in \mathbb{Z}} \hat{\mu}(n) e^{int},$$

telle que

$$\sum_{n \in \mathbb{Z}} |\hat{\mu}(n)||a_{-n}| < +\infty.$$

A-t-on nécessairement

$$\sum_{n \in \mathbb{Z}} \hat{\mu}(n) a_{-n} = 0 ?$$

Une réponse positive (dont je doute) donnerait une nouvelle preuve de l'existence d'ensembles $U(\varepsilon)$ de Zygmund de mesure pleine.

2. *Soit $f \in L^p(\mathbb{T})$, $f \sim \sum_{n \in \mathbb{Z}} \hat{f}(n) e^{int}$. Peut-on approcher f dans $L^p(\mathbb{T})$ par des polynômes trigonométriques $P = \sum_{\text{finie}} \hat{P}(n) e^{int}$ tels que $\hat{f}(m) = \hat{f}(n) \Longrightarrow \hat{P}(m) = \hat{P}(n)$?*

La question a été posée par W. Rudin [1] pour $p = 1$ (la réponse est alors négative [2]). Pour $p = 2$, la réponse positive est évidente. Pour $p = \infty$, la question n'a d'intérêt que si on suppose f continue (la réponse est négative). La question est ouverte pour $1 < p < 2$ et $2 < p < \infty$.

BIBLIOGRAPHIE

1. Rudin W., *Fourier Analysis on Groups*, Interscience, N.Y., 1962.
2. Kahane J.-P., *Idempotents and closed subalgebras of $L^1(\mathbb{T})$*, in Funct. algebras, Proc. Intern. Symp. Tulane Univ. (T. Birtel, ed.), Scott–Forestmann, Chicago, 1966 pp. 198–207.

DEPT. DE MATHÉMATIQUES
UNIVERSITÉ DE PARIS-SUD
BÂTIMENT 425
91405 ORSAY CEDEX
FRANCE

COMMENTAIRE

La réponse au second problème est négative (voir [3] pour $1 < p < 2$ et [4,5] pour $2 < p < \infty$).

BIBLIOGRAPHIE

3. Rider D., *Closed subalgebras of $L^1(\mathbb{T})$*, Duke Math. J. **36** (1969), no. 1, 105–115.
4. Oberlin D. M., *An approximation problem in $L^p[0, 2\pi]$, $2 < p < \infty$*, Studia Math. **70** (1981), no. 3, 221–224.
5. Bachelis G. F., Gilbert J. E., *Banach algebras with rider subalgebras*, Bull. Inst. Math. Acad. Sinica **7** (1979), no. 3, 333-347.

ALGEBRA AND IDEAL GENERATION
IN CERTAIN RADICAL BANACH ALGEBRAS

Marc Thomas

Let $\mathbb{C}[[z]]$ denote the algebra of formal power series over $\mathbb{C}$. We say that a sequence of positive reals $\{w(n)\}$ is *a radical algebra weight* provided the following hold:

(1) $w(0) = 1$ and $0 < w(n) \leqslant 1$ for all $n \in \mathbb{Z}_+$.
(2) $w(m + n) \leqslant w(m)w(n)$, for all $m, n \in \mathbb{Z}_+$.
(3) $\lim\limits_{n \to \infty} w(n)^{1/n} = 0$.

If these conditions hold it is routine to check that

$$\ell^1(w(n)) = \left\{ y = \sum_{n=0}^{\infty} y(n)z^n : \sum_{n=0}^{\infty} |y(n)|w(n) < \infty \right\}$$

is both a subalgebra of $\mathbb{C}[[z]]$ and a radical Banach algebra with identity adjoined. The norm is defined in the natural way: $\|y\| = \sum_{n=0}^{\infty} |y(n)|w(n)$. The multiplication is given by the usual convolution of formal power series. We shall generally refer to $l^1(w(n))$ as a radical Banach algebra and $\{w(n)\}$ as simply a "weight". Let $A = l^1(w(n))$ in all the following. Besides A itself, there are obvious proper closed ideals in A:

$$M(n) = \left\{ \sum_{k=0}^{\infty} y(k)z^k \in A : y(0) = y(1) = \ldots = y(n-1) = 0 \right\}$$

for $n = 1, 2, \ldots$, and, of course, the zero ideal. Such closed ideals are referred to as *standard ideals*. Any other closed ideals are denoted *non-standard ideals*. Note that the unique maximal ideal in A is $M(1) = M$.

We first discuss the problem of polynomial generation. Let $x = \sum_{n=1}^{\infty} x(n)z^n$ be an element of A with $x(1) \neq 0$. One says that x *generates a non-standard closed subalgebra* if the smallest closed subalgebra containing x is properly contained in M. Since this algebra is the closed linear span of polynomials in x, we could equivalently say

(4) $\overline{\mathrm{span}}\{x^{k+1}\}_{k=0}^{\infty}$ is properly contained in $M(1)$.

The requirement that $x(1) \neq 0$ is necessary, otherwise (4) is vacuous. If the weight is very well behaved there are positive results which show that non-standard closed subalgebras are not present [3]. On the other hand, it was shown [5, Theorem 3.11] that, for certain star-shaped weights, non-standard closed subalgebras exist. (A weight w is star-shaped if $\{w(n)^{1/n}\}$ is non-increasing). Hence one problem is the following.

PROBLEM 1. *Characterize the radical algebra weights w such that $\ell^1(w(n))$ has non-standard closed subalgebras.*

We next consider the problem of ideal generation. Whether $\ell^1(w(n))$ has only standard closed ideals or not is a problem whether each non-zero element x generates a standard closed ideal or not. If we let T be the operator of right translation on A, we could equivalently say [1, Lemma 4.5][1]

(5) $\overline{\mathrm{span}}\{T^k x\}_{k=0}^{\infty}$ contains a power of z,

for each non-zero x in A. If $\ln w$ is a concave function it is well known [1, Theorem 4.1][2] that all closed ideals are standard. More generally, it can be shown [4, Corollary 3.6] that if w is star-shaped and $w(n)^{1/n}$ is $O(\frac{1}{n^a})$ for some $a > 0$ then all closed ideals are standard. This is in contrast to the fact that star-shaped weights can support non-standard closed subalgebras. Apparently Šilov first posed the problem whether or not there exists any radical algebra weight w such that $\ell^1(w(n))$ contains a non-standard ideal. The answer is affirmative [6, Theorem] for certain semi-multiplicative weights [5, Definition 2.1]. These are weights where $w(m + n)$ actually equals $w(m)w(n)$ for many values of m, n in $\mathbb{Z}_+$. Hence we propose the following

PROBLEM 2. *Characterize the radical algebra weights w such that $\ell^1(w(n))$ has non-standard ideals.*

Even substantial necessary conditions on the weight w for the existence of non-standard ideal would be welcome.

Finally we remark that one can consider related radical algebras $\ell^1(\mathbb{Q}_+, w)$ built upon $\mathbb{Q}_+$ rather than $\mathbb{Z}_+$ (we again require (1)–(3) for m, n in $\mathbb{Q}_+$). Define for x non-zero in $\ell^1(\mathbb{Q}_+, w)$

$$a(x) = \inf\{\, r : x(r) \neq 0, r \in \mathbb{Q}_+ \,\}.$$

Also define

$$M = \{\, y \in \ell^1(\mathbb{Q}_+, w) : y(0) = 0 \,\}.$$

We pose the final problem.

PROBLEM 3. *Does there exist some $\ell^1(\mathbb{Q}_+, w)$ containing an element x, $a(x) = 0$, such that the closed ideal generated by x is properly contained in M?*

Preliminary results on this problem can be found in [2].

REFERENCES

1. Grabiner S., *Weighted shifts and Banach algebras of power series*, American J. Math **97** (1975), 16–42.
2. Gronbaek N., *Weighted discrete convolution algebras*, "Radical Banach Algebras and Automatic Continuity", Proceedings, Lect. Notes Math., vol. 975, 1981.
3. Söderberg D., *Generators in radical weighted ℓ^1*, Uppsala University Department of Mathematics Report 9 (1981).

---

[1]This is also a part of Lemma 1 in Nikol'ski N. K., Izvestiya AN USSR, ser. mat. **32** (1968), 1123–1137 (Russian) – *Ed.*

[2]This is also a part of Theorem 2 in Nikol'ski N. K. Vestnik LGU, ser. mat.-mech. and astron. **7** (1965), 68–77 (Russian) – *Ed.*

4. Thomas M. P., *Approximation in the radical algebra $\ell^1(w_n)$ when $\{w_n\}$ is star-shaped*, Radical Banach Algebras and Automatic Continuity, Proceedings, Lect. Notes in Math., vol. 975.
5. Thomas M. P., *A non-standard closed subalgebra of radical Banach algebra of power series*, J. London Math. Soc. **29** (1984), 153–163.
6. Thomas M. P., *A non-standard closed ideal of a radical Banach algebra of power series*, Bull. Amer. Math. Soc. **9** (1983), 331–333.

MATHEMATICS DEPARTMENT
CALIFORNIA STATE COLLEGE
AT BAKERSFIELD
9001 STOCKDALE HWY.
BAKERSFIELD, CA 93309
USA

COMMENTARY BY J. ESTERLE

There is another interesting problem concerning the radical algebras $\ell^1(\mathbb{Q}^+, w)$. It was proved in [7] that the ideal M defined above contains real semigroups $\{x^t\}_{t>0}$ such that $a(x^t) = 0$ $(t > 0)$ (the ideal M actually contains semigroups $\{x^\tau\}_{\substack{\tau \in G \\ \tau > 0}}$ for any linearly ordered group G of cardinality $\aleph_1$, such that $a(x^\tau) = 0$ $(\tau \in G, \tau > 0)$). We have the following problem:

PROBLEM 4. *Does M contain a nonzero continuous real semigroup?*

This problem seems to be open for all radical weights w over $\mathbb{Q}^+$.

REFERENCE

7. Esterle J., *Elements for a classification of commutative radical algebras*, Radical Banach Algebras and Automatic Continuity, Proceedings, Lect. Notes in Math., vol 975, pp. 1–62

CeReMaB, UA ASSOCIÉE 226
UNIVERSITÉ BORDEAUX 1
33405 TALENCE CEDEX
FRANCE

TRANSLATES OF FUNCTIONS OF TWO VARIABLES

B. Ya. Levin

In [1] and [2] the following theorem is proved: if $f \in L^1(\mathbb{R}_+)$ and $f(x) = 0$ for $x < 0$, then the system of functions $\{ f(x - h) : h \in \mathbb{R}_+ \}$ is complete in $L^1(\mathbb{R}_+)$ if and only if the following conditions are fulfilled:

1. The function

$$\mathcal{F}(z) = \int_0^\infty f(x)e^{ixz}\, dx$$

doesn't vanish in $\operatorname{Im} z \geqslant 0$.
2. There is no $\delta > 0$ such that $f(x) = 0$ a.e. on $(0, \delta)$.

PROBLEM. *Let* $f \in L^1(P_1)$, $P_1 = \{ (x,y) \in \mathbb{R}^2 : x > 0, y > 0 \}$ *and* $f \equiv 0$ *in* $\mathbb{R}^2 \setminus P_1$. *Find necessary and sufficient conditions for the system* $\{ f(x - h, y - k) : h \geqslant 0, k \geqslant 0 \}$ *to be complete in* $L^1(P_1)$.

REFERENCES

1. Nyman B., *On the one-dimension translation group and semi-group in certain function spaces*, Uppsala, 1950.
2. Gurarii V. P., Levin B. Ya., *About completeness of the system of the shifts in the space* $\mathcal{L}(0, \infty)$ *with weight*, Zap. Khark. mat. ob-va **30** (1960), no. 4. (Russian)

PHYSICAL AND TECNOLODGICAL INSTITUTE
OF THE LOW TEMPERATURE
ACADEMY OF SCIENCE OF UKRAINE
LENIN STREET 47, KHARKOV, 310164
UKRAINE

COMMENTARY BY H. HEDENMALM

In [3], the above problem was studied in extenso.

The one-dimensional analogue of the problem was solved by B. Nyman (1950), and, independently, by V. P. Gurarii and B. Ya. Levin (1964). In order to gain insight into Levin's problem, it was necessary to modify substantially the Fourier – Laplace transform technique used by Nyman – Gurarii – Levin. A new transformation $\mathcal{C}$ was constructed, with the following property: given an invariant subspace J of $L^1(\mathbb{R}_+^2)$, consisting of functions lacking common zeros of their Fourier transforms, it maps every element φ of the dual space $L^\infty(\mathbb{R}_+^2)$ annihilating J onto a corresponding entire function, $\mathcal{C}[\varphi]$. The method makes it possible to arrive at certain sufficient conditions. Necessary conditions are obtained through a simpler argument. Unfortunately, a small but nevertheless irritating gap remains between the necessary and sufficient conditions obtained. A complete solution of the problem will require deeper insights than provided in [3].

PROBLEM 11.22

A related problem for the Volterra algebra on the unit square was also considered in [3], and successfully attacked with the use of the above-mentioned transformation $\mathcal{C}$. For some further information about this resolvent-type transform, see [4].

REFERENCES

3. Hedenmalm H., *Translates of functions of two variables*, Duke Math. J. 58 (1989), 251–297.
4. Hedenmalm H., *Outer functions of several complex variables*, J. Funct. Anal. 80 (1988), 9–15.

MATEMATISKA INSTITUTIONEN
THUNBERGSVÄGEN 3
S-752 38 UPPSALA
SWEDEN

TITCHMARSH'S THEOREM FOR VECTOR FUNCTIONS

HENRY HELSON

In one version (from which others can be derived) Titchmarsh's theorem states:

If f and g are functions of $L^2(\mathbb{R}_+)$ such that $f * g$ vanishes on $(0,1)$, and if f vanishes on no interval $(0,\varepsilon)$, then g must vanish on $(0,1)$.

Here is a *proof.* Fix f, and denote by M the set of all g such that $f * g$ vanishes on $(0,1)$. M is a closed subspace invariant under shifts to the right. Beurling's theorem states that $\widehat{M}$, the space of Fourier transforms of functions in M, is exactly qH^2, where q is inner in the upper half-plane and H^2 is the Hardy space on the half-plane. Since M contains all functions vanishing on $(0,1)$, $(q(z))^{-1}\exp(iz)$ is an inner function too. The known structure of inner functions implies that $q(z) = \exp(isz)$ for some s, $0 \leqslant s \leqslant 1$. This means that M contains all functions that vanish on $(0,s)$, and it follows that f vanishes on $(0,1-s)$. Hence $s = 1$, so g must vanish on $(0,1)$. $\square$

Suppose F and G are functions in $L^2(\mathbb{R}_+)$ with values in a Hilbert space H, and suppose the expression

$$\int_0^x \big(F(t), G(x-t)\big)\, dt$$

vanishes for $0 \leqslant x \leqslant 1$. *(The integrand is the inner product in H.) What can we say about F and G?* More generally, *is there a simple characterization of the subspaces M of $L^2(\mathbb{R}_+)$ that are closed, invariant under right shifts, and contain all functions vanishing on $(0,1)$?*

By the vectorial version of Beurling's theorem (see [1]) *the problem is equivalent to describing the inner functions Q such that $Q(z)^{-1}\exp iz$ is also inner.* In the vectorial context, an inner function Q is analytic in the upper half-plane, takes values in the space of operators on H, satisfies $|Q(z)| \leqslant 1$, and has boundary values $Q(x)$ that are unitary for almost all real x. In our case $Q(x)$ has spectrum (the support of its Fourier transform) in $[0,1]$ and so is entire.

We obtain inner functions of this kind in the form $\exp izA$, where A is a constant self-adjoint operator satisfying $0 \leqslant A \leqslant I$. The corresponding subspace M is easily described. Let (H_t) be the spectral resolution of A; thus $H_t = \mathbf{0}$ for $t < 0$ and $H_t = H$ for $t > 1$. M is the set of vector functions F such that $F(t)$ lies in H_t for almost every t.

A straightforward extension of Titchmarsh's theorem would assert that the integral above vanishes for $0 \leqslant x \leqslant 1$ only if the inner product vanishes identically for such x. This is equivalent to saying the inner function of M necessarily has the form $\exp izA$. This is not true, as shown by *an example of Donald Sarason.* His example leads to a method for constructing such inner functions. Set $R(z) = (\exp(-iz/2))Q(z)$; then the unitary function $R(x)$ has spectrum in $[-\frac{1}{2}, \frac{1}{2}]$. Write $R = S + iT$ with S, T self-adjoint. The fact that R is unitary means that S and T commute at each point, and $S^2 + T^2 = I$.

Suppose H is two-dimensional and $S = rI$, $0 \leqslant r < 1$. Then on the real axis T must be $\begin{pmatrix} f & g \\ \bar{g} & -\bar{f} \end{pmatrix}$ where f and g are entire functions of exponential type at most $\frac{1}{2}$, f is real on the real axis, and $f^2 + |g|^2 = 1 - r^2 = s^2$. The choice $f(z) = s \cos bz$, $g(z) = s \sin bz$, $0 < b \leqslant \frac{1}{2}$, gives

$$Q(z) = r \left(\exp \frac{iz}{2} \right) I + s \left(\exp \frac{iz}{2} \right) \begin{pmatrix} \cos bz & \sin bz \\ \sin bz & -\cos bz \end{pmatrix}.$$

Can the structure of Q be described simply in general, or even when H is two-dimensional?

REFERENCE

1. Helson H., *Lectures on Invariant Subspaces*, Academic Press, NY–London, 1964.

DEPARTMENT OF MATHEMATICS
UNIVERSITY OF CALIFORNIA
BERKELEY, CALIFORNIA 94720
USA

INVARIANT SUBSPACES OF THE SHIFT OPERATOR
IN SOME SPACES OF ANALYTIC FUNCTIONS

F. A. SHAMOYAN

1. Let X be a Banach algebra of functions analytic in the unit disk $\mathbb{D}$ (with pointwise addition and multiplication). Assume that $X \subset C_A^{(m)}$ and let $n \overset{\text{def}}{=} \max\{m \colon X \subset C_A^{(m)}\} < \infty$. Let $E^{(n)}(f) = \{\zeta \colon \zeta \in \mathbb{T}; \ f^{(j)}(\zeta) = 0, \ 0 \leqslant j \leqslant n\}$.

DEFINITION. A closed subset E of the unit circle $\mathbb{T}$ is called D_n-set for the algebra X if for any function f, $f \in X$ with $E^{(n)}(f) \supset E$ there is a sequence $\{f_k\}_{k \geqslant 1}$ of functions in X, satisfying the following conditions:

  (i) $|f_k(z)| \leqslant C_k[\text{dist}(z, E)]^{n+1}$, $z \in \text{clos}\,\mathbb{D}$;
  (ii) $\lim\limits_k \|f_k f - f\|_X = 0$.

It follows from [1] that every Beurling–Carleson set* is a D_n-set for a number of standard algebras of analytic functions. In particular, it is true for the algebra $H_{n+1}^p = \{f \colon f^{(n+1)} \in H^p\}$, $1 < p < \infty$.

QUESTION 1. *Is every Beurling–Carleson set a D_n-set for the algebra H_{n+1}^1, $n \geqslant 0$?*

Remark. If $E = \overset{m}{\underset{k=1}{\cup}} E_k$, where $m E_k = 0$, $k = 1, \ldots, m$, and the lengths of complementary intervals of each E_k tend to zero exponentially fast then (as proved in [2]) E is a D_n-set for H_{n+1}^1 and the sequence $\{f_k\}_{k \geqslant 1}$ can be chosen not depending on f, $f \in H_{n+1}^1$.

2. Let $A^p = A^p(\mathbb{D})$ be the Bergman space in the unit disc $\mathbb{D}$ (i.e. the space of all functions in $L^p(\mathbb{D})$ analytic in $\mathbb{D}$).

QUESTION 2. *Is there a closed subspace G, $G \subset A^p$, $G \neq A^p$ invariant under the shift operator S, $S(f)(z) \overset{\text{def}}{=} z f(z)$, and not finitely generated?*

QUESTION 3. *Let G, $G \subset A^p$ be an invariant subspace of S. Assume that for every ζ, $\zeta \in \mathbb{D}$ there is a function f, $f \in G$, $f(\zeta) \neq 0$. Is it true that G is generated by one function g, i.e. $G = \vee(z^n g \colon n \geqslant 0)$?*

Remark. Questions 2 and 3 arise in the following way. Let $\mathbb{D}^2$ be the unit polydisc in $\mathbb{C}^2$ and $H^p(\mathbb{D}^2)$ be the Hardy space in $\mathbb{D}^2$ (see [3]). Denote by D_2 the "diagonal operator" in $H^p(\mathbb{D}^2)$, i.e. $D_2 f(\zeta) = f(\zeta, \zeta)$, $\zeta \in \mathbb{D}$, and denote by S_i the operator

$$S_i(f)(\zeta_1, \zeta_2) = \zeta_i f(\zeta_1, \zeta_2), \quad \zeta_i \in \mathbb{D}, \ f \in H^p(\mathbb{D}^2), \ i = 1, 2.$$

It is known that $D_2(H^p(\mathbb{D}^2)) = A^p$, $0 < p < \infty$ (see [4], [5]) and there exist $S_{1,2}$-invariant and non-finitely generated subspaces ([3], p.67). Moreover there is an invariant subspace generated by two functions in $H^2(\mathbb{D}^2)$ without common zeros in $\mathbb{D}^2$ and this subspace cannot be generated by any of its elements (see [6]).

---

*See the definition in 11.13

PROBLEM 11.24

REFERENCES

1. Shamoyan F. A., *The structure of closed ideals in certain algebras of functions analytic in the disk and smooth up to the boundary*, Dokl. Akad. Nauk Arm.SSR 60 (1975), no. 3, 133–136. (Russian)
2. Shamoyan F. A., *Construction of a certain special sequence and the structure of closed ideals in certain algebras of analytic functions*, Izv. Akad. Nauk Arm.SSR 7 (1972), no. 6, 440–470. (Russian)
3. Rudin W., *Function Theory in Polydiscs*, Benjamin, New York, 1969.
4. Horowitz C., Oberlin D., *Restrictions of functions to diagonal of $\mathbb{D}^2$*, Indiana Univ. Math. J. 24 (1975), no. 7, 767–772.
5. Shamoyan F. A., *A theorem on imbedding in spaces of n-harmonic functions, and some applications*, Dokl. Akad. Nauk Arm.SSR 62 (1976), no. 1, 10–14. (Russian)
6. Jacewicz Ch. Al., *A nonprincipal invariant subspace of the Hardy space on the torus*, Proc. Amer. Math. Soc. 31 (1972), 127–129.

OGANESYANA 24/1, 7
EREVAN AVAN 375101
ARMENIA

COMMENTARY BY H. HEDENMALM

The answer to Question 2 is yes. For $p = 2$, this follows from a result in [7], where it is shown that there are invariant subspaces J in A^2 with $J \ominus zJ$ infinite-dimensional. Such invariant subspaces necessarily must have infinitely many generators. For general p, $1 \leqslant p < +\infty$, an explicit construction will appear in [8] (see also [9]).

The answer to Question 3 is no. This is evident from the construction of an invariant subspace in [9].

REFERENCES

7. Apostol C., Bercovici H., Foiaş C., Pearcy C., *Invariant subspaces, dilation theory, and the structure of the predual of a dual algebra, I*, J. Funct. Anal. 63 (1985), 369–404.
8. Hedenmalm H., Richter S., Seip K., *Zero sets and invariant subspaces in the Bergman space*, in preparation.
9. Hedenmalm H., *An invariant subspace in the Bergman space having the codimension two property*, J. Reine Angew. Math. (to appear).

BLASCHKE PRODUCTS AND IDEALS IN C_A^∞

David L. Williams

Let A be the space of functions analytic in the open unit disc $\mathbb{D}$ and continuous in clos $\mathbb{D}$; and let $C_A^\infty = \{ f \in A : f^{(n)} \in A, \ n = 0, 1, \dots \}$. Although the sets of uniqueness for C_A^∞ have been described [1], [2], [3], [4], and the closed ideal structure of C_A^∞ is known [5], there are still some open questions concerning the relationship of Blaschke products with closed ideals in C_A^∞. I pose two problems. Let I, $I \subset C_A^\infty$, denote a closed ideal and let B denote a Blaschke product which divides some non-zero C_A^∞ function.

(1) *For which B is it true that*

$$BI \overset{\text{def}}{=} \{ Bf : f \in I \} \subset C_A^\infty ?$$

(2) *If B is the g.c.d. (greatest common divisor) of the Blaschke factors of the non-zero functions in I, when is $(\frac{1}{B})I \overset{\text{def}}{=} \{ f \in C_A^\infty : Bf \in I \}$ a closed ideal in C_A^∞?*

Note that the corresponding problem for singular inner functions are easier and are solved in section 4 of [5].

To discuss the problems for Blaschke products we need some notation. Let

$$Z^n(I) = \underset{f \in I}{\cap} \{ z \in \text{clos}\,\mathbb{D} : f^{(k)}(z) = 0, \ k = 0, \dots, n \}, \quad n = 0, 1, \dots,$$

and let

$$Z^\infty(I) = \overset{\infty}{\underset{n=0}{\cap}} Z^n(I), \quad Z(I) = \{ Z^n(I) \}_{n=0}^\infty.$$

If $I(Z(I))$ denotes the closed ideal of all f, $f \in C_A^\infty$, with $f^{(n)}(z) = 0$ for $z \in Z^n(I)$, $n = 0, 1, \dots$, then the closed ideal structure theorem says $I = SI(Z(I))$, where S is the g.c.d. of the singular inner factors of the non-zero functions in I.

DEFINITION. *A sequence $\{ z_j \} \subset \mathbb{D}$ has finite degree of contact at E, $E \subset \partial \mathbb{D}$, if there exist k, $k > 0$, and ε, $\varepsilon > 0$, such that $1 - |z_j| \geq \varepsilon \rho(z_j/|z_j|, E)^k$ for all j. (Here ρ denotes the Euclidean metric.)*

The following unpublished theorem of B. A. Taylor and the author provides solutions to problems (1) and (2) in a special case.

THEOREM.

(a) *Assume $Z^0(I) = Z^\infty(I)$. In order that $BI \subset C_A^\infty$ it is necessary and sufficient that the zeros of B have finite degree of contact at $Z^\infty(I)$. If $BI \subset C_A^\infty$, then multiplication by B is continuous on I, BI is closed, and the inverse operation is continuous.*

(b) Assume $Z^0(I) \cap \partial \mathbb{D} = Z^\infty(I)$. Let B be the g.c.d. of the Blaschke factors of the non-zero functions in I. In order that $(1/B)I$ be closed it is necessary and sufficient that the zeros of B have finite degree of contact at $Z^\infty(I)$.

The proof of sufficiency in (a) is primarily a computation of the growth of the derivatives of B near $Z^\infty(I)$. The computation has also been done by James Wells [6]. The proof of necessity in (a) requires the construction of outer functions. (One can assume without loss of generality that the g.c.d. of the singular inner factors of the non-zero functions in I is **1**.) In section 3 of [5] it is demonstrated that there is an outer function F, $F \in C_A^\infty$, vanishing to infinite order precisely on $Z^\infty(I)$, and such that $\log F(e^{i\Theta})| = -\omega(-\log \tilde{\rho}(\Theta))$, where $\tilde{\rho}$ is continuous, const $\cdot \rho(e^{i\Theta}) \leqslant \tilde{\rho}(\Theta) \leqslant$ const $\cdot \rho(E^{i\Theta})$, and ω is a positive increasing infinitely differentiable function on $\mathbb{R}$ which can be chosen so that $\lim\limits_{x \to +\infty} x^{-1}\omega(x) = +\infty$ as slowly as desired. An appeal to the closed ideal structure theorem places $F^{(n)} \in I^\infty$ for all n. Now, since BF and BF' are assumed to belong to C_A^∞, $B'F = (BF)' - BF' \in C_A^\infty$. Thus

$$\log |B'(e^{i\Theta})| = -\log |F(e^{i\Theta})| + O(1) = \omega(-\log \tilde{\rho}(\Theta)) + O(1)$$

for all choices of ω. Hence, for some $k > 0$,

$$\log |B'(e^{i\Theta})| \leqslant -k \log \rho(e^{i\Theta}, Z^\infty(I)) + O(1)$$

or $|B'(e^{i\Theta})| = O(\rho(e^\Theta, Z^\infty(I))^{-k})$. A computation shows that this implies that the zeros of B have finite degree of contact at $Z^\infty(I)$.

The last assertion follows from the closed graph theorem.

To prove sufficiency in (b), let

$$J = \{ f \in C_A^\infty : f^{(n)}(z) = 0, \; z \in Z^\infty(I), \; n = 0, 1, \dots \}.$$

Then $Z^0(J) = Z^\infty(J) = Z^\infty(I)$ and $(1/B)I \subset J$. Applying (a) to J, one concludes $(1/B)I$ is closed. To prove necessity in (b), let

$$K = \{ f \in C_A^\infty : f^{(n)}(z) = 0, \; z \in Z^0(I) \cap \partial \mathbb{D}, \; n = 0, 1, \dots \}.$$

(Again, one can ignore singular inner factors.) Then $Z^0(K) = Z^\infty(K) = Z^0(I) \cap \partial \mathbb{D}$ and by the closed ideal structure theorem $(1/B)I \supset K$. Thus $BK \subset I \subset C_A^\infty$; and so, applying (a) to K, the zeros of B have finite degree of contact at $Z^\infty(K) = Z^0(I) \cap \partial \mathbb{D}$. $\square$

Let us consider problem (1) in the more general case where $Z^0(I) \subset \partial \mathbb{D}$ but $Z^0(I) \neq Z^\infty(I)$ in the light of the above results. From the computation referred to in the proof of sufficiency in Theorem (a), it is clear that if the zeros of B have finite degree of contact at $Z^\infty(I)$, then $BI \subset C_A^\infty$; however, it is not difficult to construct examples to show that this condition is not necessary. On the other hand, Theorem (a) along with the closed ideal structure theorem implies that a necessary condition for $BI \subset C_A^\infty$ is that the zeros of B have finite degree of contact with $Z^0(I)$; however, this condition is clearly not sufficient. It appears that the sets $Z^n(I)$, $0 < n < \infty$, play a role in determining whether or not $BI \subset C_A^\infty$.

Similar remarks apply to problem (2). That is, if the zeros of B have finite degree of contact at $Z^\infty(I)$, then $(1/B)I$ is closed; and, if $(1/B)I$ is closed, then the zeros of B have finite degree of contact at $Z^0(I) \subset \partial\mathbb{D}$.

In regard to problem (2), it is not always the case that $(1/B)I$ is closed. In fact, it is possible to construct a closed ideal I where the zeros of B, the g.c.d. of the Blaschke factors of the non-zero functions in I, do not have finite degree of contact at $Z^0(I) \subset \partial\mathbb{D}$ and, hence, $(1/B)I$ is not closed.

We note that if B is a Blaschke product which divides a non-zero C_A^∞ function, then there is a Carleson set E, $E \subset \partial\mathbb{D}$, such that the zeros of B have finite degree of contact at E. In fact one can take $E = \mathrm{clos}\{\, z/|z| : B(z) = 0 \,\}$; see Theorem 1.2 [3].

References

1. Korenblum B. I., Doklady Akad. Nauk SSSR **200** (1971), no. 1, 24–27 (Russian); English transl. in Soviet Math. Dokl. (1971).
2. Caughran J. G., *Zeros of analytic function with infinitely differentiable boundary values*, Proc. Amer. Math. Soc. **24** (1970), 700–704.
3. Nelson D., *A characterization of zero sets for C_A^∞*, Mich. Math. J. **18** (1971), 141–147.
4. Taylor B. A., Williams D. L., *Zeros of Lipschitz functions analytic in the unit disc*, Mich. Math. J. **18** (1971), 129–139.
5. Taylor B. A., Williams D. L., *Ideals in rings of analytic functions with smooth boundary values*, Can. J. Math. **22** (1970), 1266–1283.
6. Wells J., *On the zeros of functions with derivatives in H^1 and H^∞*, Can. J. Math. **22** (1970), 342–347.

Department of Mathematics
Syracuse University
Syracuse, New York, 13210
USA

CLOSED IDEALS IN THE ANALYTIC GEVREY CLASS

J. Bruna

Let $\mathbb{D}$ denote the open unit disc in $\mathbb{C}$. The (analytic) Gevrey class of order α is the class of holomorphic functions f in $\mathbb{D}$ such that

$$|f^{(n)}(z)| \leqslant C_f Q_f^n n! \, n^{n/\alpha} \, .$$

The class G_α is endowed with a natural topology under which it becomes a topological algebra. So it is natural to ask for the structure of the closed ideals of G_α. For $\alpha \geqslant 1$ the class G_α is quasianalytic and so this question is trivial in this case.

For $0 < \alpha < 1$, this characterization should be along the lines of previous works on this topic ([2] and [3]) concerning the classes A^k, $k = 0, 1, \ldots, \infty$. Namely, for a closed ideal I one considers for $k = 0, 1, \ldots, \infty$

$$Z^k(I) = \{ z \in \mathbb{D} : f^{(j)}(z) = 0 \text{ for all } f \text{ in } I, \, j = 0, \ldots, k \}$$

and

$$S = \text{ greatest common divisor of the inner parts of functions in } I.$$

The precise QUESTION is then stated as follows: *is every closed ideal I characterized by the sequence $Z^k(I)$ and S_I in the sense*

$$I = \{ f \in G_\alpha : S_I \text{ divides } f \text{ and } Z^k(f) \subset Z^k(I) \} ?$$

Hruščёv's paper [1] is basic in this context for it contains the characterization of the sets of uniqueness for G_α, i.e. the sets E which can appear as $Z^\infty(I)$. The imitation of the proof of [2] or [3] for the A^k-case just gives the result in special cases which include the restriction $\alpha < 1/2$.

REFERENCES

1. Hruščёv S. V., *Sets of uniqueness for the Gevrey classes*, Ark. för mat. **15** (1977), 235–304.
2. Korenblum B. I., *Closed ideals of the ring A^n*, Funct. anal. and its applications **6** (1972), no. 3, 38–52. (Russian)
3. Taylor B. A., Williams D. L., *Ideals in rings of analytic functions with smooth boundary values*, Canad. J. Math. **22** (1970), 1266–1283.

Universitat Autònoma
de Barcelona
Secció matemàtiques
Bellaterra (Barcelona)
España

S.11.27
old

DIVISIBILITY PROBLEM IN $A(\mathbb{D})$ AND $H^\infty(\mathbb{D})$

FRANK FORELLI

We offer two problems on divisibility, one in the disc algebra $A(\mathbb{D})$, the other in $H^\infty(\mathbb{D})$.

PROBLEM 1. *For which $X \subset \overline{\mathbb{D}}$, $X \neq \emptyset$, do we have*

$$(1) \qquad A_X = \bigcap_{\varsigma \in X} A_\varsigma$$

where A_X is the ring of fractions

$$\{ f/g \colon f,g \in A(\mathbb{D}), \quad g \text{ vanishes nowhere in } X \}$$

and A_ς is the local ring of fractions

$$\{ f/g \colon f,g \in A(\mathbb{D}), \quad g(\varsigma) \neq 0 \}?$$

We might point out that (1) holds if $X \cap \mathbb{T}$ is closed.

(Proof. Put $Y = X \cap \mathbb{T}$, $\Delta = X \cap \mathbb{D}$, and let $\gamma \in$ the right side of (1). Then

$$\gamma \in \bigcap_{\varsigma \in Y} A_\varsigma.$$

It is proved in [1] that (1) holds if X is closed in $\mathbb{D}$; hence $\gamma = F/G$ where $F, G \in A(\mathbb{D})$ and G vanishes nowhere in Y. Let $\varsigma \in \Delta$; then $\gamma = f/g$ where $f, g \in A(\mathbb{D})$, $g(\varsigma) \neq 0$. We have $Fg = fg$; thus $o(G, \varsigma) \leqslant o(F, \varsigma)$ where $o(\varphi, \varsigma)$ is the order of vanishing of φ at ς. This proves that the Blaschke product B made up of the zeros of G in Δ (counting multiplicities) divides F. Put $F_1 = F/B$, $G_1 = G/B$. Then $F_1, G_1 \in A(\mathbb{D})$ and $\gamma = F_1/G_1$. $\square$).

Let R be an integral domain, F its field of fractions, and Δ the space of all prime ideals p in R, $p \neq R$. Let Q be an ideal in R; then Q is said to be an ideal of denominators if

$$Q = \{ g \in R \colon g\gamma \in R \}$$

where $\gamma \in F$. Suppose R is such that every ideal of denominators is principal; then if $X \subset \Delta$, $X \neq \emptyset$, we have

$$R_X = \bigcap_{p \in X} R_p$$

where R_X is the ring of fractions

$$\{ f/g \colon f,g \in R, \ g \notin \bigcup X \}$$

and R_p the local ring of fractions

$$\{ f/g : f, g \in R, \ g \notin p \}.$$

This is easy to prove. Now in $H^\infty(\mathbb{D})$ every ideal of denominators is principal. (Easy to prove, but not obvious.) Thus if X is any nonempty set in the maximal ideal space of $H^\infty(\mathbb{D})$, then

$$H^\infty_X = \bigcap_{\varsigma \in X} H^\infty_\varsigma$$

where

$$H^\infty_X = \{ f/g : f, g \in H^\infty, \ \hat{g} \text{ vanishes nowhere in } X \}$$
$$H^\infty_\varsigma = \{ f/g : f, g \in H^\infty, \ \hat{g}(\varsigma) \neq 0 \}.$$

This suggests that (1) might hold for every nonempty subset of $\overline{D}$, although unlike $H^\infty(\mathbb{D})$ not every ideal of denominators in $A(\mathbb{D})$ is principal. E.g. if $\varsigma \in \mathbb{T}$ and

(2) $$P_\varsigma = \{ f \in A(\mathbb{D}) : f(\varsigma) = 0 \},$$

then P_ς is an ideal of denominators, but it is not principal.

PROBLEM 2. *Let Q be a prime ideal in $H^\infty(\mathbb{D})$, $Q \neq 0, H^\infty(\mathbb{D})$. Suppose Q is finitely generated. Do we then have*

$$Q = \{ f \in H^\infty(\mathbb{D}) : f(\varsigma) = f'(\varsigma) = \cdots = f^{(n)}(\varsigma) = 0 \}$$

where $\varsigma \in \mathbb{D}$ and $n \geqslant 0$?

(Yes if Q is maximal [2].) In $A(\mathbb{D})$ we have the following.

Let P and Q be prime ideals in $A(\mathbb{D})$ with $P \neq Q$ and $P \subset Q \subset P_\varsigma$ where $\varsigma \in \mathbb{T}$ (P_ς = the right side of (2)). Then the $A(\mathbb{D})$-module Q/P is not finitely generated, i.e.

$$Q \neq P \text{ a finitely generated ideal in } A(\mathbb{D}).$$

This is a corollary to Nakayama's lemma. It suggests that the answer to Problem 2 is yes, but then the proof would have to be different.

REFERENCES

1. Forelli F., *A note on ideals in the disc algebra*, Proc. Amer. Math. Soc. **84** (1982), 389–392.
2. Tomassini G., *A remark on the Banach algebra $LH^\infty(\mathbb{D}^N)$*, Boll. Un. Mat. Ital. **2** (1969), 202–204.

DEPARTMENT OF MATHEMATICS
UNIVERSITY OF WISCONSIN
480 LINCOLN DR.
MADISON, WI 53706
USA

COMMENTARY BY R. MORTINI

1. A complete solution of Problem 1 is given in Theorem 1.4, p. 16 of [5]. This result tells that for any $X \subset \operatorname{clos}\mathbb{D}$, $X \neq \emptyset$, the ring A_X equals the intersection $\bigcap\limits_{\zeta \in X} A_\zeta$ of all its local rings of fractions. The proof is based on the solution of another problem of Forelli posed in his paper [1].

2. A complete solution of the questions on finitely generated prime ideals in H^∞ and $A(\mathbb{D})$ is given in [3]. The case of H^∞ has also been solved independently by Pamela Gorkin with essentially the same proof [4].

REFERENCES

3. Mortini R., *Finitely generated prime ideals in H^∞ and $A(\mathbb{D})$*, Math. Zeit. **191** (1986), 297–302.
4. Gorkin P., *Prime ideals in closed subalgebras of L^∞*, Michigan Math. J. **33** (1986), 315–323.
5. Mortini R., *The Chang–Marshall algebras*, Mitteilungen Math. Seminar Giessen **185** (1988), 1–76.

MATHEMATISCHES INSTITUT 1
UNIVERSITÄT KARLSRUHE
POSTFACH 6980
D-7500 KARLSRUHE 1
GERMANY

Chapter 12

APPROXIMATION and CAPACITIES

Edited by

J. Brennan
Dept. of Mathematics
University of Kentucky
Lexington, KY 40506
USA

A. Volberg
Dept. of Mathematics
Michigan State University
East Lansing, MI 48824
USA
and
Steklov Math. Institute
St. Petersburg branch
Fontanka 27
St. Petersburg, 191011
Russia

V. P. Havin
Dept. of Math. and Mechanics
St. Petersburg State University
Staryi Peterhof
St. Petersburg, 198904
Russia

INTRODUCTION

Most problems of our Collection may be viewed as approximation problems. That is why selection principles in this Chapter are even more vague and conventional than in the others. Problems collected under the above title illustrate, nevertheless, some important tendencies of modern Approximation Theory.

Some Problems below are closely related to the ideas of the preceding Chapter. That is, of course, not a mere coincidence, the approximation being really the core of spectral analysis-synthesis. An attentive reader will not be deceived by the seemingly scattered contents of items 12.3, 12.4, 12.9, S.12.33, which can be given a unified interpretation from the (broadly conceived) "spectral" point of view. What really matters is, after all, not what or by what means we approximate (by rational functions or by exponentials with prescribed frequencies, by weighted polynomials or by σ-measures within a spectral subspace), but the intrinsic sense, the aim, and the motive impelling to the approximation, i.e., singling out elementary harmonics (with respect to an action) and subsequent recovering of the object they are generated by.

The variant of spectral synthesis mentioned in Problem S.12.33 is aimed at L^p-approximation by solutions of elliptic differential equations (in particular, by analytic and harmonic functions). The same can be said about the Problems 12.9–12.12. Problem 12.10 deals also with some estimates of the derivative of a conformal mapping. Such estimates are useful in connection with "the weak invertibility" (see Chapter 11 again) and especially with the "crescent effect" discovered by M. V. Keldysh.

Problems 12.6–12.8 and 12.15–12.16 are interesting variants of the classical uniform approximation (in the spirit of Mergelyan–Vitushkin–Arakelyan).

Padé approximations, an intensively growing branch of rational approximations, is presented in Problems 12.17 and 12.18 (this direction seems to be promising in connection with some operator-theoretic aspects. See Problem 9.1).

The best approximation á la Tchebysheff, the eternal theme of Approximation Theory, emerges in Problem 12.19.

Problem 12.22 concerns some ideas arising in Complex Analysis under the influence of the Theory of Banach Algebras.

But all this explains only the first half of the title. As to the second, it is a manifestation of close connections of many modern approximation problems with potential theory. Items 12.10, 12.14–12.16, 12.21, 12.23–12.29, S.12.33 make an extensive use of various kinds of capacities though all of them have in mind (or are inspired by) some approximation theoretic problems.

"The capacitary ideology" appears here also in connection with other themes, namely with the solvability of boundary value problems for elliptic equations (see the "old" Problem 12.30, its Commentary being a new problem article), with metric estimates of capacities (12.23–12.27, 12.31) and with removable singularities of analytic functions (12.23–12.27). Sobolev spaces play an essential role in many approximation problems of this Chapter. In Problem 12.32 they are considered in their own right.

Five problems (12.23–12.27) dealing with removable singularities of bounded analytic functions (or, what is the same, with analytic capacity) formed a separate chapter in the first edition. Here we reproduce the translation of some fragments from its preface.

"Analysis became interested in sets of removable singularities of bounded analytic functions in the eighties of the last century, attracted by the very possibility to formulate problems in the new set-theoretic language. This interest being still alive today (as it is witnessed by this Chapter whose five sections have a non-void and even a fairly large intersection), modified its spirit many times during the past century. Now connected with the classification of Riemann surfaces then with extremal problems of Function Theory it was born again in the early sixties after Vitushkin's works on rational approximation.

The problem of relations between analytic capacity and length was the theme of active debates during the Yerevan Conference on Analytic Functions (September, 1960) when L. D. Ivanov pointed out in his conference talk the role which irregular plane sets (in the Besicovich sense) are likely to play in the theory of removable singularities of bounded analytic functions. But an essential new progress (namely, the proof of the Denjoy conjecture*) became possible in the last (1977) year only, after the remarkable achievement by A. Calderón, namely, after his proof of the L^p-continuity of the Cauchy singular integral operator on a smooth curve. The whole Chapter is written under the influence of the Calderón theorem. Maybe, thanks to it, the time is near when the geometric nature of singularities of bounded analytic functions will be completely understood."

The historical information concerning analytic capacity is given in 12.23 and 12.24. We should like to add the article Uryson P. S., *Sur une fonction analytique partout continue*, Fund. Math. 4 (1922), 144–150 (the Russian translation in the book Uryson P. S., *Works on Topology and Other Areas of Mathematics*, GITTL, Moscow-Leningrad, 1951, 3–100).

Capacity motives can be heard also in other Chapters. The classical logarithmic capacity appears (rather unexpectedly) in the item 1.15 devoted to the isomorphic classification of spaces of analytic functions. The analytic capacity (the main subject of Problems 12.23–12.27 influenced by the recent progress in singular integrals, see Chapter 10) takes part in the purely operator-theoretic item 8.10. The use of capacities in the Operator Theory is not at all a novelty or a surprise. Spectral capacities describing the sets carrying non-trivial spectral subspaces, the exquisite classification of the uniqueness sets for various classes of trigonometrical series (the particular case corresponding to the shift $f \mapsto zf$), metric characteristics of spectra in the classification of operators (transfinite diameters et al), all these are the everyday tools of Spectral Theory and the corresponding connections are well illustrated, e.g., by Chapter 4.

---

*See Problem 12.23

APPROXIMATION BY SOLUTIONS OF ELLIPTIC EQUATIONS

N. N. Tarkhanov

1°. Suppose that we are given a local Fréchet space $\mathcal{L}_{\mathrm{loc}}$ of distributions in $\mathbb{R}^n$ such that the embedding maps of C_{loc}^∞ into $\mathcal{L}_{\mathrm{loc}}$ and of $\mathcal{L}_{\mathrm{loc}}$ into L_{loc}^1 are both defined and continuous.

We shall often endow $\mathcal{L}_{\mathrm{loc}}$ with the "main" seminorm $\mathcal{P}$ (possibly taking the value $+\infty$).

For each integer $s \geqslant 0$ we set $\mathcal{L}_{\mathrm{loc}}^s$ to be the space of all functions $f \in \mathcal{L}_{\mathrm{loc}}$ whose derivatives up to order s belongs to $\mathcal{L}_{\mathrm{loc}}$, with the topology of convergence of these derivatives.

Let $\mathcal{K}$ be a compact set in $\mathbb{R}^n$. The space $\mathcal{L}^s(\mathcal{K})$ is defined as the quotient of $\mathcal{L}_{\mathrm{loc}}^s$ by the subspace of functions whose derivatives up to order s vanish almost everywhere on $\mathcal{K}$. It is a Fréchet space with respect to the canonical topology.

It is easy to see that $C_{\mathrm{loc}}^\infty \subset \mathcal{L}_{\mathrm{loc}}^s$. Let $\widetilde{\mathcal{L}}_{\mathrm{loc}}^s$ be the closure of C_{loc}^∞ in $\mathcal{L}_{\mathrm{loc}}^s$. As above we define the quotient $\widetilde{\mathcal{L}}^s(\mathcal{K})$.

We shall be interested in the following examples of these spaces.

Example 1. $\mathcal{L}_{\mathrm{loc}} = C_{\mathrm{loc}}$ is the space of all continuous functions on $\mathbb{R}^n$ endowed with the usual topology. In this case $\mathcal{P}(f) = \sup_{x \in \mathbb{R}^n} |f(x)|$; $\mathcal{L}_{\mathrm{loc}}^s = C_{\mathrm{loc}}^s$ is the well-known local Fréchet space of smooth functions; $\widetilde{\mathcal{L}}_{\mathrm{loc}}^s$ coincides with $\mathcal{L}_{\mathrm{loc}}^s$; $\mathcal{L}^s(\mathcal{K}) = C^s(\mathcal{K})$ is a Banach space.

Example 2. $\mathcal{L}_{\mathrm{loc}} = C_{\mathrm{loc}}^\lambda$ $(0 < \lambda \leqslant 1)$ is the space of all continuous functions on $\mathbb{R}^n$ satisfying the uniform Hölder condition of order λ on compact subsets, with the usual topology. In this case $\mathcal{P}(f) = \sup\limits_{x \in \mathbb{R}^n} \dfrac{|f(x) - f(y)|}{|x - y|^\lambda}$; $\mathcal{L}_{\mathrm{loc}}^s = C_{\mathrm{loc}}^{s,\lambda}$ is the well-known local Fréchet space of Hölder functions. If $0 < \lambda < 1$ then $\widetilde{\mathcal{L}}_{\mathrm{loc}}^s$ consists of functions f in C_{loc}^s such that $\sup_{|\alpha|=s} |D^\alpha f(x) - D^\alpha f(y)| = o\left(|x - y|^\lambda\right)$ uniformly with respect to x and y on compact subsets of $\mathbb{R}^n$, and $\widetilde{\mathcal{L}}_{\mathrm{loc}}^s = C_{\mathrm{loc}}^{s+1}$ if $\lambda = 1$; $\mathcal{L}^s(\mathcal{K}) = C^{s,\lambda}(\mathcal{K})$ and $\widetilde{\mathcal{L}}^s(\mathcal{K}) = \widetilde{C}^{s,\lambda}(\mathcal{K})$ are Banach spaces.

Example 3. $\mathcal{L}_{\mathrm{loc}} = L_{\mathrm{loc}}^q$ $(1 \leqslant q \leqslant \infty)$ is the space of functions f on $\mathbb{R}^n$ such that $|f|^q$ is summable on compact subsets with respect to the Lebesgue measure on $\mathbb{R}^n$, endowed with the usual topology. In this case $\mathcal{P}(f) = \left(\int_{\mathbb{R}^n} |f|^q \, dx\right)^{1/q}$; $\mathcal{L}_{\mathrm{loc}}^s = W_{\mathrm{loc}}^{s,q}$ is the well-known local Fréchet space of Sobolev functions; $\widetilde{\mathcal{L}}_{\mathrm{loc}}^s$ coincides with $\mathcal{L}_{\mathrm{loc}}^s$ provided $1 \leqslant q < \infty$, and $\widetilde{\mathcal{L}}_{\mathrm{loc}}^s = C_{\mathrm{loc}}^s$ if $q = \infty$; $\mathcal{L}^s(\mathcal{K}) = W^{s,q}(\mathcal{K})$ is a Banach space.

2°. Let $P \left(= \sum_{|\alpha|=p} P_\alpha D^\alpha\right)$ be a homogeneous elliptic differential operator (D. O.) of order p with constant coefficients in $\mathbb{R}^n$.

For a set $\sigma \subset \mathbb{R}^n$ we denote by $S(\sigma)$ the space of infinitely differentiable solutions of the equation $Pf = 0$ in a (variable) neighbourhood of σ. In particular, there is a natural embedding of $S(\mathcal{K})$ into $\mathcal{L}^s(\mathcal{K})$.

There are obvious conditions necessary for a function $f \in \mathcal{L}^s(\mathcal{K})$ to be approximable in $\mathcal{L}^s(\mathcal{K})$ by elements of $S(\mathcal{K})$, namely: 1) $f \in \mathcal{L}^s(\mathcal{K})$; 2) $f \in S(\overset{\circ}{\mathcal{K}})$ where $\overset{\circ}{\mathcal{K}}$ is the interior of $\mathcal{K}$, and 3) $D^\alpha(Pf)\big|_{\mathcal{K}} = 0$, $(|\alpha| \leqslant s - p)$ provided $s \geqslant p$. These conditions will be called "rough".

The following problem is motivated by the theory of rational approximation in the plane (see [1]).

PROBLEM. *For which compact sets $\mathcal{K} \subset \mathbb{R}^n$ is it possible to describe the closure of $S(\mathcal{K})$ in $\mathcal{L}^s(\mathcal{K})$ by the "rough" conditions only?*

If $s \geqslant p$ the answer is "for all $\mathcal{K}$'s" whenever $\mathcal{L}_{\text{loc}} = C_{\text{loc}}$ (see [2]), $\mathcal{L}_{\text{loc}} = C_{\text{loc}}^\lambda$, where $0 < \lambda < 1$ (see [3]), and $\mathcal{L}_{\text{loc}} = L_{\text{loc}}^q$, where $1 < q < \infty$ (see [4]). Nothing is known for $q = 1$. Moreover, for $0 \leqslant s < p$ the compact sets $\mathcal{K}$ in the Problem can be characterized equivalently by means of conditions of density of functions with compact support in $\overset{\circ}{\mathcal{K}}$ in suitable spaces of distributions in $\mathbb{R}^n$ supported by $\mathcal{K}$. These conditions are indifferent to the D. O. P, if $\mathcal{L}_{\text{loc}} = C_{\text{loc}}^\lambda$, where $0 < \lambda < 1$ (see [5]), and if $\mathcal{L}_{\text{loc}} = L_{\text{loc}}^q$, where $1 < q < \infty$ (see [6], [4]). The cases $q = 1$ and $q = \infty$ (or $\lambda = 1$) are exceptional.

3°. By analogy with well-known Vituškin's theorem (see [1]), for $0 \leqslant s < p$ it is natural to state a solution of the problem in terms of a suitable capacity. To be more precise let $\Phi(x)$ be a fundamental solution of D. O. P such that the derivatives of $\Phi(x)$ of order $|\alpha| > p - n$ are homogeneous of degree $(p - n - |\alpha|)$. Then for a set $\sigma \subset \mathbb{R}^n$ we define $c(\sigma) = \sup|f(1)|$, where the supremum is taken over distributions f with a compact support in σ, such that $\Phi * f \in \mathcal{L}_{\text{loc}}^s$ and $\sup_{|\alpha| = s} P\left(D^\alpha(\Phi * f)\right) \leqslant 1$.

If $\mathcal{L}_{\text{loc}}$ is one of the spaces of examples 1–3 and σ is open, then $c(\sigma) = \tilde{c}(\sigma)$. The capacity $\tilde{c}(\sigma)$ is defined by analogy to $c(\sigma)$, but with the additional requirement $\Phi * f \in \tilde{\mathcal{L}}_{\text{loc}}^s$.

Let us make the following observation on $c(\sigma)$ (and $\tilde{c}(\sigma)$): for the transform $Tx = kx + a$ $(k > 0)$ of $\mathbb{R}^n$ we have $c(T\sigma) = k^d c(\sigma)$. Here $d = n/q' - (p - s - \lambda)$ is the index characterizing approximations in the space of examples 1–3, and $q' = q/(q - 1)$.

CONJECTURE. *Let $\mathcal{L}_{\text{loc}}$ be one of the above spaces, and $0 \leqslant s < p$. If $d > 0$, then the following conditions for compact set $\mathcal{K} \subset \mathbb{R}^n$ are equivalent:*

1) *the closure of $S(\mathcal{K})$ in $\mathcal{L}^s(\mathcal{K})$ is equal to $S(\overset{\circ}{\mathcal{K}}) \cap \tilde{\mathcal{L}}^s(\mathcal{K})$;*

2) *$\tilde{c}(\mathcal{O} \setminus \mathcal{K}) = \tilde{c}(\mathcal{O} \setminus \overset{\circ}{\mathcal{K}})$ for all open sets $\mathcal{O} \subset \mathbb{R}^n$;*

3) *$\liminf\limits_{\delta \to +0} \dfrac{c(B(x, \delta) \setminus \mathcal{K})}{\tilde{c}(B(x, \delta) \setminus \overset{\circ}{\mathcal{K}})} > 0$ for all $x \in \partial \mathcal{K}$.*

We denote by $B(x, \delta)$ a ball centered at x and with the radius δ.

The implication 1) $\Longrightarrow$ 2) is well-known (see [5]), and the implication 2) $\Longrightarrow$ 3) is trivial. Hence it remains to prove the implication 3) $\Longrightarrow$ 1) only.

The Conjecture is true in the following cases: 1) $P = \Delta$, $\mathcal{L}_{\text{loc}} = C_{\text{loc}}$, and $s = 0$ (see [7]); 2) P is the Cauchy–Riemann operator in $\mathbb{R}^2$ and $\mathcal{L}_{\text{loc}} = C_{\text{loc}}$ (see [1]); 3) $P = \Delta$,

$\mathcal{L}_{\text{loc}} = C_{\text{loc}}$, and $s = 1$ (see [7]); 4) $\mathcal{L}_{\text{loc}} = C_{\text{loc}}^{\lambda}$, where $0 < \lambda < 1$, and $s = p-1$ (see [2]); 5) $\mathcal{L}_{\text{loc}} = L_{\text{loc}}^{q}$, where $1 < q < \infty$, and $s = p-1$ (see [4]). Under additional (necessary!) capacity conditions the implication 3) $\Longrightarrow$ 1) has been proved in [5], [10].

4°. When $\mathcal{L}_{\text{loc}} = C_{\text{loc}}^{\lambda}$ $(0 < \lambda < 1)$, the capacity $c(\sigma)$ is commensurable with the d–dimensional Hausdorff content of σ (see [5], [2]), and the capacity $\tilde{c}(\sigma)$ is commensurable with the lower d–dimensional Hausdorff content of σ (see [2]). Moreover, $c(\sigma)$ is commensurable with the $(1/q')$-th power of the $(p-s, q')$–capacity of σ provided $\mathcal{L}_{\text{loc}} = L_{\text{loc}}^{q}$, where $1 < q < \infty$ (see [5]). Thus the proof of the conjecture will in some cases result in effective geometrical description of compact sets $\mathcal{K}$ in Problem.

REFERENCES

1. Vituškin A. G., *Analytic capacity of sets in problems of approximation theory*, Uspekhi Mat. Nauk **22** (1967), 141–199. (Russian)
2. Verdera J., C^m *approximation by solutions of elliptic equations, and Calderon–Zygmund operators*, Duke Math. J. **55** (1987), 157–187.
3. O'Farell A. G., *Rational approximation in Lipschitz norms* II, Proc. Royal Irish. Acad. **79** A (1979), 104–114.
4. Tarkhanov N. N., *Approximation in Sobolev spaces by solutions of elliptic systems*, Dokl. Acad. Nauk SSSR **315** (1990), 1308–1313. (Russian)
5. Tarkhanov N. N., *Approximation on compact sets by solutions of systems with surjective symbols*, Prepr. Inst. of Physics (Krasnoyarsk) #48 M (1989), 56 pp. (Russian)
6. Havin V. P., *Approximation in the mean by analytic functions*, Dokl. Acad. Nauk SSSR **178** (1968), 1025–1028. (Russian)
7. Keldysh M. V., *On the solvability and the stability of Dirichlet's problem*, Uspekhi Mat. Nauk **8** (1941), 171–231. (Russian)
8. Paramonov P. V., *On harmonic approximation in the* C^1 *norm*, Mat. Sb. **181** (1990), 1341–1365. (Russian)
9. Mateu J., Orobitg J., *Lipschitz approximation by harmonic functions and some applications to spectral synthesis*, Prepr. Univ. Auton. de Barcelona (1988), 31 pp.
10. Tarkhanov N. N., *Uniform approximation by solutions of elliptic systems*, Mat. Sb. **133** (1987), 356–381. (Russian)

INSTITUTE OF PHYSICS
ACADEMGORODOK
KRASNOYARSK, 660036
RUSSIA

APPROXIMATION BY SMOOTH
FUNCTIONS IN SOBOLEV SPACES

PETER W. JONES

Let $G \subset \mathbb{R}^2$ be a bounded domain whose boundary is a Jordan curve. Put

$$W^{k,p}(G) = \{f : D^\alpha f \in L^p(G), \ 0 \leqslant |\alpha| \leqslant k\}.$$

This is the usual Sobolev space defined on G.

Is $C^\infty(\mathbb{R}^2)|G$ dense in $W^{k,p}(G)$, $1 \leqslant k, p < \infty$?

(The corresponding question for a disc minus a slit has a negative answer). The only thing I know is that this can be verified when $k = 1$ and $p = 2$ (use conformal mapping). To the best of my knowledge this question was first raised by C. Amick.

DEPARTMENT OF MATHEMATICS
YALE UNIVERSITY
NEW HAVEN, CONNECTICUT
USA

CHAPTER EDITORS' NOTE

J. Lewis [1] has shown that $C_0^\infty(\Omega)$ is dense in $W^{1,p}(\Omega)$ wherever $1 < p < \infty$ and Ω is a Jordan domain.

REFERENCE

1. Lewis J., *Approximation of Sobolev functions in Jordan domains*, Ark. Mat. 25 (1987), 255–264.

12.3
v.old

SPLITTING AND BOUNDARY
BEHAVIOR IN CERTAIN H^2 SPACES

Thomas Kriete

Let μ be a finite Borel measure with compact support in $\mathbb{C}$. Even for very special choices of μ the structure of $H^2(\mu)$, the $L^2(\mu)$-closure of the polynomials, can be mysterious. We consider measures $\mu = \nu + W\,dm$, where ν is carried by $\mathbb{D}$ and W is in $L^1(m)$. If $\log W$ is in $L^1(m)$, $H^2(\mu)$ is well understood and behaves like the classical Hardy space $H^2(m)$ [1]. We assume that ν is circularly symmetric, having the simple form $d\nu = G(r)r\,dr\,d\theta$, where $G > 0$ on $[0,1]$. Hastings [2] gave an example of such a measure with $W > 0$ m-a.e. such that $H^2(\mu) = H^2(\nu) \oplus L^2(W\,dm)$; we say then that $H^2(\mu)$ splits. A modification of this example will show that given any W with $\int \log W\,dm = -\infty$, G can be chosen to be positive and non-increasing on $[0,1]$ such that $H^2(\mu)$ splits. Suppose G is smooth and there exist C, $C > 0$, and d, $0 < d < 2$, so that

$$(1) \qquad G(r)^d \leqslant -G'(r) \leqslant \frac{1}{G(r)^c}$$

for $0 \leqslant r < 1$. Suppose further that for some ε, $\varepsilon > 0$,

$$(2) \qquad \int_0^1 (1-r)^{-1-\varepsilon} G(r)\,dr < +\infty.$$

THEOREM 1 ([3]). *Let G satisfy (1) and (2). Suppose that $\int_\Gamma \frac{1}{W}dm < \infty$ for some arc Γ of $\mathbb{T}$, and that $W = 0$ on a set of positive measure in $\mathbb{T} \setminus \Gamma$. Then $H^2(\mu)$ splits if and only if*

$$(3) \qquad \int_{1-\delta}^1 \log\log \frac{1}{G(r)}\,dr = \infty$$

for small δ.

This theorem settles the question of splitting only when W is well-behaved. Conditions similar to (3) were introduced by Keldyš and Džrbašjan, and have been used by several authors in the study of other closure problems (cf. [4] and [5]).

QUESTION 1. *Can W be found such that splitting occurs when the integral in (3) is finite, or even when $G \equiv 1$?*

For λ in $\mathbb{D}$ the point evaluation $p \to p(\lambda)$ is a bounded linear functional on $H^2(\mu)$ (at least for those μ that we are considering); let $E^\mu(\lambda)$ denote its norm. If $E^\nu(\lambda)$ is analogously defined, then $E^\mu(\lambda) \leqslant E^\nu(\lambda)$ (an upper bound for E^μ). It is easy to show that $H^2(\mu)$ splits if and only if $E^\mu(\lambda) = E^\nu(\lambda)$ for all λ in $\mathbb{D}$. At the other extreme, there is always an asymptotic lower bound for E^μ [6]: $\lim_{r\to 1}(1-r^2)E^\mu(re^{i\theta})^2 \geqslant 1/W(\theta)$ m-a.e. Sometimes equality holds a.e. on an arc Γ of $\mathbb{T}$:

$$(4) \qquad\qquad \lim_{r\to 1}(1-r^2)E^\mu(re^{i\theta})^2 = \frac{1}{W(\theta)} \quad \text{m-a.e. on } \Gamma$$

e.g., if $\log W$ is in $L^1(m)$, then (4) holds with $\Gamma = \mathbb{T}$. One verifies that if W does not vanish a.e. on Γ and if (4) holds, then $H^2(\mu)$ cannot split.

THEOREM 2 ([7]). *Suppose that $\int_\Gamma \log W\, dm > -\infty$ and*

$$(5) \qquad\qquad \int_0^1 \log \frac{1}{G(r)}\, dr < +\infty.$$

Then (4) holds, every f in $H^2(\mu)$ has boundary values $\tilde f(e^{i\theta})$ m-a.e. on Γ, $f = \tilde f$ m-a.e. on Γ, and $\int_J \log |f|\, dm > -\infty$ whenever J is a closed arc interior to Γ and $f \not\equiv 0$ in $H^2(\mu)$. Every zero set for $H^2(\mu)$ with no limit points outside of J is a Blaschke sequence.

The hypothesis on W is weaker than that in Theorem 1, (5) is stronger than finiteness of the integral in (3) and the conclusion is stronger than the "only if" conclusion of Theorem 1 by an unknown amount. Equation (4) can fail if the hypothesis on W is removed. Fix α, $0 < \alpha < 1$, and let

$$(6) \qquad\qquad G(r) = \exp\left(-\frac{1}{(1-r)^\alpha}\right),$$

G satisfies (5). Define $\Omega(\theta,\delta) = (\pi/\delta)\, m\{x : x \in [\theta - \delta, \theta + \delta],\ W(x) \leqslant \exp(-\delta^{-\alpha})\}$ and note that $0 \leqslant \Omega \leqslant 1$.

THEOREM 3 ([3]). *If G is as in (6), there exist constants $a, b > 0$ with $E^\mu(re^{i\theta}) > a \exp\left(b\frac{\Omega(\theta,1-r)}{(1-r)^\alpha}\right)$ for all $re^{i\theta}$ in $\mathbb{D}$.*

If $\int_\Gamma \log W\, dm > -\infty$, then $\Omega(\theta,\delta) = O(\delta^\alpha)$ as $\delta \to 0$ m-a.e. on Γ and Theorem 3 yields no information near Γ. On the other hand, for any d, $d > 1$, one can construct W, $W > 0$ a.e. with $\Omega(\theta,\delta) > (\text{const})(-\log \delta)^{-d}$ for δ small and all θ [3]. Thus (4) can fail even if (5) holds.

QUESTION 2. *Assume that the integral in (3) is finite, or even that G is given by (6), or that $G \equiv 1$. Is there a measurable set E, $E \subset \mathbb{T}$, with*

$$H^2(\mu) = H^2(\nu + \chi_{\mathbb{T}\setminus E}W\, dm) \oplus L^2(\chi_E W\, dm),$$

where the first summand consists of "analytic" functions? Might such an E contain any arc on which $\Omega(\theta,\delta)$ (or a suitable analogue) tends to zero sufficiently slowly as $\delta \to 0$? If there is no such E with $mE > 0$, exactly how can the various conclusions of Theorem 2 fail, if indeed they can?

81

QUESTION 3. *Let $W(\theta)$ be smooth with a single zero at $\theta = 0$. Assuming the integral in (3) is finite, describe the invariant subspaces of the operator "multiplication by z" on $H^2(\mu)$ in terms of the rates of decrease of $W(\theta)$ near 0 and $G(r)$ near 1.*

Perhaps more complete results can be obtained than in the similar situation discussed in [8].

Finally we want to mention that the study of other special classes may be fruitful. Recently A. L. Volberg has communicated interesting related results for measures $\nu + W dm$, where ν is supported on a radial line segment. (See [10], [13] in the reference list after Commentary. – Ed.)

REFERENCES

1. Clary S., *Quasi-similarity and subnormal operators*, Doct. Thesis, Univ. Michigan, 1973.
2. Hastings W., *A construction of Hilbert spaces of analytic functions*, Proc. Amer. Math. Soc. **74** (1979), no. 2, 295–298.
3. Kriete T., *On the structure of certain $H^2(\mu)$ spaces*, Indiana Univ. Math. J. **28** (1979), no. 5, 757–773.
4. Brennan J. E., *Approximation in the mean by polynomials on non-Caratheodory domains*, Ark. Math. **15** (1977), 117–168.
5. Mergeljan S. N., *On the completeness of systems of analytic functions*, Uspekhi Mat. Nauk **8** (1953), no. 4, 3–63 (Russian); English translation in ser. 2, Amer. Math. Soc. Translations **19** (1962), 109–166.
6. Kriete T., Trent T., *Growth near the boundary in $H^2(\mu)$ spaces*, Proc. Amer. Math. Soc. **62** (1977), 83–88.
7. Trent T., *$H^2(\mu)$ spaces and bounded evaluations*, Doct. Thesis, Univ. Virginia, 1977.
8. Kriete T., Trutt D., *On the Cesaro operator*, Indiana Univ. Math. J. **24** (1974), 197–214.

DEPARTMENT OF MATHEMATICS
UNIVERSITY OF VIRGINIA
CHARLOTTESVILLE, VIRGINIA 22903
USA

COMMENTARY

THEOREM (A. L. Vol'berg). *There exists W, $W > 0$ a.e. on $\mathbb{T}$, such that $H^2(\mu)$ splits even for $G \equiv 1$.*

The theorem gives an affirmative answer to Question 1. It may be seen from the proof that $\Omega(\theta, \delta)$ tends to zero rather rapidly for every θ. The proof follows an idea of N. K. Nikolskii [9], p. 243.

Proof. It is sufficient to construct a function W, $W > 0$ a.e. on $\mathbb{T}$ and a sequence of polynomials $\{P_n\}_{n \geqslant 1}$ such that $\lim_n (P_n|\mathbb{T}, P_n|\mathbb{D}) = (0, 1)$ in the Hilbert space $L^2(W\, dm) \oplus L^2(\mathbb{D}, dx\, dy)$.

Let $\{\delta_n\}_{n \geqslant 1}$ be any sequence of positive numbers satisfying $\sum_n \delta_n < 1$, $\delta_n \downarrow 0$, and let $\Gamma_n \stackrel{\text{def}}{=} \{e^{it} : |t| \leqslant \pi \cdot \delta_n\}$. Pick any smooth outer function h_n with the constant modulus $|h_n| = \varepsilon_n$ on $\mathbb{T} \setminus \Gamma_n$ ($\varepsilon_n \downarrow 0$) and such that $h_n(0) = 1$. The last condition implies the existence of an integer N_n such that

$$(7) \qquad \iint_{\mathbb{D}} \left| h_n(z^{N_n}) - 1 \right|^2 dx\, dy < \frac{1}{n}.$$

Consider now the set $e_n = \{\zeta \in \mathbb{T} : \zeta^{N_n} \in \Gamma_n\}$. It is clear that $m e_n = \delta_n$ and therefore $m(\cap_{n \geqslant 1} \cup_{k \geqslant n} e_k) = 0$. This implies that the increasing family of sets $S_k = \mathbb{T} \setminus \cup_{k \geqslant n} e_k = \cap_{k \geqslant n} \mathbb{T} \setminus e_k$ almost exhausts the unit circle: $\lim_k m S_k = 1$. Now we are in a position to define the weight W:

$$W(\zeta) = \begin{cases} 1, & \zeta \in S_1 \\ \min\left(1, \dfrac{1}{kC_k^2}\right), & \zeta \in S_k \setminus S_{k-1}, \ k = 2, \ldots, \end{cases}$$

where C_n stands for $\|h_n\|_\infty$.

Set $H_n(z) \stackrel{\text{def}}{=} h_n(z^{N_n})$ and note that $|H_n| = \varepsilon_n$ on S_n because $S_n \subset \mathbb{T} \setminus e_n$. Clearly $\lim_n C_n = +\infty$. These imply

$$\int_\mathbb{T} |H_n|^2 W \, dm = \int_{S_n} |H_n|^2 dm + \frac{1}{n} m(\mathbb{T} \setminus S_n) \leqslant \varepsilon_n^2 + \frac{1}{n}.$$

The last inequality together with (7) yields obviously the desired conclusion.

Theorem 1 in the text of the problem can be strengthened. Suppose that the function G satisfies some regularity conditions and there exists an arc Γ with $\int_\Gamma \frac{dm}{W} < +\infty$. Then $H^2(\mu)$ splits iff

$$\int^1 \log\log \frac{1}{G(r)} dr = +\infty,$$

$$\int_\mathbb{T} \log W \, dm = -\infty.$$

The new point here is that we do not require for W to be identically zero on a set of the positive length. See [10] for the proof.

Question 2 can also be answered affirmatively. Recall that a closed subset E of $\mathbb{T}$ satisfies (by definition) the Carleson condition if $\sum_\nu m(l_\nu) \log \frac{1}{m(l_\nu)} < +\infty$. Here $\{l_\nu\}$ stands for the family of all complementary intervals of E. Let $\mathcal{A}$ be the family of all closed E, $mE > 0$ which do not contain subsets of positive length satisfying the Carleson condition.

THEOREM (S. V. Hruščëv). *Let $E \in \mathcal{A}$. Then there exists a positive weight W such that*

$$H^2(\mu) = H^2(\nu + \chi_{\mathbb{T} \setminus E} W \, dm) \oplus L^2(\chi_E W \, dm),$$

where the first summand does not split.

It has been shown in [11] that such sets E do exist. For example, any set of Cantor type having positive Lebesgue measure and not satisfying the Carleson condition does the job.

Proof. Pick a closed set E in $\mathcal{A}$ and consider an auxiliary region $\mathcal{D}$ having the smooth boundary as it is shown on the figure:

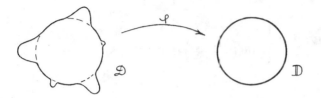

The region $\mathcal{D}$ abuts on $\mathbb{T}$ precisely at the points of E and its boundary Γ has at these points the second order of tangency. Let φ be a conformal mapping of $\mathcal{D}$ onto $\mathbb{D}$. Then φ does not distort the Euclidean distance by the Kellog's theorem (see [12], p. 411). It follows that $\varphi(E) \in \mathcal{A}$. By Theorem 4.1 in [11] there exists a sequence of polynomials $\{P_n\}_{n \geqslant 1}$ satisfying

$$\lim_n P_n = 0 \qquad\qquad \text{uniformly on } \varphi(E);$$

$$\lim_n P_n = 1 \qquad\qquad \text{uniformly on compact subsets of } \mathbb{D};$$

$$|P_n(z)| \leqslant \frac{\text{Const}}{(1 - |z|)^{1/16}}.$$

Using the Kellog's theorem again, we see that the sequence $\{f_n\}_{n \geqslant 1}$, $f_n(z) = P_n(\varphi(z))$ satisfies the following:

(8) $$\lim_n f_n = 0 \text{ uniformly on } E;$$

(9) $$\lim_n f_n = 1 \text{ uniformly on compact subsets of } \mathbb{D};$$

(10) $$|f_n(z)| \leqslant \frac{\text{Const}}{\text{dist}(z, E)^{1/8}}, \quad z \in \text{clos } \mathbb{D}.$$

Define

$$W(t) = \begin{cases} 1, & t \in E \\ (ml)^{1/4}, & t \in l, \ l \text{ being a complimentary interval of } E. \end{cases}$$

The function $g(t) = \text{dist}(t, E)^{-1/4} \cdot \sum_l (ml)^{1/4} \cdot \chi_l$ is evidently summable on $\mathbb{T}$ and dominates $|f_n|^2 \cdot W$. Together with (9) this implies

$$\lim_n \int_{\mathbb{T} \backslash E} |f_n - 1|^2 \cdot W \, dm = 0$$

by the Lebesgue theorem on dominated convergence. Besides, (9) and (10) yield

$$\lim_n \iint_{\mathbb{D}} |f_n - 1|^2 dx dy = 0$$

(see [11]). Finally, $\lim_n \int_E |f_n|^2 W \, dm = 0$, see (8).

The space $H^2(\nu + \chi_{\mathbb{T} \setminus E} W \, dm)$ does not split because $\inf_{t \in l} W(t) = m(l)^{1/4} > 0$ for every complementary interval l. This can be deduce either from Theorem 2 cited in the text of the problem or from Theorem 3.1 of [11]. □

Note that an appropriate choice of E provides the additional property of the weight W in the theorem:

$$\int_{\mathbb{T}} \left(\log \frac{1}{W} \right)^p dm < +\infty$$

for every p, $p < 1$. Pick a Cantor type set E in $\mathcal{A}$ satisfying $\sum_l l \left(\log \frac{1}{l} \right)^p < +\infty$ for $p < 1$.

The construction of the theorem can be extended for other weights G satisfying (5). Such a splitting cannot occur if $\int_E \log W \, dm > -\infty$ and E, $mE > 0$ satisfies the Carleson condition (see Theorem 3.1 in [11]).

References

9. Nikol'skii N. K., *Selected problems of weighted approximation and spectral analysis*, Trudy Mat. Inst. Steklov Akad. Nauk SSSR **120** (1974) (Russian); English transl. in Proc. Steklov Inst. Math. **120** (1974).

10. Vol'berg A. L., *The logarithm of an almost analytic function is summable*, Dokl. Akad. Nauk SSSR **265** (1982), 1297–1302 (Russian); English transl. in Soviet Math. Dokl. **26** (1982), 283–243.

11. Hruščëv S. V., *The problems of simultaneous approximation and removal of singularities of Cauchy-type integrals*, Trudy Mat. Inst. Steklov Akad. Nauk SSSR **130** (1978), 124–195 (Russian); English translation in Proc. Steklov Inst. Math. **130** (1979), 133–203.

12. Goluzin G. M., *Geometric theory of functions of a complex variable*, Transl. Math. Monographs, vol. 26, Amer. Math. Soc., Providence, 1969.

13. Vol'berg A. L., *The simultaneous approximation of polynomials on the circle and in the interior of the disk*, Zapiski Nauchn. Sem. LOMI **92** (1978), 60–84. (Russian)

Commentary by Thomas Kriete and Barbara MacCluer: A Lower Bound for Logarithmic Capacity

We can report some recent progress. As before $\mu = \nu + W \, dm$, where $d\nu = G(r) r \, dr \, d\theta$ on $\mathbb{D}$ and m is Lebesgue measure on $\mathbb{T}$.

Theorem 1 ([15]). *Suppose that G satisfies a mild regularity hypotheses, $G(r) \to 0$ as $r \to 1$, and*

$$(1) \qquad \int_{1-\delta}^1 \log \log \frac{1}{G(r)} dr < \infty \qquad (\delta \text{ small}).$$

If $\int_I \log W \, dm > -\infty$ for some non-trivial open arc I in $\mathbb{T}$, then $H^2(\mu) \neq H^2(\nu) \oplus L^2(W \, dm)$.

Going further, T. L. Miller and R. Smith have shown the following:

Theorem 2 ([16]). *Under the hypotheses of Theorem 1 we have, for every f in $H^2(\mu)$:*

(i) *For m-almost every $e^{i\theta}$ in I, $f(z) \to f(e^{i\theta})$ as $z \to e^{i\theta}$ nontangentially.*

(ii) *If $f \not\equiv 0$, then $\int_J \log |f| \, dm > -\infty$ for every closed subarc J of I.*

More recently, R. Olin and L. Yang have shown that *bounded* functions in an 'analytic direct summand' of a general $H^2(\mu)$ space agree almost everywhere with their nontangential limits, appropriately defined [17].

There is also a sufficient condition for splitting which applies to weights $G(r)$ which decrease to zero faster than $\exp[-c/(1-r)]$. For $\varepsilon > 0$ and an open arc I in $\mathbb{T}$, let $\Omega_I(\varepsilon)$ denote the proportion of I on which $W \leqslant \varepsilon$. One checks that $\log W \notin$ weak $L^1(I)$ exactly when

$$\sup_{0<\varepsilon<1} : \Omega_I(\varepsilon) \log \frac{1}{\varepsilon} = +\infty.$$

For $0 < \delta < 2\pi$, let $\Omega(\delta,\varepsilon)$ be the infimum of $\Omega_I(\varepsilon)$ over all arcs I of length δ.

THEOREM 3 ([15]). *Suppose*

$$(2) \qquad\qquad \lim_{r \to 1}(1-r)\log\frac{1}{G(r)} = +\infty.$$

If there exist positive sequences $\{\delta_n\}$ *and* $\{\varepsilon_n\}$, *both tending to zero, with*

$$(3) \qquad\qquad \lim_{n \to \infty} \Omega(\delta_n,\varepsilon_n)\log\frac{1}{\varepsilon_n} = +\infty,$$

then $H^2(\mu) = H^2(\nu) \oplus L^2(W\,dm)$.

We would like to show that (3) can be replaced by the weakest possible (by Theorem 1) assumption, namely, 'for every arc I, $\int_I \log W\,dm = -\infty$'. Or, failing that, can one weaken (3) by removing the uniformity, replacing (3) by 'for every arc I, $\log W \notin$ weak $L^1(I)$'?

There are versions of Theorem 3 [15] which apply to weights which decrease to zero more slowly than equation (2) allows, but these results are less satisfactory and need to be reconciled with Hruščëv's result [14] on G-Carleson sets. Perhaps an immediate corollary of Theorems 1 and 3 should be mentioned (Hruščëv knew the "if" direction by his methods):

COROLLARY ([15]). *Suppose that* (1) *and* (2) *both hold. If* $W = \chi_E$ *for some measurable subset* E *of* $\mathbb{T}$, *then* $H^2(\mu) = H^2(\nu) \oplus L^2(E)$ *if and only if* $m(I \setminus E) > 0$ *for every non-trivial arc* I *in* $\mathbb{T}$.

REFERENCES

14. Hruščëv S., *The problem of simultaneous approximation and removal of singularities of Cauchy-type integrals*, Trudy Math. Inst. Steklov **130** (1978), 124–195 (Russian); English transl. in Proc. Steklov Inst. Math. **130** (1979), no. 4, 133–203.

15. Kriete T., MacCluer B., *Mean-square approximation by polynomials on the unit disk*, Trans. Amer. Math. Soc. **322** (1990), 1–34.

16. Miller T., Smith R., *Nontangential limits of functions in some $P^2(\mu)$ spaces*, Indiana Univ. Math. J. **39** (1990), 19–26.

17. Olin R., Yang L., *The commutant of multiplication by z on the closure of polynomials in $L^1(\mu)$*, preprint.

ON THE SPAN OF TRIGONOMETRIC
SUMS IN WEIGHTED L^2 SPACES

HARRY DYM

Let $\Delta = \Delta(\gamma)$ be an odd non-decreasing bounded function of γ on the line $\mathbb{R}$, let $Z(\Delta) = L^2(\mathbb{R}, d\Delta)$ and let $Z^T(\Delta)$ denote the closure in $Z(\Delta)$ of finite trigonometric sums $\sum c_j e^{i\gamma t_j}$ with $|t_j| \leqslant T$. It is readily checked that $Z^{T_1}(\Delta) \subset Z^{T_2}(\Delta)$ for $T_1 \leqslant T_2$ and that $\cup_{T \geqslant 0} Z^T(\Delta)$ is dense in $Z(\Delta)$. Let

$$T_0(\Delta) = \inf \{T > 0 : Z^T(\Delta) = Z(\Delta)\}$$

with the understanding that $T_0(\Delta) = \infty$ if the equality $Z^T(\Delta) = Z(\Delta)$ is never attained. The following three examples indicate the possibilities:

(1) if $\Delta(\gamma) = \int_0^\gamma (\xi^2 + 1)^{-1} d\xi$ then $T_0 = \infty$;
(2) if $\Delta(\gamma) = \int_0^\gamma e^{-|\xi|} d\xi$ then $T_0 = 0$;
(3) if Δ is a step function with jumps of height $1/(n^2 + 1)$ at every integer n, then $T_0 = \pi$.

PROBLEM. *Find formulas for T_0, or at least bounds on T_0, in terms of Δ.*

Discussion. Let Δ' denote the Radon–Nikodym derivative of Δ with respect to Lebesgue measure. It then follows from a well-known theorem of Krein [1] that $T_0 = \infty$ as in example (1) if

$$\int_{-\infty}^{\infty} \frac{\log \Delta'(\gamma)}{\gamma^2 + 1} \, d\gamma > -\infty.$$

A partial converse due to Levinson–McKean implies that if Δ is absolutely continuous and if $\Delta'(\gamma)$ is a decreasing function of $|\gamma|$ and $\int_{-\infty}^{\infty} \frac{\log \Delta'(\gamma)}{\gamma^2 + 1} d\gamma = -\infty$ (as in example (2)), then $T_0 = 0$. A proof of the latter and a discussion of example (3) may be found in Section 4.8 of [2]. However, apart from some analogues for the case in which Δ is a step function with jumps at the integers, these two theorems seem to be the only general results available for computing T_0 directly from Δ. (There is an explicit formula for T_0 in terms of the solution to an inverse spectral problem, but this is of little practical value because the computations involved are typically not manageable).

The problem of finding T_0 can also be formulated in the language of Fourier transforms since $Z^T(\Delta)$ is a proper subspace of $Z(\Delta)$ if and only if there exists a non-zero function $f \in Z(\Delta)$ such that

$$\tilde{f}(t) = \int_{-\infty}^{\infty} e^{i\gamma t} \cdot f(\gamma) d\Delta(\gamma) = 0$$

for $|t| \leqslant T$. Thus

$$T_0 = \inf\{T > 0 : \tilde{f}(t) = 0 \text{ for } |t| \leqslant T \Rightarrow f = 0 \text{ in } Z(\Delta)\}.$$

Special cases of the problem in this formulation have been studied by Levinson [3] and Mandelbrojt [4] and a host of later authors. For an up-to-date survey of related results in the special case that Δ is a step function see [5]. The basic problem can also be formulated in $L^p(\mathbb{R}, d\Delta)$ for $1 \leqslant p \leqslant \infty$. A number of results for the case $p = \infty$ have been obtained by Koosis [6], [7] and [8].

References

1. Krein M. G., *On an extrapolation problem of A. N. Kolmogorov*, Dokl. Akad. Nauk SSSR **46** (1945), 306–309. (Russian)
2. Dym H., McKean H. P., *Gaussian Processes, Function Theory and the Inverse Spectral Problem*, Academic Press, New York, 1976.
3. Levinson N., *Gap and Density Theorems*, Colloquium Publ., 26, Amer. Math. Soc., New York, 1940.
4. Mandelbrojt S., *Séries de Fourier et Classes Quasi-analytiques*, Gauthier-Villars, Paris, 1935.
5. Redheffer R. M., *Completeness of sets of complex exponentials*, Adv. Math. **24** (1977), 1–62.
6. Koosis P., *Sur l'approximation pondérée par des polynômes et par des sommes d'exponentielles imaginaires*, Ann. Sci. Ec. Norm. Sup. **81** (1964), 387–408.
7. Koosis P., *Weighted polynomial approximation on arithmetic progressions of intervals or points*, Acta Math. **116** (1966), 223–277.
8. Koosis P., *Solution du problème de Bernstein sur les entiers*, C. R. Acad. Sci. Paris, Ser. A **262** (1966), 1100–1102.

DEPARTMENT OF MATHEMATICS
THE WEIZMANN INSTITUTE OF SCIENCE
REHOVOT, ISRAEL

NEVANLINNA EXTREMAL MEASURES
AND ZEROS OF ENTIRE FUNCTIONS

CH. BERG, H. PEDERSEN

A Nevanlinna extremal solution to an indeterminate Hamburger moment problem is a solution μ so that the polynomials are dense in $L^2(\mu)$. It is a classical fact that such a measure μ is of the form

$$(1) \qquad \mu = \sum_{\lambda \in \Lambda} \mu_\lambda \varepsilon_\lambda, \quad \mu_\lambda > 0,$$

with

$$(2) \qquad \sum_{\lambda \in \Lambda} \mu_\lambda : |\lambda^k| < \infty \quad \text{for } k = 0, 1, \dots,$$

where Λ is a countable and discrete subset of R.

There exists a real entire function of at most exponential type zero having Λ as simple roots. This function q is uniquely determined up to a constant factor, and it can be expressed in terms of the polynomials of the first and second kind associated with the moment sequence, cf. [1]. It is well-known that $1/q$ admits the following absolutely convergent partial fraction expansion

$$(3) \qquad \frac{1}{q(z)} = \sum_{\lambda \in \Lambda} \frac{1}{q'(\lambda)(z - \lambda)}, \qquad z \in C \setminus \Lambda$$

and that

$$(4) \qquad \sum_{\lambda \in \Lambda} \frac{1}{\mu_\lambda (1 + \lambda^2)(q'(\lambda))^2} < \infty,$$

$$(5) \qquad \sum_{\lambda \in \Lambda} \frac{1}{\mu_\lambda (q'(\lambda))^2} = \infty.$$

We have thus a series of necessary conditions which must be satisfied by the function q and the masses $(\mu_\lambda)_{\lambda \in \Lambda}$ of a Nevanlinna extremal measure μ. An old paper by Hamburger [3] claims that these conditions are also sufficient, and the result is presented in the Addenda and Problems to Chapter 4 of Akhiezer's monograph [1]. Unfortunately the result does not hold.

Let us state precisely what is known.

Let q be a real entire function of at most exponential type zero. Assume that q has only real and simple roots, that $\Lambda = q^{-1}(\{0\})$ is infinite and that (3) holds. Let $(\mu_\lambda)_{\lambda \in \Lambda}$ be a sequence of positive numbers so that μ defined by (1) has moments of every order.

If (4) holds then μ is easily seen to be indeterminate, cf. [1, p. 163]. The claim in [3, Theorem 4] and in [1, p. 171] is that μ is Nevanlinna extremal, if also (5) holds. While working on his master's thesis a gap in the proof in [3] was found by the second author. Inspired by this Koosis has constructed an entire function q as above of order zero so that (2)–(5) holds with $\mu_\lambda = 1/(q'(\lambda))^2$. The corresponding measure μ given by (1) is indeterminate, but there exists a function $S \in L^2(\mu) \setminus \{0\}$ annihilating the polynomials, so μ is not Nevanlinna extremal, see [4]. However more can be said. The function S is bounded so the polynomials are not even dense in $L^1(\mu)$.

Due to the importance of Nevanlinna extremal measures we state the following

PROBLEMS.

    a. *Characterize the support Λ of Nevanlinna extremal measures.*
    b. *Characterize the real entire functions q of at most exponential type zero asso-ciated with Nevanlinna extremal measures, and characterize the corresponding sequences $(\mu_\lambda)_{\lambda \in \Lambda}$.*

Comments: 1°. In [2] the following perturbation result was proved: If the support Λ of a Nevanlinna extremal measure is changed at finitely many points, then the new set supports a Nevanlinna extremal measure.

2°. Suppose q and $(\mu_\lambda)_{\lambda \in \Lambda}$ is as above satisfying (1)–(4). For $\mathcal{I}z > 0$ we define $g_z \in L^2(\mu)$ by

$$g_z(x) = \begin{cases} \dfrac{g(z)}{\mu_\lambda q'(\lambda)(z - \lambda)}, & x = \lambda \in \Lambda \\ 0, & x \notin \Lambda. \end{cases}$$

Then μ is Nevanlinna extremal if and only if g_z is in the closure of the polynomials for all z with $\mathcal{I}z > 0$.

3°. The following result is given in [4]. Let q be a real entire function of at most exponential type zero such that all zeros are real and simple and such that

$$\sum_{\lambda \in \Lambda} \left| \frac{\lambda^k}{q'(\lambda)} \right| < \infty \quad \text{for } k = 0, 1, \ldots,$$

where $\Lambda = q^{-1}(\{0\})$, and define $\mu_\lambda = 1/(q'(\lambda))^2$. Then the polynomials are not dense in $L^2(\mu)$ if and only if there exists $S \not\equiv 0$ such that

    (i) S is entire of at most exponential type zero;
    (ii) $S \in l_2(\Lambda)$;
    (iii) $\forall n \geqslant 0 : y^n S(iy)/q(iy) \to 0$ for $|y| \to \infty$.

REFERENCES

1. Akhiezer N. I., *The classical moment problem*, Oliver and Boyd, 1965.
2. Berg C., Christensen J. P. R., *Density questions in the classical theory of moments*, Ann. Inst. Fourier (Grenoble) **31** (1981), no. 3, 99–114.

3. Hamburger H., *Hermitian transformations of deficiency index (1,1), Jacobi matrices and un-determined moment problems*, Amer. J. Math. **66** (1944), 489–522.
4. Koosis P., *Measures orthogonales extrémales pour l'approximation ponderée par des polynômes*, C. R. Acad. Sci. **311** (1990), 503–506.

MATEMATISK INSTITUT
UNIVERSITETSPARKEN 5, DK-2100
COPENHAGEN Æ
DENMARK

MATEMATISK INSTITUT
UNIVERSITETSPARKEN 5, DK-2100
COPENHAGEN Æ
DENMARK

DECOMPOSITION OF APPROXIMABLE FUNCTIONS

ARNE STRAY

Let $H(\mathcal{D})$ be the space of all analytic functions in some open subset $\mathcal{D}$ of the extended complex plane $\mathbb{C}$. Let $\mathcal{D}^*$ denote the one-point compactification of $\mathcal{D}$.

If F is a relatively closed subset of $\mathcal{D}$, $A_{\mathcal{D}}(F)$ is the functions on F being uniform limits on F by sequences from $H(\mathcal{D})$. The problem of characterizing $A_{\mathcal{D}}(F)$ was raised by N. U. Arakelyan some years ago [2]. A closely related question was raised in [3].

Recently we obtained the following characterization of $A_{\mathcal{D}}(F)$ for a large class of sets $\mathcal{D}$:

$$(1) \qquad A_{\mathcal{D}}(F) = C_{na}\big(F \cup \Omega(F)\big) + H(\mathcal{D})$$

where $C_{na}\big(F \cup \Omega(F)\big)$ is the space of analytic functions on $F \cup \Omega(F)$ with a continuous extension to the Riemann sphere, and where $\Omega(F)$ is the smallest open subset of $\mathcal{D} \setminus F$ such that $\mathcal{D}^* \setminus \big(F \cup \Omega(F)\big)$ is arc-wise connected. For details see [5].

PROBLEM 1. *Obtain a decomposition like (1) for any proper nonempty open subset of the Riemann sphere.*

PROBLEM 2. *Obtain decompositions like (1) when $\mathcal{D}$ is the unit disc $\{|z| < 1\}$ and $H(\mathcal{D})$ is replaced by other function spaces in $\mathcal{D}$.*

Remark: A positive answer to Problem 1 will immediately give a solution to Problem 9.6 in [1] in light of the results about $C_{na}(F)$ in [4].

REFERENCES

1. Anderson J. M., Barth K. F., Brannan D. A., *Research Problems in Complex Analysis*, Bull. London Math. Soc. **9** (1977), 152.
2. Arakelyan N. U., *Approximation complexe et propriétés des fonctions analytiques*, Actes Congrès Intern. Math. **2** (1970), Gauthier–Villars / Paris, 595–600.
3. Brown L., Shields A. L., *Approximation by analytic functions uniformly continuous on a set*, Duke Math. Journal **42** (1975), 71–81.
4. Stray A., *Uniform and asymptotic approximation*, Math. Ann. **234** (1978), 61–68.
5. Stray A., *Decomposition of approximable functions*, Ann. of Math. (2) **120** (1984), no. 2, 225–235.

UNIVERSITY OF BERGEN
ALLEGT 55, N-5000 BERGEN
NORWAY

COMMENTARY BY THE AUTHOR

Let F be a relatively closed subset of $\mathbb{D}$. Let $\hat{F}$ denote the set $\{z \in \mathbb{D} : f(z) \leqslant \|f\|\}$, where $\|f\|$ is the supremum norm of f over F and f is an arbitrary H^∞-function. Let $0 < p < +\infty$. A function g defined on F is a uniform limit on F of a sequence of H^p-functions iff $g = u + v$ where $u \in H^p$ and v is uniformly continuous on $\hat{F}$ and analytic in the interior of $\hat{F}$. For $p = +\infty$ the problem is still open.

A PROBLEM OF UNIFORM APPROXIMATION BY FUNCTIONS ADMITTING QUASICONFORMAL CONTINUATION

V. I. Belyi

The following subalgebras of the Banach space $C(K)$ of all continuous functions on a compact set K, $K \subset \mathbb{C}$, are important in the theory of rational approximation. These are the algebra $A(K)$ of all functions in $C(K)$ holomorphic in the interior of K and the algebra $R(K)$ consisting of uniform limits of rational functions continuous on K. For $\varepsilon > 0$ let $K_\varepsilon \overset{\text{def}}{=} \{z \in \mathbb{C} : \text{dist}(z, K) < \varepsilon\}$. Consider the Beltrami equation in K_ε

$$(1) \qquad f_{\bar{z}} = \mu(z) f_z,$$

μ being a measurable function such that

$$\underset{K_\varepsilon}{\text{ess sup}} |\mu| \leqslant k < 1.$$

A continuous function f is said to be a generalized solution of (1) if its generalized derivatives (in the sense of the distribution theory) belong to L^1 locally and satisfy (1) a.e. on K_ε.

Clearly $f|K \in C(K)$ for such a solution and it is known that $f|K \in A(K)$ provided $\mu \equiv 0$ on K [1].

Fix $k < 1$ and consider the set $B_\varepsilon(K)$ of all restrictions $f|K$, where f runs over the family of generalized solutions of (1) in K_ε with $\mu \equiv 0$ on K. Let $B(K)$ be the closure of $\cup_{\varepsilon>0} B_\varepsilon(K)$ in $C(K)$. Then clearly

$$R(K) \subset B(K) \subset A(K).$$

PROBLEM 1. *Is the equality $R(K) = B(K)$ true for an arbitrary K?*

An affirmative answer to the question would entail the following problem.

PROBLEM 2. *Find necessary and sufficient conditions on K for*
(a) $B(K) = C(K)$
and for
(b) $B(K) = A(K)$.

Suppose $k = 0$. Then a complete solution of Problem 2 is given by Vitushkin's theorem [2], [3]. The case $k > 0$ corresponds to the problem of approximation by functions admitting a quasi-conformal continuation.

One of possible ways to solve Problem 1 consists in the construction of a "Swiss cheese" satisfying $R(K) \neq C(K)$, $B(K) = C(K)$.

These problems were posed for the first time at the International Conference on Approximation Theory (Varna, 1981).

PROBLEM 12.7

REFERENCES

1. Lehto O., Virtanen K. I., *Quasi-conformal Mappings in the Plane*, Springer–Verlag, 1973.
2. Vitushkin A. G., *The analytic capacity of sets in problems of approximation theory*, Uspekhi Mat. Nauk **22** (1967), no. 5, 141–199 (Russian); English transl. in Russian Math. Surveys **22** (1967), 139–200.
3. Zalcman L., *Analytic Capacity and Rational Approximation*, Lect. Notes in Math., 50, 1968.

UL. POSTYSHEVA 135, APT.45,
340055 DONETSK
UKRAINE

COMMENTARY

In [4] it is proved that $B(K) = C(K)$ for every $K \subset \mathbb{C}$.

REFERENCE

4. Maimeskul V. V., *To the question on the approximation of continuous functions by traces of generalized solutions of Beltrami equation*, Theory of functions and approximations, Proc. of 2nd Saratov Winter School, chapter 3, Saratov University, 1986, 17–19. (Russian)

TANGENTIAL APPROXIMATION

A. Boivin,   P. Gauthier

Let F be a closed subset of the complex plane $\mathbb{C}$ and let $\mathcal{F}$ and G be two spaces of functions on F. The set F is said to be a set of *tangential approximation* of functions in the class $\mathcal{F}$ by functions in the class G if for each function $f \in \mathcal{F}$ and each positive continuous ε on F, there is a function $g \in G$ with

$$\left| f(z) - g(z) \right| < \varepsilon(z), \ z \in F.$$

Carleman's theorem [1] states that the real axis is a set of tangential approximation of continuous functions by entire functions. Hence, tangential approximation is sometimes called Carleman approximation.

PROBLEM. *For given classes of functions $\mathcal{F}$ and G, characterize the sets of tangential approximation.*

Of course, the problem is of interest only for certain classes $\mathcal{F}$ and G. We shall use the following notations:

$H(\mathbb{C})$: entire functions

$M_F(\mathbb{C})$: meromorphic functions on $\mathbb{C}$ having no poles on F

$H(F)$: functions holomorphic on (some neighbourhood of) F

$U(F)$: uniform limits on F, of functions in $H(F)$

$A(F)$: functions continuous on F and holomorphic on F°

$C(F)$: continuous complex-valued functions on F

Each of these classes is included in the one below it. We consider each problem of tangential approximation which results by choosing G as one of the first three classes and choosing $\mathcal{F}$ as one of the last three. Thus each square in the following table corresponds to a problem.

G \ $\mathcal{F}$	$U(F)$	$A(F)$	$C(F)$
$H(\mathbb{C})$		[4]	[2]
$M_F(\mathbb{C})$			[3]
$H(F)$			[3]

The blank squares correspond to open problems. For partial results on the central square, see [5]. In [3], the conditions stated characterize those sets of tangential approximation for the classes $\mathcal{F} = C(F)$ and $G = M_F(\mathbb{C})$. One easily checks that these conditions are also necessary and sufficient for the case $\mathcal{F} = C(F)$ and $G = H(F)$. The first column was suggested to us by T. W. Gamelin and T. J. Lyons.

One can formulate similar problems for harmonic approximation. The most general harmonic function in a neighbourhood of an isolated singularity at a point $y \in \mathbb{R}^n$, $n \geqslant 2$ can be written in the form

$$u(x) = p_0 \cdot K(x - y) + \sum_{k=1}^{\infty} p_k(x - y)|x - y|^{-n-2k} + \sum_{k=0}^{\infty} q(x - y)$$

where

$$K(x) = \begin{cases} \log|x| & \text{if } n = 2 \\ |x|^{2-n} & \text{if } n \geqslant 3 \end{cases}$$

and p_k, q_k are homogeneous harmonic polynomials of degree k, $k \geqslant 0$. The singularity of u is said to be *non-essential* if $p_k = 0$, $k \geqslant k_0$. An *essentially harmonic function* on an open set $\Omega \subset \mathbb{R}^n$ is a function which is harmonic in Ω except possibly for non-essential singularities.

Let F be a closed set in $\mathbb{R}^n$, $n \geqslant 2$. We introduce the following notations:

$h(\mathbb{R}^n)$: functions harmonic on $\mathbb{R}^n$

$m_F(\mathbb{R}^n)$: essentially harmonic functions on $\mathbb{R}^n$ having no singularities on F

$h(F)$: functions harmonic on (some neighbourhood of) F

$u(F)$: uniform limits on F, of functions in $h(F)$

$a(F)$: functions continuous on F and harmonic on F°

$c(F)$: continuous real-valued functions on F

As in the complex case, we have a *table of problems*.

G \ $\mathcal{F}$	$u(F)$	$c(F)$	$c(F)$
$h(\mathbb{R}^n)$			[6]
$m_F(\mathbb{R}^n)$			[7]
$h(F)$			[8]

REFERENCES

1. Carleman T., *Sur un théorème de Weierstrass*, Ark. Mat. Astronom. Fys. 4 (1927), no. 20B, 1–5.
2. Keldyš M. V., Lavrent'ev M. A., *On a problem of Carleman*, Dokl. Akad. Nauk SSSR **23** (1939), no. 8, 746–748. (Russian)
   Mergeljan S. N., *Uniform approximation to functions of a complex variable*, Uspekhi Mat. Nauk **7** (1952), 31–123 (Russian); English transl. in Amer. Math. Soc. Translations 3 ser. 1 (1962), 294–391.
   Arakeljan N. U., *Uniform and tangential approximation by analytic functions*, Izv. Akad. Nauk Armjan. SSR **3** (1968), 273–286 (Russian); English transl. in Amer. Math. Soc. Translations **122** ser. 2 (1984), 85–97.

3. Nersesjan A. A., *On uniform and tangential approximation by meromorphic functions*, Izv. Akad. Nauk Armjan. SSR **7** (1972), no. 6, 405–412.
Roth A., *Meromorphe Approximationen*, Comment. Math. Helv. **48** (1973), 151–176.
Roth A., *Uniform and tangential approximations by meromorphic functions on closed sets*, Canad. J. Math. **28** (1976), 104–111.
4. Nersesjan A. A., *On Carleman sets*, Izv. Akad. Nauk. Armjan. SSR **6** (1971), no. 6, 465–471 (Russian); English transl. in Amer. Math. Soc. Translations **122** ser. 2 (1984), 99–104.
5. Boivin A., *On Carleman approximation by meromorphic functions*, Proceedings 8th Conference on Analytic Functions, Blazejewko, August, 1982.
6. Šaginjan A. A., *Uniform and tangential harmonic approximation of continuous functions on arbitrary sets*, Mat. Zametki **9** (1971), 131–142 (Russian); English transl. in Mat. Notes **9** (1971), 78–84.
7. Gauthier P. M., *Carleman approximation on unbounded sets by harmonic functions with Newtonian singularities*, Proceedings 8th Conference on Analytic Functions, Blazejewko, August, 1982.
8. Labrèche M., *De l'approximation harmonique uniforme*, Doctoral Dissertation, Université de Montréal, 1982.

MIDDLESEX COLLEGE
DEPARTMENT OF MATHEMATICS
UNIVERSITY OF WESTERN ONTARIO
LONDON, CANADA N6A 5B7

DÉPT. DE MATHÉMATIQUES ET DE STATISTIQUE
UNIVERSITÉ DE MONTRÉAL
CP 6128-A
MONTRÉAL, CANADA H3J 1J7

COMMENTARY BY THE AUTHORS

APPROXIMATION PROBLEM: For given classes of functions $\mathcal{F}$ and $\mathcal{G}$, characterize the sets of uniform (respectively, tangential) approximation.

Of course, this problem is of interest only for certain natural classes $\mathcal{F}$ and $\mathcal{G}$. In particular, we denote by $H(S)$ (respectively, $h(S)$) the class of functions on F which are restrictions of functions holomorphic (respectively, harmonic) on S, and by $A(F)$ (respectively, $a(F)$) the class of functions continuous on F and holomorphic (respectively, harmonic) on the interior of F. Since any function on F which can be uniformly approximated (or, *a fortiori*, tangentially) by functions in $H(S)$ (respectively, $h(S)$) necessarily belongs to the class $A(F)$ (respectively, $a(F)$), it is natural to take for $\mathcal{F}$ and $\mathcal{G}$ the classes $A(F)$ and $H(S)$ (respectively, $a(F)$ and $h(S)$). Since these are the only classes which we will discuss here, we shall, for brevity and when S is clear from the context, say that F is a set of uniform (respectively tangential) holomorphic approximation to mean that F is a set of uniform (respectively, tangential) approximation of functions in the class $A(F)$ by functions in the closed $H(S)$. Similarly, the phrase "F is a set of uniform (respectively, tangential) harmonic approximation" means that "F is a set of uniform (respectively, tangential) approximation of functions in the class $a(F)$ by functions in the class $h(S)$". In the previous edition of this book, for simplicity, we formulated the problem only for tangential (not uniform) approximation, and not only for the case where $S = \mathbb{C}$. Since the previous edition, we have no progress to report on the problem as it was then formulated in $\mathbb{C}$. However, there has been significant progress in the more general context of Riemann surfaces.

Harmonic approximation. Indeed, Bagby, in a series of papers, respectively with the second author (1988), alone (1988), and recently with Blanchet (submitted), has characterized closed sets of uniform and also of tangential harmonic approximation on Riemann surfaces.

Holomorphic approximation. For tangential approximation by global holomorphic functions, the problem has been solved for some time already. Of course, we may as well assume that the set is a set of uniform approximation, since tangential approximation implies uniform approximation. For such a set F in a plane domain S, a necessary condition introduced by the second author (1969) was shown by Nersesjan (1971) to be also sufficient in order that the set F be a set of tangential holomorphic approximation. The first author (1980) showed that this tangential holomorphic approximation theory carries over to arbitrary Riemann surfaces.

Uniform holomorphic approximation, however, which would seem to be easier than tangential holomorphic approximation, has turned out to be an enigma, at least on Riemann surfaces. In the case where $S = \mathbb{C}$, Arakeljan (1964) gave a complete topological characterization of those closed subsets F of $\mathbb{C}$ which are sets of uniform holomorphic approximation. The second author and Hengartner (1975) showed that the topological characterization of Arakeljan no longer holds on general Riemann surfaces. In fact Scheinberg (1978) showed that, on Riemann surfaces, no topological characterization whatsoever is possible for sets of uniform holomorphic approximation. *Thus, there remains the problem of just how to characterize sets of uniform holomorphic approximation on Riemann surfaces.* It may be that progress on this problem would contribute to our understanding of approximation on unbounded subsets in $\mathbb{C}^n$, a subject where very little is known.

A better knowledge of uniform holomorphic approximation on Riemann surfaces may also shed some light (see the second author (1980)) on the following rather well known problem.

EMBEDDING PROBLEM. *Can every open Riemann surface S be (properly) embedded into $\mathbb{C}^2$?*

The case $S = \mathbb{C}$ is of course the (only) trivial case. Nishino (see Hitotumatu (1971)) made the first major breakthrough in this problem by showing that the unit disc can be so embedded. Later, Alexander (1977) and Laufer (1973) showed that the answer is positive for any annulus. To our knowledge, these are the only cases that have been published. However, Narashiman (1975) *has conjectured that the method of Nishino actually works for any finite Riemann surface.*

REFERENCES

9. Alexander H., *Explicit imbedding of the (punctured) disc into $\mathbb{C}^2$*, Comment. Math. Helv. **52** (1977), 539–544, MR 58, 1272.

10. Arakeljan N. U., *Uniform and tangential approximation by analytic functions*, Izv. Akad. Nauk Armjan. SSR Ser. Mat. **3** (1968), 273–286, MR 43, 104. (Russian)

11. Bagby T., *A Runge theorem for harmonic functions on Riemann surfaces*, Proc. Amer. Math. Soc. **103** (1988), 160–164.

12. Bagby T., Blanchet P., *Uniform approximation by harmonic functions on closed subsets of Riemannian manifolds* (to appear).

13. Bagby T., Gauthier P. M., *Approximation by harmonic functions on closed subsets of Riemann surfaces*, J. d'Analyse Math. **51** (1988), 259–284, MR 89j, 30064.

14. Boivin A., *Carleman approximation on Riemann surface*, Math. Ann. **275** (1986), 57–70, MR 87, 30054.

15. Carleman T., *Sur un théorème de Weierstrass*, Ark. Mat. Astronom. Fys. (1927), no. 20B, 1–5, JB 53, 237.

16. Gauthier P. M., *Tangential approximation by entire functions and functions holomorphic in a disc*, Izv. Akad. Nauk Arm. SSR Ser. Mat. **4** (1969), 319–326, MR 43, 1172.

17. Gauthier P. M., *Analytic approximation on closed subsets of open Riemann surfaces*, Constructive Function Theory "77, Sofia, 1980, pp. 317–325.

18. Gauthier P. M., Hengartner W., *Uniform approximation on closed sets by functions analytic on a Riemann surface*, Approximation Theory (Z. Ciesielski and J. Musielak, eds.), Reidel, Dordrecht, 1975, pp. 63–70, MR 58, 6263.

19. Hitotumatu S., *Some recent topics in several complex variables by the Japanese school*, Proc. Rommanian-Finnish Seminar, Bucharest, 1971, MR 45, 8871.

20. Laufer H. B., *Imbedding Annuli in $\mathbb{C}^2$*, J. Analyse Math. **26** (1973), 187–215, MR 49, 10915.

21. Narashiman R., Oral communication (1975).

22. Nesresjan A. A., *Carleman sets*, Izv. Akad. Nauk Armjan. SSR Ser. Mat. **6** (1971), 465–471, MR 46, 66. (Russian)

23. Scheinberg S., *Uniform approximation by functions analytic on a Riemann surface*, Ann. Math. **108** (257–298), MR 58, 17111.

THE INTEGRABILITY OF THE
DERIVATIVE OF A CONFORMAL MAPPING

J. Brennan

Let Ω be a simply connected domain having at least two boundary points in the extended complex plane and let φ be a conformal mapping of Ω onto the open unit disk $\mathbb{D}$. In this note we pose the following:

QUESTION. *For which numbers p is*

$$\iint_\Omega |\varphi'|^p \, dx \, dy < +\infty \, ?$$

For $p = 2$ the integral is equal to the area of the disk and is therefore finite. In general, it is known to converge for $4/3 < p < 3$ and if Ω is the plane slit along the negative real axis then it obviously diverges for $p = 4/3$ and $p = 4$. These facts are consequences of the Koebe distortion theorem and were first discovered by Gehring and Hayman (unpublished) for $p < 2$ and by Metzger [1] for $p > 2$. Recently, the author has succeeded in proving that the upper bound 3 can be increased. The following theorem summarizes the known results.

THEOREM 1. *There exists a number τ, $\tau > 0$, not depending on Ω, such that*

$$\iint_\Omega |\varphi'|^p \, dx \, dy < +\infty$$

if $4/3 < p < 3 + \tau$.

For a wide class of regions, including "star-like" and "close-to-convex" domains, $p = 4$ is the correct upper bound (cf. [2], Theorem 2). Quite likely, $\iint_\Omega |\varphi'|^p \, dx \, dy < +\infty$ for $4/3 < p < 4$ in all cases but, unfortunately, the argument in [2] will not give this result.

Here is a *Sketch of the Proof of Theorem 1.* We shall assume that $x_0 \in \Omega$, $\varphi(x_0) = 0$ and we shall denote by $\delta(z)$ the Euclidean distance from the point z to $\partial\Omega$. It is easy to see, using polar coordinates, that

$$\iint_\Omega |\varphi'|^p \, dx \, dy = \int_0^1 2\pi r \int_{|\varphi|=r} |\varphi'|^{p-2} \, d\omega_r \, dr,$$

where $d\omega_r$ is harmonic measure on the curve $|\varphi| = r$ relative to x_0. Moreover, it follows from the Koebe distortion theorem that $|\varphi'(z)| \simeq K \frac{1-|\varphi(z)|}{\delta(z)}$ near $\partial\Omega$ and, consequently, $\iint_\Omega |\varphi'|^p \, dx \, dy < \infty$ if and only if

$$\int_0^1 (1-r)^{p-2} \int_{|\varphi|=r} \frac{d\omega_r}{\delta(z)^{p-2}} \, dr < +\infty.$$

Thus, Theorem 1 is now an immediate consequence of the following lemma on the growth of the integral $\int \delta(z)^{-\lambda} d\omega_r$, as $r \to 1$.

LEMMA 1. *There exists a constant ρ, $\rho > 0$, such that if $\lambda > 1/2$, then*

$$\int_{|\varphi|=r} \frac{d\omega_r}{\delta(z)^\lambda} = O\left[\frac{1}{(1-r)^{2\lambda-\rho}}\right].$$

Of course, if we could prove the lemma for all ρ, $\rho < 1$, then we could prove Theorem 1 for $4/3 < p < 4$. So far, however, this has still not been done. The proof of the lemma is based on an idea of Carleson [3], which he expressed in connection with another problem. The QUESTION is the following: *On a Jordan curve, is harmonic measure absolutely continuous with respect to α-dimensional Hausdorff measure for every α, $\alpha < 1$?* On the one hand, according to the Beurling projection theorem (cf. [4], p. 72), the question can be answered affirmatively if $\alpha \leqslant 1/2$. On the other hand, Lavrent'ev [5], McMillan and Piranian [6] and Carleson [3] have shown by measure of counterexamples that absolute continuity does not always occur if $\alpha = 1$. In addition, Carleson was able to show that the upper bound $1/2$ in Beurling's theorem can be increased. It is interesting to speculate on the extent to which it is possible to observe a similarity between the two problems. For example, it is well-known (cf. [7], p. 44) that harmonic measure is absolutely continuous with respect to 1-dimensional Hausdorff measure if there are no points ξ on $\partial\Omega$ for which

(1) $$\limsup_{\substack{z\to\xi \\ z\in\Omega}} \arg(z\to\xi) = +\infty,$$

(2) $$\limsup_{\substack{z\to\xi \\ z\in\Omega}} \arg(z\to\xi) = -\infty.$$

The QUESTION arises:

If this condition is satisfied, must $\iint |\varphi'|^p dx\,dy < \infty$ for $4/3 < p < 4$?

At this time the answer is not known. Before proceeding to the solution of the general problem it apparently remains to answer this more modest question.

To the best of my knowledge, the question about the integrability of the derivative of a conformal mapping arose in connection with several problems in approximation theory. We shall mention only one of these and then indicate an application of Theorem 1. Our problem was first posed by Keldyš in 1939 (cf. [8] and [9], p. 10) and he obtained the first results in this direction. Further progress has been achieved in the works of Džrbašjan [10], Šaginjan [11], Maz'ja and Havin [12], [13] and the author [14], [15], [2]. A complete discussion of the results obtained up to 1975 can be found in the surveys of Mergeljan [9], Mel'nikov and Sinanjan [16].

Let us assume that D, U are two Jordan domains in the complex plane, $U \subset D$, and let $\Omega = \text{Int}(D\backslash U)$. We shall denote by $H^p(\Omega)$, $p \geqslant 1$, the closure of the set of all polynomials in the space $L^p(\Omega, dx dy)$ and we shall denote by $L_a^p(\Omega)$ the subspace consisting of those functions f, $f \in L^p(\Omega)$, which are analytic in Ω. Clearly $H^p \subset L_a^p$. An interesting question concerns the possibility of equality in this inclusion. It is well-known that in order for H^p and L_a^p to coincide the determining factor is the "thinness" of the region Ω near multiple boundary points (i.e., near points of $\partial D \cap \partial U$). Here is a result which gives a quantitative description of that dependence. The proof is based in part on Theorem 1 (cf. [2] and [15], pp. 143–148).

THEOREM 2. Let $\delta(z)$ be the distance from z to $\mathbb{C} \backslash D$ and let $d\omega$ be harmonic measure on ∂U relative to the domain U. There exists an absolute constant r, $r > 0$ not depending on Ω, such that if

$$\int_{\partial U} \log \delta(z) \, d\omega(z) = -\infty,$$

then $H^p(\Omega) = L_a^p(\Omega)$ for all p, $p < 3 + r$.

The QUESTION remains: Is $p = 4$ the upper bound or is the theorem true for all p, $p < +\infty$?

REFERENCES

1. Metzger T. A., *On polynomial approximation in $A_q(D)$*, Proc. Amer. Math. Soc. **37** (1973), 468–470.

2. Brennan J., *The integrability of the derivative in conformal mapping*, J. London Math. Soc. **18** (1978), 261–272.

3. Carleson L., *On the distortion of sets on a Jordan curve under conformal mapping*, Duke Math. J. **40** (1973), 547–559.

4. McMillan J. E., *Boundary behavior under conformal mapping*, Proc. of the N. R. L. Conference on Classical Function Theory, Washington, D. C., 1970, pp. 59–76.

5. Lavrent'ev M. A., *On some boundary problems in the theory of univalent functions*, Mat. Sb. **1(43)** (1936), no. 6, 815–846 (Russian); English transl. in Amer. Math. Soc. Translations **32** ser. 2 (1963), 1–35.

6. McMillan J. E., Piranian G., *Compression and expansion of boundary sets*, Duke Math. J. **40** (1973), 599–605.

7. McMillan J. E., *Boundary behavior of a conformal mapping*, Acta Math. **123** (1939), 43–67.

8. Keldyš M. V., *Sur l'approximation en moyenne quadratique des fonctions analytiques*, Mat. Sb. **47** (1939), no. 5, 391–402.

9. Mergeljan S. N., *On the completeness of systems of analytic functions*, Uspekhi Mat. Nauk **8** (1953), 3–63 (Russian); English transl. in Amer. Math. Soc. Translations **19** ser. 2 (1962), 109–166.

10. Džrbašjan M. M., *Metric theorems on completeness and the representation of analytic functions*, Doctoral Dissertation, Erevan, 1948; Uspekhi Mat. Nauk. **5** (1950), 194–198. (Russian)

11. Šaginjan A. L., *On a criterion for the incompleteness of a system of analytic functions*, Dokl. Akad. Nauk Armjan. SSR **V** (1946), no. 4, 97–100. (Russian)

12. Maz'ja V. G., Havin V. P., *On approximation in the mean by analytic functions*, Vestnik Leningrad Univ. Math. (1968), no. 13, 62–74. (Russian)

13. Maz'ja V. G., Havin V. P., *Use of (p, l)-capacity in problems of the theory of exceptional sets*, Mat. Sb. **90** (1973), 558–591 (Russian); English transl. in Math. USSR Sbornik **19** (1973), 547–580.

14. Brennan J., *Invariant subspaces and weighted polynomial approximation*, Ark. Mat. **11** (1973), 167–189.

15. Brennan J., *Approximation in the mean by polynomials on non-Carathéodory domains*, Ark. Mat. **15** (1977), 117–168.

16. Mel'nikov M. S., Sinanjan S. O., *Questions in the theory of approximation of functions of one complex variable*, Contemporary Problems of Mathematics, vol. 4, Itogi Nauki i Tekhniki, VINITI, Moscow, 1975, pp. 143–250 (Russian); English transl. in vol. 5, 1976, pp. 688–752.

UNIVERSITY OF KENTUCKY
LEXINGTON, KENTUCKY 40506
USA

COMMENTARY BY THE AUTHOR

In recent years the integrability question has gained considerable attention and some interesting connections with other problems have been uncovered. The original question is still open, however, and the essential difficulties are as yet unresolved.

One can choose to work either on Ω or on D. Setting $f = \varphi^{-1}$, it suffices to show that

(i) $\displaystyle\iint_D |f'|^{2-p}\, dx\, dy < \infty, \qquad \frac{4}{3} < p < 4,$

or, by analogy with Lemma 1, to obtain a suitable estimate for the rate of growth of the integral means

(ii) $\displaystyle I_t(r, f') = \int_0^{2\pi} |f'(re^{i\theta})|^t\, d\theta,$

particularly in the range $-2 < t < 0$. The precise bound depends on the special nature of the function

$$\beta(t) = \inf \left\{ \beta : I_t(r, f') = O((1 - r)^{-\beta}) \quad \text{for all univalent } f \right\}$$

and our main conjecture is more or less equivalent to the assertion that $\beta(-2) = 1$. In general, $\beta(t)$ is a convex function defined for $-\infty < t < \infty$ and Pommerenke [28] (cf. also [26], [27]) has shown that

$$\beta(t) \leqslant -\frac{1}{2} + t + \sqrt{\frac{1}{4} - t + 4t^2}\,,$$

from which it follows that $1 \leqslant \beta(-2) < 1.601$ and that the integral in (i) is finite if $\frac{4}{3} < p < 3.399$. A similar result was obtained by Makarov [23], unpublished.

A more detailed analysis of the function $\beta(t)$ can be found in the article of Carleson and Makarov [19], where a kind of *phase transition* is identified, confirming the existence of a point $t_0 < 0$ such that

(iii) $\beta(t) \equiv |t| - 1$ if $t \leqslant t_0$,

(iv) $\beta(t)$ is strictly convex on $[t_0, 0]$.

Roughly speaking, for $t < t_0$ the extremal growth of I_t corresponds to the case of isolated singularities on $\partial\Omega$, while some sort of stochastic distribution of singularities is required for $t_0 < t < 0$. In general, it is known that if $t < 0$ then

$$|t| - 1 \leqslant \beta(t) < |t| - \varepsilon$$

for some absolute constant ε. Hence, the integrability conjecture would be validated if phase transition occurs at $t_0 = -2$. Moreover, a comparison could then be made between the Hausdorff dimension of an arbitrary set E in ∂D with that of its image $f(E)$ in $\partial\Omega$, which is in agreement with earlier results of Makarov [24], Lemma 5.1 (cf. also [19], Corollary 2 and [28], p. 150). This tends to support the conjecture.

As a possible source of counterexamples A. E. Eremenko has suggested a construction based on iteration theory and certain notions from thermodynamics. His idea is the following: Let

$$F(z) = z^2 + c$$

be any quadratic polynomial whose Julia set is connected; that is, any polynomial for which the constant c belongs to the Mandelbrot set $\mathcal{M}$. Because Ω_∞, the unbounded complementary component of J, is simply connected there exists a conformal map

$$\varphi: \Omega_\infty \to \Delta = \{z : |z| > 1\}.$$

In certain cases $F: J \to J$ is hyperbolic (i.e. expanding) and the average rate of growth of φ' near $\partial\Omega_\infty$ can be estimated in terms of the *pressure* of the associated dynamical system. A similar idea can be found in [21], where it appears in connection with a well-known conjecture of Littlewood on the mean spherical derivative of a polynomial.

Assuming that F is hyperbolic, fix a neighborhood W of J such that the inverse images W, $F^{-1}(W)$, $F^{-2}(W), \ldots$ contract at some specified rate to J (cf, for example, [17], Sec. 9.7). Here, and for the remainder of the discussion, F^n and F^{-n} denote the n-th iterates of F and F^{-1}, respectively. Because the Böttcher equation, (cf. [20], p. 580 and [25], p. 43),

$$\varphi \circ F = \varphi^2$$

is satisfied in Ω_∞, it follows by iteration that $\varphi \circ F^n = \varphi^{2^n}$, $n = 1, 2, \ldots$. Thus, F^{-n} has exactly 2^n univalent branches on any square $S \subset W \cap \Omega_\infty$, giving rise to the same number of disjoint neighborhoods $V_1, \ldots, V_{2^n}$, each of which is mapped onto S by F^n. In each V_j we choose a point z_j such that $F^n(z_j) = a$ coincides with the center of S. In order to determine whether or not the integral

(1)
$$I = \iint_{W\cap\Omega_\infty} |\varphi'|^p \, dx \, dy$$

diverges it suffices to estimate $|\varphi'|$ near $\partial\Omega_\infty$, where

$$(\varphi' \circ F^n)(F^n)' = 2^n \varphi^{2^n-1} \varphi'$$

is satisfied. Choosing S so that $2S$ is also contained in W, it is easily checked that

$$|(F^n)'(z)| \asymp |(F^n)'(z_j)| \quad \text{on } V_j$$

with constants independent of n. Similarly, $|\varphi'(z)| \asymp |\varphi'(z_j)|$ on V_j and

$$\iint_{V_j} |\varphi'|^p \, dx \, dy \asymp |(F^n)'(z_j)|^{p-2} 2^{-np}, j = 1, \ldots, 2^n,$$

from which it follows that

$$I \geqslant K \lim_{n\to\infty} \sum_{z:F^n(z)=a} |(F^n)'(z)|^{p-2} 2^{-np}$$

for some absolute constant K.

The *pressure* of the dynamical system $F: J \to J$ is defined as

$$P(t) = P_F(t) = \lim_{n\to\infty} \frac{1}{n} \log\left(\sum_{z:F^n(z)=a} |(F^n)'(z)|^{-t} \right), \qquad t > 0$$

and, moreover,

    (a) the limit exists and is independent of a;
    (b) $P(t)$ is a strictly convex, decreasing function of t;
    (c) $P(t_0) = 0$ exactly when $t_0 = \dim J$.

The concept itself is due to R. Bowen and D. Ruelle, at least in the present context (cf.[20] for additional references). If for some choice of $c \in \mathcal{M}$ and $p > 2$ it should happen that $P(2 - p) > p \log 2$ then the corresponding integral (3) clearly diverges. A counterexample to the integrability conjecture would therefore be obtained if $P(-2) > 4 \log 2$.

Computer experiments conducted by P. Jones at Yale University seem to indicate that the extremal situation is approached as $c \to -2$ through int $\mathcal{M}$. It is conjectured, but not yet proven, that every $c \in$ int $\mathcal{M}$ corresponds to a hyperbolic polynomial F (cf. [18], p. 127; [22], p. 209; [25], p. 116). One should keep this, as well as the inherent instability of all algorithms near $\partial \mathcal{M}$, in mind when trying to interpret the computational results. It is interesting to note, however, that for $c = -2$ the Julia set $J = [-2, 2]$ and the corresponding conformal map $\varphi \colon \mathbb{C} \setminus J \to \Delta$ exhibits essentially the same behavior as the Koebe function. This also tends to support the original conjecture.

REFERENCES

17. Beardon A. F., *Iteration of Rational Functions*, Springer-Verlag, Berlin, 1991.

18. Blanchard P., *Complex analytic dynamics on the Riemann sphere*, Bull. Amer. Math. Soc. **11** (1984), 85–141.

19. Carleson L., Makarov N. G., *Some results connected with Brennan's conjecture*, preprint, 1992.

20. Eremenko A. E., Lyubich M. Yu., *The dynamics of analytic transformations*, Algebra i Analiz **1** (1989), 1–70 (Russian); English transl. in Leningrad Math. J. **1** (1990), 563–634.

21. Eremenko A. E., *Lower estimate in Littlewood's conjecture on the mean spherical derivative of a polynomial and iteration theory*, Proc. Amer. Math. Soc. **112** (1991), 713–715.

22. Falconer K., *Fractal Geometry: Mathematical Foundations and Applications*, Wiley, 1990.

23. Makarov N. G., *Two remarks on integral means of the derivative of a univalent function*, LOMI preprint (1985).

24. Makarov N. G., *Conformal mappings and Hausdorff measures*, Ark. Mat. **25** (1987), 41–85.

25. Milnor J., *Dynamics in one complex variable: Introductory Lectures*, preprint no. 5, Inst. Math. Sci. Stony Brook, 1990.

26. Pommerenke Ch., *On the integral means of the derivative of a univalent function*, J. London Math. Soc. **32** (1985), 254–258.

27. Pommerenke Ch., *On the integral means of the derivative of a univalent function* II, Bull. London Math. Soc. **17** (1985), 565–570.

28. Pommerenke Ch., *The growth of the derivative of a univalent function*, The Bieberbach Conjecture, Math. Surveys Monograph, vol. 21, 1986, pp. 143–152.

WEIGHTED POLYNOMIAL APPROXIMATION

J. Brennan

Let Ω be a bounded simply connected domain in the complex plane $\mathbb{C}$, let $dx\,dy$ denote two-dimensional Lebesgue measure and let $w(z) > 0$ be a bounded measurable function defined on Ω. For each p, $1 \leqslant p < \infty$, we shall consider two spaces of functions:

(i) $H^p(\Omega, w\,dx\,dy)$, the closure of the polynomials in $L^p(\Omega, w\,dx\,dy)$;
(ii) $L_a^p(\Omega, w\,dx\,dy)$, the set of functions in $L^p(\Omega, w\,dx\,dy)$ which are analytic in Ω.

If w is bounded away from zero locally it is easy to see that L_a^p is a closed subspace of L^p and that $H^p \subseteq L_a^p$. It is an OLD PROBLEM to determine: for which regions Ω and weights w is $H^p(\Omega, w\,dx\,dy) = L_a^p$? Whenever this happens the polynomials are said to be complete in L_a^p.

As the problem suggests, completeness depends both on the region Ω and on the weight w. In this article, however, we shall be primarily concerned with the role of w when no restrictions are placed on Ω, save simple connectivity. The main difficulty then stems from the fact that Ω may have a nonempty inner boundary; that is, there may be points in $\partial\Omega$ which belong to the interior of Ω (viz., a Jordan domain with a cut or incision in the form of a simple arc from an interior point to a boundary point). Roughly speaking, $H^p(\Omega, w\,dx\,dy) = L_a^p$ if $w(z) \to 0$ sufficiently rapidly at every point of the inner boundary. But, this is not the only factor that must be considered and in order to avoid certain snags we shall make the additional assumption that W is constant on the level lines of some conformal mapping φ of Ω onto the open unit disk D (i.e., $w(z) = W(1 - |\varphi(z)|)$, where $w(t) \to 0$ as $t \to 0$). Put another way, w depends only on Green's function. With this requirement the problem becomes conformally invariant and every significant result going back to the early 1940's and the seminal work of Keldyš [1] makes use of this or some equivalent fact. Additional information and background material on the completeness question can be found in the survey article of Mergeljan [2], in the author's papers [3], [4] and in the references cited therein.

In the ensuing discussion Ω_∞ will denote the unbounded component of $\mathbb{C}\backslash\bar{\Omega}$ and ω is harmonic measure on $\partial\Omega$ relative to some convenient point in Ω. For weights which depend only on Green's function the author [4] has obtained the following result:

THEOREM 1. *Suppose that $w(t) \downarrow 0$ as $t \downarrow 0$ and that $\omega(\partial\Omega_\infty) > 0$. Then there exists a universal constant $r > 0$ such that*

$$\int_0 \log\log \frac{1}{w(t)}\,dt = +\infty$$

implies $H^p(\Omega, w\,dx\,dy) = L_a^p$ whenever $1 \leqslant p < 3 + r$.

Since there are only two restrictions, one on Ω and one on the L^p-class, TWO QUES-TIONS arise:

(1) Can the assumption $w(\partial\Omega_\infty) > 0$ be removed?

(2) Is the theorem true for all p, $1 \leqslant p < \infty$?

If $w(t) \downarrow 0$ in a sufficiently regular fashion then the answer to both questions is yes. For example, if $w(t) = e^{-h(t)}$ and $t^2 h(t) \uparrow +\infty$ as $t \downarrow 0$ the divergence of the integral $\int_0 \log h(t)\, dt$ is sufficient to guarantee that $H^p(\Omega, w\, dx\, dy) = L_a^p$ for all p, $1 \leqslant p < \infty$, even when $w(\partial\Omega_\infty) = 0$.

In order to give an indication of how the hypotheses are used, here is a brief outline of the proof of Theorem 1. For each $q > 1$ we shall denote by C_q the capacity naturally associated with the Sobolev space $W^{1,q}$ and Λ_α will stand for α-dimensional Hausdorff measure. A comparison of these two set functions together with their formal definitions can be found in the survey article of Maz'ja and Havin [5].

Let g be any function in $L^q(\Omega, w\, dx\, dy)$, $q = p/(p-1)$, with the property that $\int Qgw\, dx\, dy = 0$ for all polynomials Q and form the Cauchy integral

$$f(\zeta) = \int_\Omega \frac{gw(z)}{z - \zeta}\, dx\, dy, \quad z = x + iy\,.$$

Evidently, f vanishes identically in Ω_∞ and so, by "continuity", $f = 0$ a.e.–C_q on $\partial\Omega_\infty$. To establish the completeness of the polynomials we have only to prove that $f = 0$ a.e.–C_q on the rest of the boundary as well, this approach having first been suggested by Havin [6] (cf. also [3] and [4]). The argument is then carried out in two stages; one verifies that

Step 1. $f = 0$ a.e. with respect to harmonic measure on $\partial\Omega$;

Step 2. $f = 0$ a.e. with respect to the capacity C_q on $\partial\Omega$.

Moreover, in the process it will be convenient to transfer the problem from Ω to D by means of conformal mapping. With $\psi = \varphi^{-1}$ let $F = f(\psi)$. For each $\varepsilon > 0$ let $\Omega_\varepsilon = \{z \in \Omega : |\varphi(z)| < 1 - \varepsilon\}$ and put

$$f_\varepsilon(\zeta) = \int_{\Omega_\varepsilon} \frac{gw(z)}{z - \zeta}\, dx\, dy \text{ and } F_\varepsilon = f_\varepsilon(\psi)\,.$$

Thus, F and F_ε are both defined on D and F_ε is analytic near ∂D.

Step 1. By choosing $\tau > 0$ and sufficiently small we can find a corresponding $\delta > 0$ such that the following series of implications are valid for any Borel set $E \subseteq \partial\Omega$:

$$C_q(E) = 0 \overset{(i)}{\Longrightarrow} \Lambda_{\frac{1}{2} + \delta}(E) = 0 \overset{(ii)}{\Longrightarrow} \omega(E) = 0.$$

Here $q = p/(p-1)$ and $p < 3 + \tau$. The first implication (i) is essentially due to Frostman [7]. Although he considered only Newtonian capacity, his argument read-ily extends to the nonlinear capacities which enter into the completeness problem (cf. Maz'ja–Havin [5]). The second assertion (ii) is a consequence of a very deep theorem of Carleson [8]. Because $f = 0$ a.e.–C_q on $\partial\Omega_\infty$ and $\omega(\partial\Omega_\infty) > 0$ it follows that $f = 0$ on some boundary set of positive harmonic measure. Consequently, taking radial limits, $F = 0$ on a set of positive arc length on ∂D.

We may now suppose that $W(t) = e^{-h(t)}$, where $h(t) \uparrow +\infty$ as $t \downarrow 0$. Then, using the fact that $\int |\varphi'|^p dx dy < \infty$ for $p < 3 + \tau$ (cf. [9] and [10], the latter being reprinted in this collection, Problem 12.9) it is an easy matter to check that

1. $\displaystyle \int_{\partial \mathbb{D}} |F - F_\varepsilon| \, d\theta \leqslant e^{-ch(\varepsilon)}$,

2. $\displaystyle \int_{|z|=r} |F_\varepsilon| \, d\theta \leqslant K, \quad 1 - \varepsilon \leqslant r \leqslant 1$,

where c and K are constants independent of ε. Because $F = 0$ on a set of positive $d\theta$ measure on ∂D and $\int_0 \log h(t) dt = +\infty$, it follows that $F = 0$ a.e.$-d\theta$. The argument here is based on a modification of Beurling's ideas [11] and it states, in essence, that those functions on ∂D which can be sufficiently well approximated by analytic functions retain the uniqueness property of the approximating family. As a general principle, of course, this goes back to Bernstein [12]. The upshot is $f = 0$ a.e.$-d\omega$ on $\partial \Omega$ and Step 1 is complete.

Step 2. At this point we are required to show that from $f = 0$ a.e.$-d\omega$ it can be concluded that $f = 0$ a.e.$-C_q$ on $\partial \Omega$ which runs counter to the known relationship (1.1) between harmonic measure and capacity. We shall be content here to simply note that the reasoning is based on an argument from the author's article [13] and that essential use is made of the fact that Step 1 is valid for every annihilator g.

In light of what has now been said, one IMPORTANT QUESTION remains: *Is the divergence of the log log-integral necessary for completeness to occur?* That is, if $\int_0 \log \log \frac{1}{W(t)} \, dt < +\infty$, is $H^p(\Omega, w \, dx \, dy) \neq L_a^p$? In case the inner boundary of Ω contains an isolated "smooth" arc and $W(t) \downarrow 0$ as $t \downarrow 0$, the answer is yes and the proof is a simple adaptation of an argument of Domar [14].

REFERENCES

1. Keldysh M. V., *Sur l'approximation en moyenne par polynômes des fonctions d'une variable complexe*, Mat. Sb. **58** (1945), no. 1, 1–20.
2. Mergeljan S. N., *On the completeness of systems of analytic functions*, Uspekhi Mat. Nauk SSSR **8** (1953), no. 4, 3–63 (Russian); English transl. in Amer. Math. Soc. Translations **19** ser. 2 (1962), 109–166.
3. Brennan J., *Approximation in the mean by polynomials on non-Carathéodory domains*, Ark. Mat. **15** (1977), 117–168.
4. Brennan J., *Weighted polynomial approximation, quasianalyticity and analytic continuation*, J. Reine Angew. Math. **357** (1985), 23–50.
5. Maz'ja V. G., Havin V. P., *Nonlinear potential theory*, Uspekhi Mat. Nauk SSSR **27** (1972), no. 6, 67–138 (Russian); English transl. in Russian Math. Surveys **27** (1972), 71–143.
6. Havin V. P., *Approximation in the mean by analytic functions*, Dokl. Akad. Nauk SSSR **178** (1968), no. 5, 1025–1028 (Russian); English transl. in Soviet Math. Dokl. **9** (1968), 245–248.
7. Frostman O., *Potentiel d'équilibre et capacité des ensembles*, Meddel. Lunds Univ. Mat. Sem., (1935), no. 3, 1–118.
8. Carleson L., *On the distortion of sets on a Jordan curve under conformal mapping*, Duke Math. J. **40** (1973), 547–559.
9. Brennan J., *The integrability of the derivative in conformal mapping*, J. London Math. Soc. **18** (1978), 261–272.
10. Brennan J., *The integrability of the derivative of a conformal mapping*, This Volume, Problem 12.9, 101–106.
11. Beurling A., *Quasianalyticity and general distributions*, Lecture Notes, Stanford Univ., 1961.

12. Bernstein S. N., *Léçons sur les Propriétés Extrémales et la Meilleure Approximation des Fonctions Analytiques d'une Variable Réelle*, Gauthier-Villars, Paris, 1926.

13. Brennan J., *Point evaluations, invariant subspaces and approximation in the mean by polynomials*, J. Functional Analysis **34** (1979), 407–420.

14. Domar Y., *On the existence of a largest subharmonic minorant of a given function*, Ark. Mat. **3** (1958), 429–440.

UNIVERSITY OF KENTUCKY
LEXINGTON, KENTUCKY 40506
USA

COMMENTARY BY THE AUTHOR

Since this article first appeared in 1984 the author has extended and amplified the remarks immediately following QUESTIONS (1) and (2) (cf., for example, [15], [16]). As in the foregoing discussion Ω is a bounded simply connected domain and $w(z) = e^{-h\left(\log 1 / |\varphi(z)|\right)}$ is a weight depending only on Green's function and satisfying the two added conditions:

(i) $t\,h(t) \uparrow +\infty$ as $t \downarrow 0$;

(ii) $\displaystyle\int_0 \log h(t)\,dt = +\infty$.

THEOREM 2. $H^p(\Omega, w\,dx\,dy) = L_a^p$ for all p, $1 \leqslant p < \infty$, whenever w is a weight satisfying (i) and (ii).

The proof makes essential use of a strengthened form of a theorem of Vol'berg [17], the details of which can be found in [16] and [18]. It is here that the regularity condition (i) enters in an essential way as an example of Borichev and Vol'berg [19] shows. Quite likely, the approximation theorem itself is no longer true if (i) is significantly relaxed.

Theorem 2 has a natural extension and interpretation in the context of uniform weighted approximation. Beurling [11] has considered, for example, the following generalization of the classical Bernstein problem: With Ω and $w = e^{-h}$ as above, let $C_w(\Omega)$ be the Banach space of all complex valued functions f for which the product $f(z)w(z)$ is continuous on $\overline{\Omega}$ and vanishes on $\partial\Omega$, the norm being defined by

$$\|f\|_w = \sup_\Omega |f|w.$$

Evidently, the collection of functions

$$A_w(\Omega) = \left\{ f \in C_w(\Omega) : f \text{ is analytic in } \Omega \right\}$$

is a closed linear subspace of $C_w(\Omega)$. The problem is to determine whether or not the polynomials are dense in $A_w(\Omega)$. As a corollary to Theorem 2 we can easily obtain this extension of Beurling's result to arbitrary simply connected domains:

THEOREM 3. If (i) and (ii) are satisfied then the polynomials are dense in $A_w(\Omega)$.

REFERENCES

15. Brennan J., *Weighted polynomial approximation and quasianalyticity for general sets*, Centre De Recerca Matematica, Inst. D'Estudis Catalans, preprint.
16. Brennan J., *Functions with rapidly decreasing negative Fourier coefficients*, Lecture Notes in Math. Vol. 1275, 1987.
17. Vol'berg A., *The logarithm of an almost analytic function is summable*, Dokl. Akad. Nauk SSSR **265** (1982), 1297–1301. (Russian)
18. Koosis P., *The Logarithmic Integral* I, Cambridge University Press, 1988; *Vol.* II, in press.
19. Borichev A. A., Vol'berg A., *Unicity theorems for almost analytic functions*, Algebra i Analiz 1 (1989), 146–177 (Russian); English transl. in Leningrad Math. J. 1 (1990), 157–191.

12.11

SIMULTANEOUS APPROXIMATION IN L^p-NORMS

Fernando Perez-Gonzalez

Let Ω be a bounded simply connected domain in the complex plane, let dA denote two-dimensional Lebesgue measure and let $w(z) > 0$ be a bounded measurable function defined on Ω. For each p, $1 \leqslant p < \infty$, we shall consider the two spaces of functions described in problem 12.10:

(i) $H^p = H^p(\Omega, wdA)$, the closure of the polynomials in $L^p(\Omega, wdA)$;
(ii) $L_a^p = L_a^p(\Omega, wdA)$, the set of functions in $L^p(\Omega, wdA)$ which are analytic in Ω.

If w is bounded away from zero locally then L_a^p is a closed subspace of L^p and $H^p \subseteq L_a^p$. The problem of determining for which domains Ω and weights w is $H^p(\Omega, wdA) = L_a^p(\Omega, wdA)$ has been solved by J. Brennan and others (see Brennan [1], [2], [3] and references cited therein). When equality of the two spaces occurs the polynomials are said to be complete in L_a^p.

From here on we shall assume that Ω and w have been chosen in such a way that the polynomials are complete in L_a^p and X will be a relatively closed subset of Ω. We are interested in finding a geometric description of those sets X such that every function in L_a^p satisfying some auxiliary condition on X is the limit of a sequence of polynomial converging simultaneously in the L^p-norm, and in a special way on X.

Some standard notations must be recalled. For a function g defined on X we set $\|g\|_X = \sup_{z \in X} |g(z)|$. When g is bounded on X and $0 < \alpha < 1$, we say that $g \in \mathrm{Lip}(\alpha, X)$ if there is a constant $M > 0$ such that

$$\big|g(z_1) - g(z_2)\big| \leqslant M\big|z_1 - z_2\big|^\alpha$$

for all z_1, z_2 in X. Endowed with the usual norm $\|\cdot\|_{\alpha,X}$ the collection $\mathrm{Lip}(\alpha, X)$ is a Banach space. The closure of the polynomials in the $\mathrm{Lip}(\alpha, X)$-norm is a closed subspace denoted $\mathrm{lip}(\alpha, X)$.

PROBLEM 1. *Find necessary and sufficient conditions on X so that for every function $f \in L_a^p$ which is also bounded on X there exists a sequence (p_n) of polynomials such that*

(i) *$p_n \to f$ in the L^p-norm as $n \to \infty$, and*
(ii) *given any $\varepsilon > 0$ there exists an N such that $\|p_n\|_X \leqslant \|f\|_X + \varepsilon$ for all $n > N$.*

PROBLEM 2. *Find necessary and sufficient conditions on X so that for every function $f \in L_a^p$, which is also uniformly continuous on X, there exists a sequence (p_n) of polynomials such that*

(i) *$p_n \to f$ in the L^p-norm, and*
(ii) *$\|p_n - f\|_X \to 0$ as $n \to \infty$.*

PROBLEM 3. *Find necessary and sufficient conditions on X such that for every function* $f \in L_a^p \cap \mathrm{lip}(\alpha, X)$, $0 < \alpha < 1$, *there exists a sequence (p_n) of polynomials such that*
  (i) $p_n \to f$ *in the L^p-norm, and*
  (ii) $\|p_n - f\|_{\alpha, X} \to 0$ *as $n \to \infty$.*

For the usual Hardy classes $H^p(D)$ in the open unit disk D, $0 < p < \infty$, the sets X solving Problems 1 and 2 have been completely characterized by Perez-Gonzalez and Stray [6]. In fact, these sets are exactly the same, and they must satisfy the following geometric condition: Almost every point $\zeta \in X \cap \partial D$ with respect to 1-dimensional Lebesgue measure on ∂D is a non-tangential limit of a sequence (ζ_n) of points in X. These kind of questions were posed twenty years ago by L. A. Rubel and have been studied extensively by Arne Stray and the author.

Following established practice in the existing literature it is natural to refer to those subsets of Ω solving Problem 1 as *Farrell sets* for $L_a^p(\Omega, wdA)$. Those solving Problems 2 and 3 are called *Mergeljan sets* and α-*Mergeljan sets* for $L_a^p(\Omega, wdA)$, respectively.

For the weight $w \equiv 1$ such approximation problems must be considered in the Bergman space $L_a^p(\Omega, dA)$, where the corresponding functions can be very badly behaved at the boundary, causing serious difficulties to arise. This suggests that in order to attack such problems in $L_a^p(\Omega, wdA)$ one should impose some additional restrictions on w near $X \cap \partial \Omega$ in order to make the situation more manageable.

Consider for a moment a positive measure $d\nu$ supported on the closed unit disk $\overline{D}$ such that $d\nu = d\sigma + h\, d\theta$, where $d\sigma$ is carried by the interior of D, $d\theta$ is the usual angular measure on ∂D and $h \in L^1(\partial D, d\theta)$. The somewhat related question as to when $H^p(d\nu) = H^p(d\sigma) \oplus L^p(h d\theta)$ has been studied in one form or another by Hruščëv [4], Kriete [5] and Vol'berg [7]. Whenever equality occurs $H^p(d\nu)$ is said to *split*. Under the assumption that $d\nu = w(|z|)\, dA$ Theorem 7.1 in Brennan [2] gives a nice sufficient condition for $H^p(d\nu)$ to split. It should be noted here that one may always assume that $\int \log h(\theta)\, d\theta = -\infty$, since otherwise splitting can never occur. If one could obtain satisfactory results for measures $d\nu$ as above, then perhaps it would be possible to settle the corresponding questions for more general domains Ω and weights w by means of conformal mapping.

REFERENCES

1. Brennan J. E., *Approximation in the mean by polynomials on non-Carathéodory domains*, Ark. Mat. **15** (1977), 117–168.
2. Brennan J. E., *Weighted polynomial approximation, quasianalyticity and analytic continuation*, J. Reine. Angew Math. **357** (1985), 23–50.
3. Brennan J. E., *Weighted polynomial approximation and quasianalyticity for general sets*, Centre De Recerca Matematica, Inst. D'Estudis Catalans, March 1992, preprint, No. 149,.
4. Hruščëv S. V., *The problem of simultaneous approximation and removal of singularities of Cauchy-type integrals*, Trudy Mat. Inst. Steklov **130** (1978), 124–195 (Russian); English transl. in Proc. Steklov Inst. Math. **130** (1979), 133–203.
5. Kriete T., *On the structure of certain $H^2(\mu)$ spaces*, Indiana Univ. Math. J. **28** (1979), 757–773.
6. Perez Gonzalez F., Stray A., *Farrell and Mergeljan sets for E^p spaces*, $(0 < p < 1)$, Michigan Math. J. **36** (1989), 379–386.
7. Vol'berg A. L., *Simultaneous approximation by polynomials on the circle and in the interior of the disk*, Zap. Nauchn. Sem. Leningrad Otdel. Mat. Inst. Steklov (LOMI) **92** (1979), 60–84. (Russian)

DEPARTAMENTO DE ANÁLISIS MATEMÁTICO
UNIVERSIDAD DE LA LAGUNA
38271 LA LAGUNA, TENERIFE
SPAIN

12.12

A POLYNOMIAL APPROXIMATION PROBLEM

HÅKAN HEDENMALM

Let ω be a continuous, real-valued function on the open disk $\mathbb{D}$, such that

$$\left(1 - |z|\right)^2 \leqslant \omega(z) \leqslant \left(1 - |z|\right), \qquad z \in \mathbb{D}.$$

Consider the Hilbert space $L_a^2(\mathbb{D}, \omega\, dA)$ consisting of all holomorphic functions f on $\mathbb{D}$ satisfying

$$\|f\|_\omega^2 = \int_{\mathbb{D}} |f|^2 \omega\, dA < \infty,$$

where dA is a normalized area measure on $\mathbb{D}$.

PROBLEM. *Does there exist an ω such that the analytic polynomials are not dense in $L_a^2(\mathbb{D}, \omega dA)$? If so, can this happen also if we make the regularity assumption on ω that $\Delta^2 \omega \geqslant 1$ on $\mathbb{D}$?*

If the answer is negative, it would have important consequences for the invariant subspaces of the unweighted Bergman space (see [1,2]). For instance, one would know that the functions corresponding to Blaschke products for the Bergman space always generate the associated zero-type invariant subspace.

REFERENCES

1. Hedenmalm H., *A factorization theorem for square area integrable analytic functions*, J. Reine Angew. Math. **422** (1991), 45–68.
2. Hedenmalm H., *Factorization in weighted Bergman spaces* under preparation.

UPPSALA UNIVERSITET
MATEMATISKA INSTITUTIONEN
BOX 480
S-751 06 UPPSALA
SWEDEN

AN ELASTICITY PROBLEM

HÅKAN HEDENMALM

Let G be a continuous function on the closed unit disk $\overline{\mathbb{D}}$ which is analytic in the open unit disk $\mathbb{D}$, and suppose that $|G|^2 \, dA$ is a representing measure for 0:

$$u(0) = \int_{\mathbb{D}} u(z)|G(z)|^2 \, dA(z)$$

for all harmonic polynomials u. Such a function G has $|G| \geqslant 1$ on the unit circle $\mathbb{T}$. Let $U_\zeta^G(z) = U^G(z, \zeta)$ be the solution to the boundary problem

$$\begin{cases} \Delta \frac{1}{|G|^2} \Delta U_\zeta^G = \delta_\zeta & \text{on } \mathbb{D}, \\ U_\zeta^G, \quad \nabla U_\zeta^G = 0 & \text{on } \mathbb{T}, \end{cases}$$

where $\mathbb{T}$ denotes the unit circle, Δ is the Laplacian, ∇ the gradient, and δ_ζ is the Dirac mass at the arbitrary point $\zeta \in \mathbb{D}$.

CONJECTURE. *We have $U^G(z, \xi) > 0$ for all $z, \xi \in \mathbb{D}$.*

This is known to hold for $G(z) \equiv 1$. If the conjecture is true, it has implications concerning the factorization theory for the Bergman space $L_a^2(\mathbb{D})$ [1, 2].

REFERENCES

1. Hedenmalm H., *A factorization theorem for square area integrable analytic functions*, J. Reine Angew. Math. **422** (1991), 45–68.
2. Hedenmalm H., *Factorization in weighted Bergman spaces*, under preparation.

UPPSALA UNIVERSITET
MATEMATISKA INSTITUTIONEN
BOX 480
S-751 06 UPPSALA
SWEDEN

COMMENTARY BY B. KORENBLUM

Let $0 \notin \alpha = \{a_n\} \subset \mathbb{D}$ be a finite set and I_α the subspace of $L_a^2(\mathbb{D})$ consisting of functions that vanish on α. Hedenmalm's extremal function H_α is defined as a (unique) element f that realizes that maximum

$$\max\{\operatorname{Re} f(0) : f \in I_\alpha, \ \|f\| \leqslant 1\}.$$

It was proved in [1] that

(i) $|H_\alpha|^2 \, dA$ is a representing measure for 0;
(ii) H_α is a contractive divisor for I_α and an expansive multiplier for $L_a^2(\mathbb{D})$, i.e. $\|H_\alpha h\| \geqslant \|h\|$ for all $h \in L_a^2(\mathbb{D})$;
(iii) H_α is analytic on $\overline{\mathbb{D}}$.

Suppose now that $0 \notin \alpha \subset \mathbb{D}$ and $0 \notin \beta \subset \mathbb{D}$ are two finite sets, with corresponding extremal functions H_α and H_β. If the conjecture is true, it would imply that if $\alpha \subset \beta$ then H_α is a "more expansive" multiplier than H_β, i.e. $\|H_\alpha h\| \geqslant \|H_\beta h\|$ for all $h \in L^2_a(\mathbb{D})$; this relation is called domination: $H_\alpha \succ H_\beta (\succ 1)$. Thus a correspondence would be established between the partial ordering (by inclusion) of the set of zero-based subspaces of $L^2_a(\mathbb{D})$ and the partial ordering (by domination) of the set of extremal functions (also called inner functions).

It should be noted that the validity of the conjecture is by no means necessary for the validity of the implication $I_\alpha \subset I_\beta \Rightarrow H_\alpha \succ H_\beta$. My guess is that the latter is true, even for general z-invariant subspaces, but the former is not.

DEPARTMENT OF MATHEMATICS
STATE UNIVERSITY OF NEW YORK
ALBANY, NY 12222
USA

12.14

old

APPROXIMATION IN THE MEAN BY HARMONIC FUNCTIONS

Thomas Bagby

We discuss analogues of the Vituškin approximation theorem [10] for mean approximation by harmonic functions. We assume that p is fixed, $1 < p < \infty$. We let X be a compact subset of $\mathbb{R}^n$ of positive Lebesgue measure, and we assume $n \geqslant 3$. If $x \in \mathbb{R}^n$, let $B_r(x) = \{ y \in \mathbb{R}^n : |y - x| < r \}$. All functions will be real-valued. If $l \in \{0, 1\}$, let $\mathcal{P}_l$ denote the vector space of all polynomials on $\mathbb{R}^n$ which are homogeneous of degree l, with inner product

$$\{P_1, P_2\} = \sum_{|\alpha| = l} C_\alpha^{(1)} C_\alpha^{(2)} \quad \text{if } P_i(\xi) \equiv \sum_{|\alpha| = l} C_\alpha^{(i)} \xi^\alpha.$$

If $k \in \{1, 2\}$ is fixed, define the (positive) function $G_k \in L^1(\mathbb{R}^n, \mathrm{loc})$ as the inverse Fourier transform of $\widehat{C}_k(\xi) = \left(1 + |\xi|^2 \right)^{-k/2}$, and for each $A \subset \mathbb{R}^n$ define the Bessel capacity $b_{k,p'}(A) = \inf \left\{ \|f\|_{L^{p'}(\mathbb{R}^n)} : f \text{ measurable and } f \geqslant 0 \text{ on } \mathbb{R}^n, G_k * f \geqslant 1 \text{ on } A \right\}$; if $kp' < n$, there exists a constant $C > 0$ such that $C^{-1} \leqslant b_{k,p'}(B_r(0)) / r^{n/p'-k} \leqslant C$ for $0 < r \leqslant 1$. See [7], [8].

We say that X has the L^p *harmonic approximation property* (L^p h.a.p.) provided that for each $\varepsilon > 0$, and each function $f \in L^p(X)$ which is harmonic on the interior $\mathrm{int}\, X$, there exists a harmonic function u on an open neighborhood of X such that $\|u - f\|_{L^p(X)} < \varepsilon$.

Theorem 1. *If any one of the following conditions holds, then X has the L^p h.a.p.*

  (a) ([8], [2]) $p' > n$.

  (b) ([8]) $p' < n$ and there exists a constant $\eta > 0$ such that $b_{1,p'}(B_r(x) \setminus X) \geqslant \eta r^{n/p'-1}$ if $x \in \partial X$ and $0 < r \leqslant 1$.

  (c) ([4], [5]) For each $k \in \{1, 2\}$ one of the following two conditions is met: (i) $kp' > n$ or (ii) $kp' \leqslant n$ and there exists a set E_k with $b_{k,p'}(E_k) = 0$ such that $\int_0^1 r^{k-n/p'} b_{k,p'}(B_r(x) \setminus X)^p \, dr/r = \infty$ if $x \in \partial X \setminus E_k$.

See also [5, Theorem 6]; it follows from [8, Theorem 2.7] that the condition in (c) for $k = 2$ is necessary for the L^p h.a.p., but Hedberg has pointed out that the condition in (c) for $k = 1$ is not necessary (see [1, Section 2]). To characterize the sets having the L^p h.a.p. we define other capacities. We use the notation $\langle T, \varphi \rangle$ to denote the action of the distribution T of compact support on the function $\varphi \in C^\infty(\mathbb{R}^n)$. Let $E(x) = c_n/|x|^{n-2}$ be a fundamental solution for Δ. Let A be a subset of the open set $\Omega \subset \mathbb{R}^n$. If $H \in \mathcal{P}_0 \setminus \{0\}$, we define $\gamma_{p,H}(A, \Omega) = \sup_T |\langle T, H \rangle|$, where the supremum is taken over all (real) distributions T on $\mathbb{R}^n$ such that the support of T is a compact subset of A, $E * T \in L^1(\mathbb{R}^n, \mathrm{loc})$ and $\|E * T\|_{L^p(\Omega)} \leqslant 1$. If $H \in \mathcal{P}_1 \setminus \{0\}$, we define

117

$\gamma_{p,H}(A,\Omega) = \sup_T |\langle T,H \rangle|$, where the supremum is taken over all distributions T on $\mathbb{R}^n$ satisfying the following four conditions:

(i) the support of T is a compact subset of A;
(ii) $\langle T,1 \rangle = 0$;
(iii) $\langle T,P \rangle = 0$ for each $P \in \mathcal{P}_1$ satisfying $\{H,P\} = 0$;
(iv) $E * T \in L^p(\mathbb{R}^n, \text{loc})$ and $\|E * T\|_{L^p(\Omega)} \leqslant 1$.

For references to related capacities of Harvey–Polking, Hedberg, and Maz'ja, see [1, Section 2, Remark 3].

The capacity $\gamma_{p,1}$ is closely related to the Bessel capacity $b_{2,p'}$. Moreover, if $k \in \{1,2\}$, one can prove that there exists a constant $C > 0$ such that

$$(1) \qquad \gamma_{p,H}(A, B_2(x)) \geqslant C\{H,H\}^{1/2} b_{k,p'}(A)$$

for $H \in \mathcal{P}_{2-k} \setminus \{0\}$, $x \in \mathbb{R}^n$, and $A \subset B_1(x)$ a Borel set. The next result follows from the proof of [1, Theorem 2.1], with obvious changes since here our functions are real-valued; the proof is constructive, extending techniques of Lindberg [6] which are based on those of Vituškin [10].

THEOREM 2. *The following conditions are equivalent:*

(a) X has the h.a.p.
(b) *If* $H \in \mathcal{P}_0 \cup \mathcal{P}_1 \setminus \{0\}$, *and if* G *and* Ω *are open subsets of* $\mathbb{R}^n$ *satisfying* $G \Subset \Omega \Subset \mathbb{R}^n$, *then* $\gamma_{p,H}(G \setminus \text{int } X, \Omega) = \gamma_{p,H}(G \setminus X, \Omega)$.
(c) *There exist numbers* $\eta > 0$ *and* $\rho > 0$ *such that*

$$\gamma_{p,H}(B_r(x) \setminus \text{int } X, B_{2r} \setminus (X)) \leqslant \eta \gamma_{p,H}(B_r(H) \setminus X, B_{2r}(x))$$

if $H \in \mathcal{P}_0 \cup \mathcal{P}_1 \setminus \{0\}$, $x \in \mathbb{R}^n$ *and* $0 < r \leqslant 1$.

From Theorem 2 it is possible to deduce part (a) of Theorem 1 (see [1, Section 2]) and part (b) of Theorem 1; however, we have not been able to deduce part (c) of Theorem 1, and this forms a motivation for the first problem below. We also note that a motivation for Problems 1 and 3 is provided by corresponding results in the theory of mean approximation by analytic functions; see the references in the first paragraph of [1]. A number of other papers related to the present note are also given in the references in [1].

PROBLEM 1. *Can one characterize the compact sets having the L^p h.a.p. by means of conditions of Wiener type?*

Specifically, let us say that X has property $(*)$ provided that for each $k \in \{1,2\}$ one of the following two conditions is met: (i) $kp' > n$ or (ii) $kp' \leqslant n$ and there exists a set E_k with $b_{k,p'}(E_k) = 0$ such that

$$\int_0^1 r^{k-n/p'} \gamma_{p,H}(B_r(x) \setminus X, B_2(x))^p \, dr/r = \infty \quad \text{if } H \in \mathcal{P}_{2-k} \setminus \{0\}$$

and $x \in \partial X \setminus E_k$.

If X has the L^p h.a.p., then X has property $(*)$; this follows from Theorem 2, (1) and the Kellogg property [5, Theorem 2]. Our question is whether the converse holds: *if X has property $(*)$, does it have the L^p h.a.p.?* We remark that if this question were answered affirmatively, then part (c) of Theorem 1 would follow by use of (1).

PROBLEM 2. If $p = 2$ and $n \geqslant 5$, a different criterion for the L^p h.a.p. follows from work of Saak [9]. What is the relation between Saak's work and Theorem 2?

PROBLEM 3. If $H \in \mathcal{P}_0 \cup \mathcal{P}_1 \setminus \{0\}$ and Ω is an open set, then $\gamma_{p,H}(\cdot, \Omega)$ is an increasing set function defined on the subsets of Ω. Can one characterize the compact sets having the L^p h.a.p. by means of increasing set functions which are countably subadditive and have the property that all Borel sets are capacitable?

(To say that a set E is capacitable with respect to a set function γ means that $\gamma(E) = \sup\{\gamma(K) : K \text{ compact}, K \subset E\} = \inf\{\gamma(G) : G \text{ open}, G \supset E\}$). See [3], [7]. For the case $p = 2$, see [9, Lemma 2].

REFERENCES

1. Bagby T., *Approximation in the mean by solutions of elliptic equations*, Trans. Amer. Math. Soc..
2. Burenkov V. I., *On the approximation of functions in the space $W_p^r(\Omega)$ by functions with compact support for an arbitrary open set Ω*, Trudy Mat. Inst. Steklov Akad. Nauk SSSR **131** (1974), 51–63 (Russian); English transl. in Proc. Steklov Inst. Math. **131** (1974), 53–66.
3. Choquet G., *Forme abstraite du théorème de capacitabilité*, Ann. Inst. Fourier 9 (1959), Grenoble, 83–89.
4. Hedberg L. I., *Spectral synthesis in Sobolev spaces, and uniqueness of solutions of the Dirichlet problem*, Acta Math. **147** (1981), 237–264.
5. Hedberg L. I., Wolff T. H., *Thin sets in nonlinear potential theory*, Stockholm, 1982, (Rep. Dept. of Math. Univ. of Stockholm, Sweden, ISSN 0348-7652 N 24).
6. Lindberg P., *A constructive method for L^p-approximation by analytic functions*, Ark. Mat. **20** (1982), 61–68.
7. Meyers N. G., *A theory of capacities for functions in Lebesgue classes*, Math. Scand. **26** (1970), 255–292.
8. Polking J. C., *Approximation in L^p by solutions of elliptic partial differential equations*, Amer. J. Math. **94** (1972), 1231–1244.
9. Saak E. M., *A capacitary criterion for a domain with stable Dirichlet problem for higher order elliptic equations*, Mat. Sb. **100(142)** (1976), no. 2(6), 201–209 (Russian); English transl. in Math. USSR Sbornik **29** (1976), 177–185.
10. Vitushkin A. G., *The analytic capacity of sets in problems of approximation theory*, Uspekhi Mat. Nauk **22** (1967), no 6, 141–199 (Russian); English transl. in Russian Math. Surveys **22** (1967), 139–200.

INDIANA UNIVERSITY
DEPARTMENT OF MATHEMATICS
BLOOMINGTON, INDIANA 47405
USA

EDITORS' NOTE

Many years before the appearance of [6] the constructive techniques of Vitushkin were applied to the L^p-approximation by analytic functions by S. O. Sinanjan [11]. See also the survey by M. S. Melnikov and S. O. Sinanjan [12]

REFERENCES

11. Sinanyan S. O., *Approximation by analytic functions in the mean with respect to area*, Math. Sb. **69** (1966), no. 4, 546–578. (Russian)
12. Mel'nikov M. S., Sinanyan S. O., *Questions in the theory of approximations of functions of one complex variable*, Contemporary Problems of Mathematics, Itogi Nauki i Tekniki, vol. 4, VINITI, 1975, pp. 143–250. (Russian)

COMMENTARY BY YU. V. NETRUSOV

Yu. V. Netrusov answered the question in the first Problem in the affirmative. His proof (to be published in "Algebra i Analiz" (Russian), English translation in "St.Petersburg Mathematical Journal") uses Theorem 2.1 of [13].

REFERENCE

13. Meyers N., *Continuity properties of potentials*, Duke Math. J. **42** (1975), 157–166.

UNIFORM APPROXIMATION BY HARMONIC FUNCTIONS

Anthony G. O'Farrell

Let X be a compact subset of $\mathbb{R}^3$ and denote by $H(X)$ the space of all continuous real-valued functions such that there exists a sequence f_n of continuous functions, each harmonic on some neighbourhood of X (the neighbourhood may depend on n), such that $f_n \to f$ uniformly on X as $n \uparrow \infty$. The problem is to give an explicit characterization of the functions f that belong to $H(X)$.

Let C denote the Newtonian capacity on $\mathbb{R}^3$. Let f be continuous on $\mathbb{R}^3$. To avoid inessential problems, suppose that f is bounded. I conjecture the following:

CONJECTURE. $f \in H(X)$ if and only if there is a constant $\kappa_f > 0$ such that

$$\left| \int_{\mathbb{R}^3} f(x) \Delta \varphi(x) \, dx \right| \leqslant \kappa \left(d \cdot \|\nabla \varphi\|_\infty + d^2 \cdot \|\nabla^2 \varphi\|_\infty \right) C(D \sim X),$$

whenever φ is a test function, the support of φ is contained in a disk D, and $d = \operatorname{diam} D$.

This integral condition may be thought of as asserting a kind of weak harmonicity of f on X. It implies that f is actually harmonic on the interior of X. It is sufficient for $f \in H(X)$ if f is also assumed C^2. It is also sufficient if the boundary of X is smooth. It is necessary in general. These facts were established jointly by the author and J. Garnett in 1975 (unpublished). The proof depends in part on the instability of Newtonian capacity, which was proven by Lysenko and Pisarevskii [L].

There is a characterization of $H(X)$ in terms of the fine topology: $f \in H(X)$ if and only if f is continuous and f is finely-harmonic on the fine interior of X. This is due to Debiard and Gaveau. See [B].

The conjecture is motivated by the fact that a similar result holds for uniform holomorphic approximation in the plane. It is known as Vitushkin's individual function theorem. See [V].

References

[B] Bliedtner J., Hansen W., *Potential theory. An analytic and probabilistic approach to balayage*, Springer, 1986.

[L] Lysenko Yu. A., Pisarevskii B. M., *Instability of harmonic capacity and approximations of continuous functions by harmonic functions*, Mat. Sb. **76** (1968), 52–71 (Russian); English transl. in Math. USSR Sbornik **5** (1968), 53–72.

[V] Vitushkin A. G., *The analytic capacity of sets in problems of approximation theory*, Uspekhi Mat. Nauk **22** (1967), 141–199 (Russian); English transl. in Russian Math. Surveys **22** (1967), no. 6, 139–200.

MAYNOOTH COLLEGE
CO. KILDARE
IRELAND

THE UNIFORM APPROXIMATION PROBLEM FOR THE
SQUARE OF THE CAUCHY-RIEMANN OPERATOR

JOAN VERDERA

Let L be a constant coefficient, elliptic operator in $\mathbb{R}^n$. Given a compact subset of $\mathbb{R}^n$ let $h(X, L)$ be the set of continuous functions f on X such that $L(f) = 0$ on the interior of X, and let $H(X, L)$ be the uniform closure on X of the subspace of $h(X, L)$ consisting of those functions which satisfy the equation $L(f) = 0$ on some neighbourhood of X. Clearly $H(X, L) \subset h(X, L)$ and so the problem arises of characterizing those X for which we have equality in the above inclusion.

This problem was (independently) solved by Deny and Keldysh in the late forties for $L = \Delta$, the Laplace operator [2, 3]. A modern reformulation of their result states that

$$H(X, \Delta) = h(X, \Delta) \text{ is equivalent to } C(B \setminus \overset{\circ}{X}) = C(B \setminus X) \text{ for any ball } B, \text{ where } C \text{ is}$$

the classical capacity of potential theory. In the sixties Vitushkin solved the problem for $n = 2$ and $L = \bar{\partial}$, the Cauchy–Riemann operator [8]. Vitushkin's theorem asserts that $H(X, \bar{\partial}) = h(X, \bar{\partial})$ is equivalent to $\alpha(\Delta \setminus \overset{\circ}{X}) = \alpha(\Delta \setminus X)$ for any disc Δ, where α is continuous analytic capacity.

It is worth mentioning that the basic ideas and techniques are quite different in the harmonic and analytic cases. In fact no unified proof of the above two theorems is yet available. Moreover, no complete results are known for other operators besides $\bar{\partial}$ and Δ, even in the simplest case $n = 2$ and $L = \bar{\partial}^2$. We pose the following:

PROBLEM. *Show that*

$$(1) \qquad\qquad H(X, \bar{\partial}^2) = h(X, \bar{\partial}^2)$$

for any compact subset X of the plane.

The main reason we expect (1) to hold without any restriction on X is that the fundamental solution $\frac{1}{\pi} \frac{\bar{z}}{z}$ of $\bar{\partial}^2$ is bounded, and thus no capacitary set function like α or C can exist to produce necessary conditions for (1).

The study of the particular case $L = \bar{\partial}^2$ seems interesting because it could shed some light on the problem of finding an adequate general approach to deal with other operators L and (or) with norms different from the uniform.

We present now a list of known results related to our problem.

1. Any of the following assumptions on X is sufficient for (1) to hold:
   - X has no interior points [6].
   - The complement of X is finitely connected [1].
   - For some constant C one has diameter$(\Delta \setminus X) \leqslant C \sqrt{\text{area}(\Delta \setminus X)}$, for all discs Δ. (See [5].)
2. Any Dini-continuous function in $h(X, \bar{\partial}^2)$ belongs to $H(X, \bar{\partial}^2)$. (See [7].)

3. Since bounded functions are functions of bounded mean oscillation (BMO), $\frac{1}{\pi}\frac{\bar{z}}{z} \in$ BMO. It turns out that the BMO–$\bar{\partial}^2$ approximation problem (i.e., the uniform norm is replaced by the BMO norm in the above discussion) can be solved without any restriction on X. The same is true for the BMO-Δ approximation problem, because the fundamental solution $\frac{1}{2\pi}\log|z|$ of Δ is again in BMO (see [4]).

References

1. Carmona J. J., *Mergelyan's Approximation theorem for rational modules*, J. Approx. Theory **44** (1985), 113–126.
2. Deny J., *Systèmes totaux de fonctions harmoniques*, Ann. Inst. Fourier **1** (1949), 103–113.
3. Keldysh M. V., *On the solubility and stability of the Dirichlet problem*, Uspekhi Mat. Nauk **8** (1941), 171–231; Amer. Math. Soc. Trans. **51** (1966), no. 2, 1–73.
4. Mateu J., Verdera J., *BMO harmonic approximation in the plane and spectral synthesis for Hardy–Sobolev spaces*, Rev. Math. Iberoamericana **4** (1988), 291–313.
5. Paramonov P., personal communication.
6. Trent T., Wang J., *Uniform approximation by rational modules on nowhere dense sets*, Proc. Amer. Math. Soc. **81** (1981), 62–64.
7. Verdera J., preprint.
8. Vitushkin A. G., *Analytic capacity of sets in problems of approximation theory*, Uspekhi Mat. Nauk **22** (1967), 141–199 (Russian); English transl. in Russian Math. Surveys **22** (1967), 139–200.

Departement de Matematiques
Universitat Autonoma de Barcelona
08193 Bellaterra (Barcelona)
Spain

12.17
v.old

RATIONAL APPROXIMATION OF ANALYTIC FUNCTIONS

A. A. GONCHAR

1. Local approximations. Let

$$(1) \qquad f(z) = \sum_{n \geqslant 0} f_n / z^n, |z| > R_f \qquad (R_f = \varlimsup_n |f_n|^{1/n} < \infty);$$

and let f be a complete analytic function corresponding to the element f. For any $n \in \mathbb{N}$ define $\nu_n(f) = \sup\{\nu(f - r) : r \in \mathcal{R}_n\}$ where $\nu(g)$ is the multiplicity of the zero of g at ∞, $\mathcal{R}_n$ is the set of all rational functions of degree at most n.

For any n there exists a unique function π_n, $\pi_n \in \mathcal{R}_n$, such that $\nu_n(f) = \nu(f - \pi_n)$. It is called the nth diagonal Padé approximant to the series (1). Let a, $a > 1$ be an arbitrary fixed number and let $|\cdot| = a^{-\nu(\cdot)}$; then π_n is the function of the best approximation to f in $\mathcal{R}_n$ with respect to the metric: $\lambda_n(f) = |f - \pi_n| = \inf\{|f - r| : r \in \mathcal{R}_n\}$. See [1], [2] for a more detailed discussion on the Padé approximants (the definition in [2] slightly differs from the one given above).

For any powers series (1) we have

$$(2) \qquad \nu_n(f) > n, \ n \in \mathbb{N}; \quad \nu_n(f) > 2n, \ n \in \Lambda, \ \Lambda \subset \mathbb{N},$$

Λ being an infinite subset of $\mathbb{N}$ depending on f.

A functional analogue of the well-known Thue–Siegel–Roth theorem (see [3], Theorem 2, (i)) can be formulated in our case as follows: if f is an element of an algebraic nonrational function then for any x, $x > 2$, the inequality $\nu_n(f) > xn$ holds only for a finite number of indices n. From this it follows easily that in our case

$$(3) \qquad \lim n^{-1}\nu_n(f) = 2.$$

Apparently, this theorem is true for more general classes of analytic functions.

CONJECTURE 1. *If f is an element of a multi-valued analytic function f with a finite set of singular points then (3) is valid.*

In connection with Conjecture 1 we note that if

$$(4) \qquad \varlimsup_n n^{-1}\nu_n(f) = +\infty,$$

then f is a single-valued analytic function; but for any A, $A > 0$ the inequality $\varlimsup_n n^{-1}\nu_n(f) > A$ is compatible with the fact that f is multi-valued (the first assertion is contained essentially in [4], [5], the second follows from the results of Polya [6]).

124

Everything stated above can be reformulated in terms of sequences of normal indices of the diagonal Padè approximations (see [7], [1]). In essence the question is about possible lacunae in the sequence of the Hankel determinants

$$
F_n = \begin{vmatrix}
f_1 & f_2 & \cdots & f_n \\
f_2 & f_3 & \cdots & f_{n+1} \\
\cdots\cdots\cdots\cdots\cdots\cdots \\
f_n & f_{n+1} & \cdots & f_{2n}
\end{vmatrix}, \qquad n \in \mathbb{N}.
$$

Thus (3) means that the sequence $\{F_n\}$ has no "Hadamard lacunae" and (4) means that $\{F_n\}$ has "Ostrowski lacunae" (in the terminology of [8]). Apparently many results on lacunary power series (see [8]) have their analogues for diagonal Padé approximations.

2. Uniform approximation. We restrict ourselves by the corresponding approximation problems on discs centered at infinity for the functions satisfying (1). Let $R > R_f$, $E = \{z : |z| \geqslant R\}$ (f is holomorphic on E) and $\mathbb{D}_R = \{z : |z| < R\}$. Denote by $\rho_n(f)$ the test approximation of f on E by the elements of $\mathcal{R}_n$:

$$
\rho_n(f) = \inf\{\|f - r\|_E : r \in \mathcal{R}_n\},
$$

$\|\cdot\|$ is the sup–norm on E.

Let $\mathcal{F}$ be the set of all compacts, F, $F \subset \mathbb{D}_R$ (with the connected complement) such that f admits a holomorphic (single-valued) continuation on $\mathbb{C} \setminus F$. Denote by $C_{\mathbb{D}}(F)$ the Green capacity of F with respect to $\mathbb{D}_R$ (the capacity of the condenser (E, F)) and define

$$
a_f = \sup\Big\{\exp \frac{1}{C_{\mathbb{D}}(F)} : F \in \mathcal{F}\Big\}.
$$

For every f,

$$
(5) \qquad\qquad \varlimsup_n \rho_n(f)^{1/n} \leqslant a_f^{-1}.
$$

This inequality follows from the results of Walsh ([9], ch. VIII).

CONJECTURE 2. *For any f,*

$$
(6) \qquad\qquad \lim_n \rho_n(f)^{1/n} \leqslant a_f^{-2}.
$$

Inequalities (5), (6) are similar to inequalities (2). To clarify (here and further) the analogy with the local case one should pass in Section 1 from ν_n to the best approximations λ_n. In particular, equality (3) will be written

$$
\lim_n \lambda_n(f)^{1/n} \leqslant a^{-2}.
$$

CONJECTURE 3. *If f is an element of an analytic function f which has a finite set of singular points, then*

$$
(7) \qquad\qquad \lim_n \rho_n(f)^{1/n} = a_f^{-2}.
$$

If under the hypothesis of this conjecture f is a single-valued analytic function, both parts of (7) are obviously equal to zero.

Conjecture 3 can be proved for the case when all singular points of f lie on $\mathbb{R}$ (for the case of two singular points, see [10]).

In contradistinction to the local case, the question of validity of (7) remains open for the algebraic functions also.

REFERENCES

1. Perron O., *Die Lehre von den Kettenbrüchen*, II, Stuttgart, 1957.
2. Baker G. A., *Essentials of Padé Approximants*, "AP", New York, 1975.
3. Uchiyama S., *Rational approximations to algebraic functions*, Journal of the Faculty of Sciences Hokkaido University, Ser. I XV (1961), no. 3, 4, 173–192.
4. Gonchar A. A., *A local condition of single-valuedness of analytic functions*, Mat. Sb. **89** (1972), 148–164 (Russian); English transl. in Math. USSR Sbornik **18** (1972), 151–157.
5. Gonchar A. A., *On the convergence of Padé approximants*, Mat. Sb. **92** (1973), 152–164 (Russian); English transl. in Math. USSR Sbornik **21** (1973), 155–166.
6. Polya G., *Untersuchungen über Lücken and Singularitäten von Potenzreihen*, Math. Z. **29** (1929), 549–640.
7. Gonchar A. A., *On the convergence of Padé approximants for some classes of meromorphic functions*, Mat. Sb. **97** (1975), 605–627 (Russian); English transl. in Math. USSR Sbornik **26** (1975), 555–575.
8. Bieberbach L., *Analytische Fortsetzung*, Springer-Verlag, Berlin–Heidelberg, 1955.
9. Walsh J. L., *Interpolation and approximation by rational functions in the complex domain*, second ed., vol. 20, AMS Coll. Publ., 1960.
10. Gonchar A. A., *On the speed of rational approximation of some analytic functions*, Mat. Sb. **105** (1978), 147–163 (Russian); English transl. in Math. USSR Sbornik **34** (1978), 131–145.

STEKLOV MATHEMATICAL INSTITUTE
UL. VAVILOVA 42
117966 MOSKOW
RUSSIA

A CONVERGENCE PROBLEM ON RATIONAL
APPROXIMATION IN SEVERAL VARIABLES

HANS WALLIN

1. The one-variable case. Let $z \in \mathbb{C}$. Let me first give the background in the one-variable case. Let $f(z) = \sum c_j z^j$, $z \in \mathbb{C}$, be a formal power series and P/Q, $Q \not\equiv 0$, a rational function in one variable z of type (n, ν), i.e., P is a polynomial of degree $\leqslant n$ and Q of degree $\leqslant \nu$. It is in general not possible to determine P/Q so that it interpolates to f of order at least $n + \nu + 1$ at the origin (i.e., having the same Taylor polynomial of degree $n+\nu$ as f). However, given n and ν, we can always find a unique rational function P/Q of type (n, ν) such that P interpolates to fQ of order at least $n + \nu + 1$ at the origin, i.e., $(fQ - P)(z) = O(z^{n+\nu+1})$. This function P/Q, the $[n, \nu]$-*Padé approximant* to f, was first studied systematically by Padé in 1892; see [1]. In 1902 Montessus de Ballore [2] proved the following theorem which generalizes the well-known result on the circle of convergence for Taylor series.

THEOREM. *Suppose f is holomorphic at the origin and meromorphic in $|z| < R$ with ν poles (counted with their multiplicities). Then the $[n, \nu]$-Padé approximant to f, P_n/Q_n, converges uniformly to f, with geometric degree of convergence, in those compact subsets of $|z| < R$ which do not contain any poles of f.*

With the assumption in the theorem it can also be proved that P_n/Q_n diverges outside $|z| = R$ if R is chosen as large as possible [3, p. 269] and that the poles of P_n/Q_n converge to the poles of f in $|z| < R$. Furthermore, when n is sufficiently large, P_n, Q_n is the unique rational function of type (n, ν) which interpolates to f at the origin of order at least $n + \nu + 1$. Montessus de Ballore's original proof used Hadamard's theory of polar singularities (see [4]). Today several other easier proofs are known; see for instance [5], [6], [7] and [8].

Padé approximants have been used in a variety of problems in numerical analysis and theoretical physics, for instance in the numerical evaluation of functions and in order to locate singularities of functions (see [1]). One reason for this is, of course, the fact that the Padé approximants of f are easy to calculate from the power series expansion of f. In recent years there has been an increasing interest in using analogous interpolation procedures to approximate functions of several variables (see [9]). I propose the problem to investigate in which sense it is possible to generalize Montessus de Ballore's theorem to several variables.

2. The two-variable case ($z = (z_1, z_2)$; $z_1, z_2 \in \mathbb{C}$). We first generalize the definition of Padé approximants to the two-variable case. Let $f(z) = \sum c_{jk} z_1^j z_2^k$ be a formal power series and let P/Q, $Q \not\equiv 0$, be a rational function in two variables z_1 and z_2 of type (n, ν), i.e., P is a polynomial in z_1 and z_2 of degree $\leqslant n$ and Q of degree $\leqslant \nu$. By counting the number of coefficients in P and Q we see that it is always possible to determine P and Q so that, if $(Qf - P)(z) = \sum \alpha_{jk} z_1^j z_2^k$, then $\alpha_{jk} = 0$ for $(j, k) \in S$, where S, the interpolation set, is a chosen subset of $\mathbb{N} \times \mathbb{N}$ with $\frac{1}{2}(n+1)(n+2) + \frac{1}{2}(\nu+1)(\nu+2) - 1$ elements. There is no natural unique way to choose S, but it seems reasonable to assume that $\{(j,k) : j + k \leqslant n\} \subset S$ and that $(j,k) \in S \Rightarrow (l,m) \in S$ if $l \leqslant j$ and $m \leqslant k$. In this way we get a *rational approximant* P/Q of type (n, ν) to f corresponding to S. With a suitable choice of S, P/Q is unique [7, Theorem 1.1]. The definition, elementary properties and some convergence results have been considered for these and similar approximants in [9], [10] and [7]. The possibility to generalize Montessus de Ballore's theorem has been discussed in [6], [7] and [11], but the results are far from being complete.

PROBLEM 1. *In what sense can Montessus de Ballore's theorem be generalized to several variables?*

It is not clear what class of functions f one should use. We consider the following concrete situation. Let $f = F/G$, where F is holomorphic in the polydisc $\{z = (z_1, z_2) : |z_i| < R_i, i = 1, 2\}$ and G is a polynomial of degree ν, $G(0) \neq 0$. By the method described above we obtain for every n a rational approximant P_n/Q_n of type (n, ν) to f corresponding to some chosen interpolation set $S = S_n$. In what region of $\mathbb{C}^2$ does P_n/Q_n converge to f? Partial answers to this problem are given in [7] and [11] (in the latter with a somewhat different definition of the approximants). If $\nu = 1$, explicit calculations are possible and sharp results are easy to obtain [7, Section 4]. These show that in general we do not have convergence in $\{z : |z_i| < R_i, i = 1, 2\} \setminus \{z : G(z) = 0\}$. This proves that the general analogue of the Montessus de Ballore's theorem is not true. It may be added that it is easy to prove — by just using Cauchy's estimates — that there exist rational functions $r_{n\nu}$ of type (n, ν) interpolating to f at the origin of order at least $n + 1$ and converging uniformly, as $n \to \infty$, to f in compact subsets of $\{z : |z_i| < R_i\} \setminus \{z : G(z) = 0\}$.

A disadvantage, however, of $r_{n\nu}$ compared to the rational approximants defined above is that $r_{n\nu}$ is not possible to compute from the Taylor series expansion of f (see [7, Theorem 3.3]).

In the one-variable case the proof of Montessus de Ballore's theorem is essentially finished when you have proved that the poles of the Padé approximants converge to the poles of f. In the several-variable case, on the other hand, there are examples [7, Section 4, Counterexample 2], when the rational approximants P_n/Q_n do not converge in the whole region $\{z : |z_i| < R_i\} \setminus \{z : G(z) = 0\}$ in spite of the fact that the singularities of P_n/Q_n converge to the singularities of F/G. This motivates:

PROBLEM 2. *Under what conditions does Q_n converge to G ?*

The choice of the interpolation set S_n is important for the convergence. For instance, if $\nu = 1$ and $R_1 = R_2 = \infty$, we get convergence in $\mathbb{C}^2 \setminus \{z : G(z) = 0\}$ with a suitable choice of S_n [7, Section 4]. On the other hand, if we change just one point in S_n — without violating the reasonable choices of S_n indicated in the definition of the rational approximants — we get examples [7, Section 4, Counterexample 1], where we do not have convergence in any polydisc around $z = 0$.

PROBLEM 3. *How is the convergence $P_n/Q_n \to f$ influenced by the choice of the interpolation set S_n?*

Since we do not get a complete generalization of Montessus de Ballore's theorem, it is also natural to ask:

PROBLEM 4. *If the sequence of rational approximants does not converge, is there a subsequence that converges to f?* (Compare [7, Theorem 3.4]).

Finally, I want to propose the following conjecture:

CONJECTURE. $P_n/Q_n \to f$ *in* $\mathbb{C}^2 \setminus \{z : G(z) = 0\}$ *if* $R_1 = R_2 = \infty$ *and the interpolation set S_n is suitably chosen.* (Compare [11, Corollary 2] and the case $\nu = 1$ referred to just after Problem 2 above).

REFERENCES

1. Baker G. A., *Essentials of Padé Approximants*, Academic Press, New York, 1975.
2. de Montessus de Ballore R., *Sur les fractions continues algébrique*, Bull. Soc. Math. France **30** (1902), 28–36.
3. Perron O., *Die Lehre von den Kettenbrüchen*, II, Teubner, Stuttgart, 1957.
4. Gragg W. B., *On Hadamard's theory of polar singularities*, Padé Approximants and Their Applicants (Graves-Morris P. R., ed.), Academic Press, London, 1973, pp. 117–123.
5. Saff E. B., *An extension of Montessus de Ballore's theorem on the convergence of interpolation rational functions*, J. Approx. Theory **6** (1972), 63–68.
6. Chisholm J. S. R., Graves-Morris P. R., *Generalization of the theorem of de Montessus to two-variable approximants*, Proc. Royal Soc. Ser. A **342** (1975), 341–372.
7. Karlsson J., Wallin H., *Rational approximation by an interpolation procedure in several variables*, Padé and rational approximation (Saff E. B. and Varga R. S., eds.), Academic Press, New York, 1977, pp. 83–100.
8. Gonchar A. A., *On the convergence of generalized Padé approximants of meromorphic functions*, Mat. Sb. **98** (1975), no. 4, 563–577 (Russian); English transl. in Math. USSR Sbornik **27** (1975), 503–514.
9. Chisholm J. S. R., *N-variable rational approximants*, Padé and rational approximation (Saff E. B. and Varga R. S., eds.), Academic Press, New York, 1977, pp. 23–42.
10. Gonchar A. A., *A local condition for the single-valuedness of analytic functions of several variables*, Mat. Sb. **93** (1974), no. 2, 296–313 (Russian); English transl. in Math. USSR Sbornik **22** (1974), 305–322.
11. Graves-Morris P. R., *Generalization of the theorem of de Montessus using Canterbury approximant*, Padé and rational approximation (Saff E. B. and Varga R. S., eds.), Academic Press, New York, 1977, pp. 73–82.

UMEÅ UNIVERSITY
S-90187 UMEÅ
SWEDEN

COMMENTARY BY THE AUTHOR

In a recent paper, A. Cuyt (A Montessus de Ballore theorem for multivariate Padé approximants, Department of Mathematics, University of Antwerp, Belgium, 1983) considers a multivariate rational approximant P/Q to f where P and Q are polynomials of degree $n\nu + n$ and $n\nu + \nu$, respectively, such that all the terms of P and Q of degree less than $n\nu$ vanish. It is then possible to determine P and Q so that $fQ - P$ has a power series expansion where the terms of degree $\leqslant n\nu + n + \nu$ are all zero. For this approximant P/Q she proves the following theorem where $P/Q = P_n/Q_n$ and P_n and Q_n have no common non-constant factor: Let $f = F/G$ where F is holomorphic in the polydisc $\{z : |z_i| < R_i\}$ and G is a polynomial of degree $\leqslant \nu$, $G(0) \neq 0$, and assume that $Q_n(0) \neq 0$ for infinitely many n. Then there exists a polynomial $Q(z)$ of degree ν such that $\{z : G(z) = 0\} \subset \{z : Q(z) = 0\}$ and a subsequence of $\{P_n/Q_n\}$ that converges uniformly to f on compact subsets of $\{z : |z_i| < R_j\} \setminus \{z : Q(z) = 0\}$.

BADLY-APPROXIMABLE FUNCTIONS
ON CURVES AND REGIONS

LEE A. RUBEL

Let X be a compact Hausdorff space and A a uniform algebra on X: that is, A is uniformly closed, separates points, and contains the constants. For example, if $X \subset \mathbb{C}^n$ then we might take $A = P(X)$, the uniform limits on X of polynomials. We say that a function φ, $\varphi \in C(X)$, is *badly-approximable* (with respect to A) to mean

$$\|\varphi - f\|_\infty \geqslant \|\varphi\|_\infty \quad \text{for all } f \in A,$$

where $\|\cdot\|_\infty$ is the supremum norm over X. The problems discussed here concern finding concrete descriptions of the badly-approximable functions for some classical function algebras. They are the functions that it is useless to try to approximate.

In this section, we let G be a bounded domain in $\mathbb{C}$, with boundary X, and let $A^*(X)$ be the algebra of boundary values of continuous functions on $G \cup X$ that are analytic in G. In case G is the open unit disc, then $A^*(X)$ is the "disc algebra" (regarded as consisting of functions on X and not on G).

POREDA'S THEOREM ([1]). *If X consists of a simple closed Jordan curve, then φ, $\varphi \in C(X)$, is badly-approximable with respect to $A^*(X)$ if and only if φ has nonzero constant modulus, and* $\operatorname{ind}\varphi < 0$.

THEOREM A ([2]). *If φ, $\varphi \in C(X)$ has nonzero constant modulus and if $\operatorname{ind}\varphi < 0$, then φ is badly-approximable with respect to $A^*(X)$.*

Here, $\operatorname{ind}\varphi$ is the index of φ, defined as the winding number on X of φ around 0.

THEOREM B ([2]). *Each badly-approximable (with respect to $A^*(X)$) function in $C(X)$ has constant modulus on the boundary of the complement of the closure of G.*

THEOREM C ([2]). *Suppose that X consists of $N + 1$ disjoint closed Jordan curves. If φ is badly-approximable with respect to $A^*(X)$, then φ has constant modulus, and* $\operatorname{ind}\varphi < N$.

An example was given in [2] to show that the range $0 \leqslant \operatorname{ind}\varphi < N$ is indeterminate, so that one cannot tell from the winding number alone, on such domains, whether or not φ is badly-approximable.

PROBLEM 1. *Find necessary and sufficient conditions for a function φ to be badly-approximable with respect to $A^*(X)$ if X is a finite union of disjoint Jordan curves.*

Note: In the case of the annulus, $X = \{z : |z| = r \text{ or } |z| = 1\}$, where $0 < r < 1$, supposing φ is of modulus 1 on X, it is shown in [2] that φ is badly-approximable with respect to $A^*(X)$ if and only if either $\operatorname{ind}\varphi < 0$ or $\operatorname{ind}\varphi = 0$ and

$$\frac{1}{2\pi} \int_{-\pi}^{\pi} \arg\big(\varphi(e^{i\theta})\big)\, d\theta - \frac{1}{2\pi} \int_{-\pi}^{\pi} \arg\big(\varphi(re^{i\theta})\big)\, d\theta \equiv \pi \pmod{2\pi}.$$

PROBLEM 2. *The analogue of Problem 1 for $R^*(X)$, which is the limits on X of rational functions with poles off G, where one permits G to have infinitely many holes.*

PROBLEM 3. *Characterize the badly-approximable functions with respect to $P(X)$, where X is any compact set in $\mathbb{C}$.*

PROBLEM 3′. *The same as Problem 3 but in the special case $X = \operatorname{clos}\mathbb{D}$.*

Despite appearances, Problem 3′ is just about as general as Problem 3. An answer to Problem 3 could be called a "co-Mergelyan theorem" since Mergelyan's theorem [3] characterizes the "well-approximable" functions on X.

THEOREM ([4]). *If φ is badly-approximable with respect to $P(\operatorname{clos}\mathbb{D})$ then $\left\|\frac{\partial\varphi}{\partial\bar{z}}\right\|_\infty \geqslant \frac{3}{4}\|\varphi\|_\infty$ where $\|\cdot\|_\infty$ is the supremum norm over $\mathbb{D}$. The converse is false.*

PROBLEM 4. *Obtain, for sets X, $X \subset \mathbb{C}^n$, $n \geqslant 2$, any significant result about badly-approximable functions with respect to any algebra like $P(X)$, $A(X)$, or $R(X)$.*

REFERENCES

1. Poreda S. J., *A characterization of badly approximable functions*, Trans. Amer. Math. Soc. **169** (1972), 249–256.
2. Gamelin T. W., Garnett J. B., Rubel L. A., Shields A. L., *On badly approximable functions*, J. Approx. Theory **17** (1976), 280–296.
3. Rudin W., *Real and Complex analysis*, New York, 1966.
4. Kronsstadt E., Private communication, September, 1977.
5. Luecking D. H., *On badly approximable functions and uniform algebras*, J. Approx. Theory **22** (1978), 161–176.
6. Rubel L. A., Shields A. L., *Badly approximable functions and interpolation by Blaschke products*, Proc. Edinburgh Math. Soc. **20** (1976), 159–161.

DEPARTMENT OF MATHEMATICS
UNIVERSITY OF ILLINOIS AT URBANA-CHAMPAIGN
URBANA, ILLINOIS 61801
USA

AN ISOPERIMETRIC PROBLEM

Dmitry Khavinson

Let Ω be a finitely connected region on $\mathbb{C}$ whose boundary Γ consists of n simple closed analytic curves γ_j, $j = 1,\ldots,n$. Let $A = A(\Omega)$ denote the algebra of functions analytic in Ω and continuous in the closure of Ω with the norm $\|f\|_\infty = \sup_{z\in\Omega}|f(z)|$. The following quantity, called an *analytic content* of Ω, has been introduced in [4] (also, see [3]):

$$\lambda = \lambda(\Omega) = \inf_{\varphi\in A(\Omega)} \|\bar{z} - \varphi\,|_\infty.$$

It was shown in [4] that always

(1) $$\lambda(\Omega) \geqslant 2\,\text{area}\,(\Omega)/\text{perimeter}\,(\Omega).$$

From this, and the well-known Alexander estimate ([1, 2])

(2) $$\lambda(\Omega) \leqslant \left(\text{area}\,(\Omega)/\pi\right)^{1/2}$$

one obtains as a corollary the isoperimetric inequality

(3) $$P^2 \geqslant 4\pi A,$$

where P and A denote the perimeter and area of Ω respectively. More detail on the various forms of (1) and (2), and their affiliation with (3) can be found in [3, 5, 6, 7]. Both inequalities (1) and (2) are sharp, since they become equalities for disks ([3]). Moreover, it is not hard to show (see, e.g., [3]), that if equality holds in (2), Ω must be a disk of radius λ. The question concerning a complete list of extremal domains for (1) first raised in [6] still remains open. It is known that disks are not the only solutions. Equality in (1) also holds for annuli ([6]). It was conjectured in [6] that those are the only extremal domains, i.e. $\lambda = 2A/P$ if and only if Ω is a disk of radius λ or an annulus $\{z : r < |z| < R\}$ with $R - r = \lambda$. It was shown in [6] that *if the boundary of an extremal domain Ω (for which $\lambda = 2A/P$) contains a circular arc, then Ω must be either a disk or an annulus.* The argument there relies on the following characterization of extremal domains.

THEOREM A. *The following are equivalent:*

(i) $\lambda(\Omega) = 2A/P$.

(ii) $\exists\varphi \in A(\Omega)$ *such that*

(4) $$\bar{\xi}(s) - i\lambda\dot{\bar{\xi}} = \varphi(\xi(s)), \qquad (s \text{ is arclength on } \Gamma).$$

$\left(\text{Here, } \dot{\xi} = \dfrac{d\xi}{ds}\right)$. *Then, φ is the extremal function for $\lambda(\Omega)$, i.e., $\|\bar{z} - \varphi\|_\infty = \lambda$.*

(iii) *For all $f \in A(\Omega)$, the following quadrature identity holds:*

(5) $$\frac{1}{A}\int_\Omega f\,dA = \frac{1}{P}\int_\Gamma f\,ds.$$

Since the function φ in (4) is uniquely determined by its values on any subarc of Γ, then assuming that Γ contains a circular arc, we find φ from (4) and then it is easy to show that Ω must be either a disk or an annulus.

Assuming that Ω is simply connected, we can extend (5) to hold for all harmonic functions in Ω. As was noted by Kosmodem'yanskii [8], this implies that the overdetermined boundary value problem

(6)
$$\begin{cases} \Delta u = 1 \text{ in } \Omega; \\ \dfrac{\partial u}{\partial n} = \text{const on } \Gamma \\ u = 0 \text{ on } \Gamma \end{cases}$$

is solvable in Ω. A well-known theorem of Serrin (see [10,11]) then states that Ω must be a disk. This conclusion can be obtained directly from (4) by the following argument due to B. Gustafsson (private communication). Differentiating (4) with respect to s we obtain

(7)
$$1 + \overline{\lambda i (\ddot{\xi}/\dot{\xi})} = \varphi'(\xi(s))(\dot{\xi})^2 \text{ on } \Gamma.$$

Since $\arg(\ddot{\xi}/\dot{\xi}) = \pm\pi/2$, the left-hand side in (7) is real-valued. Thus, $\varphi'(\dot{\xi})^2$ is real-valued on Γ. Let $z = \alpha(w)$ be a conformal mapping of the unit disk $\mathbb{D}$ onto Ω. We have then

$$\varphi'(\dot{\xi})^2 = \varphi'(\alpha(w)) \frac{(\alpha'(w))^2}{|\alpha'(w)|^2} \left(\frac{dw}{|dw|}\right)^2$$

is real-valued on the unit circle $\{w : |w| = 1\}$. Hence, the analytic function $F(w) = \varphi'(\alpha(w))(\alpha'(w))^2 w^2$ is real-valued on the unit circle and, therefore, must be a constant. As $F(0) = 0$, $F \equiv 0$. Hence, $\varphi' \equiv 0$ and from (4) it follows that Ω is a disk of radius λ. (Incidentally, this also yields a simple independent proof of Serrin's theorem in two dimensions.) Unfortunately, this argument already fails for doubly connected domains.

If we do not assume Ω to be simply connected, then (i)–(iii) of Theorem A are equivalent to the following extension of (6): *The overdetermined problem*

(8)
$$\begin{cases} \Delta u = 1 \text{ in } \Omega \\ \dfrac{\partial u}{\partial n} = \text{const on } \Gamma \\ u|_{\gamma_j} = c_j, \quad j = 1, \ldots, n \end{cases}$$

has a solution in Ω. Thus, our conjecture is equivalent to the following extension of Serrin's theorem: *The overdetermined problem (8) can only have radially symmetric solutions.* To see that (8) is indeed equivalent to (i)–(iii) of Theorem A, consider following [9] a (unique) solution v of the Dirichlet problem

(9)
$$\begin{cases} \Delta v = 1 \text{ in } \Omega; \\ v = 0 \text{ on } \Gamma. \end{cases}$$

Then, assuming (5), we find from Green's formula that for any $u = \operatorname{Re} f$, $f \in A(\Omega)$

$$\frac{1}{A} \int_\Omega u \, dA = \frac{1}{A} \int_\Omega u \Delta v \, dA = \frac{1}{A} \int_\Gamma u \frac{\partial v}{\partial n} \, ds = \frac{1}{P} \int_\Gamma u \, ds.$$

Hence $\left(\dfrac{\partial v}{\partial n} - \dfrac{A}{P} \right) \perp \operatorname{Re} A(\Omega)$ and therefore $\exists$ real constants $c_1, \ldots, c_{n-1}$ such that

$$(10) \qquad \frac{\partial v}{\partial n} - \frac{A}{P} = \sum_{j=1}^{n-1} c_j \frac{\partial \omega_j}{\partial n},$$

where ω_j, $j = 1, \ldots, n-1$ are harmonic measures of the boundary curves γ_j. (9), (10) imply (8), and the argument is obviously reversible.

Finally, let me mention a *regularity problem* for the free boundary Γ of Ω. Everywhere above we tacitly assumed γ to be smooth, even analytic. However, I am certain that, assuming Γ to be merely rectifiable, the quadrature identity (5) alone already implies that (a) Γ *is locally real analytic*, and (b) *has at most two connected components*. Unfortunately, I cannot prove any of this, even under the additional assumption that Ω is simply connected.

References

1. Alexander H., *Projections of polynomial hulls*, J. Funct. Anal. **3** (1973), 13–19.
2. Alexander H., *On the area of the spectrum of an element of a uniform algebra*, Complex Approximation (B. Aupetit, ed.), Birkhäuser, Basel, 1980, pp. 3–12.
3. Gamelin T., Khavinson D., *The isoperimetric inequality and rational approximation*, Amer. Math. Monthly **96** (1989), 18–30.
4. Khavinson D., *Annihilating measures of the algebra $R(X)$*, J. Funct. Anal. **28** (1984), 175–193.
5. Khavinson D., *A note on Töeplitz operators*, Geometry of Banach Spaces (N. Kaltonand, E. Saab, eds.), Lecture Notes in Math., vol. 934, Springer–Verlag, 1986, pp. 89–95.
6. Khavinson D., *Symmetry and uniform approximation by analytic functions*, Proc. Amer. Math. Soc. **101** (1987), 475–483.
7. Khavinson D., Lueking D., *On an extremal problem in the theory of rational approximation*, J. Approx. Theory **50** (1987), 127–132.
8. Kosmodem'yanskii A. A., *A converse of the mean value theorem for harmonic functions*, Uspekhi Mat. Nauk **36** (1981), 175–176 (Russian); English transl. n Russian Math. Surveys **36** (1981), 159–160.
9. Marrero-Rodriguez M. I., written communication, 1991.
10. Serrin J., *A symmetry problem in potential theory*, Arch. Rational Mech. Anal. **43** (1971), 304–318.
11. Weinberger H., *Remark on the preceding paper of Serrin*, Arch. Rational Mech. Anal. **43** (1971), 319–320.

Department of Mathematical Sciences
University of Arkansas
Fayetteville, Arkansas 72701
USA

A LOWER BOUND FOR LOGARITHMIC CAPACITY

T. J. RANSFORD

Let K be a non-empty, compact subset of $\mathbb{C}$. In this article we consider a possible lower bound for $c(K)$, the logarithmic capacity of K. This quantity, notoriously difficult to compute in all but the simplest cases, turns up throughout complex analysis and potential theory. A classical background reference is Tsuji's book [T].

Given a polynomial p, let us write

$$\|p\|_\infty := \max_{z \in K} |p(z)|,$$

$$\|p\|_2 := \left(\iint_K |p(z)|^2 \, dx \, dy \right)^{1/2},$$

and let $\mathcal{P}_n$ denote the set of monic polynomials of degree n. It is a consequence of the characterization of capacity as the Chebyshev constant (see [T, Chapter III, §5]) that for each $n \geqslant 1$ we have the upper bound

$$(1) \qquad\qquad c(K)^n \leqslant \inf_{p \in \mathcal{P}_n} \|p\|_\infty$$

PROBLEM. *Is it true that for each $n \geqslant 1$ we have the lower bound*

$$(2) \qquad\qquad c(K)^n \geqslant (1/\sqrt{n\pi}) \inf_{p \in \mathcal{P}_n} \|p'\|_2 \ ?$$

The infimum in (2) is just the distance from a point in a Hilbert space to a finite-dimensional subspace. Thus, if true, the inequality (2) gives a lower bound for $c(K)$ which is easy to calculate, at least in principle. If K is 'close' to being a lemniscate then we might expect the estimate to be quite good (see (e) below) but, at the other extreme, if K has zero area then of course (2) tells us nothing.

Remarks.

(a) If $n = 1$, then (2) reduces to

$$(3) \qquad\qquad c(K) \geqslant (A(K)/\pi)^{1/2} ,$$

where $A(K)$ denotes the area of K. This 'isoperimetric' inequality is well known to be true, even with logarithmic capacity replaced by analytic capacity (see e.g. [G, p. 200]).

(b) If $n = 2$, then (2) reduces to

$$(4) \qquad\qquad c(K) \geqslant (2I(K)/\pi)^{1/4} ,$$

where $I(K)$ denotes the moment of inertia of K about its centre of mass. Even this special case remains open.

(c) The inequality (2) is true for all $n \geqslant 1$ if K is connected. To prove this we use an argument based upon [**PS**, §5.9], where it is proved in the case $n = 2$.

Filling in the 'holes' (which does not affect the capacity), we can suppose that $\mathbb{C}_\infty \setminus K$ is connected. By the Riemann mapping theorem and [**T**, Theorem III.39], there is a conformal mapping $f : \mathbb{C}_\infty \setminus \Delta_{c(K)} \to \mathbb{C}_\infty \setminus K$ such that $f(z) = z + O(1)$ near ∞. (Here and throughout, Δ_r denotes the closed disc $\{z \in \mathbb{C} : |z| \leqslant r\}$.) Choose $p \in \mathcal{P}_n$ such that $(p \circ f)(z) = z^n + O(1)$ near ∞. Take $r > c(K)$, and define K_r by $\mathbb{C}_\infty \setminus K_r = f(\mathbb{C}_\infty \setminus \Delta_r)$. Then

$$
\begin{aligned}
\iint_{K_r} |p'(z)|^2 \, dx \, dy &= \iint_{K_r} \frac{\partial}{\partial \bar{z}} \left(p'(z)\overline{p(z)} \right) \, dx \, dy \\
&= \frac{1}{2i} \int_{\partial K_r} p'(z)\overline{p(z)} \, dz \\
&= \frac{1}{2i} \int_{\partial \Delta_r} p'(f(z))\overline{p(f(z))} f'(z) \, dz \\
&= \frac{1}{2i} \int_{\partial \Delta_r} (p \circ f)'(z)\overline{(p \circ f)(z)} \, dz \\
&\leqslant n\pi r^{2n},
\end{aligned}
$$

where the final inequality comes from substituting the Laurent expansion of $p \circ f$ into the integral. The result now follows by letting r decrease to $c(K)$.

(d) The inequality (2) is also true if $K = q^{-1}(L)$, where L is compact and q is a monic polynomial of degree n. For then we have

$$
c(K)^n = c(L)
$$

(see e.g. [**F**, p. 35]), and evidently

$$
\iint_K |q'(z)|^2 \, dx \, dy = nA(L),
$$

so in this case (2) follows from (3).

(e) Equality holds in (2) if K is a disc or, more generally, if $K = q^{-1}(\Delta_r)$, where q is a monic polynomial of degree n. It is possible that these are not the extremal cases, and so it is reasonable to ask whether (2) holds with $1/\sqrt{n\pi}$ replaced by any smaller positive constant. Even this does not appear to be known.

We conclude by mentioning a reformulation of the Problem which does not explicitly involve capacity. To prove that (2) holds for general K, it actually suffices to do so for K of the form $q^{-1}(\Delta_r)$, where $r > 0$ and q is a monic polynomial of *arbitrary degree;* this follows easily from the characterization of capacity as the Chebyshev constant. Since $c(q^{-1}(\Delta_r)) = r^{1/\deg q}$, we can restate the Problem as follows.

PROBLEM$'$. *Given $n \geqslant 1$, $r > 0$, and a monic polynomial q, is it true that*

$$
(5) \qquad \inf_{p \in \mathcal{P}_n} \frac{1}{\sqrt{n\pi}} \left(\iint_{q^{-1}(\Delta_r)} |p'(z)|^2 \, dx \, dy \right)^{1/2} \leqslant r^{n/\deg q} \; ?
$$

For fixed n and q, (5) will hold if r is large enough (since $q^{-1}(\Delta_r)$ is then connected), and also if r is small enough (as can be shown by elementary asymptotic estimates). But what happens in between remains a mystery.

REFERENCES

[F]   Fisher S. D., *Function Theory on Planar Domains*, Wiley, New York, 1983.

[G]   Gamelin T. W., *Uniform Algebras*, Chelsea, New York, 1984.

[PS] Pólya G., Szegö G., *Isoperimetric Inequalities in Mathematical Physics*, Annals of Mathematical Studies, vol. 27, Princeton University Press, 1951.

[T]   Tsuji M., *Potential Theory in Modern Function Theory*, Chelsea, New York, 1975.

DEPARTMENT OF PURE MATHEMATICS
AND MATHEMATICAL STATISTICS,
16 MILL LANE,
CAMBRIDGE CB2 1SB,
UNITED KINGDOM

EXOTIC JORDAN ARCS IN $\mathbb{C}^n$

G. M. Henkin

Let γ be a simple (non-closed) Jordan arc in $\mathbb{C}^n (n \geqslant 2)$, $P(\gamma)$ be the closure in $C(\gamma)$ of polynomials in complex variables, $O(\gamma)$ be the uniform closure on γ of the algebra of functions holomorphic in a neighbourhood of γ. Denote by $A(\gamma)$ a uniform algebra on γ such that $P(\gamma) \subset A(\gamma) \subset C(\gamma)$ and let $h_A(\gamma)$ be its spectrum (maximal ideal space). For an arbitrary compact set K in $\mathbb{C}^n$ the spectrum $h_A(K)$ depends essentially on the choice of the subalgebra $A(K)$. Until recently, however, it seemed plausible that for Jordan arcs the spectrum $h_A(\gamma)$ depends on γ only.

Consider also the algebra $R(\gamma)$ of uniform limits on γ of rational functions with poles off γ, and the algebra $H(\gamma)$ which is the closure in $C(\gamma)$ of the set of all functions holomorphic in a pseudoconvex neighbourhood of γ. Then we obviously have

$$P(\gamma) \subset R(\gamma) \subset H(\gamma) \subset O(\gamma).$$

In 1968 A. Vitushkin (see [1, 2]) discovered the first example of a rationally convex but not polynomially convex arc γ in $\mathbb{C}^2$. In other words, in this example

$$h_R(\gamma) = \gamma, \text{ but } h_P(\gamma) \neq \gamma.$$

In 1974 the author (see [2], p. 116; [3], p. 174) found an example of a Jordan arc γ in $\mathbb{C}^2$ which, being holomorphically convex, cannot, nevertheless, coincide with an intersection of holomorphically convex domains, i.e., $h_O(\gamma) = \gamma$, but $h_H(\gamma) \neq \gamma$.

A curious problem remains, however, unsolved. Namely, whether $h_R(\gamma) = h_H(\gamma)$ for every Jordan arc.

CONJECTURE 1. *There exists a Jordan arc γ in $\mathbb{C}^2$ satisfying*

$$h_R(\gamma) \neq \gamma, \quad h_H(\gamma) = \gamma.$$

Consider now the algebra $A(K, S)$ of all functions continuous on the Riemann sphere S and holomorphic outside a compact set K, $K \subset S$. To prove Conjecture 1 it is sufficient, for example, to prove the following statement which simultaneously strengthens the classical results of J. Wermer (see [4], [6]) and R. Arens (see [5], [6]).

CONJECTURE 2. *There exists a Jordan arc γ in S such that $A(\gamma, S)$ contains a finitely generated subalgebra with the spectrum S.*

All known exotic Jordan arcs in $\mathbb{C}^n$ are of positive two-dimensional Hausdorff measure. It would be very interesting therefore to prove that there is no exotic arc of zero two-dimensional Hausdorff measure.

PROBLEM. *Suppose that a simple (non-closed) Jordan arc γ in $\mathbb{C}^n$ has zero two-dimensional Hausdorff measure. Is γ polynomially convex (i.e. $h_P(\gamma) = \gamma$)?*

Recall that H. Alexander [7] has proved that every rectifiable simple arc in $\mathbb{C}^n$ is polynomially convex.

REFERENCES

1. Vitushkin A. G., *On a problem of Rudin*, Dokl. Akad. Nauk SSSR **213** (1973), no. 1, 14–15 (Russian); English transl. in Soviet Math. Dokl. **14** (1973), 1618–1619.

2. Henkin G. M., Chirka E. M., Contemporary Problems of Mathematics, vol. 4, VINITI, Moscow, 1975, pp. 13–142 (Russian); English transl. in vol. 5, 1976, pp. 612–687.

3. Wells R. O., *Function theory on differentiable submanifolds*, Contributions to Analysis, a collection of papers dedicated to Lipman Bers, Academic Press, 1974, pp. 407–441.

4. Wermer J., *Polynomial approximation on an arc in* $\mathbb{C}^3$, Ann. Math. **62** N 2 (1955), 269–270.

5. Arens R., *The maximal ideals of certain function algebras*, Pacific J. Math. **8** (1958), 641–648.

6. Gamelin Th. W., *Uniform Algebras*, Prentice Hall, New Jersey, 1969.

7. Alexander H., *Polynomial approximation and hulls in sets of finite linear measure in* $\mathbb{C}^n$, Amer. J. Math. **93** (1971), no. 1, 65–74.

ANALYSE COMPLEXE ET GÉOMÉTRIE
UNIVERSITÉ PARIS VI
4, PLACE JUSSIEU
75252 PARIS CEDEX 05
FRANCE

REMOVABLE SETS FOR BOUNDED ANALYTIC FUNCTIONS*

DONALD E. MARSHALL**

Suppose E is a compact subset of an open set V, $V \subset \mathbb{C}$. Then E is said to be removable, or a Painlevé null set [1], if every bounded analytic function on $V \backslash E$ extends to be analytic on V. This is easily seen to be a property of the set E and not V. Painlevé [2] asked for a necessary and sufficient condition for a compact set E to be removable. The corresponding problem for harmonic functions has been answered in terms of logarithmic capacity and transfinite diameter. Ahlfors [3] has restated the question in terms of the following extremal problem. Let

$$\gamma(E) = \sup \left\{ \lim_{z \to \infty} \left| z \big(f(z) - f(\infty) \big) \right| : f \text{ is analytic on } \mathbb{C} \backslash E \text{ and } |f(z)| \leqslant 1 \right\}$$

be the analytic capacity of E. Then E is removable if and only if $\gamma(E) = 0$. A geometric solution to this problem would have applications in rational approximation and cluster-value theory. See, for example, [4] and [5]. Also [6] contains an interesting historical account.

It is known that Hausdorff measure is not "fine" enough to characterize removable sets. Painlevé (and later Besicovitch [7]) proved that if the one-dimensional Hausdorff measure, $H^1(E)$, is zero then $\gamma(E) = 0$. It is also classical that if $H^{1+\varepsilon}(E) > 0$, for some $\varepsilon > 0$, then $\gamma(E) > 0$. However, examples [8] and [9] show that it is possible for $H^1(E) > 0$ and $\gamma(E) = 0$.***

If l_θ is the ray from the origin with argument θ, let $|P_\theta(E)|$ denote the Lebesgue measure of the orthogonal projection of E on l_θ. Let

$$CR(E) = \int_0^{2\pi} \left| P_\theta(E) \right| d\theta / \pi .$$

This quantity first arose in connection with the solution of the Buffon needle problem as given by Crofton [11] in 1868. If the diameter of E is less than 1, it is the probability of E falling on a system of parallel lines one unit apart. See [12] for an interesting geometric interpretation. Vitushkin [4] asked if $CR(E) = 0$ is equivalent to $\gamma(E) = 0$. It is not hard to see that if $H^1(E) = 0$, then $CR(E) = 0$. Marstrand [13] has proved that if $H^{1+\varepsilon}(E) > 0$ then $CR(E) > 0$. In order to answer Vitushkin's question, one thus needs to consider only sets of Hausdorff dimension 1.

A special case is the following theorem asserted by Denjoy [14] in 1909.

---

*see also Commentary after Problem 12.27
**Research supported in part by National Foundation Grant No. MCS 77-01873.
***See also pp. 346–348 of the book [10]. – Ed.

THEOREM. *If E is a compact subset of a rectifiable curve Γ, then $\gamma(E) = 0$ if and only if $H^1(E) = 0$.*

Although his proof has a gap, Ahlfors and Beurling [1] noted that it is correct if Γ is a straight line. They extended this result to analytic curves Γ. Ivanov [16] proved it for curves slightly smoother than C^1. Davie [17] proved that it sufficed to assume Γ is a C^1 curve. Recently, A. P. Calderón [18] proved that the Cauchy integral operator, for C^1 curves, is bounded on L^p, $1 < p < \infty$. Denjoy's conjecture is a corollary of this theorem. Here is an *Outline of the Proof*:

Let $\mathcal{D}$ be a finitely connected planar domain bounded by C, a union of rectifiable arcs $C_1, \ldots, C_k$. Let F_n map the unit disk conformally onto C_h^c and let $C_k^r = F_n(|z| = r)$. We say that f, analytic in $\mathcal{D}$, is in $E^2(\mathcal{D})$ if and only if $\sup_{r > r_0} \int_{\cup_n C_n^r} |f(z)|^2 |dz| < \infty$ and define $\|f\|_{E^2}^2 = \int_C |f(z)|^2 \frac{|dz|}{2\pi}$, where C is traced twice if it is an arc.

LEMMA 1 ([19]). *If C consists of finitely many analytic curves, then*

$$\gamma(C)^{1/2} = \sup \left(|f'(\infty)| : f \in E^2(\mathcal{D}), \|f\|_{E^2} \leqslant 1 \right) .$$

In this classical paper, Garabedian introduces the dual extremal problem: $\inf(\|g\|_{E^1} : g \in E^1(\mathcal{D}), g(\infty) = 1)$ to obtain the above relation. It was noticed by Havinson [15] that the result remains true for rectifiable arcs. If $g \in L^2(C)$, let

$$G(z) = \int_C \frac{g(\zeta)}{\zeta - z} \frac{d\zeta}{2\pi i}.$$

LEMMA 2 ([20]). *If C is the union of finitely many C^1 curves and if the Cauchy integral has boundary values G^*, $G^* \in L^2(C)$, then $G \in E^2(\mathcal{D})$.*

This follows by writing $G = \sum G_n$ where each G_n is analytic off one of the contours in C. Then use the well-known fact that $\int_{|z|=r} |f(z)|^2 |dz|$ increases with r if $f \in H^2(\mathbb{D})$.

LEMMA 3 ([21]). *Let C be a C^1 curve. If for all g, $g \in L^2(C)$, we have $G \in E^2(\mathcal{D})$, then the length and capacity of a subset E of C are simultaneously positive or zero.*

This follows by approximating the set E by a subset $\tilde{E}$ of C consisting of finitely many subarcs, then applying Lemma 1 to the characteristic function of $\tilde{E}$.

Thus by Calderón's theorem, Denjoy's conjecture is true for C^1 curves. Davie's result finishes the proof. Incidentally an older theorem of [25], p. 267, immediately implies Davie's result. $\square$

About the same time that Besicovitch rediscovered Painlevé's theorem (see above), he proved one of the fundamental theorems of geometric measure theory. A set E is said to be *regular* if it is contained in a countable union of rectifiable curves. A set E is said to be *irregular* if

$$\lim_{r \to 0} \sup \frac{H^1(E \cap B(x, r))}{r} \neq \lim_{r \to 0} \inf \frac{H^1(E \cap B(x, r))}{r} \quad \text{for } H^1 - \text{a.a. } x, \ x \in E,$$

where $B(x, r) = \{y : |y - x| \leqslant r\}$. Besicovitch [22] proved that if $H^1(E) < \infty$, then $E = E_1 \cup E_2$ where E_1 is regular and E_2 is irregular. Later [23], he showed that if

E is irregular, then the orthogonal projection of E in almost all directions has zero length. Thus if $H^1(E) < \infty$ and $CR(E) > 0$ there is a rectifiable curve Γ so that the length of $E \cap \Gamma$ is positive. Since Denjoy's conjecture is true, $\gamma(E)$ is positive whenever $H^1(E) < \infty$ and $CR(E) > 0$.

All examples where the analytic capacity is known concur with Vitushkin's conjecture. For instance, let $\tilde{E}$ be the cross product of the Cantor set, obtained by removing middle halves, with itself. It is shown in [9] that $\gamma(E) = 0$. For each x, $x \in E$, one can find annuli centered on x which are disjoint from $\tilde{E}$ and proportional in size to their distance from x. Thus $\tilde{E}$ is irregular and $CR(E) = 0$. We remark that the projection of $\tilde{E}$ on a line with slope $\frac{1}{2}$ is a full segment. Another relevant example is the cross product of the usual Cantor tertiary set with itself, call it F. The Hausdorff dimension of F is greater than one so that $CR(F) > 0$ and $\gamma(F) > 0$. However every subset $\tilde{F}$ of F with $H^1(\tilde{F}) < \infty$ is irregular and hence satisfies $CR(\tilde{F}) = 0$. This shows we cannot easily reduce the problem to compact sets E with $H^1(E) < \infty$.

If $\gamma(E) > 0$, one possible approach to prove $CR(E) > 0$ is to consider the set

$$\tilde{E} = \left\{ \frac{1 + e^{i\theta}}{2} z : z \in E,\ 0 \leqslant \theta \leqslant 2\pi \right\}.$$

A point ξ is not in $\tilde{E}$ if and only if the line passing through z and whose distance to the origin is $|z|$, misses the set E. It is not hard to see $CR(E) > 0$ if and only if $\tilde{E}$ has positive area. Uy [24]* has recently shown that a set F has positive area if and only if there is a Lipschitz continuous function which is analytic on $\mathbb{C} \setminus F$. So one might try to construct such a function for the set $\tilde{E}$. A related question was asked by A. Beurling. He asked, if $\gamma(E) > 0$ and if E has no removable points, then must the part of the boundary of the normal fundamental domain (for the universal covering map) on the unit circle have positive length? This was shown to fail in [26].

Finally, I would like to mention that I see no reason why $CR(E)$ is not comparable to analytic capacity. In other words, does there exist a constant K with $\frac{1}{K} \cdot CR(E) \leqslant \gamma(E) \leqslant K \cdot CR(E)$? If this were true, it would have application to other problems. For example, it would prove that analytic capacity is semi-subadditive.

References

1. Ahlfors L. V., Beurling A., *Conformal invariants and function-theoretic null sets*, Acta Math. **83** (1950), 101–129.
2. Painlevé P., *Sur les lignes singulières des fonctions analytiques*, Ann. Fac. Sci. Toulouse **2** (1888).
3. Ahlfors L. V., *Bounded analytic functions*, Duke Math. J. **14** (1947), 1–11.
4. Vitushkin A. G., *The analytic capacity of sets in problems of approximation theory*, Uspekhi Mat. Nauk **22** (1967), no. 6, 141–199 (Russian); English transl. in Russian Math. Surveys **22** (1967), 139–200.
5. Zalcman L., *Analytic capacity and Rational Approximation*, Lect. Notes Math., vol. 50, Springer, Berlin, 1968.
6. Collingwood E. P., Lohwater A. J., *The Theory of Cluster Sets*, Cambridge U. P., Cambridge, 1966.
7. Besicovitch A., *On sufficient conditions for a function to be analytic and on behavior of analytic functions in the neighborhood of non-isolated singular points* Proc. London Math. Soc. **32** (1931), no. 2, 1–9.

---

*See [27] for a short proof. – Ed.

Chapter 12. APPROXIMATION and CAPACITIES

8. Vitushkin A. G., *An example of a set of positive length, but zero analytic capacity*, Dokl. Akad. Nauk SSSR **127** (1959), 246–249. (Russian)
9. Garnett J., *Positive length but zero analytic capacity*, Proc. Amer. Math. Soc. **24** (1970), 696–699.
10. Ivanov L. D., *The variation of sets and functions*, Nauka, Moscow, 1975.
11. Crofton M. W., *On the Theory of Local Probability*, Philos. Trans. Roy. Soc. **177** (1968), 181–199.
12. Sylvester J. J., *On a funicular solution of Buffon's "Problem of the needle" in its most general form*, Acta Math. **14** (1891), 185–205.
13. Marstrand J. M., *Fundamental geometrical properties of plane sets of fractional dimensions*, Proc. London Math. Soc. **4** (1954), 257–302.
14. Denjoy A., *Sur les fonctions analytiques uniformes à singularités discontinues*, C. R. Acad. Sci. Paris **149** (1909), 258–260.
15. Havinson S. Ya., *Analytic capacity of sets, joint nontriviality of various classes of analytic functions and the Schwarz lemma in arbitrary domains*, Mat. Sb. **54** (1961), no. 1, 3–50 (Russian); English transl. in Amer. Math. Soc. Translations **43** ser. 2 (1964), 215–266.
16. Ivanov L. D., *On the analytic capacity of linear sets*, Uspekhi Mat. Nauk **17** (1962), 143–144. (Russian)
17. Davie A. M., *Analytic capacity and approximation problems*, Trans. Amer. Math. Soc. **171** (1972), 409–444.
18. Calderón A. P., *Cauchy integrals on Lipschitz curves and related operators*, Proc. Nat. Acad. Sci. USA **74** (1977), 1324–1327.
19. Garabedian P. R., *Schwarz's lemma and the Szegö kernel function*, Trans. Amer. Math. Soc. **67** (1949), 1–35.
20. Havin V. P., *Boundary properties of integrals of Cauchy type and harmonic conjugate functions in domains with rectifiable boundary*, Mat. Sb. **68** (1965), 499–517 (Russian); English transl. in Amer. Math. Soc. Translations **74** ser. 2 (1968), 40–60.
21. Havin V. P., Havinson S. Ya., *Some estimates of analytic capacity*, Dokl. Akad. Nauk SSSR **138** (1961), 789–792 (Russian); English transl. in Soviet Math. Dokl. **2** (1961), 731–734.
22. Besicovitch A., *On the fundamental geometric properties of linearly measurable plane sets of points*, I, Math. Ann. **98** (1927), 422–464; II, Math. Ann. **115** (1938), 296–329.
23. Besicovitch A., *On the fundamental geometric properties of linearly measurable plane sets of points*, III, Math. Ann. **116** (1939), 349–357.
24. Uy N., *Removable sets of analytic functions satisfying a Lipschitz condition*, Ark. Mat. **17** (1979), 19–27.
25. Federer H., *Geometric Measure Theory*, Springer–Verlag, Berlin, 1969.
26. Marshall D. E., *Painlevé null sets*, Colloq. d'Analyse Harmonique et Complexe, Univ. Aix-Marseill I, Marseill, 1977.
27. Hruščëv S. V., *A simple proof of a theorem on removable singularities of analytic functions satisfying a Lipschitz condition*, Zapiski Nauchn. Sem. LOMI **113** (1981), 199–203 (Russian); English transl. in J. Soviet Math. **22** (1983), 1829–1832.

DEPARTMENT OF MATHEMATICS
UNIVERSITY OF WASHINGTON
SEATTLE, WASHINGTON 98195
USA

ON PAINLEVE NULL SETS*

W. K. HAYMAN

Suppose that E is a compact plane set and that N s an open neighbourhood of E. A set is called a *Painlevé null* set (or P. N. set) if every function regular and bounded in $N \setminus E$ can be analytically continued onto E. In this case we also say that E has *zero analytic capacity*.

The problem of the structure of P. N. sets has a long history. Painlevé proved that if E has linear (i.e., one-dimensional Hausdorff) measure zero, then E is a P. N. set, though it seems that this result was first published by Zoretti [1]. Painlevé's theorem has been rediscovered by various people including Besicovitch [2] who proved that if f is continuous on E, as well as regular outside E, and if E has finite linear measure, then f can be analytically continued onto E. Denjoy 3] conjectured that if E lies on a rectifiable curve, then E is a P. N. set if and only if E has linear measure zero. He proved this result for linear sets. Ahlfors and Beurling [4] proved Denjoy's conjecture for sets on analytic curves and Ivanov [5] for sets on sufficiently smooth curves. Davie [6] has shown that it is sufficient to prove Denjoy's conjecture for C^1 curves. On the other hand, Havin and Havinson [7] and Havin [8] showed that Denjoy's conjecture follows if the Cauchy integral operator is bounded on L^2 for C^1 curves. This latter result has now been proved by Calderón [9] so that Denjoy's conjecture is true. I am grateful to D. E. Marshall [10] for informing me about the above results.

Besicovitch [11] proved that every compact set E of finite linear measure is the union of two subsets E_1, E_2. The subset E_1 lies on the union of a finite or countable number of rectifiable Jordan arcs. It follows from the above result that E_1 is not a P. N. set unless E_1 has linear measure zero. The set E_2 on the other hand meets every rectifiable curve in a set of measure zero, has projection zero in almost all directions and has a linear density at almost none of its points. The sets E_1 and E_2 were called respectively *regular* and *irregular* by Besicovitch [11]. Since irregular sets behave in some respects like sets of measure zero, I have tentatively conjectured [12, p. 231] that they might be P. N. sets. Vitushkin [13] and Garnett [14] have given examples of irregular sets which are indeed P. N. sets, but the complete conjecture is still open.

A more comprehensive conjecture is due to Vitushkin [15, p. 147]. He conjectures that E is a P. N. set if and only if E has zero projection in almost all directions.

It is not difficult to see that a compact set E is a P. N. set if and only if for every bounded complex measure distributed on E, the function

$$(1) \qquad F(z) = \int_E \frac{d\mu(\zeta)}{z - \zeta}$$

---

*see also Commentary after Problem 12.27

is unbounded outside E.* Thus E is certainly not a P. N. set if there exists a positive unit measure μ on E such that

$$\int_E \frac{d\mu(\zeta)}{|z - \zeta|}$$

is bounded outside E, i.e., if E has positive linear capacity [16, p. 73]. This is certainly the case if E has positive measure with respect to some Hausdorff function h, such that

$$\int_0^1 \frac{h(r)dr}{r^2} < \infty$$

([17]). Thus in particular E is not a P. N. set if E has Hausdorff dimension greater than one. While a full geometric characterization of P. N. sets is likely to be difficult, there still seems plenty of scope for further work on this intriguing class of sets.

REFERENCES

1. Zoretti L., *Sur les fonctions analytiques uniformes qui possèdent un ensemble parfait discontinu de points singuliers*, J. Math. Pures Appl. **6** N 1 (1905), 1–51.
2. Besicovitch A., *On sufficient conditions for a function to be analytic and on behavior of analytic functions in the neighborhood of non-isolated singular points*, Proc. London Math. Soc. **32** N 2 (1931), 1–9.
3. Denjoy A., *Sur les fonctions analytiques uniformes à singularités discontinues*, C. R. Acad. Sci. Paris **149** (1909), 258–260.
4. Havinson S. Ya., *Analytic capacity of sets, joint nontriviality of various classes of analytic functions and the Schwarz lemma in arbitrary domains*, Mat. Sb. **54** (1961), no. 1, 3–50 (Russian); English transl. in Amer. Math. Soc. Translations **43** ser. 2 (1964), 215–266.
5. Ivanov L. D., *On a conjecture of Denjoy*, Uspekhi Mat. Nauk **18** (1964), 147–149. (Russian)
6. Davie A. M., *Analytic capacity and approximation problems*, Trans. Amer. Math. Soc. **171** (1972), 409–444.
7. Havin V. P., Havinson S. Ya., *Some estimates of analytic capacity*, Dokl. Akad. Nauk SSSR **138** (1961), 789–792 (Russian); English transl. in Soviet Math. Dokl. **2** (1961), 731–734.
8. Havin V. P., *Boundary properties of integrals of Cauchy type and harmonic conjugate functions in domains with rectifiable boundary*, Mat. Sb. **68** (1965), 499–517 (Russian); English transl. in Amer. Math. Soc. Translations **74** ser. 2 (1968), 40–60.
9. Calderón A. P., *Cauchy integrals on Lipschitz curves and related operators*, Proc. Nat. Acad. Sci. USA **74** (1977), 1324–1327.
10. Marshall D. E., *The Denjoy Conjecture*, preprint, 1977.
11. Besicovitch A., *On the fundamental geometric properties of linearly measurable plane sets of points*, I, Math. Ann. **98** (1927), 422–464; II, Math. Ann. **115** (1938), 296–329.
12. Hayman W. K., Kennedy P. B., *Subharmonic Functions*, Vol. 1., Academic Press, London – New York, 1976.
13. Vitushkin A. G., *An example of a set of positive length but zero analytic capacity*, Dokl. Akad. Nauk SSSR **127** (1959), 246–249. (Russian)
14. Garnett J., *Positive length but zero analytic capacity*, Proc. Amer. Math. Soc. **24** (1970), 696–699.
15. Vitushkin A. G., *The analytic capacity of sets in problems of approximation theory*, Uspekhi Mat. Nauk **22** (1967), no. 6, 141–199 (Russian); English transl. in Russian Math. Surveys **22** (1967), 139–200.
16. Carleson L., *Selected problems on exceptional sets*, Van Nostrand Math. Stud., N 13, Van Nostrand, Toronto, 1967.
17. Forstman O., *Potentiel d'équilibre et capacité des ensembles avec quelques applications à la théorie des fonctions*, Medded. Lunds. Univ. Mat. Sem. **3** (1935), 1–118.

*See ed. note at the end of the section. – Ed.

18. Vitushkin A. G., *On a problem of Denjoy*, Izv. Akad. Nauk SSSR **28** (1964), no. 4, 745–756. (Russian)
19. Val'skii R. É., *Remarks on bounded functions representable by an integral of Cauchy–Stieltjes type*, Sibirsk. Mat. Zh. **7** (1966), no. 2, 252–260 (Russian); English translation in Siberian Math. J. **7** (1966), 202–209.

DEPARTMENT OF MATHEMATICS
IMPERIAL COLLEGE
SOUTH KENSINGTON
LONDON SW7 ENGLAND

EDITORS' NOTE (1984)

As far as we know the representability of *all* functions bounded and analytic off E and vanishing at infinity by "Cauchy potentials" (1) is guaranteed when E has finite Painlevé length whereas examples show that this is no longer true for an arbitrary E ([18], [19]). We think the QUESTION of *existence of potentials* (1) *bounded* on $\mathbb{C} \setminus E$ (*provided E is not a P. N. set*) is one more interesting problem. (See also §5 of [4]).

12.25
v.old

ANALYTIC CAPACITY AND RATIONAL APPROXIMATION*

M. S. MEL'NIKOV, A. G. VITUSHKIN

Let E be a bounded subset of $\mathbb{C}$ and $B(E, 1)$ be the set of all functions f in $\hat{\mathbb{C}}$ analytic on $\hat{\mathbb{C}} \setminus E$ and with $f(\infty) = 0$, $\sup_{\mathbb{C}} |f| \leqslant 1$. Put $A(E, 1) = \{ f \in B(E, 1) : f$ is continuous on $\mathbb{C} \}$. The number

$$\gamma(E) \overset{\text{def}}{=} \sup_{f \in B(E, 1)} \lim_{z \to \infty} |z f(z)|$$

is called the analytic capacity of E. The number

$$\alpha(E) \overset{\text{def}}{=} \sup_{f \in A(E, 1)} \lim_{z \to \infty} |z f(z)|$$

is called the analytic C-capacity of E.

The analytic capacity has been introduced by Ahlfors [1] in connection with the Painlevé problem to describe sets of removable singularities of bounded analytic functions. Ahlfors [1] has proved that these sets are characterized by $\gamma(E) = 0$. However, it would be desirable to describe removable sets in metric terms.

CONJECTURE 1. *A compact set E, $E \subset \mathbb{C}$, has zero analytic capacity iff the projection of E onto almost every direction has zero length* ("almost every" means "a.e. with respect to the linear measure on the unit circle"). *Such an E is called irregular provided its linear Hausdorff measure is positive.*

If the linear Hausdorff measure of E is finite and $\gamma(E) = 0$, then the average of the measures of the projections of E is zero. This follows from Calderon's result [2] and the well-known theorems about irregular sets (see [3], p. 341–348). The connections between capacity and measures is described in detail in [4].

The capacitary characteristics are most efficient in the approximation theory [5], [6], [7], [8]. A number of approximation problems lead to an unsolved question of the semi-additivity of the analytic capacity:

$$\gamma(E \cup F) \leqslant c \left[\gamma(E) + \gamma(F) \right],$$

where c is an absolute constant and E, F are arbitrary disjoint compact sets.

Let $A(K)$ denote the algebra of all functions continuous on a compact set K, $K \subset \mathbb{C}$, and analytic in its interior. Let $R(K)$ denote the uniform closure of rational functions with poles off K and, finally, let $\partial^o K$ be the inner boundary of K, i.e. the set of boundary points of K not belonging to the boundary of a component of $\mathbb{C} \setminus K$. Sets K satisfying $A(K) = R(K)$ were characterized in terms of the analytic capacity [6]. To obtain geometric conditions of the approximability, a further study of capacities is needed.

---

*see also Commentary after Problem 12.27

CONJECTURE 2. *If $\alpha(\partial^o K) = 0$, then $A(K) = R(K)$.*

The affirmative answer to the question of semiadditivity would yield a proof of this conjecture. Since $\alpha(E) = 0$ provided E is of finite linear Hausdorff measure, this would also lead to the proof of the following statement.

CONJECTURE 3. *If the linear Hausdorff measure of $\partial^o K$ is zero (K being a compact subset of $\mathbb{C}$), then $A(K) = C(K)$.*

The last equality is not proved even for K's with $\partial^o K$ of zero linear Hausdorff measure.

It is possible however that the semiadditivity problem can be avoided in the proof of Conjecture 3.

The semiadditivity of the capacity has been proved only in some special cases ([9], [10–13]), e.g., for sets E and F separated by a straight line. For a detailed discussion of this and some other relevant problems see [14].

REFERENCES

1. Ahlfors L. V., *Bounded analytic functions*, Duke Math. J. **14** (1947), 1–11.
2. Calderón A. P., *Cauchy integrals on Lipschitz curves and related operators*, Proc. Nat. Acad. Sci. USA **74** (1977), 1324–1327.
3. Ivanov L. D., *The variation of sets and functions*, Nauka, Moscow, 1975. (Russian)
4. Garnett J., *Analytic capacity and measure*, Lect. Notes in Math., 297, Springer, Berlin, 1972.
5. Vitushkin A. G., *The analytic capacity of sets in problems of approximation theory*, Uspekhi Mat. Nauk **22** (1967), no. 6, 141–199 (Russian); English transl. in Russian Math. Surveys **22** (1967), 139–200.
6. Mel'nikov M. S., Sinanjan S. O., *Questions in the theory of approximation of functions of one complex variable*, Contemporary Problems of Mathematics, vol. 4, Itogi Nauki i Tekhniki, VINITI, Moscow, 1975, pp. 143–250 (Russian); English transl. in vol. 5, 1976, pp. 688–752.
7. Zalcman L., *Analytic capacity and rational approximation*, Lect. Notes in Math., 50, Springer, Berlin, 1968.
8. Gamelin T. W., *Uniform Algebras*, Prentice Hall, N. J., 1969.
9. Davie A. M., *Analytic capacity and approximation problems*, Trans. Amer. Math. Soc **171** (1972), 409–444.
10. Mel'nikov M. S., *An estimate of the Cauchy integral along an analytic curve*, Mat. Sb. **71** (1966), no. 4, 503–514 (Russian); English transl. in AMS Translations **80** ser. 2 (1969), 243–255.
11. Vitushkin A. G., *Estimates of the Cauchy integral*, Mat. Sb. **71** (1966), no. 4, 515–534 (Russian); English transl. in AMS Translations **80** ser. 2 (1969), 257–278.
12. Shirokov N. A., *On a property of analytic capacity*, Vestnik Leningrad Univ., Mat. (1971), no. 19, 75–82. (Russian)
13. Shirokov N. A., *Some properties of analytic capacity*, Vestnik Leningrad Univ., Mat. **1** (1972), no. 1, 77–86. (Russian)
14. Besicovitch A., *On sufficient conditions for a function to be analytic and on behavior of analytic functions in the neighborhood of non-isolated singular points*, Proc. London Math. Soc. **32** (1931), no. 2, 1–9.

MECHMAT FACULTET
MOSCOW STATE UNIVERSITY
LENINSKIE GORY, MOSCOW 117234
RUSSIA

LENINSKII PR. 69-3, 418
MOSCOW, RUSSIA

12.26
v.old

ON SETS OF ANALYTIC CAPACITY ZERO*

L. D. IVANOV

Let K be a compact plane set and $A_\infty(K)$ the space of all functions analytic and bounded outside K endowed with the sup-norm. Define a linear functional L on $A_\infty(K)$ by the formula

$$L(f) = \frac{1}{2\pi i} \int_{|z|=r} f(z)dz,,$$

with $r > \max\{|z| : z \in K\}$. The norm of L is called the *analytic capacity of K*. We denote it by $\gamma(K)$. The function γ is invariant under isometries of $\mathbb{C}$. Therefore it would be desirable to have a method to compute it in terms of Euclidean distance. E. P. Dolženko has found a simple solution of a similar question related to the so-called α-capacity [1]. But for γ the answer is far from being clear. I would like to draw attention to three conjectures.

CONJECTURE 1. *There exists a positive number C such that for any compact set K,*

$$\gamma(K) \geqslant C \int_{\mathbb{T}} \mu(K, \zeta)\, dm(\zeta),$$

where $\mu(K, \zeta)$ denotes the length of the projection of K onto the line through 0 and $\zeta \in \mathbb{T}$.

CONJECTURE 2. *There exists a positive number C such that for any compact set K,*

$$\gamma(K) \leqslant C \int_{\mathbb{T}} \mu(K, \zeta)\, dm(\zeta).$$

These conjectures are in agreement with known facts about analytic capacity. For example, it follows immediately from Conjecture 1 that $\gamma(K) > 0$ if K lies on a continuum of finite length and has positive Hausdorff length. In turn, Conjecture 2 implies that $\gamma(K) = 0$ provided the Favard length of K equals to zero. At last, let K be a set of positive Hausdorff h–measure (a survey of literature on the Hausdorff measures can be found in [3]). If $\int_0 \frac{h(t)}{t^2} dt < \infty$, then the Favard length of K is positive. This ensures the existence of a compact K_1, $K_1 \subset K$, such that $\mu_h(K_1) > 0$ and the function f,

$$f(z) \overset{\text{def}}{=} \int_{K_1} \frac{d\mu_h(z)}{\zeta - z},$$

is continuous on $\mathbb{C}$, μ_h being the Hausdorff h–measure. Hence $\gamma(K) \geqslant \gamma(K_1) > 0$ which also easily follows from Conjecture 2.

---

*see also Commentary after Problem 12.27

CONJECTURE 3. *For any increasing function* $h\colon (0,+\infty) \to (0,+\infty)$ *with* $\int_0 \frac{h(t)}{t^2} dt = \infty$, *there exists a set* K *satisfying* $\mu_h(K) > 0$ *and* $\gamma(K) = 0$.

To corroborate this conjecture, I shall construct a function h and a set E such that $\lim_{t \to 0} \frac{h(t)}{t} = 0$, $\mu_h(K) > 0$, but $\gamma(E) = 0$. Assign to any sequence $\varepsilon = \{\varepsilon_n\}$, $\varepsilon_n \downarrow 0$, a compact set $E(\varepsilon)$. Namely, let $Q_0(\varepsilon) = [0,1]$. If $Q_n(\varepsilon)$ is the union of 2^n disjoint segments Δ_n of length $\delta_n(\varepsilon)$, then $Q_{n+1}(\varepsilon)$ is the union of 2^n sets $\Delta_n^j \setminus \hat{\Delta}_n^j$, $\hat{\Delta}_n^j$ being the interval of length $\frac{1}{2}\delta_n(\varepsilon)(1 - \varepsilon_n)$ concentric with the segment Δ_n^j. Put

$$E(\varepsilon) = \bigcap_{n \geqslant 1} Q_n^2(\varepsilon), \quad Q_n^2(\varepsilon) \overset{\text{def}}{=} Q_n(\varepsilon) \times Q_n(\varepsilon),$$

and let ε^c be a constant sequence, $(\varepsilon^c)_n = c$. Finally, let $E = E(\varepsilon^0)$.

It is known (see [3]) that $\gamma(E) = 0$. This implies the existence of a function φ such that

$$\lim_{t \to 0} \varphi(t) = 0 \quad \text{and} \quad \gamma(E(\varepsilon)) < \varphi(t)$$

for any sequence ε satisfying $\varepsilon_1 < t$. Then $E(\varepsilon)$ has the desired properties for properly chosen ε, as will be shown later. To choose ε, pick numbers $\alpha_1 > 0$ and $n_1 \in \mathbb{N}$ such that $\varphi(\alpha_1) < \frac{1}{2}$, $\gamma(Q_{n_1}^2(\varepsilon^{\alpha_1})) < \frac{1}{2}$ and $(1 + \alpha_1)^{n_1} > 2$. Set $\varepsilon_j = \alpha_1$ for $j = 1, 2, \ldots, n_1$. Proceeding by induction, pick α_{k+1} to provide the inequality $\varphi(\alpha_{k+1}) < \beta_k = \frac{1}{k+2} \prod_{j=1}^{k}(1 + \alpha_j)^{-n_j}$ and next pick n_{k+1} such that

$$(1 + \alpha_{k+1})^{n_{k+1}} > 2 \quad \text{and} \quad \gamma\left[Q_{n_{k+1}}^2(\varepsilon^{\alpha_{k+1}})\right] < \beta_k.$$

Set now $\varepsilon_j = \alpha_{k+1}$ for $j = N_k + 1, \ldots, N_{k+1}$ ($N_s \overset{\text{def}}{=} n_1 + \cdots + n_s$).

The sequence ε defines a function ψ_ε equal to 4^{-n} at $\delta_n(\varepsilon)$, $n \in \mathbb{N}$ and linear on each segment $[\delta_j(\varepsilon), \delta_{j-1}(\varepsilon)]$. It is easy to verify that $\lim_{t \to 0+}(\psi_\varepsilon(t))/t = 0$ and $E(\varepsilon)$ has positive ψ_ε-measure. It remains to check only that $\gamma[E(\varepsilon)] = 0$.

For this purpose, let $f \in A_\infty(E(\varepsilon))$ and let

$$f_n^j(z) = \frac{1}{2\pi i} \int_\gamma \frac{f(\zeta)d\zeta}{\zeta - z}, \quad z \notin E(\varepsilon) \cap (\Delta_n^{j_1} \times \Delta_n^{j_2}),$$

where contour γ embraces $E(\varepsilon) \cap (\Delta_n^{j_1} \times \Delta_n^{j_2})$ and separates it from z. The set $Q_n^2(\varepsilon)$ being the union of 4^n squares $\Delta_n^{j_1} \times \Delta_n^{j_2}$ with the side $\delta_n(\varepsilon)$ and lying at the distance at least $2\delta_n(\varepsilon)\frac{1-\varepsilon_n}{1+\varepsilon_n}$ one from another, it is clear that f_n^j are uniformly bounded and

$$|L(f)| \leqslant 4^n \cdot \sup_{m,j} \|f_m^j\| \cdot \gamma[E(\varepsilon) \cap (\Delta_n^{j_1} \times \Delta_n^{j_2})].$$

This implies

$$|L(f)| \leqslant 4^n \cdot \sup_{m,j} \|f_m^j\| \frac{\prod_{j=1}^k (1 + \alpha_j)^{n_j}}{4^{N_k}} \cdot \gamma\left[Q_{n_{k+1}}^2(\varepsilon^{\alpha_{k-1}})\right]$$

$$\leqslant \frac{1}{k+1} \sup_{m,j} \|f_m^j\|$$

and finally $L(f) = 0$. $\square$

Chapter 12. APPROXIMATION and CAPACITIES

REFERENCES

1. Dolženko E. P., *On the removal of singularities of analytic functions*, Uspekhi Mat. Nauk **18** (1963), no. 4, 135–142 (Russian); English translation in Amer. Math. Soc. Translations **97** ser. 2 (1971), 33–41.
2. Calderón A. P., *Cauchy integrals on Lipschitz curves and related operators*, Proc. Nat. Acad. Sci. USA **74** (1977), 1324–1327.
3. Rogers C. A., *Hausdorff Measures*, Cambridge University Press, Cambridge, 1970.
4. Garnett J., *Positive length but zero analytic capacity*, Proc. Amer. Math. Soc. **24** (1970), 696–699.

COMMENTARY BY V. YA. EIDERMAN

Suppose that the function $h(r), r \geqslant 0$, is continuous and increasing, and that $h(0) = 0$ (such function are called measuring functions). It is well known that if $\mu_h(K) = 0$ with $h(r) = r$, then $\gamma(K) = 0$. On other hand, if

$$(1) \qquad \int_0 h(t)/t^2 dt < \infty$$

and $\gamma(K) = 0$, then $\mu_h(K) = 0$.

The following question arises: is it possible to reduce the gap between the conditions on h in these implications? The example $K = [a, b]$ shows that in the first of them we cannot take $h(r) = o(r), r \to 0+$. Hence, it is natural to investigate the necessity of condition (1) and to consider Conjecture 3. (An analogous problem for the classical capacity was considered in [5].)

We will show that this Conjecture 3 is not true. Let $C(K)$ be the "classical" capacity corresponding to the kernel $k(r) = 1/r$.

THEOREM. *If $C(K) = 0$ and the measuring function $h(r)$ is such that*

$$(2) \qquad \limsup \psi(r) = \alpha > 0 \ \text{ as } \ r \to 0+, \qquad \psi(r) = r^2/h(r),$$

then $\mu_h(K) = 0$.

LEMMA. *There exists a measuring function $h(r)$ satisfying (2) and such that*

$$(3) \qquad \int_0 h(t)/t^2 dt = \infty.$$

The theorem and the lemma stated above are particular cases of Theorem 1.1 and Lemma 1.1 in [5].

So, there exists a measuring function h (which is even infinitely differentiable) such that (3) holds, and if $\mu_h(K) > 0$ then $C(K) > 0$. Since $C(K) \leqslant \gamma(K)$ (see [6, p. 71]), we have the implication $(\mu_h(K) > 0 \implies \gamma(K) > 0)$. Hence, Conjecture 3 fails to be true.

But the following question arises: is Conjecture 3 true under some additional regularity conditions on $h(r)$? To answer this question, it is very important to solve Garnett's problem [6, p. 95]:

PROBLEM. Let K be a plane Cantor set, and let σ_n be the side of squares at stage n of the construction of K. Then $\gamma(K) > 0$ if and only if $\sum_n 4^{-n}/\sigma_n < \infty$.

(*This is equivalent to* $\gamma(K) = 0 \iff C(K) = 0$.)

There is a proof of this statement ([6, p. 88, Theorem 2.2]) under the assumption that σ_n/σ_{n-1} is a non-increasing sequence. From this Theorem J. Garnett deduces Theorem 2.7 [6, p. 94], which states that Conjecture 3 is true under the following additional condition:

(4) the function $a(t) = th'(t)/h(t)$ is increasing.

It should be noted that there is a gap in the proof of Theorem 2.2 in [6]. To fill this gap it suffices, for instance, to supplement Lemma 2.6 [6, p. 92] with the following additional statement: the constant δ in this Lemma 2.6 does not depend on γ. (I am not sure that this statement is true.)

Finally, we note that the solution of the above Problem gives also a proof of Conjecture 3 under the following additional assumption: there exist some constants $r_0 > 0$, $b \in (0, 1)$, such that

(5) $$h(r_1)/h(r_2) \geqslant b(r_1/r_2)^2, \quad 0 < r_1 \leqslant r_2 \leqslant r_0$$

((3) and (4) imply (5)).

I am grateful to Professor J. Garnett for useful discussion on Theorem 2.2 [6].

REFERENCES

5. Eiderman V. Ya., *On the comparison of Hausdorff measure and capacity*, Algebra i Analiz **3** (1991), no. 6, 174–189 (Russian); English translation in St.Petersburg Math. J. **3** (1992), no. 6, 1367–1381.
6. Garnett J., *Analytic capacity and measure*, Lect. Notes in Math. **297** (1972), Springer, Berlin etc.

VOLODARSKOGO ST. 26/32, APT. 165
MOSCOW 109172
RUSSIA

ESTIMATES OF ANALYTIC CAPACITY

JOSEF KRÁL

Let μ denote the Lebesgue measure on $\mathbb{R}^2 \equiv \mathbb{C}$. In what follows we let z range over $\mathbb{C}$ and r over the interval $(0,1) \subset \mathbb{R}$. We always suppose that $p \geqslant 1$ and $\lambda > 0$. E will be used as a generic notation for compact subsets of $\mathbb{C}$. For any locally integrable (complex-valued) function f on $\mathbb{C}$, we denote by

$$f_{r,z} = \frac{1}{\pi r^2} \int_{\mathbb{D}(r,z)} f \, d\mu$$

its mean value over the disc $\mathbb{D}_{(r,z)} = \{\zeta : \zeta \in \mathbb{C}, |\zeta - z| < r\}$. $\mathcal{L}^{p,\lambda}$ will stand for the class of all functions f on $\mathbb{C}$ that are locally integrable to the power p and satisfy

$$\|f\|_{p,\lambda} \equiv \sup_{r,z} \left\{ r^{-\lambda} \int_{\mathcal{D}(r,z)} |f - f_{r,z}|^p \right\}^{1/p} < +\infty$$

(cf. [1] for references on related function spaces). Investigation of removable singularities for holomorphic functions in these classes gives rise naturally to the corresponding capacities $\gamma_{p,\lambda}$ defined as follows (compare [2]). If $A_{p,\lambda}(E,1)$ denotes the class of all $f \in \mathcal{L}^{p,\lambda}$ that are holomorphic in the complement of E (including ∞) and satisfy the conditions

$$f(\infty) = 0, \quad \|f\|_{p,\lambda} \leqslant 1,$$

then

$$\gamma_{p,\lambda}(E) = \sup\{|f'(\infty)| : f \in A_{p,\lambda}(E,1)\},$$

where $f'(\infty) = \lim_{z \to \infty} z f(z)$.

It is an important feature of these capacities that they admit simple metrical estimates which reduce to those of Melnikov (cf. Chapter V in [3]) for a special choice of the parameters when they yield Dolzenko's result on removable singularities in Hölder classes. Writing $\operatorname{diam} M$ for the diameter of M, $M \subset \mathbb{C}$, and defining for α, $\alpha \geqslant 0$, and ε, $\varepsilon > 0$,

$$I_\alpha^\varepsilon(E) = \inf \sum_n (\operatorname{diam} M_n)^\alpha,$$

where the infimum is taken over all sequences of sets M_n, $M_n \subset \mathbb{C}$ with $\operatorname{diam} M_n < \varepsilon$ such that $E \subset \cup_n M_n$, we may state the following inequalities (cf. [4]).

THEOREM 1. *Let* $2 \leqslant p + \lambda < \tilde{\lambda} p + 2$, $\tilde{\lambda} = \min(2,\lambda)$ *and define* $\alpha = 1 + \dfrac{\lambda - 2}{p}$. *Then there are constants* c *and* k *such that*

(1) $$c \, I_\alpha^1(E) \leqslant \gamma_{p,\lambda}(E) \leqslant k \, I_\alpha^1(E)$$

for all E.

PROBLEM 1. *What are the best values of the constants c, k occurring in* (1)?

Theorem 1 is of special interest in the case $\lambda = 2$, because it characterizes removable singularities of holomorphic functions of bounded mean oscillation (cf. [5]) as those sets E whose linear measure

$$I(E) = \lim_{\varepsilon \to +0} I_1^\varepsilon(\varepsilon)$$

vanishes. This is in agreement with the example of Vitushkin (cf. [6]). The capacity of E corresponding to the broader class of functions of bounded mean oscillation may be positive even though E has zero analytic capacity corresponding to bounded functions which is defined by

$$\gamma(E) = \sup\{|f'(\infty)| : f \in A(E, 1)\},$$

where now $A(E, 1)$ is the class of all functions holomorphic off E and vanishing at ∞ whose absolute value never exceeds 1. Nevertheless, by the so-called DENJOY CONJEC-TURE (which follows from combination of results in [7], [8], [9], [10]), the equivalence

$$\gamma(E) = 0 \Leftrightarrow I(E) = 0$$

is true for E situated on a rectifiable curve. The upper estimate of $\gamma(\cdot)$ by means of $I(\cdot)$ is generally valid (cf. [11]) while the lower estimates of $\gamma(E)$ by means of a multiple of $I(E)$ is possible only for E situated on sets Q of a special shape.

PROBLEM 2. *Let $Q \subset \mathbb{C}$ be a compact set. Find geometric conditions on Q guaranteeing the existence of a constant c such that*

(2) $$\gamma(E) \geqslant c\, I(E), \quad E \subset Q.$$

The following theorems, 2 and 3, may serve as sample results.

THEOREM 2 (cf. [12]). *Let $Q = \psi(\langle 0, 1\rangle)$, where $\psi \colon \langle 0, 1\rangle \to \mathbb{C}$ is simple and continuously differentiable, $|\psi'| = 1$ and*

$$\sup_r \int_0^1 \frac{|\psi'(t) - \psi'(r)|^2}{|t - r|}\, dt \equiv \varepsilon < +\infty.$$

Then (2) *holds and c can be computed by means of s (see also §7 in* [11]).

THEOREM 3 (cf. [13]). *If Q has only a finite number of components and*

$$V(Q) \stackrel{\text{def}}{=} \sup_{z \in Q} \int_{\mathbb{T}} N_z^Q(\zeta)\, dm(\zeta) < +\infty,$$

where $N_z^Q(\zeta)$ is the number of points in $Q \cap \{z + t\zeta : t > 0\}$ then (2) *holds with $c = (2\pi)^{-1}(2V(Q) + 1)^{-1}$.*

If Q is a straight-line segment, then $V(Q) = 0$ and Pommerenke's equality $\gamma(E) = \frac{1}{4}l(E)$ holds (cf. [13]). This leads to the following:

PROBLEM 3. *Is it possible to improve c in Theorem 3 to $\dfrac{1}{4}(2V(Q) + 1)^{-1}$?*

Remark. It was asserted in [4] that Theorem 3 holds with this value of the constant. Dr. J. Matyska kindly pointed out that there was a numerical error in the original draft of the corresponding proof in [14].

Chapter 12. APPROXIMATION and CAPACITIES

REFERENCES

1. Peetre J., *On the theory of $\mathcal{L}^{p,\lambda}$-spaces*, J. Funct. Anal. 4 (1969), 71–87.
2. Harvey R., Polking J., *A notion of capacity which characterizes removable singularities*, Trans. Amer. Math. Soc. 169 (1972), 183–195.
3. Mel'nikov M. S., Sinanjan S. O., *Questions in the theory of approximation of functions of one complex variable*, Contemporary Problems of Mathematics, vol. 4, Itogi Nauki i Tekhniki, VINITI, Moscow, 1975, pp. 143–250 (Russian); English transl. in vol. 5, 1976, pp. 688–752.
4. Král J., *Analytic capacity*, Proc. Conf. "Elliptische Differentialgleichungen", Rostock, 1977.
5. John F., Nirenberg L., *On functions of bounded mean oscillations*, Comm. Pure Appl. Math. 14 (1961), 415–426.
6. Vitushkin A. G., *An example of a set of positive length, but zero analytic capacity*, Dokl. Akad. Nauk SSSR 127 (1959), 246–249. (Russian)
7. Calderón A. P., *Cauchy integrals on Lipschitz curves and related operators*, Proc. Nat. Acad. Sci. USA 74 (1977), 1324–1327.
8. Davie A. M., *Analytic capacity and approximation problems*, Trans. Amer. Math. Soc. 171 (1972), 409–444.
9. Havin V. P., Havinson S. Ya., *Some estimates of analytic capacity*, Dokl. Akad. Nauk SSSR 138 (1961), 789–792 (Russian); English transl. in Soviet Math. Dokl. 2 (1961), 731–734.
10. Havin V. P., *Boundary properties of integrals of Cauchy type and harmonic conjugate functions in domains with rectifiable boundary*, Mat. Sb. 68 (1965), 499–517 (Russian); English transl. in Amer. Math. Soc. Translations 74 ser. 2, (1968), 40–60.
11. Garnett J., *Analytic capacity and measure*, Lect. Notes Math., 297, Springer, Berlin, 1972.
12. Ivanov L. D., *On a conjecture of Denjoy*, Uspekhi Mat. Nauk 18 (1964), 147–149. (Russian)
13. Pommerenke Ch., *Über die analytische Kapazität*, Arch. Math. 11 (1960), 270–277.
14. Fuka J., Král J., *Analytic capacity and linear measure*, Czechoslovak Math. J. 28 (103) (1978), no. 3, 445–461.

MATEMATICKÝ ÚSTAV ČSAV,
ŽITNÁ 25, 11567, PRAHA 1,
ČESKÁ REPUBLIKA

COMMENTARY ON PROBLEMS 12.23 – 12.27

Each of the problems 12.23 – 12.27 makes reference to, or is in some way connected with, a well-known conjecture of Vitushkin [1] on characterizing the sets of zero analytic capacity. In keeping with the notation established in 12.23, E is a compact subset of $\mathbb{C}$, l_θ is the ray from the origin making angle θ with the positive real axis, $|P_\theta(E)|$ denotes the Lebesgue measure of the orthogonal projection of E onto l_θ,

$$CR(E) = \frac{1}{\pi} \int_0^{2\pi} |P_\theta(E)| \, d\theta,$$

and $\gamma(E)$ is the analytic capacity of E. Vitushkin conjectured that $\gamma(E) = 0$ if and only if $CR(E) = 0$ (i.e. if and only if $|P_\theta(E)| = 0$ for almost all θ). More recently, however, Mattila [2] has disproved that conjecture, but his argument gives no indication as to which implication (or if both) is incorrect. On the other hand, it is now known that there exists a compact set E_0 such that $CR(E_0) = 0$ but $\gamma(E_0) > 0$ (cf. [3] and [4]).

REFERENCES

1. Vitushkin A. G., *The analytic capacity of sets in problems of approximation theory*, Uspekhi Mat. Nauk **22** (1967), 141–199 (Russian); English transl. in Russian Math. Surveys **22** (1967), 139–200.
2. Mattila P., *Smooth maps, null-sets for integral geometric measure and analytic capacity*, Ann. Math. **123** (1986), 303–309.
3. Murai T., *Comparison between analytic capacity and the Buffon needle probability*, Amer. Math. Soc. Trans. **304** (1987), 501–514.
4. Jones P., Murai T., *Positive analytic capacity but zero Buffon needle probability*, Pacific J. Math. **133** (1988), 99–114.

ANALYTIC CAPACITY AND THE SZEGÖ KERNEL FUNCTION

Takafumi Murai

We are concerned with analytic capacity and the Szegö kernel function. Let $H^\infty(\Omega)$ denote the Banach space of bounded analytic functions in a planar domain Ω with the supremum norm $\|\cdot\|_{H^\infty}$. For $\zeta \in \Omega$, we define $c(\zeta; \Omega) = \sup|f'(\zeta)|$, where the supremum is taken over all $f \in H^\infty(\Omega)$, $\|f\|_{H^\infty} \leqslant 1$. (The derivative at infinity is defined by $f'(\infty) = \lim_{z\to\infty} z\{f(\infty) - f(z)\}$). The analytic capacity of a compact set E in the complex plane is defined by $\gamma(E) = c(\infty; \Omega_E)$, where Ω_E denotes the component of E^c containing ∞. Analytic capacity plays an important role in function theory (e.g. [3, 4]).

1. The arc-length variation. Let Ω be a domain containing ∞ bounded by a finite number of analytic Jordan curves and let $\zeta \in \Omega$, $\zeta \neq \infty$. Then there uniquely exists a pair $\big(K(z, \bar\zeta), L(z, \zeta)\big)$ of meromorphic functions on Ω (as functions of z) such that

$$K(z, \bar\zeta), \ L(z, \zeta) - \frac{1}{z - \zeta} \ \text{are analytic in } \Omega, \quad K(\infty, \bar\zeta) = L(\infty, \zeta) = 0,$$

$$\frac{1}{i} L(z, \zeta)\, dz = \overline{K(z, \bar\zeta)}\, |dz| \ \text{on the boundary } \partial\Omega \text{ of } \Omega.$$

Here the orientation of dz is chosen so that Ω lies to the left. The function $K(z, \bar\zeta)$ is the Szegö kernel function, i.e.

$$f(z) = \frac{1}{2\pi} \int_{\partial\Omega} K(z, \bar\zeta) f(\zeta)\, |d\zeta|$$

for any analytic function f in a neighborhood of $\overline\Omega$ such that $f(\infty) = 0$. For three mutually distinct numbers $w, z, \zeta \in \Omega - \{\infty\}$, we define

$$Dc(z, \zeta; \Omega) = |L(z, \zeta)|^2 - |K(z, \bar\zeta)|^2,$$

$$D^2c(w, z, \zeta; \Omega) = 2\mathcal{R}\Big[\{L(w, z)\overline{K(w, \bar\zeta)} - \overline{K(w, \bar z)}L(w, \zeta)\}\overline{L(z, \zeta)}$$

$$- \{L(w, z)\overline{L(w, \zeta)} - \overline{K(w, \bar z)}K(w, \bar\zeta)\}\overline{K(z, \bar\zeta)}\Big].$$

The definition of these functions is based on variational formulae of analytic capacity.

Conjecture 1. $D^2c(w, z, \zeta; \Omega) \leqslant 0$.

Conjecture 2. $Dc(z, \zeta; \Omega) \leqslant 1$.

If Conjecture 1 is true, then analytic capacity is subadditive and Conjecture 2 is also true. If Conjecture 2 is true, then $\gamma(A \cup B) \leqslant \gamma(A) + \gamma(B)$ for any continuum A. Conjecture 1 is true for simply and doubly connected domains Ω. See [3, 4].

2. Cauchy potentials. The Cauchy potential of a complex plane measure μ is defined by

$$\mathcal{C}\mu(z) = \frac{1}{\pi} \int \frac{1}{\zeta - z} \, d\mu(\zeta) \quad (z \notin \text{(the support of } \mu)).$$

For a compact set E in the complex plane, we define

$$\gamma_0(E) = \sup \left\{ \frac{1}{\pi} \left| \int d\mu \right|; \quad \mathcal{C}\mu \in H^\infty(\Omega_E), \, \|\mathcal{C}\mu\|_{H^\infty} \leqslant 1 \right\}.$$

Then $\gamma_0(E) \leqslant \gamma(E)$.

PROBLEM 3. *Does there exists a compact set E_0 such that $\gamma(E_0) = 1$ and $\gamma_0(E_0)$ is equal to a given non-negative number less than 1?*

See [2, p. 99]. For any compact set E, there uniquely exists $f_E \in H^\infty(\Omega_E), \|f_E\|_{H^\infty} \leqslant 1$ such that $f'_E(\infty) = \gamma(E)$. This is called the Ahlfors function of E and satisfies $f_E(\infty) = 0$. Let $\Lambda(\cdot)$ denote the 1-dimensional Hausdorff measure. If $\Lambda(E) < \infty$, then any function f in $H^\infty(\Omega_E)$ vanishing at infinity is expressed as a Cauchy potential. As an information about this problem, we here note

PROPOSITION 4. *There exists a perfect set E_1 such that $\Lambda(E_1) = \infty$ and f_{E_1} is expressed as a Cauchy potential. There exists a perfect set E_2 such that f_{E_2} cannot be expressed as a Cauchy potential.*

To show the method of the proof, we take

$$I_\delta = [-\delta, \delta] \cup (i[-1, 1] + \delta) \cup (i[-1, 1] - \delta),$$
$$J_\delta = ([-\delta, \delta] + i) \cup (i[-1, 1] + \delta) \cup (i[-1, 1] - \delta) \quad (\delta > 0),$$

where, in general, $zE + \zeta = \{zw + \zeta; \ w \in E\}$. Evidently, both Ahlfors functions $f_{I_\delta}, f_{J_\delta}$ converge to the Ahlfors function $f_{I_0} = \cdot - \sqrt{(\cdot - i)(\cdot + i)}$ of $I_0 = i[-1, 1]$ as $\delta \downarrow 0$. Let

$$V(f_{I_\delta}) = \sup \frac{1}{2\pi} \left| \int_{\partial I_\delta} f_{I_\delta} \, g \, dz \right|,$$

where the supremum is taken over all continuous functions g on I_δ such that $|g(z)| \leqslant 1$ ($z \in I_\delta$), and ∂I_δ denotes the boundary of I_δ^c having two sides. The quantity $V(f_{I_\delta})$ shows the total variation of a measure μ_δ/π such that $f_{I_\delta} = \mathcal{C}\mu_\delta$. Since $f_{I_\delta}(\delta + i)$, $f_{I_\delta}(-\delta + i)$ converge to $f_{I_0}(i) = i$ as $\delta \downarrow 0$, we have $\lim_{\delta \downarrow 0} \max|f_{I_\delta}(z) - i| = 0$, where the maximum is taken over all points z on the subarc of ∂I_δ with endpoints $\pm\delta + i$ containing $+0i$ ($= \lim_{\eta \downarrow 0} \eta i \in \partial I_\delta$). Since I_δ is symmetric with respect to the real line, a simple calculation shows that

$$\lim_{\delta \downarrow 0} V(f_{I_\delta}) = \frac{1}{2\pi} \int_{\substack{z>0, \\ z \in \partial I_0}} |f_{I_0}(z) - i| \, |dz| + \frac{1}{2\pi} \int_{\substack{z<0, \\ z \in I_0}} |f_{I_0}(z) + i| \, |dz| = \frac{4\sqrt{2}}{3\pi}.$$

Analogously,

$$\lim_{\delta \downarrow 0} V(f_{J_\delta}) = \frac{1}{2\pi} \int_{\partial I_0} |f_{I_0}(z) - i| \, |dz| = \frac{8}{3\pi} \quad \left(> \frac{4\sqrt{2}}{3\pi} \right).$$

The slits $[-\delta, \delta]$ and $[-\delta, \delta] + i$ are small, however, this shows that even a small slit is not negligible in the computation of $V(\cdot)$ for the Ahlfors functions.

Modifying I_δ and J_δ, we can construct the required perfect sets in Proposition 4. In the construction, it is necessary to remove a countable union of small segments from a given segment, and a delicate estimation of the Ahlfors functions is required in this step. Take, for example, $K_\delta = K \cup \{I_0 - i(-\delta, \delta)\}$ $(0 \leqslant \delta < 1)$, where K is a continuum disjoint from I_0. Then f_{K_δ} locally uniformly converges to f_{K_0} as $\delta \downarrow 0$, $f_{K_0}(\Omega_{K_0})$ covers the open unit disk twice and $f_{K_\delta}(\Omega_{K_\delta})$ $(\delta \neq 0)$ covers the open unit disk three times. This shows that the convergence $f_{K_0} = \lim_{\delta \downarrow 0} f_{K_\delta}$ is complicated near the boundary. Let Ω be a domain containing ∞ bounded by a finite number of analytic Jordan curves $\{C_k\}_{k=1}^{n}$. Let $g(\cdot, \zeta)$ denote the Green's function of Ω with a pole at ζ and let u_k denote the harmonic function in Ω such that $u_k = 1$ on C_k and $u_k = 0$ on $\partial\Omega - C_k$. Ahlfors [1] shows that

$$\log\left|f_{\Omega^c}(z)\right| = \sum_{j=1}^{n} g(z, \zeta_j), \quad \sum_{j=1}^{n} u_k(\zeta_j) = 1 \quad (1 \leqslant k \leqslant n),$$

where $\{\zeta_j\}_{j=1}^{n}$ are zeros of f_{Ω^c}. This property is applicable to construct the required perfect sets E_1 and E_2.

REFERENCES

1. Ahlfors L., *Bounded analytic functions*, Duke Math. J. **14** (1947), 1–11.
2. Garnett J., *Analytic Capacity and Measure*, Lecture Notes in Math., 297, Springer–Verlag, Berlin, 1972.
3. Murai T., *Analytic capacity for arcs*, ICM90 Kyoto, Springer–Verlag, Tokyo, 1991.
4. Murai T., *Analytic capacity (Theory of the Szegö kernel function)*, Sūgaku, 1991 (to appear).

DEPARTMENT OF MATHEMATICS
SCHOOL OF SCIENCE
NAGOYA UNIVERSITY
NAGOYA, 464-01
JAPAN

ON THE VARIATION OF ANALYTIC CAPACITY

Jaroslav Zemánek

The analytic capacity of a compact set E in the complex plane is defined by

$$\gamma(E) = \sup_f |f'(\infty)|,$$

where the supremum is taken over all functions f holomorphic outside E and vanishing at ∞, with absolute value not exceeding 1, and

$$f'(\infty) = \lim_{z \to \infty} z f(z)$$

is the coefficient at z^{-1} in the Laurent expansion of $f(z)$ at ∞.

In view of the prominent role of this concept in complex analysis (see, for instance, [1], [3], [6], [7], and the references given there), it would be interesting to determine the quantitative behaviour of $\gamma(E)$ as a function of E. To measure the deviation of compact sets, it is natural to use the Hausdorff distance

$$\chi(E, F) = \max\{\sup_{x \in E} \inf_{y \in F} |x - y|, \sup_{y \in F} \inf_{x \in E} : |x - y|\}.$$

The function γ is then discontinuous at any E with $\gamma(E) > 0$, because E can be approximated by finite nets that have zero analytic capacity. However, it follows from the very basic facts ([3, I.1.2 and I.1.5], see also [7, p. 16]) that the analytic capacity is upper semi-continuous in general. That is,

$$\lim_{n \to \infty} \chi(E_n, E) = 0 \text{ implies that } \limsup_{n \to \infty} \gamma(E_n) \leqslant \gamma(E).$$

We note that no inclusion between the sets is assumed here. In particular, the function γ is continuous at any E with $\gamma(E) = 0$.

PROBLEM 1. *Let E be a compact set with $\gamma(E) = 0$ and let $\varepsilon > 0$ be given. Do there exist positive constants α and β such that*

$$|\gamma(E) - \gamma(F)| \leqslant \alpha \chi(E, F)^\beta$$

for any compact set F with $\chi(E, F) < \varepsilon$? Is it possible to choose the constants α and β independently of E and/or ε?

Next, let us consider the restriction of the function $\sim$ to the connected sets. Then some other elegant characterizations of $\gamma(E)$ are available (see, for instance [5, Theorem III.26] and [2, Theorem VII.3.3]), and the continuity of γ on this class is a consequence of [4, Theorem 1]. Given a compact connected set E, we define, for each $t > 0$,

$$\varphi_E(t) = \sup_F \{|\gamma(E) - \gamma(F)| : \chi(E, F) \leqslant t\},$$

where F varies over compact connected sets. This is a finite number by [3, I.1.2], and we have

$$\lim_{t \to 0_+} \varphi_E(t) = 0$$

by [4, Theorem 1].

PROBLEM 2. *Describe the behaviour of the function $\varphi_E(t)$. Is it possible to obtain estimates similar to those in Problem 1?*

REFERENCES

1. Gamelin T. W., *Uniform algebras*, Prentice-Hall, 1969.
2. Goluzin G. M., *Geometric theory of functions of a complex variable*, Amer. Math. Soc., 1969.
3. Mel'nikov M. S., Sinanjan S. O., *Questions in the theory of approximation of functions of a one complex variable*, Contemporary Problems of Mathematics, vol. 4, Itogi Nauki i Tekhniki, VINITI, Moscow, 1975, pp. 143–250 (Russian); English transl. in vol. 5, 1976, pp. 688–752.
4. Shirokov N. A., *On a property of analytic capacity*, Vestnik Leningrad Univ., Mat. **1** (1971), 75–82. (Russian)
5. Tsuji M., *Potential theory in modern function theory*, Maruzen, 1959.
6. Vitushkin A. G., *The analytic capacity of sets in problems of approximation theory*, Uspekhi Mat. Nauk **22** (1967), 141–199 (Russian); English transl. in Russian Math. Surveys **22** (1967), 139–200.
7. Zalcman L., *Analytic capacity and rational approximation*, Lecture Notes in Math., vol. 50, Springer - Verlag, 1968.

INSTITUTE OF MATHEMATICS
POLISH ACADEMY OF SCIENCES
00-950 WARSZAWA, P.O. BOX 137
POLAND

ADDED IN PROOF

Professor Christian Pommerenke observed that a local estimate with the exponent $\beta = 1/2$ can be obtained in Problem 2, by using Theorems 2 and 7 of his paper it Über die Kapazität der Summe von Kontinuen, Math. Ann. **139** (1959), 127–132.

ÜBER DIE REGULARITÄT EINES RANDPUNKTES
FÜR ELLIPTISCHE DIFFERENTIALGLEICHUNGEN

V. G. Maz'ya

In den letzten Jahren wurde dem Kreis von Fragen die um das klassische Kriterium von Wiener über die Regularität eines Rundpunktes in Bezug auf harmonische Funtionen gruppiert sind, viel Aufmerksamkeit geschenkt [1,2]. Nach dem Satz von Wiener ist die Stetigkeit im Punkt 0, $0 \in \partial\Omega$, der Lösung des Dirichletproblems für die Laplace-Gleichung im n-dimensionalen Gebiet Ω ($n > 2$) unter der Bedingung, daß auf $\partial\Omega$ eine in 0 stetige Funktion gegeben ist, äquivalent zur Divergenz der Reihe

$$\sum_{k \geqslant 1} 2^{(n-2)k} \operatorname{cap}(C_{2^{-k}} \setminus \Omega) \, .$$

Hierbei ist $C_\rho = \{x : x \in \mathbb{R}^n, \ \rho/2 \leqslant |x| \leqslant \rho\}$ und cap K die harmonische Kapazität der kompakten Menge K.

Diese Behauptung wurde (manchmal nur der Teil der Hinlänglichkeit) auf verschiedene Klassen von linearen und quasilinearen Gleichungen zweiter Ordnung ausgedehnt (eine Charakterisierung dieser Untersuchungen und eine Bibliographie kann man im Buch [3] finden). Was die Gleichungen höherer als zweiter Ordnung betrifft, so gab es für sie bis zur letzten Zeit keine Resultate, die analog zum Satz von Wiener sind. In der Arbeit [4] des Autors wird das Verhalten der Lösung des Dirichletproblems für die Gleichung $\Delta^2 u = f$ mit homogenen Randbedingungen, wobei $f \in C_0^\infty(\Omega)$ ist, in der Umgebung eines Randpunktes untersucht. In [4] wird gezeigt, daß für $n = 5, 6, 7$ die Bedingung

$$(1) \qquad \sum_{k \geqslant 1} 2^{k(n-4)} \operatorname{cap}_2(C_{2^{-k}} \setminus \Omega) = \infty \, ,$$

wobei cap_2 die sogenannte biharmonische Kapazität ist, die Stetigkeit der Lösung im Punkt 0 garantiert. Für $n = 2, 3$ folgt die Stetigkeit der Lösung aus dem Einbettungssatz von S. L. Sobolev, aber im Fall $n = 4$, der ebenfals in [4] analysiert wird, hat die Bedingung für die Stetigkeit eine andere Gestalt.

HYPOTHESE 1. *Die Bedingung $n < 8$ ist nicht wesentlich.*

Dem Author ist nur ein Argument für diese Annahme bekannt. Für alle n ist die Lösung der betrachteten Aufgabe für einen beliebigen Kugelsektor im Eckpunkt stetig. Die Einschränkung $n < \infty$ tritt nur bei einem der Lemmata auf, auf denen die Beweise in [4] beruhen. Sie ist aber notwendig für diese Lemma. Es geht hierbei um die Eigenschaft des Operators Δ^2, positiv mit dem Gewicht $|x|^{4-n}$ zu sein. Diese Eigenschaft erlaubt es, für $n = 5, 6, 7$ die folgende Abschätzung der Greenschen Funktion des biharmonischen Operators in einem beliebigen Gebiet anzugeben:

$$|G(x, y)| \leqslant c(n)|x - y|^{4-n}$$

wobei $x, y \in \Omega$ und $c(n)$ eine nur von n abhängige Konstante ist.

HYPOTHESE 2. *Die Abschätzung (2) gilt auch für $n \geqslant 8$.*

Es versteht sich, daß man analoge Fragen auch für allgemeinere Gleichungen stellen kann. Ich möchte die Aufmerksamkeit des Lesers aber auf eine Aufgabe lenken, die auch für den Laplace-Operator nicht gelöst ist. Nach [5], [6] genügt eine harmonische Funktion, deren verallgemeinerte Randwerte einer Hölder-Bedingung im Punkt 0 genügen, derselben Bedingung in diesem Punkt, falls

$$(3) \qquad \varliminf_{N \to \infty} N^{-1} \sum_{N \geqslant k \geqslant 1} 2^{k(n-2)} \operatorname{cap}(C_{2^{-k}} \backslash \Omega) > 0 \,.$$

Es wäre interessant, folgende Annahme zu rechtfertigen oder zu widerlegen.

HYPOTHESE 3. *Die Bedingung (3) ist notwendig.*

Wir wenden uns zum Schluß nichtlinearen elliptischen Gleichungen zweiter Ordnung zu. Wie in [7] gezeigt wurde, ist der Punkt 0 regulär für die Gleichung

$$\operatorname{div}\left(|\operatorname{grad} u|^{p-2} \operatorname{grad} u\right) = 0, \qquad 1 < p < n,$$

falls

$$(4) \qquad \sum_{k \geqslant 1} \left[2^{k(n-p)} p\text{-cap}(C \backslash \Omega) \right]^{1/(p-1)} = \infty$$

ist, wobei $p\text{-cap}(K) = \inf \left\{ \| \operatorname{grad} u \|_{L^p(\mathbb{R}^n)}^p : u \in C_0^\infty(\mathbb{R}^n), \ u \geqslant 1 \text{ in } K \right\}$ ist. Diese Resultat wurde unlängst in der Arbeit [8] auf die sehr allgemeine Klasse von Gleichungen $\operatorname{div} \vec{A}(x, u, \operatorname{grad} u) = B(x, u, \operatorname{grad} u)$ übertragen. Da die Bedingung (4) für $p = 2$ mit dem Kriterium von Wiener zusammenfällt, ist es natürlich, folgende Hypothese aufzustellen.

HYPOTHESE 4. *Die Bedingung (4) ist notwendig.*

In [9] wurden Beispiele behandelt, die zeigen, daß die Bedingung (4) in einem gewissen Sinne genau ist. Für die Hypothese 4 sprechen auch neuere Ergebnisse über die Stetigkeit nichtlinearer Potentiale [10], [11].

KOMMENTAR DES VERFASSERS (1984)

"Anscheinend ist noch keine der formulierten Aufgaben gelöst" — mit diesen Worten hatte der Autor vor, die Erörterung des obenangeführten Textes zu beginnen. Aber als der Kommentar fast fertig war, hörte dieser Satz auf, wahr zu sein. Es tauchte folgendes **Gegenbeispiel zur Hypothese 3** auf.

Es sei $\mathcal{B}_\nu = \{ x \in \mathbb{R}^n : |x| < \rho_\nu \}$, $n > 2$, $\log_2 \log_2 \rho_\nu^{-1} = \nu$ und Ω die Vereinigung der Kugelschichten $\mathcal{B}_\nu \backslash \bar{\mathcal{B}}_{\nu+1}$, $\nu = 0, 1, \ldots$, die durch Öffnungen ω_ν in den Sphären $\partial \mathcal{B}_\nu$, $\nu \geqslant 1$, verbunden sind Die Öffnung ω_ν stellt eine geodätische Kugel mit beliebigem Mittelpunkt und dem Radius $2^{-1/\rho_\nu}$ dar.

Es ist klar, daß die Kapazität der Menge $C_{2-k}\backslash\Omega$ nur für $k = 2^\nu$, $2^{\nu-1}$ von Null verschieden ist und daß für diese k die Ungleichung $\mathrm{cap}(C_{2-k}\backslash\Omega) \geqslant c2^{-k(n-2)}$ gilt. Deshalb ist für $N > 1$,

$$c_1 \log_2 N \leqslant \sum_{n \geqslant k \geqslant 1} 2^{k(n-2)} \, \mathrm{cap}(C_{2-k}\backslash\Omega) \leqslant C_2 \log_2 N \,.$$

Folglich divergiert die Reihe von Wiener für das betrachtete Gebiet, aber die Bedingung (3) ist nicht erfüllt.

Wir zeigen, daß trotzdem eine beliebige in Ω harmonische Funktion, deren Randwerte der Hölder-Bedingung im Punkt 0 genügen, ebenfalls der Hölder-Bedingung im Punkt 0 genügt.

Es sei u Lösung des Dirichletproblems $\Delta u = 0$ in Ω, $u - \varphi = 0$ auf $\partial\Omega$, wobei φ eine stetige Funktion ist, die der Bedingung $\varphi(x) = O(|x|^\beta)$, $\beta > 0$, genügt. Man kann annehmen, daß $1 \geqslant \varphi(x) \geqslant 0$ ist. Wir bezeichnen mit x einen beliebigen Punkt des Gebiets Ω und mit ν eine Zahl, für die $\rho_{\nu-1} \geqslant |x| \geqslant \rho_\nu$ ist. Es sei $\varphi_\nu = 0$ außerhalb von $\mathrm{clos}(\mathcal{B}_{\nu-1})$ und $\varphi_\nu = \varphi$ in $\mathcal{B}_{\nu-1}$. Ferner sei u_ν eine harmonische Funktion in Ω, die auf $\partial\Omega$ mit φ_ν übereinstimmt. Wegen $0 \leqslant \varphi_\nu \leqslant c\rho_{\nu-1}^\beta$ auf $\partial\Omega$, ist $0 \leqslant u_\nu(x) \leqslant c\rho_{\nu-1}^\beta \leqslant c|x|^{\beta/2}$. Wir führen eine in der Kugel $\mathcal{B}_{\nu-2}$ harmonische Funktion h_ν ein, die auf $\partial\mathcal{B}_{\nu-2}$ gleich $u - u_\nu$ ist. Offenbar ist $0 \leqslant h_\nu \leqslant 1$ auf $\omega_{\nu-2}$ und $0 \leqslant h_\nu \leqslant c\rho_{\nu-2}^\beta$ auf $\partial\mathcal{B}_{\nu-2}\backslash\omega_{\nu-2}$. Stellt man h_ν in der Gestalt eines Poisson-Integrals dar, so erhält man hieraus die Ungleichung $0 \leqslant h_\nu \leqslant c\left(\rho_{\nu-2}^\beta + 2^{-(n-1)/\rho_{\nu-2}}\rho_{\nu-2}^{1-n}\right)$ in $\mathcal{B}_{\nu-1}$. Da nach dem Maximumprinzip $0 \leqslant u - u_\nu \leqslant h_\nu$ ist, erhält man $0 \leqslant u(x) - u_\nu(x) \leqslant c|x|^{\beta/4}$. Folglich genügt die Funktion u im Punkt 0 der Hölder-Bedingung mit dem Exponenten $\beta/4$.

Modifiziert man das konstruierte Beispiel, so kann man leicht zeigen, daß man auf Grund der Geschwindigkeit des Wachstums der Partialsummen der Reihe von Wiener keine unteren Abschätzungen einer harmonischen Funktion mit Null-Randbedingungen in der Nähe des Punktes 0 machen kann. Indem man nämlich die Art und Weise der Konvergenz der Radien ρ_ν und der Durchmesser der Öffnungen ω_ν gegen Null vorschreibt, kann man eine beliebig schnelle Konvergenz der Funktion u gegen Null im Punkt 0 bei beliebig langsamer Divergenz der Reihe von Wiener erreichen. Somit ist die Hypothese 3 widerlegt, aber desto interessanter bleibt die Frage nach den Rand charakterisierenden notwendigen und hinreichenden Bedingungen für die Hölder-Stetigkeit einer beliebigen harmonischen Funktion mit Hölder-stetigen Randwerten.

Was die anderen Fragen betrifft, die vor fünf Jahren gestellt wurden, so gibt es auf sie bisher noch keine Antwort*. In der letzten Zeit wurden mit ihnen verbundene neue Informationen gewonnen, die, wenn man es richtig betrachtet, nicht so sehr in die Tiefe wie in die Breite gehen.

Im Zusammenhang mit den Hypothesen 1 und 2 erwähnen wir die Arbeit [12], in der die grundlegenden Ergebnisse des Artikels [4] (die ausführlich in [13] dargestellt sind) auf die erste Randwertaufgabe für die polyharmonische Gleichung $(-\Delta)^m u = f$ übertragen wurden. Leider verlangte auch hier die Methode, die auf der Eigenschaft des

---

*Anmerkung bei der Korrektur: I. W. Skrypnik teilte soeben auf der Tagung "Nichtlineare Probleme der Mathematischen Physik" (13 April, LOMI, Leningrad) mit, daß er die Notwendigkeit der Bedingung (4) für $p \geqslant 2$ bewiesen hat. Damit ist die Hypothese 3 teilweise gestützt.

Operators $(-\Delta)^m$, $m > 2$, beruht, positive mit dem Gewicht $\Gamma(x-y)$ zu sein, wobei Γ die Fundamentallösung ist, alle Dimensionen mit Ausnahme der folgenden drei zu opfern: $n = 2m, 2m + 1, 2m + 2$. Als fragwürdigen Ausgleich gestattet uns dies, die folgenden beiden Hypothesen zu formulieren, die sich an die Hypothesen 1 and 2 anschließen.

HYPOTHESE 1′. *Für $m > 2$, $n \geqslant 2m + 3$ ist die Gleichung*

$$\sum_{k \geqslant 1} 2^{k(n-2m)} \operatorname{cap}_m(C_{2-k} \backslash \Omega) = \infty \,,$$

wobei cap_m die m-harmonische Kapazität ist, hinreichend für die Stetigkeit der Lösung des Dirichletproblems mit Nullrandbedingungen der Gleichung $(-\Delta)^m u = f$, $f \in C_0^\infty(\Omega)$ im Punkt 0.

HYPOTHESE 2′. *Für $m > 2$, $n \geqslant 2m + 3$ gilt für die Greensche Funktion G_m des Operators $(-\Delta)^m$ die Abschätzung*

$$|G_m(x,y)| \leqslant c|x - y|^{2m-n} \,,$$

wobei die Konstante c von n und m, aber nicht vom Gebiet abhängt.

In der letzten Zeit wurden neue Erkenntnisse über das Verhalten der Lösung der ersten Randwertaufgabe für stark elliptische Gleichungen der Ordnung $2m$ in der Nähe konischer Punkte erhalten. Im allgemeinen (s. [14]) haben die Hauptglieder der Asymptotik solcher Lösungen in der Umgebung des Eckpunktes des Konus die Gestalt

$$(5) \qquad c|x|^\lambda \sum_{k=0}^n (\log |x|)^k \, \varphi_k(x'/|x|) \,.$$

Dabei is λ Eigenwert des Dirichletproblems für einen gewissen polynomial vom Spektralparameter abhängigen elliptischen Operator in dem Gebiet, das durch den Konus auf der Einheitssphäre ausgeschnitten wird. Die Funktion (5) hat genau dann ein endliches Dirichlet-Integral, wenn $\Re\lambda > m - n/2$. Sie ist des weiteren stetig und genügt sogar einer Hölder-Bedingung, falls $\Re\lambda > 0$ ist. Wenn im Band $0 > \Re\lambda > m - n/2$ Eigenwerte des genannten Operators existieren, dann besitzt die Ausgangsrandwertaufgabe verallgemeinert Lösungen, die in einer beliebigen Umgebung des Eckpunktes des Konus unbeschränkt sind, und von einer Regularität nach Wiener kann man selbst bei einem konischen Punkt nicht reden. Es zeigt sich ([15], [16]), daß wir auf solche unerwarteten Erscheinungen schon bei stark elliptischen Gleichungen zweiter Ordnung mit konstanten Koeffizienten

$$(6) \qquad P_2(D_x)u \equiv \sum_{i,j=1}^n a_{ij} \frac{\partial^2 u}{\partial x_i \partial x_j} = 0, \quad n > 3,$$

stoßen, falls nicht alle Koeffizienten reell sind. In [16] (eine ausfürliche Darstellung erscheint evtl. in Matematicheskii Sbornik) wird das homogene Dirichletproblem außerhalb eines dünnen Konus $K_\varepsilon = \{x = (y, x_n) \in \mathbb{R}^n : x_n > 0, \ yx_n^{-1} \in \omega_\varepsilon\}$ untersucht, wobei ε ein kleiner Parameter, $\omega_\varepsilon = \{y \in \mathbb{R}^{n-1} : y\varepsilon^{-1} \in \omega\}$ and ω ein Gebiet im $\mathbb{R}^{n-1}$ ist.

Es wird die Lösung $u(x) = |x|^{\lambda(\varepsilon)}\varphi(\varepsilon, x/|x|)$ des stark elliptischen Systems $P_{2m}(D_x)u(\varepsilon, x) = 0$ betrachtet, wobei $P_{2m}(\xi)$ eine Matrix mit homogenen Polynomen der Ordnung $2m$ als Elementen und $\lambda(\varepsilon) = o(1)$ für $\varepsilon \to +0$ ist. Hauptergebnis ist eine asymptotische Formel für den Eigenwert $\lambda(\varepsilon)$, welche für den einfachsten Fall der Gleichung (6) die Gestalt

$$\lambda(\varepsilon) = \varepsilon^{n-3}\left\{\frac{n-3}{n-2}\frac{|S^{n-2}|}{|S^{n-1}|}\operatorname{cap}_{P_2(D_y,0)}(\omega)\frac{\left(\det|\bar{a}_{jk}\|_{j,k=1}^{n}\right)^{(n-3)/2}}{\left(\det|\bar{a}_{jk}\|_{j,k=1}^{n-1}\right)^{(n-2)/2}} + o(1)\right\}$$

hat. Hierbei ist $|S^k|$ die Oberfläche der $(k+1)$-dimensionalen Einheitskugel und $\operatorname{cap}_{P_2(D_y,0)}$ eine komplexwertige Funktion des Gebiets ω, welche eine Verallgemeinerung der harmonischen Kapazität darstellt:

$$\operatorname{cap}_{P_2(D_y,0)}(\omega) = \frac{1}{(n-3)|S^{n-2}|}\int_{\mathbb{R}^{n-1}\setminus\omega}\sum_{j,k=1}^{r-1}a_{jk}\frac{\partial w}{\partial y_k}\frac{\partial \bar{w}}{\partial y_j}\,dy,$$

wobei w eine im Unendlichen verschwindende Lösung der Gleichung $P_2(D_y, 0)w(y) = 0$ in $\mathbb{R}^{n-1}\setminus\omega$ ist, die auf $\partial\omega$ gleich 0 ist Nach [16] kann man die Koeffizienten d_{jk} so wählen, daß die Ungleichung $0 > \Re\lambda > (2-n)/2$ erfüllt ist. Im Fall $n = 3$ gilt

$$\lambda(\varepsilon) = (2|\log\varepsilon|)^{-1}(1 + o(1)) \quad \text{für } \varepsilon \to +0.$$

Folglich erfüllt jede verallgemeinerte Lösung die Hölder-Bedingung, falls der Öffnungswinkel des Konus K_ε genügend klein ist. Es ist nicht ausgeschlossen, daß die Forderung nach einem kleinen Öffnungswinkel unwesentlich ist. Dies ist gleichbedeutend mit folgendem Satz.

HYPOTHESE 5. *Für $n = 3$ ist ein konischer Punkt für einen beliebigen elliptischen Operator $P_2(D_x)$ mit komplexen Koeffizienten regulär nach Wiener.*

Für den biharmonischen Operator im $\mathbb{R}^n$ und für die Systeme von Lamé und Stokes im $\mathbb{R}^3$ wurden derartige Ergebnisse in [17], [18] erhalten.

REFERENCES

1. Wiener N., *The Dirichlet problem*, J. Math. and Phys. **3** (1924), 127–146.
2. Wiener N., *Certain notions in potential theory*, J. Math. and Phys. **3** (1924), 24–51.
3. Landis E. M., *Second-order equations of elliptic and parabolic type*, Nauka, M., 1971. (Russian)
4. Maz'ja V. G., *On the behavior near the boundary of solutions of the Dirichlet problem for the biharmonic operator*, Dokl. Akad. Nauk SSSR **18** (1977), no 4, 15–19 (Russian); English transl. in Soviet Math. Dokl. **18** (1977), 1152–1155.
5. Maz'ja V. G., *Regularity at the boundary of solutions of elliptic equations and conformal mapping*, Dokl. Akad. Nauk SSSR **152** (1963), no. 6, 1297–1300 (Russian); English transl. in Soviet Math. Dokl. **152** (1963), 1547–1551.
6. Maz'ja V. G., *Behavior near the boundary of solutions of the Dirichlet problem for a second-order elliptic equation in divergent form*, Mat. Zametki **2** (1967), 209–220 (Russian); English transl. in Math. Notes **2** (1967), 610–617.

7. Maz'ja V. G., *Continuity at a boundary point of solutions to quasi-linear elliptic equations*, Vestnik Leningrad Univ., Math. **25** (1970), 42–55; *Correction*, Vestnik Leningrad Univ., Math. **1** (1972), 160. (Russian)

8. Gariepy R., Ziemer W. P., *A regularity condition at the boundary for solutions of quasilinear elliptic equations*, Arch. Rat. Mech. Anal. **67** (1977), no. 1, 25–39.

9. Krol I. N., Maz'ja V. G., *On the absence of continuity and Hölder continuity of solutions of quasilinear elliptic equations near a nonregular boundary*, Trudy Moskovsk. Mat. Obshch. **26** (1972), 73–93 (Russian); English transl. in Trans. Moscow Math. Soc.

10. Hedberg L., *Non-linear potentials and approximation in the mean by analytic functions*, Math. Z. **129** (1972), 299-319.

11. Adams D. R., Meyers N., *Thinness and Wiener criteria for non-linear potentials*, Indiana Univ. Math. J. **22** (1972), 169–197.

12. Maz'ja V. G., Donchev T., *On Wiener regularity at a boundary point for a polyharmonic operator*, C.R. Acad. Bulgare Sci. **36** (1983), no. 2, 177–179 (Russian); English transl. in Amer. Math. Soc. Translations, ser. 2 **137** (1987), 53–55.

13. Maz'ja V. G., *Behaviour of solutions to the Dirichlet problem for the biharmonic operator at the boundary point*, Equadiff IV, Lect. Notes Math. **703** (1979), 250–262.

14. Kondrat'ev V. A., *Boundary problems for elliptic equations in domains with conical or angular points*, Trudy Moscow Mat. Obsc. **16** (1967), 209–292 (Russian); English transl. in Trans. Moscow Math. Soc. **16** (1967), 227–313.

15. Maz'ja V. G., Nazarov S. A., Plamenevskii B. A., *Absence of De Georgi-type theorems for strongly elliptic equations*, Zapiski Nauch. Sem. LOMI **115** (1982), 156–168 (Russian); English transl. in J. Soviet Math. **28** (1985), 726–734.

16. Maz'ja V. G., Nazarov S. A., Plamenevskii B. A., *On homogeneous solutions of the Dirichlet problem in the exterior of a slender cone*, Dokl. Akad. Nauk SSSR **266** (1982), no. 2, 281–284 (Russian); English transl. in Soviet Math. Dokl. **26** (1982), 320-323.

17. Maz'ja V. G., Plamenevskii B. A., *The maximum principle for the biharmonic equations in a region with conical points*, Izv. Vyssh. Uchebn. Zaved. Mat. **25** (1981), no. 2, 52–59 (Russian); English transl. in Soviet Math. (Iz. VUZ) **25** (1981), 61–70.

18. Maz'ja V. G., Plamenevskii B. A., *99–120*, Dinamika Sploshnoi Sredy, Vyp. 50, Novosibirsk, 1981 (Russian); English transl. in Amer. Math. Soc. Translations, ser. 2 vol. 123, 1984, pp. 109–123.

DEPARTMENT OF MATHEMATICS
LINKÖPING UNIVERSITY
S-58183 LINKÖPING
SWEDEN

THE EXCEPTIONAL SETS
ASSOCIATED WITH THE BESOV SPACES

DAVID R. ADAMS

For α real and $0 < p, q < \infty$, we will use Stein's notation Λ_α^{pq} for the familiar Besov spaces of distributions on $\mathbb{R}^n$; see [8] and [9] for details. The purpose of this note is to generally survey and point out open questions concerning the general problem of determining all the inclusion relations between the classes $\mathcal{E}_{\alpha,p,q}$, $\alpha > 0$, of exceptional sets naturally associated with the spaces Λ_α^{pq} for various choices of the parameters α, p, q; c.f. [4]. These exceptional sets can be described as sets of Besov capacity zero. Let $A_{\alpha,p,q}(K) = \inf\{|u|_{\alpha,p,q}^p : u \in \mathcal{D} \text{ and } u \geqslant 1 \text{ on } K\}$, $\mathcal{D}$ some fixed smooth dense class in the spaces Λ_α^{pq}, $|u|_{\alpha,p,q}$ the norm (quasi-norm) of u in Λ_α^{pq}, K compact. $A_{\alpha,p,q}$ is extended to all subsets of $\mathbb{R}^n$ as an outer capacity. Then $E \in \mathcal{E}_{\alpha,p,q}$ iff $A_{\alpha,p,q}(E) = 0$. Thus OUR PROBLEM is: given an arbitrary compact set K such that $A_{\alpha,p,q}(K) = 0$, for which β, r, s does it follow that $A_{\beta,r,s}(K) = 0$? When this holds, we will write $A_{\beta,r,s} \ll A_{\alpha,p,q}$. The symbol $\approx$ will mean that both directions, $\ll$ and $\gg$ hold.

Now when $1 \leqslant p, q < \infty$, there is quite a bit that can be said about this problem. First of all, one can restrict attention to $1 \leqslant p \leqslant n/\alpha$. Functions in Λ_α^{pq} for $p > n/\alpha$ are all equivalent to continuous functions and hence $A_{\alpha,p,q}(E) > 0$ iff $E \neq \emptyset$. Continuity also occurs, for example, when $p = n/\alpha$ and $q = 1$. Secondly, in the range $1 < p \leqslant n/\alpha$, $1 \leqslant q < \infty$, there appear to be presently four methods for obtaining inclusion relations. They are:

1. If $\Lambda_\alpha^{pq} \subset \Lambda_\beta^{rs}$ (continuous embedding), then clearly $A_{\beta,r,s} \ll A_{\alpha,p,q}$ Such embeddings occur, but not very often. However, since $\mathcal{L}_{\alpha+\varepsilon}^p \subset \Lambda_\alpha^{pq} \subset \mathcal{L}_{\alpha-\varepsilon}^p$, $\varepsilon > 0$, with $\mathcal{L}_\alpha^p$ denoting the usual class of Bessel potentials of L^p functions on $\mathbb{R}^n$ (see [9]), and since the inclusion relations for the exceptional sets associated with the Bessel potentials are all known [3], it is easy to see that $A_{\beta,r,s} \ll A_{\alpha,p,q}$ when $\beta r < \alpha p$ (no additional restrictions on s and q) and that the reverse implication is false.

2. Using the min-max theorem, it is possible to give a dual formulation of the Besov capacities: $A_{\alpha,p,q}(K)^{1/p} = \sup\{\langle \mu, 1\rangle : \mu \in (\Lambda_{-\alpha}^{p'q'})^+, \text{ supp}\,\mu \subset K \text{ and } |\mu|_{-\alpha,p',q'} \leqslant 1\}$, $\frac{1}{p'} = 1 - \frac{1}{p}$. Thus it suffices to prove inclusion relations between the positive cones in the dual spaces. This method is facilitated by the characterization: $\mu \in (\Lambda_{-\alpha}^{p'q'})^+$ iff

$$\int_0^1 \left\{ \left[r^{\alpha p - n} \mu(B(x,r)) \right]^{q'/p'} \right\} \frac{dr}{r} < \infty$$

for $p, q > 1$, and

$$\sup_{0 < r \leqslant 1} \int \left[r^{\alpha p - n} \mu(B(x,r)) \right] d\mu(x) < \infty$$

for $p > 1$, $q = 1$.

3. The reason that one cannot expect all inclusion relations to follow from the first two methods is the simple fact that the capacitary extremals (in the primal and dual problems) generally have additional regularity. One can take advantage of this by comparing the Besov capacities $A_{\alpha,p,p}$ to the Bessel capacities $B_{\alpha,p}$, i.e., the capacities associated with $\mathcal{L}_\alpha^p$. Recently, P. Nilsson observed that the positive cones in $\Lambda_{-\alpha}^{p'p'}$ and $\mathcal{L}_{-\alpha}^{p'}$ coincide. Hence $A_{\alpha,p,p} \approx B_{\alpha,p}$. And again since all the inclusion (relies on the) relations for the Bessel exceptional sets are known (a result that relies on the regularity of the Bessel extremals), it follows that the corresponding relations carry over to $A_{\alpha,p,p}$. (The equivalence of $A_{\alpha,p,p}$ and $B_{\alpha,p}$ for all $p > 1$ has been known since T. Wolff's recent proof of the Kellogg property in non-linear potential theory: see [5], also [1]).

4. It is possible to apply the method of smooth truncation to the class of Bessel potentials of non-negative functions that belong to the mixed norm space $L^{pq}(\mathbb{R}^n \times \mathbb{R}^n)$ to obtain still further inclusion relations. This is due to the fact that the Besov capacities can be viewed as restrictions of such mixed norm Bessel capacities to subsets of $\mathbb{R}^n$. We refer the reader to [2].

In addition to the above relations, it is also possible to show that $A_{\alpha,1,1} \approx H^{n-\alpha} =$ Hausdorff $(n - \alpha)$–dimensional measure $(0 < \alpha < n)$.

We summarize the results of 1 through 4 in the following diagrams; the cross indicates $A_{\alpha,p,q}(K) = 0$ and the shaded region the pairs $\left(\frac{1}{r}, \frac{1}{s}\right)$ for which $A_{\beta,r,s}(K) = 0$, $\beta r = \alpha\rho$, as a consequence.

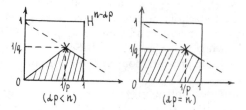

QUESTIONS. *That $H^{n-\alpha p}(K) = 0$ implies $A_{\alpha,p,q}(K) = 0$ for $p \leqslant q$ is quite easy. Is this still true for $q < p$? Do these diagrams represent all inclusion relations? If so, how does one account for the difference in the cases $\alpha p < n$ and $\alpha p = n$?*

When $0 < p < 1$, $0 < q < \infty$, very little seems to be known. One obvious thing to try is to compare $A_{\alpha,p,p}$ with $B_{\alpha,p}$ — the latter is now defined using Bessel potentials of the real Hardy spaces H^p on $\mathbb{R}^n$, $0 < p < 1$. This seems to be a good idea in view of all the recent developments on the structure of Hardy spaces, especially the atomic decomposition. Indeed, it is just such an approach that leads to $H^{n-\alpha p} \ll B_{\alpha,p}$, $\alpha p < n$, $0 < p \leqslant 1$, and then via trace theorems [6] to $H^{n-\alpha p} \ll A_{\alpha,p,p}$. However, it is not presently known if $A_{\alpha,p,p} \approx B_{\alpha,p}$ holds for $0 < p < 1$ though it is probably true. One of the main difficulties now is that the obvious dual capacity is no longer equivalent to the primal one (and the min-max theorem does not apply since the spaces in question are no longer locally convex). This all does, however, suggest comparing the Besov capacities to yet another class of capacities, namely those naturally associated with the Lizorkin–Triebel spaces F_α^{pq}; see [8] and [7]. And since it is known that F_α^{p2} coincides with Bessel potentials of H^p ($0 < p < \infty$), the F-capacities are a natural extension of the Bessel capacities. Thus we might expect some rather interesting results here in view of the things discussed above. However, it should be noted that for fixed α and p, the F capacities agree with $B_{\alpha,p}$ whenever q satisfies $1 \leqslant p \leqslant q < 2$ or $2 \leqslant q \leqslant p$. Hence the F-diagram summarizing the inclusion relations for the F-exceptional sets will be considerably different than that for the Λ-exceptional sets $\mathcal{E}_{\alpha,p,q}$.

QUESTION. *What does it look like?*

REFERENCES

1. Adams D. R., *On the exceptional sets for spaces of potentials*, Pac. J. Math. **52** (1974), 1–5.
2. Adams D. R., *Lectures on L^p-potential theory*, Umeå Univ. Reports, 1981.
3. Adams D. R., Meyers N. G., *Bessel potentials: Inclusion relations among classes of exceptional sets*, Ind. U. Math. J. **221** (1973), 873–905.
4. Aronszajn N., Mulla F., Szeptycki P., *On spaces of potentials connected with L^p classes*, Ann. Inst. Fourier **13** (1962), 211–306.
5. Hedberg L. I., Wolff T., *Thin sets in nonlinear potential theory*, Ann. Inst. Fourier **33** (1983).
6. Jawerth B., *The trace of Sobolev and Besov spaces, $0 < p < 1$*, Studia Math. **62** (1978), 65–71.
7. Jawerth B., *Some observations on Besov and Lizorkin-Triebel spaces*, Math. Scand. **40** (1977), 94–104.
8. Peetre J., *New Thoughts on Besov Spaces*, Duke Univ. Press 1976.
9. Stein E., *Singular Integrals and Differentiability Properties of Functions*, Princeton U. Press, 1970.

DEPARTMENT OF MATHEMATICS
UNIVERSITY OF KENTUCKY
LEXINGTON KY 40506-0027
USA

COMMENTARY BY THE AUTHOR

My knowledge of the status of the questions in "The Exceptional Sets Associated with the Besov Space", as of 1986, was summarized in [10]. The basic question is to find all the so called "relations" between the capacities naturally associated with the Besov spaces $\Lambda_\alpha^{p,q}(\mathbb{R}^n)$ and the Triebel–Lizorkin spaces $F_\alpha^{p,q}(\mathbb{R}^n)$, for all values of $\alpha > 0$ and $0 < p, q < \infty$. Also, usually included in the classification scheme are various forms of the Hausdorff measure (capacity) H^h, for $h = h(t)$ a nondecreasing measure function of $t \geqslant 0$; $h(0) = 0$. Even in [10] several of the original questions remain open, especially with regard to the capacities for $\Lambda_\alpha^{p,q}$ and $\Lambda_\beta^{r,s}$ when $\alpha p = \beta r < n$ and $p/q' < r/s'$, $p/q < r/s$. Here $p, q, r, s > 1$ and $1/q' = 1 - 1/q$.

When $0 < p = q \leqslant 1$, the capacities for $F_\alpha^{p,q}$ are the same as those associated with α-Riesz potentials of functions (distributions) that belong to the Hardy spaces H^p. A conjecture here is that $\mathrm{cap}(\cdot; F_\alpha^{p,q}) \approx H^{n-\alpha p}$, the Hausdorff measure for the measure function $h(t) = t^{n-\alpha p}$, $\alpha p < n$. This has been verified at least for $n/(n + \alpha) < p \leqslant 1$, by J. Orobitg, [15]. This result is a consequence of the characterization of H^p given for example in [14], because this formulation allows truncation, i.e. one can restrict attention to, say, Riesz potentials of elements $h \in H^p$ for which $I_\alpha * h \geqslant 0$. For a treatment of the case $p = 1$, see [11].

Some recent progress has been made in the Besov case for $0 < p < 1$; see [16]. Other related results are contained in [12] and [13].

It might be remarked that it seems to be the case that if one is to find all the relations for these capacities, then one must deal directly with the extremal measures that arise, i.e. prove regularity results for them, as was done in [10]. It does not appear to be enough to look for all the embeddings between the various spaces involved. Also dealing with the case $0 < p < 1$ is made especially difficult since there is no longer a "dual" form of the capacity as there is in the case $p \geqslant 1$; the unit ball in L^p is no longer convex when $0 < p < 1$, or, to say it another way, the min-max theorem fails to apply when $0 < p < 1$; see [10]. In convexity theory one often refers to a resulting "duality gap" when such convexity assumptions fail to be satisfied.

REFERENCES

10. Adams D. R., *The classification problem for the capacities associated with the Besov and Triebel-Lizorkin spaces*, Approx. and Function Spaces, Banach Center Pub., vol. 22, PWN Polish Scientific Publishers, Warsaw, 1989, pp. 9–24.
11. Adams D. R., *A note on the Choquet integral with respect to Hausdorff capacity*, Function Spaces and Applications, Lecture Notes in Mathematics, 1302, Springer–Verlag, 1988, pp. 115–124.
12. Jawerth B., Perez C., Welland G., *The positive cone in Triebel–Lizorkin spaces and the relation among potential and maximal operators*, Contemporary Math. (M. Milman, ed.), Amer. Math. Soc., Providence, 1989.
13. Ahern P. Exceptional sets for holomorphic functions, Mich. Math. J. **35** (1988), 29–41.
14. Strichartz R. S., H^p Sobolev spaces, preprint.
15. Orobitg J., Personal communication, July 1988.
16. Netrusov Yu. V., *Metric estimates for the capacities of sets in Besov spaces*, Trudy Math. Inst. Steklov **190** (1989), 159–185 (Russian); English transl. in Proc. Steklov Inst. Math. **190** (1992), 167–192.

COMMENTARY BY YU. V. NETRUSOV

Some of the questions above are answered in [10], [16]–[18].

1) If $q < p$, then the implication $H^{n-\alpha p}(K) = 0 \implies A_{\alpha,p,q}(K) = 0$ does not hold (see the counterexample in [16]).

2) For $0 < \alpha p < n$, $1 \leqslant p,q < \infty$ the diagram preceding "Questions" describes all possible inclusions (see [16]). But for $\alpha p = n$, $1 \leqslant p,q < \infty$ it doesn't work. The right answer is given in [10], [16].

3) It is shown in [10], [17] that the capacities in $F_{p,q}^\alpha$ with different q's $(0 < q \leqslant \infty)$ are equivalent.

4) The case $\min(p,q) < 1$ was considered in [16], [18]. In particular, it was shown that if $\alpha p = \beta r < n$ and if one of the following conditions is satisfied:

   a) $p,r \leqslant 1$, $s,q \neq \infty$

   or

   b) $q,s < 1$,

then $A_{\alpha,p,q}(K) \asymp A_{\beta,r,s}(K)$, $p/r = q/s$.

REFERENCES

17. Netrusov Yu. V., *Singularities of functions from Besov and Lizorkin–Triebel spaces*, Trudy Mat. Instituta AN SSSR **187** (1989), 162–178 (Russian); English transl. in Proc. Steklov Inst. Math.
18. Netrusov Yu. V., *Estimates of capacities assosiated with Besov spaces*, Zapiski nauchn. semin. LOMI **201** (1992), 124–156 (Russian); English transl. in J. Soviet Math.

12.32
old

COMPLEX INTERPOLATION BETWEEN SOBOLEV SPACES

Peter W. Jones

Put $W^{k,p}(\mathbb{R}^n) = \{f : D^\alpha f \in L^p(\mathbb{R}^n),\ 0 \leqslant \alpha \leqslant k\}$, the usual Sobolev space. The space $W^{k,\infty}$ seems to be poorly understood. Problem 1.6 gives one example of this. Another example is furnished by considering the complex method of interpolation, $(\cdot, \cdot)_\theta$. Let $1 \leqslant p_0 < \infty$.

QUESTION. Is $\left(W^{k,p_0}(\mathbb{R}^n), W^{k,\infty}(\mathbb{R}^n)\right)_\theta = W^{k,p}(\mathbb{R}^n)$, $\frac{1}{p} = \frac{1-\theta}{p_0}$?

This is easy and true when $n = 1$. Using Wolff's theorem [1] it is easy to show that a positive answer for one value of p_0 is equivalent to a positive answer for all values of p_0. The question is also easy to answer if one replaces the L^∞ endpoint by a BMO endpoint. The corresponding problem for the real method of interpolation is solved in [2].

References

1. Wolff T., *A note on interpolation spaces*, Lecture Notes in Math., 908, Springer–Verlag, 1982.
2. De Vore R., Scherer K., *Interpolation of linear operators on Sobolev spaces*, Ann. Math. **109** (1979), 583–599.

DEPARTMENT OF MATHEMATICS
YALE UNIVERSITY
NEW HAVEN, CONN.
USA

S.12.33
v.old

SPECTRAL SYNTHESIS IN SOBOLEV SPACES

L. I. HEDBERG

Let X be a Banach space of functions (function classes) on $\mathbb{R}^d$. We have in mind the Sobolev spaces W_s^p, $1 \leqslant p < \infty$, $s \in \mathbb{Z}_+$ or the spaces obtained from Sobolev spaces by interpolation (Bessel potential spaces L_s^p, $s > 0$, and Besov spaces $B_s^{p,q}$, $s > 0$). Then the dual space X' is a space of distributions. We say that a closed set K in $\mathbb{R}^d$ admits X-*spectral synthesis* if every T in X' that has support in K can be approximated arbitrarily closely in X' by linear combinations of measures and derivatives of order $< s$ of measures with support in K.

PROBLEM. *Do all closed sets admit X-spectral synthesis for the above spaces?*

The Problem can also be given a dual formulation. If ν is a measure with support in K such that a partial derivative $D^k\nu$ belongs to X', then one can define $\int D^k f \, d\nu$ for all f in X. Then K admits X-spectral synthesis if every f such that $\int D^k f \, d\nu = 0$ for all such ν and all such multi-indices k can be approximated arbitrarily closely in X by test functions that vanish on some neighborhood of K.

The Problem is of course analogous to the famous spectral synthesis problem of Beurling, but in the case of W_s^2 this terminology was introduced by Fuglede. He also observed that the so-called fine Dirichlet problem in a domain $\mathcal{D}$ for an elliptic partial differential equation of order $2s$ always has a unique solution if and only if the complement of $\mathcal{D}$ admits W_s^2-spectral synthesis. See [1; IX, 5.1].

In the case of W_1^p the Problem appeared and was solved in the work of V. P. Havin [2] and T. Bagby [3] in connection with the problem of approximation in $L^{p'}$ by analytic functions. For W_1^2 the solution appears already in the work of Beurling and Deny [4]. In fact, in these spaces all closed sets have the spectral synthesis property. This result, which can be extended to L_s^p, $0 < s < 1$, depends mainly on the fact that these spaces are closed under truncations.

When $s > 1$ this is no longer true, and the Problem is more complicated. Using potential theoretic methods the author [5] has given sufficient conditions for sets to admit spectral synthesis in $W_s^p(\mathbb{R}^d)$, $s \in \mathbb{Z}_+$. These conditions are so weak that they are satisfied for all closed sets if $p > \max(d/2, 2 - 1/d)$, thus in particular if $p = 2$ and $d = 2$ or 3. There are also some still unpublished results for L_s^p and $B_s^{p,p}$ showing for example that sets that satisfy a cone condition have the spectral synthesis property.

Otherwise, for general spaces the author is only aware of the work of H. Triebel [6], where he proved, extending earlier results of Lions and Magenes, that the boundary of a C^∞ domain admits spectral synthesis for L_s^p and $B_s^{p,p}$.

REFERENCES

1. Schulze B.-W., Wildenhain G., *Methoden der Potentialtheorie für elliptische Differentialgleichungen beliebiger Ordnung*, Akademie–Verlag, Berlin, 1977.

2. Havin V. P., *Approximation in the mean by analytic functions*, Dokl. Akad. Nauk SSSR **178** (1968), 1025–1028 (Russian); English transl. in Soviet Math. Dokl. **9** (1968), 245–248.
3. Bagby T., *Quasi topologies and rational approximation*, J. Funct. Anal. **10** (1972), 259–268.
4. Beurling A., Deny J., *Dirichlet spaces*, Proc. Nat. Acad. Sci. **45** (1959), 208–215.
5. Hedberg L. I., *Two approximation problems in function spaces*, Ark. Mat. **16** (1978), 51–81.
6. Triebel H., *Boundary values for Sobolev-spaces with weights. Density of $D(\Omega)$*, Ann. Sc. Norm. Sup. Pisa **3** (1973), no. 27, 73–96.

DEPARTMENT OF MATHEMATICS
UNIVERSITY OF STOCKHOLM
BOX 6701
S-11385 STOCKHOLM, SWEDEN

COMMENTARY BY THE AUTHOR

For the Sobolev spaces W_s^p, $1 < p < \infty$, $s \in \mathbb{Z}_+$, the problem has been solved. In fact, all closed sets admit spectral synthesis for these spaces. See L. I. Hedberg [7], L. I. Hedberg and T. H. Wolff [8], and concerning the Dirichlet problem, also T. Kolsrud [9].

REFERENCES

7. Hedberg L. I., *Spectral synthesis in Sobolev spaces, and uniqueness of solutions of the Dirichlet problem*, Acta Math. **147** (1981), 237–264.
8. Hedberg L. I., Wolff T. H., *Thin sets in nonlinear potential theory*, Ann. Inst. Fourier (Grenoble) **33** (1983), no. 4, 161–187.
9. Kolsrud T., *A uniqueness theorem for higher order elliptic partial differential equations*, Math. Scand. **51** (1982), 323–332.

EDITORS' NOTE

1) The works [7] and [8] are of importance not only in connection with the Problem but in a much wider context representing an essential breakthrough in the general nonlinear potential theory.

2) When $d = 1$ some details concerning the problem of synthesis in $W_s^p(\mathbb{R})$, $W_s^p(\mathbb{T})$ are contained in the following papers:

Kahane J.-P., Séminaire N. Bourbaki (1966, Nov.).

Akutowicz E. G., C. R. Acad. Sci. **256** (1963), no. 25, 5268–5270.

———, Ann. Scient. École Norm. Sup. **82** (1965), no. 3, 297–325.

———, Ill. J. Math. **14** (1970), no. 2, 198–204.

Osadčii N. M., Ukrain. Mat. Zh. **26** (1974), 669–670 (Russian); English transl. in Ukranian Math. J. **26** (1974), 548–549.

3) If $X \subset W_s^p(\mathbb{R}^d)$, $p > n$, then the spectral synthesis holds; every ideal of X is divisorial, i.e., is the intersection of primary ideals. However, the identification of divisors generating closed ideals is a non-trivial task. This problem is the theme of articles by L. G. Hanin:

Hanin L. G., *A geometric classification of ideals in algebras of differentiable functions of two variables*, Studies in the Theory of Functions of Several Real Variables, Yaroslavl', YGU, 1982, pp. 122–144. (Russian)

———, *A geometric classification of ideals in algebras of differentiable functions*, Dokl. Akad. Nauk SSSR **254** (1980), 303–307 (Russian); English transl. in Soviet Math. Dokl. **22** (1980), 370–375.

COMMENTARY BY YU. V. NETRUSOV

The following theorem by Yu. V. Netrusov gives a positive solution of the Problem.

THEOREM. *Let $l > 0$, $1 \leqslant p < \infty$, $m \in \mathbb{Z}_+$, $l - 1 - \frac{1}{p} < m < l$; let X (resp., $X(j)$, $j = 0, 1, \ldots, m$) be one of the following spaces:*

a) $B^l_{p,\theta}$ *or* $F^l_{p,\theta}$ *(resp.,* $B^{l-j}_{p,\theta}$ *or* $F^{l-j}_{p,\theta}$*),* $0 < \theta < +\infty$
   or

b) *the closure of the set of all C^∞-functions in $B^l_{p,\theta}$ or $F^l_{p,\infty}$ (resp., in $B^{l-j}_{p,\theta}$ or in $F^{l-j}_{p,\theta}$).*

Suppose $A \subset \mathbb{R}^n$, $f \in X$ and for any $\alpha = (\alpha_1, \ldots, \alpha_n) \in \mathbb{Z}^n_+$, $\alpha_1 + \cdots + \alpha_n = |\alpha| \leqslant m$

$$D^\alpha f(x) = 0 \text{ quasi everywhere on } A$$

(with respect to the capacity in $X(|\alpha|)$). Then for any $\varepsilon > 0$ there exists an $f_\varepsilon \in X$ such that

$$\|f - f_\varepsilon\|_X \leqslant \varepsilon \text{ and } f_\varepsilon \text{ vanishes on an open set, containing } A.$$

The proof will be published in the Russian journal "Algebra and Analysis" (English translation in "St. Petersburg Mathematical Journal").

Chapter 13

ORTHOGONAL POLYNOMIALS

Edited by

Paul Nevai
Department of Mathematics
The Ohio State University
231 West 18th Avenue
Columbus, Ohio 43210–1174
U.S.A.

email: nevai@mps.ohio-state.edu *and* neva@ohstpy.bitnet

INTRODUCTION

It is probably correct to say that orthogonal polynomials returned to the mainstream of classical analysis, approximation theory, and special functions in 1974 when Dick Askey delivered a series of ten lectures on orthogonal polynomials and special functions at Virginia Polytechnic Institute. Since then a new generation of specialists grew up who have completely transformed the classical theory into a contemporary subject with a variety of new results and ideas with numerous applications both within and outside mathematics.

There have been 3 international conferences (Bar-le-Duc in 1984, Segovia in 1987, and Erice in 1990), a NATO Advanced Study Institute (Columbus, 1989), yearly conferences in Spain, and a number of special sessions at AMS and SIAM meetings dedicated exclusively to orthogonal polynomials and special functions.

An (almost) up-to-date account of (almost) all aspects of orthogonal polynomials is given in the 20 survey papers collected in [N5].

Some of the significant achievements of the past decades were the detailed study of characterizations and classification of classical and not so classical orthogonal polynomials, orthogonal polynomials associated with exponential weights, the relationship between orthogonal polynomials, continued fractions, and Padé approximations, the relationship between Jacobi matrices, difference equations, and orthogonal polynomials, various asymptotic properties of real and complex orthogonal polynomials, applications of potential theory to the asymptotic behavior of orthogonal polynomials and their zeros, orthogonal polynomials associated with root systems and quantum groups, the connections between orthogonal polynomials and digital signal processing, birth and death processes, scattering theory, Julia sets and function iterations, functional analysis, operator theory, statistics, stochastic processes, just to name a few.

In recent years the attention has started to turn to orthogonal polynomials in several variables, matrix and operator valued orthogonal polynomials, to orthogonal polynomials depending on parameters, and to some special type of orthogonality such as Sobolev orthogonality, as well.

There are quite a few problems in orthogonal polynomials which look rather innocent, and, nevertheless, whose solution so far has been out of reach. In this short discussion I will concentrate on a few such problems which can be stated and understood without any special knowledge of orthogonal polynomials, and whose connection to orthogonal polynomials may be evident only for the experts who then may become handicapped by their vast knowledge of the area. This seems to lessen their chances to solve them. Believe me, this negative attitude is based on my own personal experiences. One of my favorites is about difference equations.

A. A nonlinear difference equation. Given $C \in \mathbb{R}$, $x_0 = 0$, and $x_1 > 0$, let the sequence $\{x_n\}_{n=2}^{\infty}$ be generated by

$$n = x_n(x_{n+1} + x_n + x_{n-1}) + Cx_n, \qquad n = 1, 2, \ldots . \tag{a1}$$

This material is based upon work supported by the National Science Foundation under Grant No DMS–9024901, and by NATO under Grant No. CRG.870806

Though it is not necessary to know, I mention that such a sequence arises when solving the extremal problem

$$\min \int_{\mathbb{R}} (x^n + \text{ lower degree terms })^2 \exp(-x^4) \, dx \ .$$

It was proved in [BN] that there is a unique $x_1 = x_1(C) > 0$ such that the sequence $\{x_n\}_{n=2}^{\infty}$ is positive. As it turns out, the proof of this requires some (perhaps ad hoc) nonelementary methods from the moment problem and orthogonal polynomials combined with a drop of Fourier analysis. Naturally, since the problem seems to have nothing to do with orthogonal polynomials, one would like to get a proof independent of them. For $C \geqslant 0$ this was done in [N2]. For $C < 0$, this seems to be much more difficult. I believe the difficulty is caused by the unpredictably oscillating nature of all but one solution. I have posed this problem several times, but, alas, no complete proof has been obtained yet (apart from the one in [BN] and related results in [Ma1] and [BLN]).

The obvious shortcoming of an approach based on orthogonal polynomials is that it makes generalizations difficult if not impossible. For instance, we have no way of handling equations similar to (a1) such as

$$n = x_n(x_{n+1} + 2x_n + x_{n-1}) + Cx_n, \qquad n = 1, 2, \ldots ,$$

or

$$n = x_n(x_{n+1} + x_n + x_{n-2}) + Cx_n, \qquad n = 1, 2, \ldots ,$$

or their higher order variants. The intriguing questions relate to the positivity, to the uniqueness, and to the nonoscillating nature of solutions of generally unstable nonlinear difference equations.

B. An extremal problem. Given a positive Borel measure α on (a, b), the minimum in

$$E_n(a, b, \alpha, 2) = \min \left[\int_a^b (x^n + \text{ lower degree terms })^2 \, d\alpha(x) \right]^{\frac{1}{2}} \tag{b1}$$

is realized by the nth degree monic orthogonal polynomial associated with α. From the point of view of approximation theory, we are looking at the degree of best approximation of x^n by lower degree polynomials. Normally, such an extremal problem is viewed as one depending on the variable n with the interval (a, b) and the measure α fixed. However, in number of interesting cases the latter may vary as well. Hence the question arises as to the dependence of $E_n(a, b, \alpha, 2)$, or, in a more general setting, of $E_n(a, b, \alpha, p)$ on a, b, and α, where

$$E_n(a, b, \alpha, p) = \min \left[\int_a^b (x^n + \text{ lower degree terms })^p \, d\alpha(x) \right]^{\frac{1}{p}}, \qquad 0 < p \leqslant \infty. \tag{b2}$$

For instance, the behavior of

$$\min \int_{-1}^{1} (x^n + \text{ lower degree terms })^2 \exp(-cx^2) \, dx \tag{b3}$$

for $0 \leqslant c \leqslant \infty$ is just one of the few very interesting examples (cf. [Ga] and [CR]). For $c = 0$, (b3) leads to Legendre polynomials, whereas $c \to \infty$ brings the Hermite polynomials to the scene.

C. Ratio asymptotics. According to a result of Rahmanov, E_n given by (b1) satisfies

$$\lim_{n \to \infty} \frac{E_n(a, b, \alpha, 2)}{E_{n-1}(a, b, \alpha, 2)} = \frac{b - a}{4} \tag{c1}$$

as long as $\alpha' > 0$ a.e. in the finite interval (a, b). There are several proofs of this (cf. [Ra], [MNT1], and [N6]), but none of them are based on the direct definition of E_n in (b1). Instead, they all use the fact that E_n is the reciprocal of the leading coefficient of the orthonormal polynomial associated with α, and then (c1) is derived from its analogue for orthogonal polynomials on the unit circle. The proof of the latter uses some identities involving orthogonal polynomials on the unit circle which do not carry over to the real line. It would be very desirable to find a direct proof of (c1). Moreover, I firmly believe that (c1) can be generalized to

$$\lim_{n \to \infty} \frac{E_n(a, b, \alpha, p)}{E_{n-1}(a, b, \alpha, p)} = \left(\frac{b - a}{4} \right)^{\frac{2}{p}} \tag{c2}$$

for all $0 < p < \infty$ provided $\alpha' > 0$ a.e. in the *finite* interval (a, b). In addition, I suspect that (c1) and (c2) remain true for $E_n(\mathbb{R}, \alpha, p)$ instead of $E_n(a, b, \alpha, p)$ where

$$E_n(\mathbb{R}, \alpha, p) = \min \left[\int_{\mathbb{R}} (x^n + \text{ lower degree terms })^p \, d\alpha(x) \right]^{\frac{1}{p}}, \qquad 0 < p \leqslant \infty,$$

provided that $\alpha' > 0$ a.e. in the *finite* interval (a, b) and that for every $\epsilon > 0$, the set $\mathbb{R} \setminus (a - \epsilon, b + \epsilon)$ is finite.

Lubinsky, Magnus and Van Assche, and Totik look at ratio asymptotics from a different point of view of examining how such properties can be passed from one measure to another (see [To]).

D. Comparative asymptotics. Given two measure α and β, even if $E_n(a, b, \alpha, p)$ and $E_n(a, b, \beta, p)$ (cf. (b2)) are not computable, there is a good chance that one of them can be estimated in terms of the other as long as α and β are absolutely continuous with respect to each other and their ratio behaves reasonably. There have been several studies showing that this expectation is not unreasonable (cf. [N1], [MNT3], and [Lo]). A typical result in this direction would be that if $d\beta = g \, d\alpha$ then

$$\lim_{n \to \infty} \frac{E_n(a, b, \beta, p)}{E_n(a, b, \alpha, p)} = D(g)$$

as long as, say, $\alpha' > 0$ a.e. in the interval (a, b) and g is reasonably well behaved, say, positive and continuous. Here D is independent of α; it is related to the geometric mean (and to the Szegő function) of g.

E. Christoffel functions. Recently, with Máté and Totik [MNT4], we proved that, given a measure μ on the interval $[0, 2\pi)$ satisfying Szegő's condition $\int_0^{2\pi} \log \mu' > -\infty$, we have

$$\lim_{n \to \infty} n \omega_n(\mu, e^{it}) = \mu'(t) \tag{e1}$$

for almost every $t \in [0, 2\pi)$. Here the *Christoffel function* $\omega_n(\mu)$ is defined by

$$\omega_n(\mu, z) = \min \frac{1}{2\pi |P(z)|^2} \int_0^{2\pi} \left| P(e^{it}) \right|^2 \, d\mu(t), \tag{e2}$$

180

where the minimum is taken for all polynomials P of degree less than n. Using this result, we also showed that a similar conclusion holds under a weaker assumption on μ. Namely, let $I \subset [0, 2\pi)$ be an arbitrary interval, and assume that $\int_I \log \mu' > -\infty$ and $\mu'(t) > 0$ for almost every $t \in [0, 2\pi)$. Then (e1) holds for almost every $t \in I$. In fact, this conclusion holds under an even slightly weaker condition. The open problem is what is the precise condition for (e1). Vili Totik came up with the following conjecture about possible extension of (e1) to measures on more general sets.

CONJECTURE (Totik). *Let μ be a positive Borel measure in the complex plane. Assume the support of μ is the boundary of a connected open set containing a neighborhood of infinity. Assume, further, that this boundary is the union of finitely many smooth Jordan arcs, and on each of these arcs $\int \log \mu'(u) |du| > -\infty$, where μ' is the Radon–Nikodym derivative of μ with respect to arc length on this arc. Then, writing*

$$\omega_n(\mu, z) = \min \frac{1}{|P(z)|^2} \int_{\mathrm{supp}(\mu)} |P|^2 \, d\mu$$

in analogy with (e2), where the minimum is taken for all polynomials P of degree less than n, and denoting by ν the equilibrium measure of the support of μ, we have

$$\lim_{n \to \infty} n\omega_n(\mu, z) = \frac{d\mu(z)}{d\nu(z)}$$

ν-*almost everywhere.*

There is a good possibility that (e1) and even maybe Conjecture 1 can be extended to L^p spaces for $0 < p < \infty$.

F. Linear difference equations; boundedness of solutions. There has been plenty of research regarding difference equations of the form

$$x_{n+m} + a_{1n}x_{n+m-1} + \cdots + a_{m-1,n}x_{n+1} + a_{mn}x_n = b_n \tag{f1}$$

and their matrix analogues

$$X_{n+1} + A_n X_n = B_n . \tag{f2}$$

The question I want to raise is to find conditions assuring the boundedness of *all* solutions of (f1) or (f2). Of course, one would expect that if (f1) and (f2) are slight perturbations of equations with bounded solutions, then they will inherit this property. Hence, the question becomes as to find the right perturbation conditions. If

$$X_{n+1} + (A + \mathcal{E}_n)X_n = 0 \tag{f3}$$

with

$$\sum_{n=1}^{\infty} \|\mathcal{E}_n\| < \infty , \tag{f4}$$

and if *all* solutions of the unperturbed equation $X_{n+1} + AX_n = 0$ are bounded then *all* solutions of (f3) are bounded as well (cf. [Mi, p. 23]. I expect that (f4) can be replaced by the significantly weaker

$$\sum_{n=1}^{\infty} \|\mathcal{E}_{n+1} - \mathcal{E}_n\| < \infty \qquad \text{and} \qquad \lim_{n \to \infty} \|\mathcal{E}_n\| = 0 . \tag{f5}$$

Some special cases of this have been proved in, for instance, [MN1], [MNT2], and [GVA]. Interestingly, all the proof of all the theorems involving analogues of (f5) (except the trivial cases) require a real variable setting, whereas theorems with conditions such as (f4) work with both real and complex (or even operator-valued) coefficients in (f1)–(f4). Therefore, I see a high priority in considering the complex case (cf. [Ma2]). For instance, I would very much like to see a proof that all solutions of

$$x_{n+2} + a_{1n}x_{n+1} + a_{0n}x_n = 0, \qquad a_{0n} \in \mathbb{C}, \ a_{1n} \in \mathbb{C}, \tag{f6}$$

are bounded as long as

$$\lim_{n\to\infty} a_{0n} = a_0 \in \mathbb{R} \setminus \{0\} \qquad \text{and} \qquad \lim_{n\to\infty} a_{1n} = a_1 \in \mathbb{R},$$

$$\sum_{n=1}^{\infty} \left[|a_{0,n+1} - a_{0n}| + |a_{1,n+1} - a_{1n}| \right] < \infty,$$

and all the solutions of

$$x_{n+2} + a_1 x_{n+1} + a_0 x_n = 0$$

are bounded. Nonhomogeneous analogues of (f6) are even more interesting. Based on my experience with Jacobi polynomials, I have a hunch that conditions of the type (f4) are the right ones when the unperturbed equation has solutions with polynomial growth, and we expect similar behavior for the perturbation. As far as I know such problems would be equally interesting for differential equations as well.

G. Linear difference equations; growth of solutions. It is nice to have bounded or polynomially growing solutions of difference (or differential equations). However, in many contexts, a different kind of growth is equally useful. A sequence $\{x_n\}_{n=1}^{\infty}$ is said to have *sub-exponential growth*, if for each $0 < p < \infty$,

$$\lim_{n\to\infty} \frac{|x_n|^p}{\sum_{k=1}^{n} |x_k|^p} = 0. \tag{g1}$$

Similarly, a function f grows sub-exponentially in $\mathbb{R}^+$ if

$$\lim_{t\to+\infty} \frac{|f(t)|^p}{\int_0^t |f(u)|^p \, du} = 0 \tag{g2}$$

for every $0 < p < \infty$. It is clear that every bounded sequence $\{x_n\}_{n=1}^{\infty}$ satisfies (g1). Orthogonal polynomials with subexponential growth were first studied in [NTZ], whereas somewhat weaker results in a much more general setting were obtained in [LN]. John Zhang [Zh] proved that, given $0 < p < \infty$ and a norm $\| \cdot \|$ on complex 2×2 matrices, there is a constant $C(p, \| \cdot \|)$ depending only on p and the norm $\| \cdot \|$ such that

$$\sup_{\substack{n \in \mathbb{N} \\ A \in \mathbb{C}^{2\times 2} \\ \rho(A) \leqslant 1}} \sup_{X_0 \neq 0} \frac{n \, \|A^n X_0\|^p}{\sum_{k=1}^{n} \|A^k X_0\|^p} \leqslant C(p, \| \cdot \|) \tag{g3}$$

where $\rho(A)$ denotes the spectral radius of A. I expect that (g3) remains true for higher dimensions as well, and then one can prove subexponential growth of solutions of

$$X_{n+1} + (A + \mathcal{E}_n)X_n = 0 \tag{g4}$$

with $\rho(A) \leqslant 1$ and $\lim_{n\to\infty} \|\mathcal{E}_n\| = 0$. Note, that if $\mathcal{E}_n \equiv 0$, then $\rho(A) \leqslant 1$ guarantees polynomial growth of all solutions of (g4). As far as I know, differential equations have not been studied yet from the point of view of subexponential growth. This surprises me since there are a number of fascinating applications of this notion (cf. [N1], [MNT3], [NTZ], [LN], and [Zh]).

H. Jacobi matrices. This is like a benevolent Pandora's box. I will momentarily open it only. If $J(\{a_n, b_n\})$ is a tridiagonal matrix given by

$$J = \begin{pmatrix} b_0 & a_1 & 0 & 0 & 0 & \cdots \\ a_1 & b_1 & a_2 & 0 & 0 & \cdots \\ 0 & a_2 & b_2 & a_3 & 0 & \cdots \\ 0 & 0 & a_3 & b_3 & a_4 & \cdots \\ \vdots & \vdots & \vdots & \vdots & \ddots & \cdots \end{pmatrix}$$

then J is called a Jacobi-type matrix. If, in addition, $a_j > 0$ and $b_j \in \mathbb{R}$, then J is called a Jacobi matrix.

Let H be infinite-dimensional and separable Hilbert space, and let J be a bounded linear self-adjoint operator on H. Assume there is a cyclic vector ψ for J, that is, $\overline{\text{Span}(\{J^n\psi\}_{n=1}^\infty)} = H$, and let $\{\psi_n\}_{n=0}^\infty$ be the Gram–Schmidt orthogonalization of $\{J^{n+1}\psi\}_{n=0}^\infty$. Then J has a tridiagonal Jacobi matrix representation of the form (f1) such that $\psi_n = p_n(J)\psi$ where

$$x p_n(x) = a_{n+1} p_{n+1}(x) + b_n p_n(x) + a_n p_{n-1}(x), \qquad n = 0, 1, \dots,$$

with $p_{-1} = 0$ and $p_0 = 1$. If $J = \int \lambda dE_\lambda$ is the spectral resolution of J and if α is defined by $\alpha(B) = \|E(B)\psi\|^2$ for every Borel set B, then J is unitarily equivalent to a multiplication operator on $L^2(\alpha, \mathbb{R})$ and $\int_\mathbb{R} p_n p_m d\alpha = \delta_{nm}$.

Now the problem is to find properties of α in terms of J, that is, in terms of the recursion coefficients, and vice versa.

Given a Jacobi matrix or a Jacobi-type matrix J, we define the spectrum of J as

$$\text{sp}(J) = \bigcap_{N=1}^\infty \overline{\bigcup_{n=N}^\infty \bigcup_{k=1}^n \{\Lambda_{kn}\}}$$

where Λ_{kn} are the eigenvalues of the truncated matrix $J_{(N)}$ consisting of the $N \times N$ section of the northeast corner of J. Since $\text{sp}(J)$ is closed, it can be written as $\text{sp}(J) = A \cup B$ where A consists of all the limit points of $\text{sp}(J)$. With this notation, we will call A the *essential (or continuous) spectrum* of J and B the *discrete (or point) spectrum* of J, respectively.

Given two Jacobi (or Jacobi-type) matrices $J_1 = J(\{a_{n,1}, b_{n,1}\})$ and $J_2 = J(\{a_{n,2}, b_{n,2}\})$, they are called *compact perturbations* of each other if

$$\lim_{n\to\infty} \left[|a_{n,1} - a_{n,2}| + |b_{n,1} - b_{n,2}| \right] = 0 \tag{h1}$$

holds. By H. Weyl's theorem, the essential spectrum does not change under a compact perturbation. The question, how the point spectrum changes, seems to be a much more delicate issue. Presently there are (more or less) satisfactory results only for the case when the Jacobi matrix J is a compact perturbation of a constant Jacobi matrix.

EXAMPLE: If $a_n = 1 + \frac{1}{n}$ & $b_n = 0$ then

$$\text{supp}(\alpha) = [-2, 2] \cup \{\infty \ \# \text{ of mass-points}\}.$$

If $a_n = 1 + \frac{A}{n^2}$ & $b_n = 0$ then

supp$(\alpha) = [-2, 2] \cup \{$depending on A there is a finite or infinite # of mass-points$\}$.

If $a_n = 1 + \frac{1}{n^3}$ & $b_n = 0$ then

$$\text{supp}(\alpha) = [-2, 2] \cup \{\text{finite # of mass-points}\}.$$

It is quite plausible to expect that if $J_0 = J(\{a_{n,0}, b_{n,0}\})$ is a Jacobi matrix with compact spectrum and finite discrete spectrum, and if the Jacobi matrix $J = J(\{a_n, b_n\})$ is a compact perturbation of J_0 such that

$$\sum_{n=1}^{\infty} n\big[|a_n - a_{n,0}| + |b_n - b_{n,0}|\big] < \infty$$

holds, then the discrete spectrum of J is also finite.

As a matter of fact, I expect that the following generalization holds as well.

CONJECTURE. *Let J_0 be a Jacobi matrix with compact spectrum and finite discrete spectrum, and let J be a (real or complex) Jacobi-type compact perturbation of J_0 such that*

$$\sum_{n,m=1}^{\infty} n\big| (Q(J))_{mn} - (Q(J_0))_{mn} \big| < \infty \tag{h2}$$

with a suitable polynomial Q. Then the discrete spectrum of J is finite.

Since $Q(J)$ is always a banded matrix whenever Q is a polynomial, the summation with respect to m in (h2) is over a finite range of indexes only. A good starting point for this conjecture is [MN2] where this is proved when J_0 is a constant Jacobi matrix and Q is a Chebyshev polynomial. In the following proposition, $J(Chebyshev)$ means the Jacobi matrix with 0's on the main diagonal and with $\frac{1}{2}$'s on the sub- and superdiagonals.

THEOREM. [MN2] *Let $k \geqslant 1$, and let T_k be the Chebyshev polynomial of degree k. If*

$$\sum_{l=n}^{\infty} \sum_{j=l-k}^{l+k} \big| [[T_k(J(\alpha)) - T_k(J(Chebyshev))]]_{jl} \big| < \frac{k}{36(n + 2k)}$$

for every sufficiently large n, then

$$\text{supp}(\alpha) = [-1, 1] \cup \{\text{finite # of mass-points}\}.$$

The above inequality describes a quantitative perturbation which is stronger than the trace class. Our result was based on ideas stemming from the works of Bargmann, Geronimo and Case, Nikishin, et al. Another closely related result is the following

THEOREM. [GVA] *Let N be a fixed positive integer. Let the Jacobi matrix J be compact perturbation (cf. (h2)) of the N-periodic Jacobi matrix $J_0 = J(\{a_{n,0}, b_{n,0}\})$. If*

$$\sum_{n=0}^{\infty} \big[|a_n - a_{n,0}| + |b_n - b_{n,0}|\big] < \infty$$

then the spectral measure α *is absolutely continuous on the essential spectrum of* J. *In addition,* α' *is also continuous and positive inside the essential spectrum of* J. *If*

$$\sum_{n=0}^{\infty} n\left[|a_n - a_{n,0}| + |b_n - b_{n,0}|\right] < \infty$$

then the discrete spectrum of J *is finite.*

This again leads to a number of problems as to the right conditions guaranteeing certain spectral properties of compact perturbations of Jacobi matrices

Whoops ... time is up. In what follows, in addition to items which were referenced above, is a short selection of books and conference proceedings published in the recent past on orthogonal polynomials. Many of them contain problem sections. In particular, [N4] is a good place to look for unsolved research problems.

References

[ADMR] Alfaro M., Dehesa J. S., Marcellán F. J., Rubio de Francia J. L., Vinuesa J., eds., *Orthogonal Polynomials and Their Applications*, Lecture Notes in Mathematics, Vol. 1329, Springer Verlag, Berlin, 1988.

[As] Askey R., *Orthogonal Polynomials and Special Functions*, Regional Conference Series in Applied Mathematics, Vol. 21, SIAM, Philadelphia, 1975.

[BN] Bonan S. S. and Nevai P., *Orthogonal polynomials and their derivatives, I*, J. Approx. Theory **40** (1984), 134–147.

[BLN] Bonan S. S., Lubinsky D. S. and Nevai P., *Orthogonal polynomials and their derivatives, II*, SIAM J. Math. Anal **18** (1987), 1163–1176.

[BDMR] Brezinski C., Draux A., Magnus A. P. and Ronveaux A., eds., *Polynômes Orthogonaux et Applications, Proceedings, Bar-le-Duc, 1984*, Lecture Notes in Mathematics, Vol. 1171, Springer Verlag, Berlin, 1985.

[BGR] Brezinski C., Gori L. and Ronveaux A., eds., 1991.

[CT] Chihara T. S., *An Introduction to Orthogonal Polynomials*, Gordon and Breach, New York–London–Paris, 1978.

[CR] Chin R. C. Y., *A domain decomposition method for generating orthogonal polynomials for a Gaussian weight on a finite interval*, J. Comp. Phys. **99** (1992), 321–336.

[Ga] Gautschi W., *Computational aspects of orthogonal polynomials*, Orthogonal Polynomials: Theory and Practice, NATO ASI Series C: Mathematical and Physical Sciences, Vol. 294, P. Nevai, ed., Kluwer Academic Publishers, Dordrecht–Boston–London, 1990 pp. 181–216.

[GVA] Geronimo J. S. and Van Assche W., *Approximating the weight function for orthogonal polynomials on several intervals*, J. Approx. Theory **65** (1991), 341–371.

[Lo] López G. L., *Convergence of Padé approximants of Stieltjes type meromorphic functions and comparative asymptotics for orthogonal polynomials*, Math. USSR Sb. **64** (1989), 207–227.

[Lu1] Lubinsky D. S., *A survey of general orthogonal polynomials for weights on finite and infinite intervals*, Acta Appl. Math. **10** (1987), 237–296.

[Lu2] Lubinsky D. S., *Strong Asymptotics for Extremal Errors and Polynomials Associated with Erdős–type Weights*, Pitman Research Notes in Mathematics Series, Longman Scientific & Technical, Vol. 202, Harlow, United Kingdom, 1988.

[LN] Lubinsky D. S. and Nevai P., *Sub-exponential growth of solutions of difference equations*, J. London Math. Soc. (to appear).

[LS] Lubinsky D. S. and Saff E. B., *Strong asymptotics for extremal polynomials associated with weights on* $\mathbb{R}$, Lecture Notes in Mathematics, vol. 1305, Springer-Verlag, Berlin, 1988.

[Ma1] Magnus A. P., *On Freud's equations for exponential weights*, J. Approx. Theory **46** (1986), 65–99.

[Ma2] Magnus A. P., *Toeplitz matrix techniques and convergence of complex weight Padé approximants*, J. Comput. Appl. Math. **19** (1987), 23–38.

Chapter 13. ORTHOGONAL POLYNOMIALS

[MN1]   Máté A. and Nevai P., *Orthogonal polynomials and absolutely continuous measures*, Approximation Theory IV, Chui C. K., Schumaker L. L., and Ward J. D., eds., Academic Press, New York, 1983, pp. 611–617.

[MN2]   Máté A. and Nevai P., *Eigenvalues of finite band-width Hilbert space operators and their application to orthogonal polynomials*, Can. J. Math. **16** (1989), 106–122.

[MNT1] Máté A., Nevai P. and Totik V., *Asymptotics for the leading coefficients of orthogonal polynomials on the unit circle*, Constr. Approx. **1** (1985), 63–69.

[MNT2] Máté A., Nevai P. and Totik V., *Asymptotics for orthogonal polynomials defined by a recurrence relation*, Constr. Approx. **1** (1985), 231–248.

[MNT3] Máté A., Nevai P. and Totik V., *Extensions of Szegő's theory of orthogonal polynomials, II & III*, Constr. Approx. **3** (1987), 51–72 & 73–96.

[MNT4] Máté A., Nevai P. and Totik V., *Szegő's extremum problem on the unit circle*, Ann. Math. **134** (1991), 433–453.

[Mi]   Miller K. S., *Linear Difference Equations*, W. A. Benjamin, Inc., New York, 1968.

[N1]   Nevai P., *Orthogonal Polynomials*, vol. 213, Mem. Amer. Math. Soc., Providence, Rhode Island, 1979.

[N2]   Nevai P., *Orthogonal polynomials associated with* $\exp(-x^4)$, Canadian Math. Soc. Conf. Proc. **3** (1983), 263–285.

[N3]   Nevai P., *Géza Freud, orthogonal polynomials and Christoffel functions. A case study*, J. Approx. Theory **48** (1986), 3–167.

[N4]   Nevai P., *Research problems in orthogonal polynomials*, Approximation Theory VI, Vol. II, Chui C. K., Schumaker L. L. and Ward J. D., eds., Academic Press, Inc., Boston, 1989, pp. 449–489.

[N5]   Nevai P., ed., *Orthogonal Polynomials: Theory and Practice*, NATO ASI Series C: Mathematical and Physical Sciences, Vol. 294, Kluwer Academic Publishers, Dordrecht–Boston–London, 1990.

[N6]   Nevai P., *Weakly convergent sequences of functions and orthogonal polynomials*, J. Approx. Theory **65** (1991), 322–340.

[NTZ]   Nevai P., Totik V. and Zhang J., *Orthogonal polynomials: their growth relative to their sums*, J. Approx. Theory **67** (1991), 215–234.

[Ra]   Rahmanov E. A., *On the asymptotics of the ratio of orthogonal polynomials*, Math. USSR–Sb. **33** (1977), 199–213; *II*, ibid. **46** (1983), 105–117.

[SaT]   Saff E. B. and Totik V., *Logarithmic Potentials with External Fields*, in preparation.

[StT]   Stahl H. and Totik V., *General Orthogonal Polynomials*, Encyclopedia of Mathematics, Vol. 43, Cambridge University Press, Cambridge, 1992.

[To]   Totik V., *Orthogonal polynomials with ratio asymptotics*, Proc. Amer. Math. Soc. **114** (1992), 491–495.

[VA1]   Van Assche W., *Asymptotics for orthogonal polynomials*, Lecture Notes in Mathematics, vol. 1265, Springer-Verlag, Berlin, 1987.

[VA2]   Van Assche W., *Constructive methods in the analysis of orthogonal polynomials*, Dissertation, Katholieke Universiteit Leuven, Heverlee (Leuven), Belgium, 1992.

[Zh]   Zhang J., *Relative growth of linear iterations and orthogonal polynomials on several intervals*, J. Lin. Alg. Appl. (to appear).

ZEROS OF ORTHOGONAL POLYNOMIALS ON
REGULAR N-GONS

M. Eiermann, H. Stahl

Let $D \subset \mathbb{C}$ be a bounded domain and denote by $L^2(D)$ the Hilbert space of all functions f which are analytic in D and fulfill $\iint_D |f(z)|^2 \, db_z < \infty$, where db_z is the two–dimensional Lebesgue measure. We consider orthogonal polynomials p_m, $\deg(p_m) = m$, with respect to the associated inner product (cf. [2] or [4, §1.1])

$$\langle p, q \rangle := \int \int_D p(z)\overline{q(z)} \, db_z.$$

The difference between real orthogonal polynomials and these complex ones is substantial. It is for instance well-known that the polynomials p_m do in general not fulfill a finite recurrence relation (cf. [3]). In contrast to the real case not much is known about the location of the zeros of p_m. Fejér's theorem (cf. [2, p. 19] or [4, §2.1]) states that all roots of p_m are contained in the convex hull cov(D) of D. In case of a non-convex domain D, one knows that the number of zeros of p_m which lie outside the polynomial convex hull $P_C(\bar{D})$ of $\bar{D}$ is bounded for $m \to \infty$ (cf. [4, Theorem 2.1.1]). One further knows that the asymptotic distribution of the zeros (for $m \to \infty$) is related to the equilibrium distribution $\omega_{\bar{D}}$ on $\bar{D}$ (cf. [4, Remark 2 to Theorem 3.1.4]). More precise informations about the locations of the zeros are only available for very special domains D, e.g., if D is an elliptic region, where the p_m turn out to be suitably transformed Chebyshev polynomials of the second kind (cf. [1]).

Here we will be concerned with convex domains having a polygonal boundary, and especially with the N–gon G_N which we normalize to have its vertices at the N th roots of unity. For G_N, it follows from standard symmetry arguments that the associated orthogonal polynomials p_m are real and that they have the form $p_m(z) = z^k q_j(z^N)$, where $\deg(q_j) = j$ and $m = jN + k$ with $k \in \{0, 1, \ldots, N - 1\}$. Thus, p_m has a root of order k at zero and the remaining roots are symmetric with respect to the axes $\arg(z) = 2\pi k/N$ ($k \in \{0, 1, \ldots, N - 1\}$).

The figures below show the zeros of the orthogonal polynomials $\{p_m\}_{m=1}^{50}$ associated with G_3 and G_5, respectively. The corresponding picture for G_4 is similar to that of G_3 and on the other hand, the pictures for G_N ($N > 5$) are similar to that of G_5[†]. Based on these numerical studies and backed by theoretical considerations, we make the following two conjectures.

This material is based upon work supported by the National Science Foundation under Grant No DMS–9024901, and by NATO under Grant No. CRG.870806

[†]The authors can send more material to interested readers.

Regular 3-gon: Regular 5-gon:

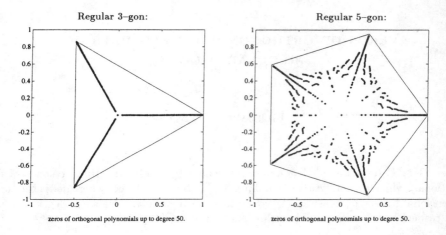

zeros of orthogonal polynomials up to degree 50.   zeros of orthogonal polynomials up to degree 50.

zeros of orthogonal polynomials up to degree 50 zeros of orthogonal polynomials up to degree 50

CONJECTURE 1. *For $N = 3$ and $N = 4$, the zeros of the orthogonal polynomials p_m are located exactly on the diagonals*

$$\Gamma_k := \{z \,:\, |z| < 1 \text{ and } \arg(z) = 2\pi k/N\} \qquad (k = 0, 1, \dots, N)$$

of G_N. For $N \geqslant 5$ however, there are zeros of p_m (m sufficiently large) which are not in $\bigcup_{0 \leqslant k \leqslant N} \Gamma_k$.

For the formulation of the next conjecture, let us subdivide $\{p_m\}_{m \geqslant 0}$ into N disjoint subsequences, namely

$$\left\{p_{jN+k}(z)\right\}_{j \geqslant 0} =: \left\{z^k q_{j,k}(z)\right\}_{j \geqslant 0} \qquad (k = 0, 1, \dots, N-1),$$

where $q_{j,k}$ is a real polynomial of degree j. If Conjecture 1 is valid, then $q_{j,k}$ has all its roots in the interval $(0, 1)$ (for $N = 3$ and $N = 4$). Moreover, on the basis of our numerical computations, we make the following

CONJECTURE 2. *For $N = 3$ and $N = 4$, the zeros of $q_{j+1,k}$ interlace the zeros of $q_{j,k}$ for each $k \in \{0, 1, \dots N-1\}$. I. e., if for a fixed k, $\lambda_\nu^{(j)}$ ($\nu = 1, 2, \dots, j$) denote the zeros of $q_{j,k}$, then*

$$0 < \lambda_1^{(j+1)} < \lambda_1^{(j)} < \lambda_2^{(j+1)} < \lambda_2^{(j)} < \dots < \lambda_j^{(j+1)} < \lambda_j^{(j)} < \lambda_{j+1}^{(j+1)} < 1$$

and asymptotically, we have

$$\frac{1}{j} \sum_{\nu=1}^{j} \delta_{\lambda_\nu^{(j)}} \xrightarrow{\;*\;} \mu_N \qquad (\text{for } j \longrightarrow \infty),$$

where δ_z, $z \in \mathbb{C}$, is Dirac's measure, μ_N is a probability measure on $[0,1]$ which is determined by the identity

$$\log |\varphi_N(z^N)| = \int \log |z - x| \, d\mu_N(x) + \log c_N \qquad (\text{for } |z| > 1)$$

with φ_N denoting the outer mapping function of $\mathbb{C}_\infty \setminus G_N$ ($N = 3, 4$) onto the exterior of the unit disk with $\varphi(\infty) = \infty$ and $c_N = \text{cap}(G_N)$. ($\xrightarrow{*}$ denotes weak convergence of measures).

The numerical examples which led to Conjecture 1 suggest that, also for $N \geqslant 5$, only the vertices of G_N attract zeros of p_m (for $m \to \infty$). This phenomenon is formulated in the following conjecture which we believe to be valid not only for regular N-gons, but for arbitrary convex domains D with a bounded polygonal boundary.

CONJECTURE 3. Let Z_m be the set of zeros of the orthogonal polynomial p_m associated with a bounded convex domain D whose boundary is a polygon and let $\{z_1, z_2, \ldots, z_N\} \subset \partial D$ be the set of vertices of the polygon. Then

$$\bigcap_{n=1}^{\infty} \overline{\bigcup_{m > n} Z_m} \cap \partial D = \{z_1, z_2, \ldots, z_N\}.$$

We consider orthogonal polynomials for special regions — like regular N-gons — because we believe that a better understanding of their zeros will lead to more insight into the asymptotic behaviour of the zeros of more general orthogonal polynomials. We add that in all three conjectures, the question arises what happens if in the inner product $\langle \cdot, \cdot \rangle$ a positive non-constant density function is introduced. For which class of such functions do the asymptotic statements of the conjectures remain true?

REFERENCES

1. Davis P. J., *Interpolation and Approximation*, Blaisdell, 1963.
2. Gaier D., *Vorlesungen über Approximation im Komplexen.*, Birkhäuser Verlag, 1980.
3. Lempert L., *Recursion for orthogonal polynomials on complex domains*, Fourier Analysis and Approximation Theory (G. Alexis, P. Turán, eds.), Vol. II. Colloquia Mathematica Societatis János Bolyai 19, North-Holland, 1978, pp. 481–494.
4. Stahl H., Totik V., *General Orthogonal Polynomials*, Encyclopedia of Mathematics, Cambridge University Press (to appear).

IBM SCIENTIFIC CENTER
TIERGARTENSTR. 15
D–6900 HEIDELBERG
GERMANY

AND

INSTITUT FÜR PRAKTISCHE MATHEMATIK
UNIVERSITÄT KARLSRUHE
D–7500 KARLSRUHE
GERMANY

TECHNISCHE FACHHOCHSCHULE BERLIN
LUXEMBURGER STR. 10
D–1000 BERLIN 65
GERMANY

POLYNOMIALS ORTHOGONAL IN A SOBOLEV SPACE

A. Iserles,   L. L. Littlejohn

Let $d\varphi$ and $d\psi$ be two positive measures and $\lambda \geqslant 0$ a constant. We define a Sobolev inner product

$$(f,g)_\lambda := \int_\mathcal{R} f(x)\bar{g}(x)d\varphi(x) + \lambda \int_\mathcal{R} f'(x)\bar{g}'(x)d\psi(x) \tag{1}$$

and consider monic polynomials $p_n^{(\lambda)}$, $n = 0, 1, \ldots$, which are orthogonal with respect to (1). We denote by $p_n \equiv p_n^{(0)}$, $n = 0, 1, \ldots$, and q_n, $n = 0, 1, \ldots$, the monic orthogonal polynomials with respect to the measures $d\varphi$ and $d\psi$ respectively. Clearly, each $p_n^{(\lambda)}$ can be written as a linear combination of $p_1, \ldots, p_n$,

$$p_n^{(\lambda)}(x) = \sum_{\ell=1}^{n} \alpha_\ell^{(n)}(\lambda)p_\ell(x), \quad n = 1, 2, \ldots .$$

It has been proved by Iserles, Koch, Nørsett and Sanz-Serna [4] that the aforementioned representation assumes the form

$$p_n^{(\lambda)}(x) = \sum_{\ell=1}^{n-1} \alpha_\ell(\lambda)p_\ell(x) + \sigma_n(\lambda)p_n(x), \quad n = 2, 3, \ldots , \tag{2}$$

where $\alpha_1, \alpha_2, \ldots$ are independent of n, as long as $\{d\varphi, d\psi\}$ are a *coherent pair*. There are several equivalent ways of defining coherence, the most useful being

1. There exist nonzero constants $C_1, C_2, \ldots$ and a sequence $\{d_k\}_{k=1}^{\infty}$ such that

$$C_m C_n \int_\mathcal{R} p'_m(x)p'_n(x)d\psi(x) = d_{\min\{m,n\}}, \quad m, n, = 1, 2, \ldots .$$

2. There exist nonzero constants $C_1, C_2, \ldots$ such that

$$q_n = C_{n+1}p'_{n+1} - C_n p'_n, \quad n = 1, 2, \ldots .$$

This material is based upon work supported by the National Science Foundation under Grant No DMS–9024901, and by NATO under Grant No. CRG.870806

The coefficients α_ℓ in (2) are themselves polynomials. Recalling

$$R_n(x) := \frac{\alpha_{n+1}(x)}{x}, \quad n = 0, 1, \ldots,$$

it follows that each R_n is itself an nth degree polynomial. Moreover the sequence $\{R_n\}_{n=0}^{\infty}$ obeys a three–term recurrence relation, hence, by the Favard theorem (Chihara, [2]) it is orthogonal with respect to some positive measure.

The paper of Iserles *et el.* [4] identifies $d\psi$ such that $\{d\varphi, d\psi\}$ is coherent for $d\varphi$ being either a *Jacobi* $(d\varphi(x) = (1 - x)^\alpha (1 + x)^\beta \, dx, \, \alpha, \beta > -1, \, -1 < x < 1)$ or a *Laguerre* $(d\varphi(x) = x^\alpha e^{-x} \, dx, \, \alpha > -1, \, x > 0)$ measure. Moreover, it proves that there are no positive measures $d\psi$ such that $\{d\varphi, d\psi\}$ is coherent when $d\varphi(x) = e^{-x^2}, \, x \in \mathcal{R}$ (the *Hermite* measure). However, a general characterization of coherent pairs, or even an example of coherence when $d\varphi$ is not a classical measure, are not available at present and we pose them thereby as an interesting open problem.

An application of coherence, to numerical solution of partial differential equations of evolution by pseudospectral methods, will be discussed in Iserles *et al.* [5].

Another, related, open problem is concerned with the characterization of polynomials that are orthogonal simultaneously with respect to L_2 and Sobolev inner products. A result of Sonine (1887, [8]), later independently shown by Hahn (1935, [3]), is the following:

THEOREM 1. *Suppose that $\{\omega_n(x)\}$ is an orthogonal polynomial sequence (OPS) with respect to some positive measure φ on the real line $\mathcal{R}$ and $\{d\omega_n(x)/dx\}$ is also an OPS with respect to some positive measure σ on $\mathcal{R}$. Then, up to a linear change of variable, $\{\omega_n(x)\}$ is*

$$\left.\begin{array}{lll} \text{either} & \text{the Jacobi OPS } \left\{P_n^{(\alpha,\beta)}(x)\right\} \text{ for some } \alpha > -1 \text{ and } \beta > -1 \\[2mm] \text{or} & \text{the Laguerre OPS } \left\{L_n^{(\alpha)}(x)\right\} \text{ for some } \alpha > -1 \\[2mm] \text{or} & \text{the Hermite OPS } \{H_n(x)\}. \end{array}\right\} \quad (3)$$

□

The key step in Hahn's proof involves showing that ω_n, $n \in \mathcal{Z}^+$, satisfies a second–order differential equation

$$a_2(x)y''(x) + a_1(x)y'(x) + a_0(n)y(x) = 0, \quad (4)$$

where a_1, a_2 are independent of n and a_0 is independent of x. He then appeals to Bochner's classification result (Bochner, 1929, [1]) which states that the only OPS's (orthogonal with respect to a positive measure on $\mathcal{R}$) which are solutions to an equation of the form (4) are the classical polynomials given in (3) above.

In terms of weighted Sobolev inner products, we can restate Theorem 1 in a slightly different way,

THEOREM 1. *The only OPS's that are orthogonal on the real line $\mathcal{R}$ with respect to some positive measure φ and also orthogonal with respect to a Sobolev inner product of the form*

$$(f, g) = \int_{\mathcal{R}} f(x)\bar{g}(x)\,\mathrm{d}\varphi(x) + \int_{\mathcal{R}} f'(x)\bar{g}'(x)\,\mathrm{d}\sigma(x),$$

where σ is a positive measure on $\mathcal{R}$, are the classical orthogonal polynomials of (3). $\square$

The orthogonality of the derivatives of the classical orthogonal polynomials can, of course, be easily seen from the differential equation (4) that they satisfy. For example, the Laguerre polynomials $\left\{L_n^{(\alpha)}(x)\right\}$ are orthogonal on $(0, \infty)$ with respect to the weight function $w(x) = x^\alpha e^{-x}$ and $y = L_n^{(\alpha)}$, $n \in \mathcal{Z}^+$, is a solution of

$$xy''(x) + (1 + \alpha - x)y'(x) + ny(x) = 0.$$

Using the identity

$$\int_0^\infty L_n^{(\alpha)}(x)L_m^{(\alpha)}(x)w(x)\,\mathrm{d}x = \frac{\Gamma(n + \alpha + 1)}{n!}\delta_{n,m},$$

we see that

$$\begin{aligned}
\frac{\Gamma(n + \alpha + 1)}{(n - 1)!}\delta_{n,m} &= \int_0^\infty \left[-x\frac{\mathrm{d}^2 L_n^{(\alpha)}(x)}{\mathrm{d}x^2} - (1 + \alpha - x)\frac{\mathrm{d}L_n^{(\alpha)}(x)}{\mathrm{d}x}\right]L_m^{(\alpha)}(x)x^\alpha e^{-x}\,\mathrm{d}x \\
&= -\int_0^\infty \left[x^{\alpha+1}e^{-x}\frac{\mathrm{d}L_n^{(\alpha)}(x)}{\mathrm{d}x}\right]'L_m^{(\alpha)}(x)\,\mathrm{d}x \\
&= -\left. x^{\alpha+1}e^{-x}\frac{\mathrm{d}L_n^{(\alpha)}(x)}{\mathrm{d}x}L_m^{(\alpha)}(x)\right|_0^\infty \\
&\quad + \int_0^\infty \frac{\mathrm{d}L_n^{(\alpha)}(x)}{\mathrm{d}x}\frac{\mathrm{d}L_m^{(\alpha)}(x)}{\mathrm{d}x}x^{\alpha+1}e^{-x}\,\mathrm{d}x \\
&= \int_0^\infty \frac{\mathrm{d}L_n^{(\alpha)}(x)}{\mathrm{d}x}\frac{\mathrm{d}L_m^{(\alpha)}(x)}{\mathrm{d}x}x^{\alpha+1}e^{-x}\,\mathrm{d}x;
\end{aligned}$$

i.e. $\{\mathrm{d}L_n^{(\alpha)}(x)/\mathrm{d}x\}$ is an OPS on $(0, \infty)$ with respect to the weight function $x^{\alpha+1}e^{-x}$.

During the past few years there has been a concentrated effort to study orthogonal polynomial solutions to higher-order differential equations. This interest stems from the self-adjoint operators generated from these differential equations. In 1940, H. L. Krall [7] extended Bochner's result and classified all OPS's (with respect to a positive measure) that satisfy fourth-order differential equations of the form

$$a_4(x)y''''(x) + a_3(x)y'''(x) + a_2(x)y''(x) + a_1(x)y'(x) + a_0(n)y(x) = 0. \tag{5}$$

Besides recovering the classical orthogonal polynomials given in (3),[1] Krall showed that there are only three more OPS that satisfy (5); they have been subsequently named

---

[1] These polynomials satisfy a fourth–order equation that is obtained by 'squaring' (4).

and studied as the Legendre type, Laguerre type and Jacobi type polynomials (see (A. M. Krall, [6])).

Interestingly enough, these polynomial sets are orthogonal with respect to an L_2 inner product

$$(f,g) = \int_{\mathcal{R}} f(x)\bar{g}(x)\mathrm{d}\varphi(x) \tag{6}$$

and a Sobolev inner product of the form

$$(f,g)_S = \int_{\mathcal{R}} f(x)\bar{g}(x)\mathrm{d}\varphi(x) + \int_{\mathcal{R}} f'(x)\bar{g}'(x)\mathrm{d}\psi(x) + \int_{\mathcal{R}} f''(x)\bar{g}''(x)\mathrm{d}\eta(x), \tag{7}$$

where ψ and η are positive measures.

In view of Theorem 1, it is natural to ask if the only OPS's that are simultaneously orthogonal with respect to the inner products (3) and (7) are the three classical OPS's, the Legendre type OPS, the Laguerre type OPS and the Jacobi type OPS. The classification of all sequences of polynomials that are simultaneously orthogonal with respect to inner products of the form (6) and (7) would provide a generalization of the well-known Sonine–Hahn classification theorem.

References

1. Bochner S., *Uber Sturm–Liouvillesche Polynomsysteme*, Math. Zeit. **29** (1929), 730–736.
2. Chihara T. S., *An Introduction to Orthogonal Polynomials*, Gordon and Breach, New York, 1978.
3. Hahn W., *Uber die Jacobischen Polynome und zwei verwandte Polynomklassen*, Math. Zeit. **39** (1935), 634–638.
4. Iserles A., Koch P. E., Nørsett S. P. and Sanz-Serna J. M., *On polynomials orthogonal with respect to certain Sobolev inner products*, J. Approx. Th. **65** (1991), 151–175.
5. Iserles A., Koch P. E., Nørsett S. P. and Sanz-Serna J. M. *Polynomials orthogonal in a Sobolev space and pseudospectral methods* (1991) (to appear).
6. Krall A. M., *Orthogonal polynomials satisfying a certain fourth-order differential equations*, Proc. Roy. Soc. Edinburgh **87** (1981), 271–288.
7. Krall H. L., *On orthogonal polynomials satisfying a certain fourth-order differential equation*, The Pennsylvania State College Studies (1940), no. 6.
8. Sonine N. Ja., *Uber die angenäherte Berechnung der bestimmten Integrale und über die dabei vorkommenden ganzen Functionen*, Warsaw Univ. Izv. **18** (1887), 1–76. (Russian)

DAMTP
University of Cambridge
Cambridge, England

Department of Mathematics
Utah State University
Logan, Utah, USA

13.3

RATIONAL VERSUS POLYNOMIAL APPROXIMATION
OF ENTIRE FUNCTIONS

D. S. LUBINSKY

Comparison of the efficiency of rational versus polynomial approximation is a classical theme in approximation theory. To elaborate, let $\mathcal{P}_m$ denote the polynomials of degree at most m, and let $\mathcal{R}_{m,n}$ denote the rational functions of type (m, n). That is, $R \in \mathcal{R}_{m,n}$ means $R = P/Q$, where $P \in \mathcal{P}_m$, $Q \in \mathcal{P}_n$ and Q is not identically zero. If $K \subset C$ is compact, and $f : K \to C$, we set

$$E_{m,n}(f; K) := \inf\{\|f - R\|_{L_\infty(K)} : R \in \mathcal{R}_{m,n}\}.$$

That is, $E_{m,n}(f; K)$ is the error in approximation of f on K by rational functions of type (m, n). In particular, $E_{m,0}(f; K)$ is the error in approximation of f on K by polynomials of degree at most m.

When f has some sort of singularity on K, or at a finite distance to K, it is known under fairly general conditions that at least for a subsequence of integers, $E_{n,n}(f; K)$ decays faster than $E_{n,0}(f; K)$. In the presence of natural boundaries, this effect is less well understood. For recent and relevant variations on this theme, see [2, 6, 7, 9, 10, 12].

Paradoxically, when f is entire, very little has been said, except for special functions such as $\exp(z)$ [1, 3, 5, 8, 11]. In this direction, we offer the following conjecture, which asserts that $\mathcal{R}_{n,n}$ approximates more than geometrically faster than $\mathcal{P}_n$:

CONJECTURE 1. *Let $f(z)$ be entire. There exists an infinite sequence S of integers, such that for every compact set K with positive logarithmic capacity,*

$$\lim_{\substack{n \to \infty \\ n \in S}} \left(E_{n,n}(f; K)/E_{n,0}(f; K) \right)^{1/n} = 0.$$

Note that the sequence S is independent of the particular set K. The conjecture has been proved under a regularity condition on the Maclaurin series coefficients of f [5].

THEOREM. *Let $f(z) := \sum_{j=0}^{\infty} a_j z^j$ be entire, with $\lim_{j \to \infty} a_{j+1}/a_j = 0$. Then Conjecture 1 is true for f.*

The proof of the theorem depends on a result about distribution of poles of diagonal rational approximants [4].

When one recalls that rational functions of type (n, n) have $2n + 1$ free parameters, while polynomials of degree at most n have only $n + 1$, one might expect to compare

This material is based upon work supported by the National Science Foundation under Grant No DMS–9024901, and by NATO under Grant No. CRG.870806

$\mathcal{R}_{n,n}$ favorably with $\mathcal{P}_{2n}$, or at least with $\mathcal{P}_{[\lambda n]}$, some $\lambda > 1$. Here $[x]$ denotes the greatest integer $\leqslant x$. This is, at first surprisingly, not possible in general:
Let

$$f(z) := \sum_{j=0}^{\infty} \exp(-2^j) z^j$$

and $K := \{z : |z| \leqslant r\}$, some $r > 0$. Results in [3] show that

$$\lim_{n \to \infty} \left(E_{n,n}(f;K) / E_{[\lambda n],0}(f;K) \right)^{1/n} = \infty,$$

for each $\lambda > 1$. So $\mathcal{P}_{[\lambda n]}$ fares better than $\mathcal{R}_{n,n}$ for each $\lambda > 1$. Since this entire function is of order zero, we offer:

CONJECTURE 2. *Let f be entire of finite positive, or infinite, order. Then there exists $\lambda > 1$ such that for every compact K with positive logarithmic capacity,*

$$\liminf_{n \to \infty} \left(E_{n,n}(f;K) / E_{[\lambda n],0}(f;K) \right)^{1/n} = 0.$$

REFERENCES

1. Braess D., *On the Conjecture of Meinardus on Rational Approximation of e^x*, II, J. Approx. Theory **40** (1984), 375–379.
2. Goncar A. A., Rahmanov E. A., *Equilibrium Distributions and Degree of Rational Approximation of Analytic Functions*, Math. USSR Sbornik **62** (1989).
3. Levin A. L., Lubinsky D. S., *Rows and Diagonals of the Walsh Array for Entire Functions with Smooth Maclaurin Series Coefficients*, Constr. Approx. **3** (1990), 257–286.
4. Lubinsky D. S., *Distribution of Poles of Diagonal Rational Approximants to Functions of Fast Rational Approximability*, Constr. Approx. **7** (1991), 501–519.
5. Lubinsky D. S., *Rational Versus Polynomial Approximation of Entire Functions*, IMACS Annals on Computing and Applied Mathematics **9** (1991), 81–86.
6. Parfenov O. G., *Estimates of the Singular Numbers of the Carleson Imbedding Operator*, Math. USSR Sbornik **59** (1988), 497–514.
7. Petrushev P. P., Popov V. A., *Rational Approximation of Real Functions*, Encyclopaedia of Mathematics and its Applications, vol. 28, Cambridge University Press, Cambridge, 1988.
8. Saff E. B., *On the Degree of Best Rational Approximation to the Exponential Function*, J. Approx. Theory **9** (1973), 97–101.
9. Stahl H., *A Note on Three Conjectures by Gonchar on Rational Approximation*, J. Approx. Theory **50** (1987), 3–7.
10. Stahl H., *General Convergence Results for Padé Approximants*, Approximation Theory VI (Chui C. K., et al., eds.), Academic Press, San Diego, 1989, pp. 605–634.
11. Trefethen L. N., *The Asymptotic Accuracy of Rational Best Approximants to e^z on a Disc*, J. Approx. Theory **40** (1984), 380–383.
12. Walsh J. L., *Interpolation and Approximation in the Complex Domain*, Amer. Math. Soc. Colloq. Publns. **20** (1969), Amer. Math. Soc., Providence.

DEPARTMENT OF MATHEMATICS
UNIVERSITY OF WITWATERSRAND
P.O. WITS 2050
JOHANNESBURG
SOUTH AFRICA

REFINED ASYMPTOTICS FOR
FREUD'S RECURRENCE COEFFICIENTS

A. P. MAGNUS

1. Freud's weights and coefficients.

Let $\{p_n(x) = p_n(x; d\mu(x))\}_{n=0}^{\infty}$ be the orthonormal polynomials with respect to an even measure (i.e., $\int_{-\infty}^{\infty} x^{2m-1} d\mu(x) = 0, m = 1, 2, \dots$) on $\mathbb{R}$. These polynomials then satisfy the recurrence relations

$$a_{n+1}p_{n+1}(x) = xp_n(x) - a_n p_{n-1}(x), \qquad n = 0, 1, \dots \qquad (a_0 = 0) \qquad (1)$$

One wishes to relate the behaviour of $d\mu(x)$ for large x to the behaviour of the recurrence coefficients a_n for large n (Freud's program [2], see also § 4.18 of [11]). Quite a number of dramatic achievements have been made recently, using advanced orthogonal polynomials theory (Christoffel functions), functional spaces theory and potential theory, see [3, 4, 5 (Appendix), 6, 7, 11, 13, 14 (Chap. 4)] as landmarks and surveys.

Freud remarked in [2] how one can use the identity

$$\frac{n}{a_n} = \int_{-\infty}^{\infty} p_n'(x) p_{n-1}(x) e^{-Q(x)} dx$$

$$= \int_{-\infty}^{\infty} p_n(x) p_{n-1}(x) Q'(x) e^{-Q(x)} dx, \quad n = 1, 2, \dots \quad (2)$$

when the polynomials p_n are orthonormal with respect to $\exp(-Q(x))dx$ (*Freud's weight*). Indeed, if Q is a polynomial, repeated applications of (1) in the right-hand side of (2) yields a polynomial in $a_n, a_{n\pm1}, \dots$, so that (2) turns as a set of equations for the recurrence coefficients $\{a_n\}$. For instance, $Q(x) = x^4$ gives

$$4a_n^2(a_{n-1}^2 + a_n^2 + a_{n+1}^2) = n, \qquad n = 1, 2, \dots, \qquad (3)$$

a well worked example ([1, 2, 8, 9, 10, 11, 12 (pp. 470–471)]). Moreover, such equations allowed Freud and followers to establish the asymptotic behaviour of a_n for various exponential weights (exponentials of polynomials [1, 2, 8, 9, 10]), and to arrive naturally to a conjecture for non polynomial exponentials:

$$\text{if } d\mu(x) = \exp\left(-|x|^{\alpha}\right) dx, \quad \text{then } a_n \sim \left(\frac{n}{C(\alpha)}\right)^{1/\alpha}, \qquad (4)$$

This material is based upon work supported by the National Science Foundation under Grant No DMS–9024901, and by NATO under Grant No. CRG.870806

when $n \to \infty$, with $C(\alpha) = \dfrac{2^\alpha \Gamma((\alpha+1)/2)}{\Gamma(1/2)\Gamma(\alpha/2)}$, for $\alpha > 0$. The proof of (4) appears as a special case of very powerful investigations which led to

if $d\mu(x) = \exp(-Q(x))dx$, then $a_n \sim \tilde{a}_n$, with $\tilde{a}_n \int_0^\pi Q'(2\tilde{a}_n \cos\theta) \cos\theta \, d\theta = n\pi$ (4')

where Q is even, continuous and convex on $\mathbb{R}$, $Q'(x) > 0$ for $x > 0$, among other conditions. The (positive) root $\tilde{a}_n$ of the equation of (4') is called the *Lubinsky–Mhaskar–Rahmanov–Saff's number*. The method of proof of (4') can even give asymptotic estimates of $a_1 a_2 \ldots a_n$ ([3, 4, 5, 6, 7, 13]). Now, let us try to explore the subject further with Freud's equations.

2. Freud's functionals and equations.
Let us consider only positive sequences $\mathbf{a} = \{a_n\}_1^\infty$ with $\sum_1^\infty 1/a_n = \infty$ so as to be sure that the related moment problem, amounting to finding $d\mu$ such that

$$\cfrac{1}{z - \cfrac{a_1^2}{z - \cfrac{a_2^2}{z - \cdots}}} = \int_{-\infty}^\infty (z-x)^{-1} d\mu(x), \qquad \forall z \notin \mathbb{R} \tag{5}$$

has a unique solution. Then the *Freud's functional* related to Q and $\mathbf{a}$ is defined as

$$F_n(Q; \mathbf{a}) = a_n \int_{-\infty}^\infty p_n(x) p_{n-1}(x) Q'(x) \, d\mu(x), \tag{6}$$

where the $p_n(x) = p_n(x; d\mu(x))$'s are the orthonormal polynomials related to the measure $d\mu$ solving the moment problem (5). Thus, Freud's remark becomes: $\mathbf{a}$ is truly the sequence of recurrence coefficients related to the measure $\exp(-Q(x))dx$ on $\mathbb{R}$ if

$$F_n(Q; \mathbf{a}) = n, \qquad n = 1, 2, \ldots \tag{7}$$

making the *Freud's equations* for $\mathbf{a}$ (when Q is an even function, or else one also must consider the functionals $G_n = \int_{-\infty}^\infty p_n^2(x) Q'(x) d\mu(x)$ [9]). The functionals F_n are linear in Q, but nonlinear in $\mathbf{a}$, so with $Q(x) = x^4$ one recovers the example (3).

3. Asymptotic expansions.
When Q is a polynomial, (6) is explicit in a finite number of neighbors of a_n, and *asymptotic expansions* can be studied: so (4) has been completed as

$$a_n \sim \left(\frac{n}{C(\alpha)}\right)^{1/\alpha} \sum_{k=0}^\infty \frac{A_k}{n^{2k}} \tag{8}$$

with $A_0 = 1$, when α is an even integer ([10], see also [1]).

The PROBLEM now is to extend (8) when α is not an even integer, still using (6). Making the assumption that $F_n(Q; \mathbf{a})$ still depends essentially on the close neighbors of a_n (technically, that $\partial F_n(Q; \mathbf{a})/\partial a_k \to 0$ when $|n - k| \to \infty$, see further for more on $\partial F_n/\partial a_k$), we approximate $F_m(Q; \mathbf{a})$ for m near n by $F_m(Q; \mathbf{\mathring{a}}^{(n)})$, where $\mathbf{\mathring{a}}^{(n)}$ is the constant sequence $\ldots = \mathring{a}_{n-2}^{(n)} = \mathring{a}_{n-1}^{(n)} = \mathring{a}_n^{(n)} = \mathring{a}_{n+1}^{(n)} = \ldots = \tilde{a}_n$ (as we already know that $a_{n+k}/a_n \to 1$ when $n \to \infty$). The measure $\mathring{\mu}^{(n)}$ is then $d\mathring{\mu}^{(n)}(x) = (1/(\pi \tilde{a}_n))\sqrt{1 - (x/(2\tilde{a}_n))^2} \, dx$ on $[-2\tilde{a}_n, 2\tilde{a}_n]$. This does not mean that $d\mathring{\mu}^{(n)}$ is close to $d\mu$, but that $F_m(Q; \mathbf{\mathring{a}}^{(n)})$ (probably) is close to $F_m(Q; \mathbf{a})$ for m close to n $\ldots$, and a lot of other trial measures would be as good. Proceeding with the computations ($\mathring{p}_m(x)$ is the Chebyshev polynomial $U_m(x/(2\tilde{a}_n))$), one finds

$$F_m(Q; \mathbf{\mathring{a}}^{(n)}) = \tilde{a}_n \int_{-2\tilde{a}_n}^{2\tilde{a}_n} U_m\left(\frac{x}{2\tilde{a}_n}\right) U_{m-1}\left(\frac{x}{2\tilde{a}_n}\right) Q'(x) \frac{1}{\pi \tilde{a}_n} \sqrt{1 - \left(\frac{x}{2\tilde{a}_n}\right)^2} \, dx$$

$$= \frac{\tilde{a}_n}{\pi} \int_0^\pi [\cos\theta - \cos(2m+1)\theta] Q'(2\tilde{a}_n \cos\theta) \, d\theta \qquad (9)$$

If Q is reasonably smooth, the $(2m+1)^{\text{th}}$ Fourier coefficient of $Q'(2\tilde{a}_n \cos\theta)$ may be neglected for large m and we recover the $LMRS$ approximation $\tilde{a}_n$ for a_n from (9). For instance, with $Q(x) = |x|^\alpha$:

$$F_m(|x|^\alpha; \mathbf{\mathring{a}}^{(n)}) = \frac{\tilde{a}_n}{\pi} \int_0^{\pi/2} [\cos\theta - \cos(2m+1)\theta]\alpha(2\tilde{a}_n \cos\theta)^{\alpha-1} \, d\theta$$

$$= \frac{2^\alpha \alpha \tilde{a}_n^\alpha}{\pi} \left[\frac{\Gamma((\alpha+1)/2)\Gamma(1/2)}{2\Gamma(1+\alpha/2)} - \frac{\Gamma(\alpha/2)\Gamma((\alpha+1)/2)\Gamma(1/2)}{2\Gamma(1+m+\alpha/2)\Gamma(\alpha/2-m)} \right]$$

$$= \tilde{a}_n^\alpha C(\alpha) \left[1 - (-1)^m \sin(\pi\alpha/2) \frac{\Gamma(1+m-\alpha/2)\Gamma(\alpha/2)\Gamma(1+\alpha/2)}{\pi\Gamma(1+m+\alpha/2)} \right]$$

$$\sim \tilde{a}_n^\alpha C(\alpha) \left[1 - (-1)^m \sin(\pi\alpha/2) \frac{\Gamma(\alpha/2)\Gamma(1+\alpha/2)}{\pi m^\alpha} \right]$$

This suggests an $(-1)^n/n^\alpha$ term in the expansion:

CONJECTURE. *The recurrence coefficients of* $\exp(-|x|^\alpha)$ *satisfy the asymptotic expansion*

$$a_n \sim \left(\frac{n}{C(\alpha)}\right)^{1/\alpha} \left(\sum_{k=0}^\infty \frac{A_k}{n^{i_k}} + (-1)^n \sum_{k=0}^\infty \frac{B_k}{n^{j_k}} \right) \qquad (10)$$

with $0 = i_0 < i_1 < \ldots$, $A_0 = 1$, $\alpha = j_0 < j_1 < \ldots$ $(\alpha > 1)$.

For more accurate predictions, one relates errors on $\mathbf{a}$ to errors on $\mathbf{F}(\mathbf{a})$ ($\mathbf{F}(\mathbf{a})$ is the sequence $\{F_m(Q; \mathbf{a})\}$) by $\mathbf{F}(\mathbf{a}) - \mathbf{F}(\mathbf{\mathring{a}}^{(n)}) \sim \mathbf{J}(\mathbf{a} - \mathbf{\mathring{a}}^{(n)})$, where J is the Jacobian matrix of the partial derivatives $\partial F_m(Q; \mathbf{a})/\partial a_k$. The elements of this matrix (computed at $\mathbf{\mathring{a}}^{(n)}$) are

$$J_{m,k} = \partial F_m(Q; \mathbf{\mathring{a}}^{(n)})/\partial \mathring{a}_k^{(n)}$$

$$= 2\mathring{a}_m \int_{\mathbb{R}} \int_{\mathbb{R}} p_m(x)p_{m-1}(y)p_k(x)p_{k-1}(y)\frac{Q'(x) - Q'(y)}{x - y} \, d\mathring{\mu}^{(n)}(x) \, d\mathring{\mu}^{(n)}(y) \qquad [9]$$

$$= \frac{8\tilde{a}_n}{\pi^2} \int_0^\pi \int_0^\pi \sin(m+1)\theta \sin m\psi \sin(k+1)\theta \sin k\psi \frac{Q'(2\tilde{a}_n \cos\theta) - Q'(2\tilde{a}_n \cos\psi)}{2\tilde{a}_n(\cos\theta - \cos\psi)} \, d\theta d\psi$$

Here again, keeping only the lowest order Fourier coefficient:

$$J_{m,k} \sim \frac{2\tilde{a}_n}{\pi^2} \int_0^\pi \int_0^\pi \cos(m-k)\theta \cos(m-k)\psi \frac{Q'(2\tilde{a}_r \cos\theta) - Q'(2\tilde{a}_n \cos\psi)}{2\tilde{a}_n(\cos\theta - \cos\psi)} d\theta \, d\psi$$

leaving a Toeplitz matrix of symbol $\mathring{\Phi}(\varphi) = \sum_{-\infty}^{\infty} J_{p,0} \exp(ip\varphi)$ such that

$$\frac{1}{\pi} \int_0^\pi \cos p\varphi \, \mathring{\Phi}(\varphi) \, d\varphi = \frac{2\tilde{a}_n}{\pi^2} \int_0^\pi \int_0^\pi \cos p\theta \cos p\psi \frac{Q'(2\tilde{a}_n \cos\theta) - Q'(2\tilde{a}_n \cos\psi)}{2\tilde{a}_n(\cos\theta - \cos\psi)} d\theta \, d\psi$$

i.e., $$\mathring{\Phi}(\varphi) = \frac{2\tilde{a}_n}{\pi} \int_0^\pi \frac{Q'(2\tilde{a}_n \cos\theta) - Q'(2\tilde{a}_n \cos(\theta+\varphi))}{2\tilde{a}_n(\cos\theta - \cos(\theta+\varphi))} d\theta$$

$\mathbf{J}^{-1}$ is approximately the Toeplitz matrix of symbol $1/\mathring{\Phi}$, so that for $|x|^c$:

$$a_n - \mathring{a}_n^{(n)} \sim \sum_{p=-\infty}^{\infty} (\mathbf{J}^{-1})_{n,n+p} \big(F_{n+p}(Q;\mathbf{a}) - F_{n+p}(Q;\mathring{\mathbf{a}}^{(n)}) \big)$$

$$\sim \sum_{-\infty}^{\infty} (\mathbf{J}^{-1})_{p,0} \left(n+p - n + n(-1)^{n+p} \sin(\pi\alpha/2) \frac{\Gamma(\alpha/2)\Gamma(1+\alpha/2)}{\pi(n+p)^\alpha} \right)$$

$$\sim (-1)^n \sin(\pi\alpha/2) \frac{\Gamma(\alpha/2)\Gamma(1+\alpha/2)}{\pi\mathring{\Phi}(\pi)} n^{1-\alpha}$$

suggesting $B_0 = (\alpha-1)\sin(\pi\alpha/2)(\Gamma(\alpha/2))^2/(2\pi)$ in (10), using

$$\mathring{\Phi}(\pi) = \pi^{-1}(2\tilde{a}_n)^{\alpha-1} \int_0^{\pi/2} (\cos\theta)^{\alpha-2} d\theta = \alpha(2\tilde{a}_n)^{\alpha-1} \Gamma((\alpha-1)/2)/(\Gamma(1/2)\Gamma(\alpha/2)).$$

Now, some horribly wrong mistake must have occurred somewhere, because very high accuracy (up to 200 digits, on the IBM 3090 of the University) calculations of instances of a_n for various values of α, followed by severe extrapolation devices designed to exhibit B_0, lead to the

PROBLEM. Show that $j_0 = \alpha$ and $B_0 = (\alpha-1)\sin(\pi\alpha/2)(1 - 1/\alpha)^\alpha (\Gamma(\alpha/2))^2/(2\pi)$ in (10) when $\alpha > 1$.

Where does this $(1 - 1/\alpha)^\alpha$ come from ?

REFERENCES

1. Bauldry W. C., Máté A., Nevai P., *Asymptotics for the solutions of systems of smooth recurrence equations*, Pacific J. Math. **133** (1988), 209–227.
2. Freud G., *On the coefficients of the recursion formulae of orthogonal polynomials*, Proc. Roy. Irish Acad. Sect. A(1) **76** (1976), 1–6.
3. Knopfmacher A., Lubinsky D. S., Nevai P., *Freud's conjecture and approximation of reciprocals of weights by polynomials*, Constr. Approx. **4** (1988), 9–20.
4. Lubinsky D. S., *A survey of general orthogonal polynomials for weights on finite and infinite intervals*, Acta ApplicandæMathematicæ **10** (1987), 237–296.

5. Lubinsky D. S., *Strong asymptotics for extremal errors and polynomials associated with Erdős-type weights*, Pitman Res. Notes Math. **202** (1989), Longman.
6. Lubinsky D. S., Mhaskar H. N., Saff E. B., *A proof of Freud's conjecture for exponential weights*, Constr. Approx. **4** (1988), 65–83.
7. Lubinsky D. S., Saff E. B., *Strong Asymptotics for Extremal Polynomials Associated with Weights on ℝ*, Lecture Notes Math., vol. 1305, Springer, 1988.
8. Magnus A. P., *A proof of Freud's conjecture about orthogonal polynomials related to $|x|^\rho \exp(-x^{2m})$ for integer m*, Lecture Notes Math., **1171**, Springer, 1985, pp. 362–372.
9. Magnus A. P., *On Freud's equations for exponential weights*, J. Approx. Theory **46** (1986), 65–99.
10. Máté A., Nevai P., Zaslavsky T., *Asymptotic expansion of ratios of coefficients of orthogonal polynomials with exponential weights*, Trans. Amer. Math. Soc. **287** (1985), 495–505.
11. Nevai P., *Géza Freud, orthogonal polynomials and Christoffel functions. A case study*, J. Approx. Theory **48** (1986), 3–167.
12. Nevai P., *Research problems in orthogonal polynomials,*, Approximation Theory VI, vol. 2, Academic Press, 1989, pp. 449–489.
13. Rahmanov E. A., *On asymptotic properties of polynomials orthogonal on the real axis*, Math. Sb. **119** (161) (1982), 163–203 (Russian); Math. USSR Sb. **47** (1984), 155–193. (English)
14. Van Assche W., *Asymptotics For Orthogonal Polynomials*, Lect. Notes Math., vol. 1265, Springer, 1987.

INSTITUT MATHÉMATIQUE
UNIVERSITÉ CATHOLIQUE DE LOUVAIN
CHEMIN DU CYCLOTRON, 2
B–1348 LOUVAIN-LA-NEUVE
BELGIUM

FAST DECREASING POLYNOMIALS AND POTENTIALS
IN THE PRESENCE OF AN EXTERNAL FIELD

V. TOTIK

PROBLEM 1. *Are there polynomials p_n of degree $n = 1, 2, \ldots$ such that*

$$p_n(0) = 1 \text{ and } |p_n(x)| \leqslant Ce^{-nx^2} \text{ for } x \in [-1, 1]? \tag{1}$$

Such fast decreasing polynomials imitate the "Dirac delta" and can be used as good polynomial kernel functions or to create well localized polynomial "partitions of unity". For a general estimate on fast decreasing polynomials see [2].

It can be shown [4] that if $\alpha < 1$, then there are polynomials with the properties

$$p_n(0) = 1 \text{ and } |p_n(x)| \leqslant Ce^{-\alpha nx^2} \text{ for } x \in [-1, 1],$$

while this cannot happen for $\alpha > 1$. Furthermore, the answer is YES if we replace (1) by

$$p_n(0) = 1 \text{ and } |p_n(x)| \leqslant e^{-n(x^2 + o(1))} \text{ for } x \in [-1, 1],$$

where the o is uniform in n and $x \in [-1, 1]$.

In general, if ϕ is an even function such that $\phi(\sqrt{x})$ is increasing and concave on $[0, 1]$, then the necessary and sufficient condition for the existence of polynomials with the properties

$$p_n(0) = 1 \text{ and } |p_n(x)| \leqslant e^{-n(\phi(x) + o(1))} \text{ for } x \in [-1, 1] \tag{2}$$

is that the relation

$$\frac{2}{\pi} \int_0^1 \frac{\phi(x)}{x^2 \sqrt{1 - x^2}} \, dx \leqslant 1$$

holds.

When we drop the concavity of $\phi(\sqrt{x})$, then a necessary and sufficient condition is that $\phi^*(0) = 0$, where ϕ^* is the smallest superharmonic majorant of ϕ defined as the infimum of all superharmonic functions h which majorize ϕ on $[-1, 1]$ and are bounded from below around infinity by const $- \log |z|$. This, however, is an implicit condition that is very difficult to verify.

The above problems are closely connected with logarithmic potentials with external fields (see [1], [3]). In fact, if we take the logarithm of both sides of (2), then we can see that the "weighted" potential

$$\frac{1}{n} \log \frac{1}{|p_n(x)|} + \phi(x)$$

This material is based upon work supported by the National Science Foundation under Grant No DMS–9024901, and by NATO under Grant No. CRG.870806

essentially takes its maximum at $x = 0$. This can happen if and only if $x = 0$ belongs to the support of the extremal measure in the energy problem in the presence of the external field ϕ: minimize

$$\iint \log \frac{1}{|z - t|} \, d\mu(t) \, d\mu(z) - 2 \int \phi(t) \, d\mu(t)$$

for all probability measures on $[-1, 1]$. Now if ϕ is even, then using symmetry, the support turns out to be one or two intervals provided $\phi(\sqrt{x})$ is concave on $[0, 1]$, and then we can explicitly determine this support. If, however, this condition is not satisfied, e.g. $\phi(x) = c|x|^\beta$ with $\beta > 2$, then the support can be wild (can be a Cantor type set) and no method is available for determining if 0 belongs to the support or not. Even the following is not known:

PROBLEM 2. *Let $\beta > 2$. Determine explicitly the largest possible constant C such that there are polynomials p_n of degree $n = 1, 2, \ldots$ with*

$$p_n(0) = 1 \text{ and } |p_n(x)| \leqslant C e^{-cn|x|^\beta} \text{ for } x \in [-1, 1].$$

REFERENCES

1. Gonchar A. A., Rahmanov E. A., *Equilibrium measure and distribution of zeros of extremal polynomials*, Mat. Sb. **125** (1984), no. 167, 117–127. (Russian)
2. Ivanov K. G., Totik V., *Fast decreasing polynomials*, Constructive Approximation 6 (1990), 1–20.
3. Mhaskar H. N., Saff E. B., *Where does the sup norm of a weighted polynomial live?*, Constructive Approx. **1** (1985), 71–91.
4. Totik V., *Fast decreasing polynomials via potentials*, manuscript.

BOLYAI INSTITUTE
ARADI VÉRTANÚK TERE 1
SZEGED 6720
HUNGARY

AND

DEPARTMENT OF MATHEMATICS
UNIVERSITY OF SOUTH FLORIDA
TAMPA, FL 33620
USA

13.6

FREUD'S CONJECTURE FOR ORTHOGONAL POLYNOMIALS
WITH NON-SYMMETRIC EXPONENTIAL WEIGHTS

WALTER VAN ASSCHE

Let $p_n(x)$ $(n = 0, 1, 2, \ldots)$ be a sequence of orthonormal polynomials on the real line with respect to a probability measure μ,

$$\int_{-\infty}^{\infty} p_n(x)p_m(x)\, d\mu(x) = \delta_{m,n}, \qquad m, n \geqslant 0.$$

It is well known that these polynomials satisfy a three-term recurrence relation

$$xp_n(x) = a_{n+1}p_{n+1}(x) + b_np_n(x) + a_np_{n-1}(x), \qquad n \geqslant 0,$$

with $a_n > 0$ and $b_n \in \mathbb{R}$. Freud's conjecture relates the asymptotic behavior of the tail of μ with the asymptotic behavior of the recurrence coefficients:

FREUD'S CONJECTURE [1]. *Suppose that μ is absolutely continuous with weight function $w(x) = e^{-|x|^\alpha}$ $(-\infty < x < \infty, \alpha > 0)$, then*

$$\lim_{n \to \infty} \frac{a_n}{n^{1/\alpha}} = \frac{1}{2}\left(\frac{\sqrt{\pi}\,\Gamma\left(\frac{\alpha}{2}\right)}{\Gamma\left(\frac{\alpha+1}{2}\right)}\right)^{1/\alpha}.$$

Because of symmetry it follows also that $b_n = 0$ for all $n \in \mathbb{N}$. A more general statement has been proved by Lubinsky, Mhaskar and Saff [2] for weight functions of the form $w(x) = \exp(-2Q(x))$ $(-\infty < x < \infty)$:

THEOREM (Lubinsky, Mhaskar and Saff). *Suppose Q is even, continuous and $Q'(x)$ exists for $x > 0$, while $xQ'(x)$ remains bounded as $x \to 0+$. Further, assume that $Q'''(x)$ exists for x large enough and for some $C > 0$ and $\alpha > 0$*

$$Q'(x) > 0, \qquad x \text{ large enough,}$$
$$x^2|Q'''(x)|/Q'(x) \leqslant C, \qquad x \text{ large enough,}$$

and

$$\lim_{x \to \infty} \frac{xQ''(x)}{Q'(x)} = \alpha - 1.$$

This material is based upon work supported by the National Science Foundation under Grant No DMS-9024901, and by NATO under Grant No. CRG.870806

The author is a Research Associate of the Belgian National Fund for Scientific Research

If c_n is the positive root of

$$n = \frac{2}{\pi} \int_0^1 \frac{c_n x Q'(x)(c_n x)}{\sqrt{1-x^2}} \, dx,$$

then

$$\lim_{n \to \infty} a_n/c_n = \frac{1}{2}.$$

Again the symmetry implies $b_n = 0$ for all non-negative integers. The weight function $w(x) = \exp(-2Q(x))$ can still be multiplied by some factors that do not change the exponential behavior near ∞ (such as generalized Jacobi factors) (see [2]). The case $Q(x) = |x|^\alpha/2$ satisfies all the necessary conditions.

The previous result is basically a result about the asymptotic behavior of the recurrence coefficients a_n. The fact that Q is even implies that the weight function is symmetric, which in turn forces the recurrence coefficients b_n to vanish. Small perturbations of the even weight $\exp(-2Q(x))$, such as a multiplication by a non-symmetric positive function ψ with $\lim_{|x| \to \infty} \psi(x) = 1$, are possible but this will lead to

$$\lim_{n \to \infty} b_n/c_n = 0,$$

which one can still consider as an asymptotically symmetric case.

In [4] and [5] exponential weights of the form $\exp(-2Q(x))$ on $(-\infty, \infty)$ are considered with

$$Q(x) \sim \begin{cases} A x^\alpha L^\alpha(x), & x > 0, \\ B|x|^\alpha L^\alpha(|x|), & x < 0, \end{cases} \tag{1}$$

where L is a slowly varying function i.e., a positive and measurable function on $\mathbb{R}^+$ such that for every $t > 0$

$$\lim_{x \to \infty} \frac{L(xt)}{L(x)} = 1.$$

The asymptotic behavior of the zeros of the corresponding polynomials turns out to be the Nevai–Ullman density v with support $[b - 2a, b + 2a]$ and parameter α, where

$$a = \frac{1+c}{4} \left(\frac{\sqrt{\pi}}{A} \frac{\Gamma(\alpha)}{\Gamma(\alpha + \frac{1}{2})} F(\frac{1}{2}, \alpha; \alpha + \frac{1}{2}; -c) \right)^{1/\alpha}, \tag{2}$$

$$b = \frac{1-c}{2} \left(\frac{\sqrt{\pi}}{A} \frac{\Gamma(\alpha)}{\Gamma(\alpha + \frac{1}{2})} F(\frac{1}{2}, \alpha; \alpha + \frac{1}{2}; -c) \right)^{1/\alpha}, \tag{3}$$

with $0 < c \leqslant 1$ the unique solution of

$$c^\alpha \frac{F(\frac{1}{2}, \alpha; \alpha + \frac{1}{2}; -c)}{F(\frac{1}{2}, \alpha; \alpha + \frac{1}{2}; -1/c)} = \frac{A}{B}. \tag{4}$$

This Nevai–Ullman density is just a Mellin convolution of the arcsin density on the interval $[b - 2a, b + 2a]$ and the beta density $f(t) = \alpha t^{\alpha-1}$ on $[0, 1]$. The asymptotic behavior of the zeros corresponds to the asymptotic behavior of the n-th root of the orthogonal polynomials. On the other hand, the asymptotic behavior of the recurrence coefficients corresponds to the asymptotic behavior of the ratio of two consecutive orthogonal polynomials. By Cesàro's lemma ratio asymptotics implies root asymptotics, and therefore if the asymptotic behavior of the recurrence coefficients is known, then the asymptotic zero distribution follows from this. We have made a study of the asymptotic distribution of zeros of orthogonal polynomials defined by the three-term recurrence relation in [3, §4.3] and these results suggest the following

CONJECTURE. *Suppose that the weight function* $w(x) = \exp(-2Q(x))$ *satisfies* (1), *with* $0 < A/B \leqslant 1$ *when* $\alpha > \frac{1}{2}$ *and* $\cos(\pi\alpha) < A/B \leqslant 1$ *when* $0 < \alpha \leqslant \frac{1}{2}$, *then the recurrence coefficients have the asymptotic behavior*

$$\lim_{n\to\infty} a_n/c_n = a, \quad \lim_{n\to\infty} b_n/c_n = b, \tag{5}$$

where $c_n = n^{1/\alpha}L^*(n^{1/\alpha})$, *with* L^* *a de Bruijn conjugate for* L *i.e., a slowly varying function such that* $L^*(x)L(xL^*(x)) \sim 1$ *as* $x \to \infty$, *and* a, b *given by* (2)–(4). *When* $0 < \alpha \leqslant \frac{1}{2}$ *and* $0 < A/B \leqslant \cos(\pi\alpha)$, *then* $c = 1$ *in* (2)–(3).

Notice that the limits in (5) satisfy $b^2 - 4a^2 < 0$. In [3] we have also encountered asymptotic zero distributions given by a Nevai–Ullman density on $[b - 2a, b + 2a]$ with parameter α and $b^2 - 4a^2 > 0$.

PROBLEM. *Which measures* μ *give rise to the Nevai-Ullman density on* $[b - 2a, b + 2a]$ *with* $b^2 - 4a^2 > 0$?

Example 1. The Meixner–Pollaczek polynomials $M_n(x)$ are orthogonal on $(-\infty, \infty)$ with weight function

$$w(x) = \left| \Gamma\left(\frac{\eta + ix}{2}\right) \right|^2 \exp(-x \arctan \delta),$$

where $\eta > 0$ and $\delta \in \mathbb{R}$. This is a non-symmetric weight as in (1), and the recurrence coefficients have the behavior as conjectured.

Example 2. The Meixner polynomials $m_n(x)$ are orthogonal on the integers:

$$\sum_{k=0}^{\infty} m_i(k)m_j(k)(\beta)_k \frac{c^k}{k!} = (1 - c)^{-\beta}c^{-i}i!(\beta)_i\delta_{i,j},$$

where $\beta > 0$ and $0 < c < 1$. The recurrence formula is

$$cm_{n+1}(x) = [(c - 1)x + (1 + c)n + c\beta]m_n(x) - n(n + \beta - 1)m_{n-1}(x)$$

so that the recurrence coefficients for the orthonormal polynomials are

$$b_n = \frac{(1+c)n + \beta c}{1 - c}, \quad a_n = \frac{\sqrt{cn(n + \beta - 1)}}{1 - c},$$

and asymptotically we have

$$a = \lim_{n\to\infty} a_n/n = \frac{\sqrt{c}}{1 - c}, \quad b = \lim_{n\to\infty} b_n/n = \frac{1 + c}{1 - c}.$$

Note that $b^2 - 4a^2 = 1 > 0$. This example shows that the negative binomial measure gives rise to a zero distribution on $[b - 2a, b + 2a]$ with $b^2 - 4a^2 > 0$.

Chapter 13. ORTHOGONAL POLYNOMIALS

REFERENCES

1. Freud G., *On the coefficients in the recursion formulae of orthogonal polynomials*, Proc. Roy. Irish Acad. Sect. A **76** (1976), 1–6.
2. Lubinsky D. S., Mhaskar H. N. and Saff E. B., *A proof of Freud's conjecture for exponential weights*, Constr. Approx. **4** (1988), 65–83.
3. Van Assche W., *Asymptotics for orthogonal polynomials*, Lecture Notes in Mathematics, vol. 1265, Springer-Verlag, Berlin, 1987.
4. Van Assche W., *Norm behavior and zero distribution for orthogonal polynomials with nonsymmetric weights*, Constr. Approx. **5** (1989), 329–345.
5. Van Assche W., *Orthogonal polynomials on non-compact sets*, Acad. Analecta, Med. Konink. Acad. Wetensch. Lett. Sch. Kunsten België **51** (1989), no. 2, 1–36.

DEPARTMENT OF MATHEMATICS
KATHOLIEKE UNIVERSITEIT LEUVEN
CELESTIJNENLAAN 200 B
B-3001 HEVERLEE (LEUVEN)
BELGIUM
email: fgaee03@cc1.kuleven.ac.be

Chapter 14

UNIQUENESS, MOMENTS, NORMALITY

Edited by

J. Brennan
Dept. of Mathematics
University of Kentucky
Lexington, KY 40506
USA

A. Volberg
Dept. of Mathematics
Michigan State University
East Lansing, MI 48824
USA
and
Steklov Math. Institute
St. Petersburg branch
Fontanka 27
St. Petersburg, 191011
Russia

V. P. Havin
Dept. of Math. and Mechanics
St. Petersburg State University
Staryi Peterhof
St. Petersburg, 198904
Russia

INTRODUCTION

Problems collected in this chapter are variations on the following theme: a "sufficiently analytic" function vanishing "intensively enough" is identically zero. The words in quotation makers get an exact meaning in accordance with every concrete situation. For instance, dealing with the uniqueness of the solution of a moment problem we often exploit traces of the analyticity of the function $\alpha \to \int x^\alpha d\mu(x)$.

The theme is wide. It encompasses such phenomena as the quasi-analyticity and the uniqueness of the moment problem, and borders on normal families (see e.g., Problem 14.5), various refinements of the maximum principle and approximation. Its importance hardly needs any explanation. The Uniqueness marks (more or less explicitly) all contents of this book. After all, every linear approximation problem (and the book abounds in such problems) is a dual reformulation of a uniqueness problem.

Every problem of this Chapter (except for 14.3 and 14.7) deals not only with "the pure uniqueness" but with other topics as well. Problem 14.1 is connected not only with zeros of some function classes but with a moment problem and with Fourier-Laplace transforms of measures; in Problem 14.6 the uniqueness in analytic Gevrey classes is considered in connection with peak sets for Hölder analytic functions. "Old" Problems 14.8 and 14.10 deal (from different points of views) with differential and differential-like operators (both have evoked a great interest, see respective commentary). Problem 14.4 has certain relation to spectral operators and to the "anti-locality" of some convolution operators (in contrast with "the locality" of convolutions discussed in 14.10). Problem 14.5 is a quantitative variation on the title theme and 14.13 gravitates towards spectral analysis-synthesis of Chapter 11. Problem 14.11 is aimed at approximation properties of exponentials and concerns also some aspects of quasianalyticity, as does Problem 14.12. Problem S.14.19 deals with an interesting "perturbation" of the "$\int \log |f| > -\infty$"-theorem.

The theme of this chapter emerges in some Problems of other chapters.

SOME OPEN PROBLEMS IN THE THEORY OF
REPRESENTATIONS OF ANALYTIC FUNCTIONS

M. M. Djrbashyan

I. Denote by Ω the set of functions ω, satisfying the following conditions:

1. $\omega(x) > 0$, $\omega \in C([0,1])$;
2. $\omega(0) = 1$, $\int_0^1 \omega(x)\,dx < +\infty$;
3. $\int_0^1 |1 - \omega(x)| x^{-1} dx < +\infty$.

In the factorization theory of meromorphic functions in the unit disc $\mathbb{D}$, developed in [1], the following theorem on solvability of the Hausdorff moment problem, proved in [2], played an important rôle: the Hausdorff moment problem

$$\mu_n = \int_0^1 x^n d\alpha(x) \quad (n = 0, 1, 2, \dots)$$

where

$$\mu_0 = 1, \quad \mu_n = \left(n \int_0^1 \omega(x) x^{n-1} dx \right)^{-1} \quad (n = 1, 2, \dots)$$

and $\omega \in \Omega$, $\omega \uparrow$, has a solution in the class of nondecreasing and bounded functions α on [0,1].

Assuming $\omega_j \in \Omega$ $(j = 1, 2)$, consider the Hausdorff moment problem of the form

$$(1) \qquad \lambda_n = \int_0^1 x^n d\beta(x) \quad (n = 0, 1, 2, \dots)$$

where

$$(2) \qquad \lambda_0 = 1, \quad \lambda_n = \left(\int_0^1 \omega_1(x) x^{n-1} dx \right) \left(\int_0^1 \omega_2(x) x^{n-1} dx \right)^{-1} \quad (n = 1, 2, \dots).$$

CONJECTURE. *The moment problem* (1)–(2) *has a solution in the class of nondecreasing and bounded functions on* [0, 1], *or at least in the class of functions β with bounded variation $V_{[0,1]}(\beta) < +\infty$ provided the functions ω_j are monotone on* [0, 1] *and ω_1/ω_2 is nonincreasing on* [0, 1].

The proof of this conjecture, which is true in the special case $\omega_1(x) \equiv 1$, would lead to important results on embeddings of classes $N\{\omega_j\}$ $(j = 1, 2)$ of meromorphic functions in $\mathbb{D}$, considered in [1].

II. Denote by Ω_∞ the set of functions ω, satisfying the following conditions:

1. the function ω is continuous and nonincreasing on $[0, \infty)$, $\omega(x) > 0$;
2. the integrals

$$\Delta_k = k \int_0^\infty \omega(x) x^{k-1} dx, \quad (k = 1, 2, \dots)$$

are finite.

Putting $\Delta_0 = 1$, consider entire functions of z:

$$W_\omega^{(\infty)}(z; \xi) = \int_{|\xi|}^\infty \frac{\omega(x)}{x} dx -$$

$$\sum_{k=1}^\infty \left(\xi^{-k} \int_0^{|\xi|} \omega(x) x^{k-1} dx - \bar{\xi}^{-k} \int_{|\xi|}^\infty \omega(x) x^{-k-1} dx \right) \frac{z^k}{\Delta_k} \quad (0 < |\xi| < \infty)$$

and

$$A_\omega^{(\infty)}(z; \xi) = \left(1 - \frac{z}{\xi} \right) e^{-W_\omega^{(\infty)}(z; \xi)} .$$

Let, finally, $\{z_k\}_1^\infty$ $(0 < |z_k| \leqslant |z_{k+1}| < \infty)$ be an arbitrary sequence of complex numbers such that

(3)
$$\sum_{k=1}^\infty \int_{|z_k|}^\infty \frac{\omega(x)}{x} dx < \infty .$$

CONJECTURE. *Under condition (3) the infinite product*

$$\pi_\omega(z; z_k) = \prod_{k=1}^\infty A_\omega^{(\infty)}(z; z_k)$$

converges on any compact set, not containing points of $\{z_k\}_1^\infty$, provided ω satisfies the additional condition

$$\frac{d \log \omega(x)}{d \log x} \downarrow -\infty, \quad x \uparrow +\infty .$$

The validity of this conjecture for some special cases and in particular for

$$\omega(x) = \frac{\rho^{\sigma^\mu}}{\Gamma(\mu)} \int_x^\infty e^{-\sigma t^\rho} t^{\mu \rho - 1} dt ,$$

where ρ $(0 < \rho < \infty)$, μ $(0 < \mu < \infty)$ and σ $(0 < \sigma < \infty)$ are arbitrary parameters, was provided in [3].

III. Let μ be a complex function on $[0, \infty)$ such that

(4)
$$V_\mu(r) \overset{\text{def}}{=} \int_0^\infty r^t |d\mu(t)| < \infty, \quad r \geqslant 0 .$$

Then it is obvious that the function

$$f_\mu(z) = \int_0^\infty z^t \, d\mu(t)$$

is regular on the Riemann surface

$$G_\infty = \{z : |\operatorname{Arg} z| < \infty, \quad 0 < |z| < \infty\}$$

and that

$$\sup_{|\varphi| < \infty} |f_\mu(re^{i\varphi})| \leqslant V_\mu(r), \quad r \in [0, \infty) \, .$$

In view of this the following problem arises.

QUESTION. *Let f be an analytic function on G_∞ satisfying*

$$\widetilde{M}_f(r) \overset{\text{def}}{=} \sup_{|\varphi| < \infty} |f(re^{i\varphi})| < \infty, \quad r \in [0, \infty) \, .$$

What supplementary condition is to be imposed on f to ensure the representation

(5)
$$f(z) = \int_0^\infty z^t \, d\mu_f(t), \quad z \in G_\infty \, ,$$

with a function μ_f satisfying (4)?

IV. Denote by $H^p(\alpha)$ $(0 < p < \infty, -1 < \alpha < \infty)$ the class of all analytic functions f on $\mathbb{D}$ such that

$$\iint_{\mathbb{D}} (1 - r^2)^\alpha \, |f(re^{i\theta})|^p \, r \, dr \, d\theta < \infty \, .$$

Let $f \in H^p(\alpha)$ and $\{\alpha_j\}_1^\infty$ $(0 < |\alpha_j| \leqslant |\alpha_{j+1}| < 1)$ be the sequence of zeros of f enumerated in accordance with their multiplicity. Denote by $n(t)$ the number of α_j's in the disc $|z| \leqslant t$ $(0 < t < 1)$ and let

$$N_\alpha(r) = \int_0^r \frac{n(t)}{t} \, dt \quad (0 < r < 1).$$

It is well-known (cf. [4]) that if $f \in H^p(\alpha)$ and $f \not\equiv 0$, then

(6)
$$\int_0^1 (1 - r)^\alpha e^{pN_\alpha(r)} \, dr < \infty \, .$$

In this connection it is natural to state the following:

CONJECTURE. *Let $\{\alpha_j\}_1^\infty$ be an arbitrary sequence in $\mathbb{D}$ satisfying (6). Then there exists a sequence of numbers $\{\theta_j\}_1^\infty$ $(0 \leqslant \theta_j < 2\pi)$ and a function f_*, $f_* \not\equiv 0$, $f_* \in H^p(\alpha)$, such that*

$$f_*(\alpha_j e^{j\theta_j}) = 0, \quad (j = 1, 2, \dots) \, .$$

Note that a statement equivalent to (6) and some other results about zeros of functions of the class $H^p(\alpha)$ (if $\alpha \geqslant 0$) were obtained in [5] much later than [4].

REFERENCES

1. Džrbašjan M. M., *Theory of factorization of functions meromorphic in the disk*, Mat. Sb. **79** (1969), no. 4, 517–615 (Russian); English transl. in Math. USSR–Sbornik **8** (1969), 493–591.
2. Džrbašjan M. M., *A generalized Riemann-Liouville operator and some of its applications*, Izv. Akad. Nauk SSSR **32** (1968), 1075–1111 (Russian); English transl. in Math. USSR–Izvestija **2** (1968), 1027–1064.
3. Džrbašjan M. M., *On an infinite product*, Izv. Akad. Nauk Armjan. SSR **13** (1978), 177–208 (Russian); English transl. in Soviet Math. Dokl. **19** (1978), 621–625.
4. Džrbašjan M. M., *On the problem of the representability of analytic functions*, Soobsc. Inst. Mat.-Mech. Akad. Nauk Armjan. SSR **2** (1948), 3–39. (Russian)
5. Horowitz C., *Zeros of functions in the Bergman spaces*, Duke Math. J. **41** (1974), 693–710.

MATHEMATICAL INSTITUTE
BAREKAMUTYAN 24B
EREVAN-19, 375019
ARMENIA

CHAPTER EDITORS' COMMENTARY

1. It was shown by S. V. Hruščëv that the mere finiteness of $\widetilde{M}_f(r)$ does not imply the representation (5) (see the previous edition of the Problem Book).

2. I. V. Ostrovskii proved in [6] that the first conjecture is true (under a somewhat stronger condition on w_1/w_2; this condition is shown in [6] to be sharp).

REFERENCE

6. Ostrovskii I. V., *A Hausdorff moment problem*, Izv. Akad. Nauk Armjan. SSR **25** (1990), 91–96. (Russian)

KERNEL FUNCTIONS, MOMENTS,
AND COMPOSITION OPERATORS

T. Kriete,   B. MacCluer

Let $G(r)$ be a positive continuous function on $(0,1)$ for which $G(|z|)$ is integrable with respect to area measure dA on the open unit disk $\mathbb{D}$ in the complex plane. We assume $G(r)$ is non-increasing for r near 1. Let A_G^2 denote the weighted Bergman space (a Hilbert space) consisting of all functions f analytic on $\mathbb{D}$ for which the norm

$$\|f\| = \left\{ \int_{\mathbb{D}} |f(z)|^2 G(|z|) \, dA(z) \right\}^{1/2}$$

is finite. In terms of the moments $p_n = \int_0^1 r^{2n+1} G(r) \, dr$, the norm can be written as

$$\|f\|^2 = 2\pi \sum_{n=0}^{\infty} p_n |a_n|^2$$

where $\{a_n\}_{n=0}^{\infty}$ are the Taylor coefficients of f. The space A_G^2 has reproducing kernels $\{k_w : w \in \mathbb{D}\}$ given by

$$k_w(z) = \frac{1}{2\pi} \sum_{n=0}^{\infty} \frac{1}{p_n} (\bar{w}z)^n,$$

so that $f(w) = \langle f, k_w \rangle$ for f in A_G^2 and w in $\mathbb{D}$. We note that $\|k_w\|^2 = k_w(w)$.

Let $\varphi : \mathbb{D} \to \mathbb{D}$ be analytic and consider the composition operator C_φ on A_G^2, given by $C_\varphi f = f \circ \varphi$. It is well known that every C_φ is bounded on A_G^2 when G is one of the *standard weights*: $G(r) = (1-r)^\alpha$, $\alpha \geq 0$ [3]. For more general G, boundedness can fail. The broad question we want to consider is how to characterize boundedness.

If ξ lies on the unit circle $\mathbb{T}$, let $|\varphi'(\xi)|$ denote the modulus of the angular derivative of φ at ξ, which can be written as

$$|\varphi'(\xi)| = \liminf_{z \to \xi} \frac{1 - |\varphi(z)|}{1 - |z|};$$

we have $0 < |\varphi'(\xi)| \leq \infty$. Moreover, let us say that G is a *fast* weight provided

$$\lim_{r \to 1} \frac{G(r)}{(1-r)^\alpha} = 0$$

for all $\alpha > 0$.

THEOREM 1 [2]. *If G is fast and $|\varphi'(\xi)| < 1$ for some ξ in $\mathbb{T}$, then C_φ is not a bounded operator on A_G^2.*

Thus for fast weights G, automorphisms of $\mathbb{D}$ (excepting rotations) do not induce bounded composition operators on A_G^2. In the positive direction, we have the following. Let $M(r) = \max_\theta |\varphi(re^{i\theta})|$.

THEOREM 2 [2]. *If*

$$
\text{(1)} \qquad\qquad \limsup_{r \to 1} \frac{G(r)}{G(M(r))} < \infty,
$$

then C_φ is bounded on A_G^2.

QUESTION 1. *Is (1) necessary for boundedness?*

We believe the answer is 'yes'. When $G(r) = (1-r)^\alpha$, (1) is always true and amounts to the well-known fact that $\inf_{\xi \in \mathbb{T}} |\varphi'(\xi)| > 0$.

QUESTION 2. *If (1) fails, must*

$$
\text{(2)} \qquad\qquad \limsup_{r \to 1} \frac{\|k_{M(r)}\|}{\|k_r\|} = \infty \ ?
$$

Since $C_\varphi^* k_w = k_{\varphi(w)}$, a 'yes' answer to Question 2 implies a 'yes' to Question 1. Moreover, Theorem 2 implies that (2) fails when (1) holds. It is conceivable that Question 2 even has a 'yes' answer when $M(r)$ is replaced by any nice function increasing to 1 as $r \to 1$.

Two examples are known to us. If $G(r) = (1 - r)^\alpha$, it is well known that

$$
\text{(3)} \qquad\qquad \|k_r\|^2 \sim \frac{1}{(1-r)^{\alpha+2}} = \frac{1}{(1-r)^2 G(r)} \quad \text{as } r \to 1.
$$

Here $\sim$ indicates the ratio of the quantities is bounded above and away from zero. In this example, (2) cannot occur. On the other hand, some delicate estimates of Trent [4] show that for the fast weights

$$
G(r) = (1 - r)^\alpha \exp\left(-B\frac{1}{1-r}\right), \quad \alpha \text{ real}, \quad B > 0,
$$

one has

$$
\text{(4)} \qquad\qquad \|k_r\|^2 \sim \frac{1}{(1-r)^3 G(r)} \quad \text{as } r \to 1.
$$

It follows from (4) that in this instance, Question 2 has a 'yes' answer.

Some further heuristic evidence for an affirmative answer to Question 2 is provided by a condition [2] which is equivalent (for fast weights satisfying a mild regularity hypothesis) to compactness of the operator C_φ, namely

$$
\text{(5)} \qquad\qquad \lim_{r \to 1} \frac{G(r)}{G(M(r))} = 0.
$$

When (5) holds, the compactness for C_φ easily yields

$$\lim_{r \to 1} \frac{\|k_{M(r)}\|}{\|k_r\|} = 0.$$

In this implication, then, we merely want to replace $\complement$ (in both places) by ∞ (and lim by lim sup). Perhaps an indirect solution can be based on this. Failing that, one can follow Trent and try to estimate the moments p_n and then $\|k_r\|$ directly. This seems difficult. A general estimate on p_n in terms of G has been found by Dynkin [1, p. 338], but it seems to contain too much slack to be of use here.

Estimates (3) and (4) suggest a final problem.

QUESTION 3. *What, in general, can be said about the function* $G(r)\|k_r\|^2$?

REFERENCES

1. Koosis P., *The Logarithmic Integral I*, Cambridge Studies in Advanced Mathematics 12, Cambridge University Press, 1988.
2. Kriete T., MacCluer B., *Composition operators on large weighted Bergman spaces*, Indiana Univ. Math. J. (to appear).
3. MacCluer B., Shapiro J., *Angular derivatives and compact composition operators on the Hardy and Bergman spaces*, Can. J. Math. **38** (1986), 878–906.
4. Trent T., *A measure inequality*, preprint.

DEPARTMENT OF MATHEMATICS
MATHEMATICS-ASTRONOMY BUILDING
UNIVERSITY OF VIRGINIA
CHARLOTTESVILLE, VA 22903
USA

DEPARTMENT OF MATHEMATICS
UNIVERSITY OF RICHMOND
RICHMOND, VA 23173
USA

SETS OF UNIQUENESS FOR ANALYTIC FUNCTIONS
WITH FINITE DIRICHLET INTEGRAL

V. Havin,   S. Hruščëv

Let $\mathcal{X}$ be a class of functions analytic in $\mathbb{D}$. A closed subset E of the closed disc clos $\mathbb{D}$ is said to be a uniqueness set for $\mathcal{X}$ (briefly $E \in \mathcal{E}(\mathcal{X})$) if

$$f \in \mathcal{X}, \quad f|E \equiv 0 \Rightarrow f \equiv 0.$$

(It is assumed that $f(\zeta) \overset{\text{def}}{=} \lim_{r \to 1-0} f(r\zeta)$ at $\zeta \in \mathbb{T} \cap E$). The structure of $\mathcal{E}(\mathcal{X})$–sets is well understood for many important classes $\mathcal{X}$ (see [1], [2]; [3] contains a short survey). The same cannot be said about the family $\mathcal{E}(\mathcal{D}_A)$, $\mathcal{D}_A$ being the space of all functions analytic in $\mathbb{D}$ with finite Dirichlet integral $\frac{1}{\pi} \iint_{\mathbb{D}} |f'(z)|^2 \, dx \, dy < +\infty$. The description of $\mathcal{E}(\mathcal{D}_A)$ seems to have to do not only with the Beurling–Carleson condition (see (1) below) but with capacity characteristics of sets.

We propose two conjectures concerning subsets of $\mathbb{T}$ in $\mathcal{E}(\mathcal{D}_A)$.

Associate with every closed set $F \subset \mathbb{T}$ a (unique) closed set F^*, $F^* \subset F$ so that $\text{cap}(F \setminus F^*) = 0$ (cap stands for the capacity corresponding to the logarithmic kernel) and every non-empty relatively open (in $\mathbb{T}$) part of F^* has positive logarithmic capacity.

CONJECTURE 1. *A closed subset F of $\mathbb{T}$ does not belong to $\mathcal{E}(\mathcal{D}_A \cap C_A)$* ($C_A$ stands for the disc-algebra) *iff*

1. $mF = 0$;
2. $\displaystyle\int_{\mathbb{T}} \log \text{dist}(\zeta, F^*) \, dm(\zeta) > -\infty.$

The difficulty of this problem is caused by the fact that functions in $\mathcal{D}_A$ possess no local smoothness on $\mathbb{T}$.

The conjecture agrees with all boundary uniqueness theorems for $\mathcal{D}_A$ we are aware of. These are two.

THEOREM (Carleson [4]). *Suppose that $F \subset \mathbb{T}$, clos $F = F$, $mF = 0$ and for some $\alpha > 0$ the inequality $C_\alpha (F \cap I(\zeta, \delta)) > \text{const } \delta$ holds for an arbitrary $\zeta \in F$ and every $\delta > 0$, $I(\zeta, \delta)$ being the arch of length 2δ centered at ζ. Then $F \in \mathcal{E}(\mathcal{D}_A)$ iff*

$$(1) \qquad\qquad \int_{\mathbb{T}} \log \text{dist}(\zeta, F) \, dm(\zeta) = -\infty.$$

Here C_α denotes the capacity corresponding to the kernel $|x-y|^{-\alpha}$. A set F satisfying the conditions of the Carleson theorem coincides with F^* because $C_\alpha(E) > 0$ implies $\text{cap}(E) > 0$.

THEOREM (Maz'ya–Havin [6]). *Suppose that $F \subset \mathbb{T}$, $\operatorname{clos} F = F$ and that there exists a family Ω of mutually disjoint open arcs I satisfying $F \subset \bigcup_{I \in \Omega} I$ and*

$$(2) \qquad \sum_{I \in \Omega} (mI) \cdot \log \frac{\operatorname{cap}(F \cap I)}{m(I)} = +\infty.$$

Then $F \in \mathcal{E}(\mathcal{D}_A)$.

Evidently (2) implies $\sum_{I \in \Omega} mI \cdot \log \frac{1}{m(I)} = +\infty$. Any family $\{l_\nu\}$ of open mutually disjoint arcs, which almost covers an arc I (i.e., $m(I \setminus \bigcup_\nu l_\nu) = 0$, satisfies $mI \cdot \log \frac{1}{mI} \leqslant \sum_\nu l_\nu \log \frac{1}{l_\nu}$. This remark is an easy consequence of the sub-additivity of $x \log \frac{1}{x}$. Therefore (2) implies the divergence of the series $\sum_\nu l_\nu \log \frac{1}{l_\nu}$, $\{l_\nu\}$ being the family of complementary intervals of F^*, provided $mF = 0$.

To state the second conjecture consider a class $\mathcal{Y}$ of non-negative functions defined on $\mathbb{T}$. A closed subset E is said to belong to $\mathcal{L}(\mathcal{Y})$ if

$$h \in \mathcal{Y}, \quad h|E \equiv 0 \Rightarrow \int_{\mathbb{T}} \log h \, dm = -\infty.$$

Let $\mathcal{D}_+$ be the set of all traces on $\mathbb{T}$ of non-negative functions from $W_2^1(\mathbb{C})$ (i.e., the functions in $L^2(\mathbb{C})$ with square summable generalized gradient).

CONJECTURE 2.

$$(3) \qquad \mathcal{E}(\mathcal{D}_A) = \mathcal{L}(\mathcal{D}_+).$$

Equality (3) (if true) permits us to separate the difficulties connected with the analyticity of functions of $\mathcal{D}_A$ from those of purely real character (such as the investigation of $\int \log h \, dm$ for non-negative h's in W_2^1).

The inclusion $\mathcal{L}(\mathcal{D}_+) \subset \mathcal{E}(\mathcal{D}_A)$ is obvious because $|f| \, | \, \mathbb{T} \in \mathcal{D}_+$ for $f \in \mathcal{D}_A$. The proofs of the theorems cited above are based precisely upon this inclusion (and upon Jensen's inequality). Here is another remark suggesting that (3) is a "right" analogue for the Beurling–Carleson theorem. This theorem asserts that $\mathcal{E}(\operatorname{Lip}_A(\alpha)) = \mathcal{L}(\operatorname{Lip}_A(\alpha))$. Here $\operatorname{Lip}_A(\alpha) \overset{\text{def}}{=} C_A \cap \operatorname{Lip}(\alpha)$ stands for the space of functions in the disc-algebra satisfying the Lipschitz condition of order α. The well-known Carleson formula for the Dirichlet integral of an analytic function [7] permits conjecture 2 to be reformulated.

Suppose that for a given $E \subset \mathbb{T}$ there exists a non-zero h in $L_+^2(\mathbb{T})$ satisfying

$$\iint\limits_{\mathbb{T} \times \mathbb{T}} \frac{|h(\xi) - h(\zeta)|^2}{|\xi - \zeta|^2} \, dm(\xi) \, dm(\zeta) < +\infty,$$

$$(4) \qquad \int_{\mathbb{T}} \log h \, dm > -\infty,$$

$$(5) \qquad h|E \equiv 0,$$

then there exists a function $h \in L_+^2(\mathbb{T})$ satisfying (4), (5) and

$$(6) \qquad \iint\limits_{\mathbb{T} \times \mathbb{T}} \frac{\left(h^2(\xi) - h^2(\zeta)\right)\left(\log h(\xi) - \log h(\zeta)\right)}{|\xi - \zeta|^2} \, dm(\xi) \, dm(\zeta) < +\infty.$$

Some estimates of the Carleson integral in (6) are given in [8]. The sets of $\mathcal{E}(\mathcal{D}_A)$ located in $\mathbb{D}$ have been considered in [4] and [5].

Chapter 14. UNIQUENESS, MOMENTS, NORMALITY

REFERENCES

1. Beurling A., *Ensembles exceptionnels*, Acta Math. **72** (1940).
2. Carleson L., *Sets of uniqueness for functions regular in the unit circle*, Acta Math. **87** (1952), 325–345.
3. Hruščëv S. V., *Sets of uniqueness for the Gevrey classes*, Arkiv Mat. **6** (1977), 253–304.
4. Carleson L., *On the zeros of functions with bounded Dirichlet integrals*, Math. Z. **56** (1952), no. 3, 289–295.
5. Shapiro H. S., Shields A. L., *On the zeros of functions with finite Dirichlet integral and some related function spaces*, Math. Z. **80** (1962), 217–229.
6. Havin V. P., Hruščëv S. V., *Use of (p, l)-capacity in problems of the theory of exceptional sets*, Mat. Sb. **90** (1973), 558–591 (Russian); English transl. in Math. USSR Sbornik **19** (1973), 547–580.
7. Carleson L., *A representation formula for the Dirichlet integral*, Math. Z. **73** (1960), no. 2, 190–196.
8. Alexandrov A. B., Džrbašjan A. E., Havin V. P., *On a formula of Carleson for the Dirichlet integral*, Vestnik Leningrad Univ., Math. **19** (1979), 8–14. (Russian)

DEPARTMENT OF MATHEMATICS AND MECHANICS
ST. PETERSBURG STATE UNIVERSITY
STARYI PETERHOF
ST. PETERSBURG, 198904
RUSSIA

EULER INTERNATIONAL MATHEMATICAL INSTITUTE
PESOCHNAYA NAB., 10
ST. PETERSBURG, 197022
RUSSIA

COMMENTARY

A description of $\mathcal{E}(\mathcal{D}_A)$ can be found in [9] which, unfortunately, is difficult to apply. Conjecture 1 has been disproved by an ingenious counter-example of L. Carleson [10]. Conjecture 2 remains open.

It is interesting to note that the closely related problem of the description of interpolating sets for $\mathcal{D}_A$ has been solved in [11]. Namely, a closed set $E \subset \mathbb{T}$ is said to be an interpolation set if $(\mathcal{D}_A \cap C_A)|E = C(E)$. Then E is an interpolation set iff cap $E \equiv 0$.

REFERENCES

9. Malliavin P., *Sur l'analyse harmonique des certaines classes de séries de Taylor*, Symp. Math. Inst. Naz. Alto Mat. **22** (1977), London – N.Y., 71–91.
10. Carleson L., *An example concerning analytic functions with finite Dirichlet integrals*, Zap. Nauchn. Semin. LOMI **92** (1979), 283–287.
11. Peller V. V., Hruščëv S. V., *Hankel operators, best approximations and stationary Gaussian processes*, Uspekhi Mat. Nauk **37** (1982), 53–124 (Russian); English transl. in Russian Math. Surveys **37** (1982), 61–144.

ANALYTIC FUNCTIONS STATIONARY ON A SET,
THE UNCERTAINTY PRINCIPLE FOR CONVOLUTIONS,
AND ALGEBRAS OF JORDAN OPERATORS

V. Havin,   B. Jöricke,   N. Makarov

1. The statement of the problem. We say a Lebesgue measurable function φ defined on the circle $\mathbb{T}$ is stationary on the set E, $E \subset \mathbb{T}$, if there exists a function ψ absolutely continuous on $\mathbb{T}$ and such that

$$\left. \begin{array}{c} \psi(\zeta) = \varphi(\zeta) \\ \dfrac{d}{d\theta}\, \psi(e^{i\theta}) = 0 \end{array} \right\} \quad \text{a.e. on } E.$$

A measurable set E, $E \subset \mathbb{T}$, is said to have the property S (in which case we write $E \in (S)$) if there is not non-constant function in $H^1(\mathbb{T})$ stationary on E.

PROBLEM. *Give a description of sets of the class* (S).

We mean a description yielding an answer to the following:

QUESTION 1. *Does every E with mes $E > 0$ belong to (S)?*

There are natural modifications of the Problem, e.g., we may ask the following:

QUESTION 2. *Suppose $E \subset \mathbb{T}$, mes $E > 0$. Is there a non-constant f in the disc-algebra A (i.e., $f \in C(\mathbb{T})$, $\hat{f}(n) = 0$ for all $n \in \mathbb{Z}_-$) stationary on E? What about $f \in A$ satisfying a Lipschitz condition of order less than one?*

Note that every $f \in A$ satisfying a first-order Lipschitz condition and stationary on a set of positive length is constant.

Using a theorem of S. V. Hruščev [1], it is not hard to prove [2, 3] that a closed set E, $E \subset \mathbb{T}$, has the property S if mes $E > 0$ and if, moreover,

(C) $$\sum |l| \log \frac{1}{|l|} < +\infty,$$

the sum being taken over the set of all complementary arcs l of E and $|l| \stackrel{\text{def}}{=} \text{mes}\, l$.

QUESTION 3. *Suppose $E \in (S)$. Does E contain a closed subset E_1 of positive length satisfying (C)?*

This question may be, of course, modified in the spirit of Question 2. A deep theorem by S. V. Hruščev (deserving to be known better than it is, [1], Th. 4.1 on p. 133) suggests the positive answer.

We like our Problem in its own right and feel it is worth solving because it is nice in itself. But there are two "exterior" reasons to look for its solution.

2. The uncertainty principle for convolutions. Let K be a distribution in $\mathbb{R}^n$, X a class of distributions (in $\mathbb{R}^n$). Suppose the convolution $K * f$ has a sense for every $f \in X$. The set $E \in \mathbb{R}^n$ is called a (K, X)–set if

$$f \in X, \quad (K * f)\big|E = 0, \quad f\big|E = 0 \Rightarrow f = 0.$$

(The exact meaning of the convolution $K * f$ and of the restrictions $f\big|E$, $(K * f)\big|E$ becomes clear in concrete situations, see [2], [4].)

If the class of (K, X)–sets is sufficiently large (e.g., contains all non-void open sets) then we may say that the operator $f \rightarrow K * f$ obeys "the uncertainty principle", namely, the knowledge of both restrictions $f\big|E$ and $(K * f)\big|E$ determines f uniquely. For example, every set $E \subset \mathbb{R}$ of positive length is an $(\mathcal{H}, L^2)$–set, $\mathcal{H}$ being the Hilbert transform (i.e., $(\mathcal{H} * f)(t) \stackrel{\text{def}}{=}$ p.v. $\frac{i}{\pi} \int_{-\infty}^{+\infty} \frac{f(u)du}{t-u}$). Other examples can be seen in [2], [4], [5]. There are interesting situations (e.g., $K(x) = |x|^{-\alpha}$ ($x \in \mathbb{R}^n$, $0 < \alpha < n$), X a suitable class of distributions) when we only know that there are many (K, X)–sets but have no satisfactory characterization of such sets. The most interesting is, may be, the case of $K(x) = |x|^{-n+1}$ ($x \in \mathbb{R}^n$, $n \geqslant 2$) closely connected with the Cauchy problem for the Laplace equation.

In an attempt to obtain a large class of relatively simple K's obeying the above "uncertainty principle" and to understand this principle better, the kernels with the so-called semirational symbols have been introduced in [2].

Consider a Lebesgue measurable function $k \colon \mathbb{R} \to \mathbb{C}$ and put $\mathcal{D}_k = \{f \in L^2(\mathbb{R}) : k\hat{f} \in L^2(\mathbb{R})\}$, $\hat{f}$ being the Fourier transform of f, and define $K * f$ for $f \in \mathcal{D}_k$ by the identity $(K * f)\hat{} = \hat{f}_k$. We call k the symbol of the operator $f \to K * f$. The function k is called semirational if there exists a rational function r such that $k\big|(-\infty, 0) = r\big|(-\infty, 0)$, $k(\xi) \neq r(\xi)$ a.e. on a neighbourhood of $+\infty$.

In [2] it was proved that every closed set $E \subset \mathbb{R}$ of positive length satisfying (C) (where l runs through the set of bounded complementary intervals of E) is a $(K, \mathcal{D}_k)$–set provided k is semirational (a simpler proof is seen in [3]). It is not known whether condition (C) can be removed. An interesting (and typical) example of a convolution with a semirational symbol is the operator K,

$$(Kf)(t) \stackrel{\text{def}}{=} (\mathcal{H} * f)(t) + ce^{-\varepsilon t} \int_{-\infty}^{t} e^{\varepsilon \tau} \hat{f}(\tau) d\tau \quad (t \in \mathbb{R}, f \in L^2, \varepsilon > 0),$$

a perturbation of the Hilbert transform. We do not know whether every set E with mes $E > 0$ is a (K, L^2)–set (though we know it is when $c = 0$ or when E satisfies (C)).

All this is closely connected with our Problem (or better to say with its slight modification).

DEFINITION.

(1) *A Lebesgue measurable function φ on the line $\mathbb{R}$ is said to be n-stationary on the set E, $E \subset \mathbb{R}$, if there are functions $\psi_1, \ldots, \psi_n \in W_2^1(\mathbb{R})$ (i.e. $\in L^2(\mathbb{R})$, absolutely continuous and with the $L^2(\mathbb{R})$-derivative) such that $\varphi\big|E = \psi_1\big|E$, $\psi_1'\big|E = \psi_2\big|E, \ldots, \psi_{n-1}'\big|E = \psi_n\big|E, \psi_n'\big|E = 0$.*

(2) *A set E, $E \subset \mathbb{R}$ is said to have the property S_n (or $E \in (S_n)$) if*

$$\varphi \in H^2(\mathbb{R}), \quad \varphi \text{ } n\text{-stationary on } E \Rightarrow \varphi \equiv 0.$$

It is not hard to see that

$$E = \text{clos}\, E,\ E \in (C),\ \text{mes}\, E > 0 \Rightarrow E \in (S_n),\ n = 1, 2, \dots;$$

and that if $E \in \cap_{n=1}^{\infty} S_n$, then E is a $(K, \mathcal{D}_k)$–set for every semirational k [3]. Moreover, if there exists a $\varphi \in H^2(\mathbb{R})$, $\varphi \neq 0$, stationary on the set E, then E is not a $(K, \mathcal{D}_k)$–set for a semirational k (which may be even chosen so that k agrees with a linear function on $(-\infty, 0)$).

Another circle of problems where n-stationary analytic functions emerge in a compulsory way is connected with

3. Jordan operators.

We are going to discuss Jordan operators (J.o.) T of the form

(*) $$T = U + Q$$

where U is unitary, $Q^{n+1} = 0$, $UQ = QU$ (in this case we say T is of order n). It is well-known that the spectrum of any such T lies on $\mathbb{T}$ so that T is invertible. Denote by $\mathcal{R}(T)$ the weakly closed operator algebra spanned by T and the identity I. We are interested in conditions ensuring the inclusion

(**) $$T^{-1} \in \mathcal{R}(T).$$

Example. Let E be a Lebesgue measurable subset of $\mathbb{T}$ and H be the direct sum of $(n+1)$ copies of $L^2(\mathbb{T} \setminus E)$. The operator $J = J(E, n)$ defined by the $(n + 1) \times (n + 1)$–matrix

$$\begin{pmatrix} z & & & \\ I & z & & 0 \\ & I & \ddots & \\ 0 & & I & z \end{pmatrix}$$

(z being the operator of multiplication by the complex variable z) is a J.o. of order n. It is proved in [3] that

$$J^{-1} \in \mathcal{R}(J) \Leftrightarrow E \in (S_n').$$

Here (S_n') denotes the class of subsets of $\mathbb{T}$ defined exactly as (S_n) in section 2 but with W_2^1 replaced by the class of all functions absolutely continuous on $\mathbb{T}$.

The special operator $J = J(E, n)$ is of importance for the investigation of J.o. in general, namely [3], if T is our J.o. (*) of order n and $\mathcal{E}_U$ stands for the spectral measure of U. Then

$$\mathcal{E}_U(E) = 0,\ J^{-1} \in \mathcal{R}(J) \Rightarrow (**).$$

Therefore, if $E \in \cap_{n=1}^{\infty}(S_n')$ (in particular if $\text{mes}\, E > 0$ and $E \in (C)$) then (**) holds whenever $\mathcal{E}_U(E) = 0$.

Recall that for a unitary operator T (i.e., when $T = U$, $Q = 0$ in (*)) the inclusion (**) is equivalent to the vanishing of $\mathcal{E}_U$ on a set of positive length. A deep approximation theorem by Sarason [6] yields spectral criteria of (**) for a normal T. Our questions concerning sets with the property S and analogous questions on classes $(S_n), (S_n')$ are related to the following difficult

PROBLEM. *Which spectral conditions ensure (**) for $T = N + Q$ where N is normal and Q is a nilpotent commuting with N?*

REFERENCES

1. Hruščëv S. V., *The problem of simultaneous approximation and removal of singularities of Cauchy-type integrals*, Trudy Mat. Inst. Steklov Akad. Nauk SSSR **130** (1978), 124–195 (Russian); English transl. in Proc. Steklov Inst. Math. **130** (1979), 133–203.
2. Jöricke B., Havin V. P., *The uncertainty principle for operators commuting with translations* I, Zapiski Nauchn. Sem. LOMI **92** (1979), 134–170; II, – ibid. **113** (1981), 97–134 (Russian); English transl. of part II in J. Soviet Math. **22** (1983), 1758–1783.
3. Makarov N. G., *On stationary functions*, Vestnik Leningrad Univ., Math. (1985), no. 22, 7–14. (Russian)
4. Havin V. P., Jöricke B., *On a class of uniqueness theorem for convolutions*, Lect. Notes in Math., vol. 864, 1981.
5. Havin V. P., *The indeterminacy principle for one-dimensional M. Riesz potentials*, Dokl. Akad. Nauk SSSR **264** (1982), no. 3, 559–563 (Russian); English transl. in Soviet Math. Dokl. **25** (1982), 694–698.
6. Sarason D., *Weak-star density of polynomials*, J. Reine und Angew. Math. **252** (1972), 1–15.

DEPARTMENT OF MATHEMATICS AND MECHANICS
ST. PETERSBURG STATE UNIVERSITY
STARYI PETERHOF
ST. PETERSBURG, 198904
RUSSIA

MAX-PLANCK-INSTITUT FÜR MATHEMATIK
GOTTFRIED-CLAREN-STR. 26
W-5300 BONN 3
GERMANY

DEPARTEMENT OF MATHEMATICS   STEKLOV MATHEMATICAL INSTITUTE
CALIFORNIA INSTITUTE OF TECHNOLOGY   ST. PETERSBURG BRANCH
253-37 PASADENA   AND   FONTANKA 27
CALIFORNIA 91125   ST. PETERSBURG, 191011
USA   RUSSIA

COMMENTARY BY C. J. BISHOP

In the above article, Havin, Jöricke, and Makarov ask whether there exists a set E in the unit circle and function $f \in H^1$ stationary on E (i.e. is there a function φ of bounded variation on the circle so that $f = \varphi$ on E and $\varphi' = 0$ on E). In [7] I prove that there is such a set and function by constructing the Riemann surface corresponding to the function.

REFERENCE

7. Bishop C. J., *An element of the disk algebra stationary on a set of positive length*, Algebra i Analiz **1** (1989), no. 3, 83–88; Leningrad Math. J. **1** (1990), 647–652.

DEPARTMENT OF MATHEMATICS
STATE UNIVERSITY OF NEW YORK
STONY BROOK
NEW YORK 11794
USA

PROBLEM IN THE THEORY OF FUNCTIONS

PAUL KOOSIS

In 1966 I published the following theorem:

There exists a constant $a > 0$ such that any collection of polynomials Q of the form

$$Q(z) = \prod_k \left(1 - \frac{z^2}{z_k^2}\right)$$

with

$$\sum_1^\infty \frac{1}{n^2} \log^+ |Q(n)| \leqslant a$$

is a normal family in the complex plane.

See [1], the theorem is on page 273.

This result can easily be made to apply to collections of polynomials of more general form provided that the sum from 1 to ∞ in its statement is replaced by one over all the non-zero integers. One peculiarity is that the constant $a > 0$ really must be taken quite small for the asserted normality to hold. If a is large enough, the theorem is false.

The result's proof is close to 40 pages long, and I think very few people have been through it. Can one find a shorter and clearer proof? This is my question.

Let me explain what I am thinking of. Take any fixed ρ, $0 < \rho < \frac{1}{2}$ and let $\mathcal{D}_\rho$ be the slit domain

$$\mathbb{C} \setminus \bigcup_{-\infty}^\infty [n - \rho, n + \rho].$$

If Q is any polynomial, write

$$Q(n, \rho) = \sup \left\{ |Q(x)| \,;\, n - \rho \leqslant x \leqslant n + \rho \right\}.$$

By direct harmonic estimation in $\mathcal{D}_\rho$ one can find without too much trouble that

$$\log |Q(z)| \leqslant K_\rho(z) \sum_{-\infty}^\infty \frac{\log^+ Q(n, \rho)}{1 + n^2},$$

where $K_\rho(z)$ depends only on z and ρ. (This is proved in the first part of the paper cited above.) A natural idea is to try to obtain the theorem by making $\rho \to 0$ in the above formula. This, however, cannot work because $K_\rho(z)$ tends to ∞ as $\rho \to 0$ whenever z is not an integer. The latter must happen since the set of integers has logarithmic capacity zero.

For polynomials, the estimate provided by the formula is too crude. The formula is valid if, in it, we replace $\log|Q(z)|$ by *any* function subharmonic in $\mathcal{D}_\rho$ having sufficiently slow growth at ∞ and some mild regularity near the slits $[n - \rho, n + \rho]$. Polynomials, however, are single-valued in $\mathcal{D}_\rho$. This single-valuedness imposes constraints on the subharmonic function $\log|Q(z)|$ which somehow work to diminish $K_\rho(z)$ to something bounded (for each fixed z) as $\rho \to 0$, provided that the sum figuring in the formula is sufficiently small. The PROBLEM here is to *see quantitatively how the constraints cause this diminishing to take place.*

The phenomenon just described can be easily observed in one simple situation. Suppose that $U(z)$ is subharmonic in $\mathcal{D}_\rho$, that $U(z) \leqslant A|z|$ there, and that $[U(z)]^+$ is (say) continuous up to the slits $[n-\rho, n+\rho]$. If $U(x) \leqslant M$ on each of the intervals $[n-\rho, n+\rho]$, then

$$U(z) \leqslant M + \frac{A}{\pi} \log \left| \frac{\sin \pi z}{\sin \pi \rho} + \sqrt{\frac{\sin^2 \pi z}{\sin^2 \pi \rho} - 1} \right|.$$

This estimate is best possible, and the quantity on the right blows up as $\rho \to 0$. However, if $U(z) = \log|F(z)|$ where $F(z)$ is a single-valued entire function of exponential type $A < \pi$, we have the better estimate

$$U(z) \leqslant M + C_A + A|y|$$

with a constant C_A independent of ρ. The improved result follows from the theorem of Duffin and Schaeffer. It is no longer true when $A \geqslant \pi$ — consider the functions $F(z) = L \sin \pi z$ with $L \to \infty$.

The whole idea here is to see how harmonic estimation for functions analytic in multiply connected domains can be improved by taking into account those functions' single-valuedness.

REFERENCE

1. Koosis P., *Weighted polynomial approximation on arithmetic progressions of intervals or points,* Acta Math. **116** (1966), 224–277.

McGILL UNIVERSITY   AND   UCLA
MONTREAL   LOS ANGELES
CANADA   USA

PEAK SETS FOR LIPSCHITZ CLASSES

E. M. Dyn'kin

The Lipschitz class A_α, $0 < \alpha \leqslant 1$, consists of all functions f analytic in $\mathbb{D}$, continuous on clos $\mathbb{D}$ and such that

$$(1) \qquad |f(\zeta_1) - f(\zeta_2)| \leqslant \text{const} \, |\zeta_1 - \zeta_2|^\alpha, \quad \zeta_1, \zeta_2 \in \mathbb{T}.$$

A closed set E, $E \subset \mathbb{T}$, is called *peak set* for A_α (in symbols, $E \in \mathcal{P}_\alpha$), if there exists a function f, $f \in A_\alpha$ (the so-called *peak function*) such that

$$|f| < 1 \text{ on } \mathbb{T} \setminus E; \quad f|E \equiv 1.$$

The PROBLEM is *to describe the structure of $\mathcal{P}_\alpha$-sets.*

B. S. Pavlov [1] discovered a necessary condition; this condition was rediscovered by H. Hutt [2] in a more complicated way. Write $E_\delta = \{\zeta \in \mathbb{T} : \text{dist}(\zeta, E) \leqslant \delta\}$.

THEOREM 1 (Pavlov [1], Hutt [2]). *If* $E \in \mathcal{P}_\alpha$, *then*

$$(2) \qquad m(E_\delta) = O(\delta^\alpha), \quad \delta \to +0.$$

COROLLARY. *$\mathcal{P}_1$-sets are finite.*

Sketch of the Proof. Let f be a peak function for E and $g = 1 - f$. Then $g \in A_\alpha$, $\text{Re}\, g \geqslant 0$ in $\mathbb{D}$ and $g|E \equiv 0$. Hence $\text{Re}\, 1/g \geqslant 0$ in $\mathbb{D}$ and Herglotz' theorem [3] says that $1/g$ is the Cauchy integral of a finite measure. Now condition (2) follows from the weak type estimate [4]

$$m\{\zeta : \zeta \in \mathbb{T}, \, |1/g| > t\} \leqslant \text{const} \cdot t^{-1}, \quad t > 0,$$

and from the evident inequality $|g(\zeta)| \leqslant \text{const} \cdot \text{dist}(\zeta, E)^\alpha$. $\square$

Until quite recently only some simple examples of $\mathcal{P}_\alpha$-sets were known [2]; condition (2) holds for these examples "with a reserve". But recent S. V. Hruščev's results [5] on zero sets for Gevrey classes permit us to obtain a very exact sufficient condition. Define the *Gevrey class* G_α as the class of all analytic functions in $\mathbb{D}$ such that $\sup_{\zeta \in \mathbb{D}} |f^{(n)}(\zeta)| \leqslant (\text{const})^{n+1} \cdot (n!)^{1+\frac{1}{\alpha}}$, $n = 0, 1, \dots$. A set E, $E \subset \mathbb{T}$ is called a *zero set* for G_α (or $E \in Z(G_\alpha)$), if there is a non-zero function f, $f \in G_\alpha$, with $f^{(n)}|E = 0$, $n = 0, 1, \dots$. Hruščëv [5] has completely investigated $Z(G_\alpha)$-sets and gave a lot of examples.

THEOREM 2. $Z(G_\alpha) \subset \mathcal{P}_\alpha$.

Sketch of the Proof. It has been shown in [5] that every non-empty E in $Z(G_\alpha)$ defines a positive function U on $\mathbb{T}$ with the following list of properties:

(a) $|(\log f)'| \leqslant \text{const } U^{1+\frac{1}{\alpha}}$;
(b) $\text{dist}(\zeta, E)^{-\alpha} \leqslant U(\zeta), \zeta \in \mathbb{T}$;
(c) $0 < c_1 < U(\zeta_1)/U(\zeta_2) < c_2 < +\infty$

provided $\zeta_1, \zeta_2 \in \mathbb{T} \backslash E$ and $|\zeta_1 - \zeta_2| < \frac{1}{2} \text{dist}(\zeta_1, E)$. Here $f = \exp\{-U - i\widetilde{U}\}$ stands for the outer function with the modulus $\exp(-U)$ on $\mathbb{T}$. It is easy to see that $f \in G_\alpha$, $f^{(n)}|E \equiv 0, n = 0, 1, \ldots$ (see [5] for details).

Set $g \overset{\text{def}}{=} (\log f)^{-1}$ in clos $\mathbb{D}$. Then

$$\text{Re } g(z) \leqslant 0, \ z \in \text{clos } \mathbb{D}; \quad |g(\zeta)| \leqslant U(\zeta)^{-1} \leqslant \text{dist}(\zeta, E)^\alpha.$$

Let us prove now that $g \in A_\alpha$. Obviously (a) holds if either ζ_1 and ζ_2 lie in different complementary intervals of E or

$$|\zeta_1 - \zeta_2| > \frac{1}{5} \max\left\{U(\zeta_1)^{-1/\alpha}, U(\zeta_2)^{-1/\alpha}\right\}.$$

If $|\zeta_1 - \zeta_2| < \frac{1}{5} U(\zeta_1)^{-1/\alpha}$, then by (b), $|\zeta_1 - \zeta_2| < \frac{1}{5} \text{dist}(\zeta_1, E)$ and by (a) and (c),

$$|g(\zeta_1) - g(\zeta_2)| \leqslant \text{const } |\zeta_1 - \zeta_2| U(\zeta_1)^{1+\frac{1}{\alpha}} U(\zeta_2)^{-2} \leqslant \text{const } |\zeta_1 - \zeta_2|^\alpha.$$

It is clear that $h = (1 + g)(1 - g)^{-1} \in A_\alpha$ and h is a peak function for E.

Theorem 2 gives a number of examples of $\mathcal{P}_\alpha$-sets and the following conjecture seems now plausible.

CONJECTURE. $\mathcal{P}_\alpha = Z(G_\alpha)$.

There exist some (not very clear) connections between the free interpolation sets for A_α and G_α [6]; these connections corroborate our conjecture. The A_α-function from Theorem 2 is a "logarithm" of some G_α-function. Probably, it is a general rule and it is possible to link two interpolation problems in a direct way.

A possible way to prove our conjecture is a conversion of the proof of Theorem 2. The "strongly vanishing function" $f = \exp\left(-\frac{1}{1-h}\right)$ is of course not in G_α for an arbitrary peak function $h \in A_\alpha$. But in Theorem 2 such a function f is not arbitrary; it is extremal in some sense [5]. Perhaps, it is possible to obtain $f \in G_\alpha$ for some extremal peak function h. The extremal functions are often analytic on $\mathbb{T} \setminus E$, and such an f may be a smooth function.

The following necessary condition may be a first step to the conjecture: if $E \in \mathcal{P}_\alpha$ then $\text{dist}(\cdot, E)^{-\alpha} \in L^1(\mathbb{T})$.

The description of $\mathcal{P}_\alpha$-sets is interesting for the investigation of the singular spectrum in the Friedrichs model [1, 7].

PROBLEM 14.6

REFERENCES

1. Pavlov B. S., *Uniqueness theorems for functions with positive imaginary part*. Prob. Math. Phys., no. 4; Spectral Theory, Wave Processes (1970), Izdat. Leningrad Univ., 118–125. (Russian)
2. Hutt H., *Some results on peak and interpolation sets of analytic functions with higher regularity*, Thesis, Uppsala Univ. Dep. Math., 1976.
3. Privalov I. I., *Raudeigenschaften Analytischer Funktionen*, Deutscher Verlag der Wissenschaften, Berlin, 1956.
4. Zygmund A., *Trigonometric series*, Cambridge Univ. Press, London, New York, 1969.
5. Hruščëv S. V., *Sets of uniqueness for the Gevrey classes*, Ark. Mat. **15** (1977), no 2, 256–304.
6. Dyn'kin E. M., *Free interpolation sets for Hölder classes*, Mat. Sb. **109** (1979), 107–128 (Russian); English transl in Math. USSR Sbornik **37** (1980), 97–117.
7. Pavlov B. S., Faddeev L. D., This collection, Problem S.6.11.

DEPT. OF MATHEMATICS
TECHNION
32000, HAIFA
ISRAEL

14.7
v.old

A PROBLEM BY R. KAUFMAN

R. Kaufman

Let G be a bounded Lipschitz domain and $C^r(G)$ $(r \geqslant 1)$ the class of functions analytic in G, with r^{th} derivative uniformly continuous on G. *Do the classes $C^r(G)$ have the same zero-sets in clos G, or on ∂G?*

University of Illinois at Urbana-Champaign
Department of Mathematics
Urbana, IL 61801
USA

QUASI-ANALYTICITY OF FUNCTIONS
WITH RESPECT TO A DIFFERENTIAL OPERATOR

V. I. MATSAEV

Suppose Ω is a domain in $\mathbb{R}^n$, E is a closed subset of Ω, $\{m_n\}$ is a sequence of positive numbers such that $\sum_n m_n^{*-1/n} = \infty$ (m_n^* is the best monotone majorant for m_n), L is a differential operator of order l with $C^\infty(\Omega)$ coefficients. A function $f \in C^\infty(\Omega)$ is said to belong to the class $\mathcal{L}(m_n)$ if the following inequalities are satisfied:

$$\|L^k f\|_{L^2(\Omega)} \leqslant Cm_k^l, \ k = 0, 1, 2, \ldots \qquad (C = C_f).$$

Denote by Ω_0 the maximal set among the subsets of Ω enjoying the following property: If a function f_1, $f_1 \in C^\infty(\Omega)$, has a zero of infinite multiplicity on E and satisfies the equation $Lf_1 = 0$, then $f_1|\Omega_0 = 0$.

I conjecture that under an appropriate definition of the order l of the operator L, the following is true.

CONJECTURE. *If f belongs to the class $\mathcal{L}(m_n)$ and has a zero of infinite multiplicity on E, then $f|\Omega_0 = 0$.*

In other words, functions quasi-analytic with respect to the operator L behave with respect to the uniqueness theorem as solutions of the homogeneous equation $Lf = 0$.

We will *Sketch the Proof* of the Conjecture in two cases:

(a) Suppose L is an elliptic operator such that the operator $D_t^l - L_x$ is elliptic, $E = \{x_0\}$ is a point of Ω, and $m_n = n!$. Consider the solution $g(t, x)$ of the problem $(D_t^l - L_x)g = 0$, $g(0, x) = f(x)$, $D_t^j g(0, x) = 0$ $(j = 1, \ldots, l - 1)$. It exists for small t:

$$g(t, x) = \sum_{k=0}^\infty \frac{t^{kl}}{(kl)!} L_x^k f.$$

The function $g(t, x)$ has zero of infinite order at $t = x = 0$, hence $g(t, x) \equiv 0$ and $f(x) \equiv 0$.

(b) $L = D_{x_1}^2 + D_{x_2}^2 - D_{x_3}^2$, $\Omega = \mathbb{R}^3$, E is a two-dimensional smooth surface, $f \in C^\infty(\mathbb{R}^3)$ and satisfies $\|L^k f\|_{L^2(\Omega)} \leqslant Cm_k$, $k = 0, 1, 2, \ldots$ (so that here $l = 1$). Denote by L_0 the closure of the operator L defined on the set of functions vanishing to infinite order on E. It is clear that L_0 is a symmetric operator. Suppose the vector φ, $\varphi \neq 0$ is such that $L_0^* \varphi = i\varphi$. Consider for $\operatorname{Im} \lambda > 0$ the vector-valued function $\lambda \to \varphi_\lambda = \varphi + (\lambda - i)\tilde{R}_\lambda \varphi$ where $\tilde{R}_\lambda$ is the resolvent of a self-adjoint extension of the operator L_0. Then $L_0^* \varphi_\lambda = \lambda \varphi_\lambda$ and

$$\left|\lambda^k(\varphi_\lambda, f)\right| = \left|(L_0^{*k}\varphi_\lambda, f)\right| = \left|(\varphi_\lambda, L_0^k\varphi)\right| \leqslant Cm_k, \ k = 0, 1, 2, \ldots.$$

Hence, $(\varphi_\lambda, f) = (\varphi, f) = 0$. Similarly $(\psi, f) = 0$ where $L_0^* \psi = -i\psi$. Hence f belongs to an invariant subspace of the operator L_0 on which L_0 is self-adjoint. By the theorem of Gelfand and Kostjučenko [1] f belongs to the linear span of the generalized eigenfunctions of the operator L_0, i.e., of solutions of the equation $L_0 g = \lambda g$ vanishing on Ω_0. Hence $f|\Omega_0 = 0$. $\square$

REFERENCES

1. Gelfand I. M.,Kostjučenko A. G., *Expansion in eigenfunctions of differential and other operators*, Dokl. Akad. Nauk SSSR **103** (1955), 349–352. (Russian)
2. Berezanskii Ju. M., *Expansions in eigenfunctions of self-adjoint operators*, Akad. Sci. Ukrain. SSR, Mat. Inst., Naukova Dumka, Kiev, 1965. (Russian)

SCHOOL OF MATHEMATICAL SCIENCES
TEL AVIV UNIVERSITY
RAMAT AVIV, TEL AVIV 69978
ISRAEL

COMMENTARY

The Conjecture is confirmed now in various particular cases. Here are some quite recent (unpublished) results.

Let Ω be a domain in $\mathbb{R}^n$, A a differential operator of order l with $C^\infty(\Omega)$–coefficients, $p \in (1, +\infty]$, $M = \{M_k\}_{k=0}^{+\infty}$ a sequence of positive numbers. Put $m_k = \frac{M_k}{k!}$,

$$\mathcal{A}_p(M, \Omega) \stackrel{\text{def}}{=} \left\{ f \in C^\infty(\Omega) : \|A^j f\|_{L^P(\Omega)} \leqslant C_f Q_f^{jl} M_j^l, \; j = 0, 1, \dots \right\}.$$

Suppose $m_k m_p \leqslant C m_{k+p}$ $(k, p = 1, 2, \dots)$. Consider a family $d = (d_1, \dots, d_n)$ of positive numbers, $d_1 \leqslant d_2 \leqslant \cdots \leqslant d_n$ and a compact set K, $K \subset \Omega$. Put

$$C(M, d, K) \stackrel{\text{def}}{=} \left\{ f \in C^\infty(\Omega) : \|\mathcal{D}^\alpha f\|_{C(K)} \leqslant C_{f,K} \, Q_{f,K}^{|\alpha|} \, m_{[d_1\alpha_1 + \cdots + d_n\alpha_n]} \Gamma \left(1 + \sum_1^n \alpha_j d_j \right) \right\},$$

$[t]$ being the integer part of t. When $d \equiv 1$ we write $C(M, K)$ instead of $C(M, d, K)$.

THEOREM 1 (A. G. Chernyavskii). *Suppose that*

(a) $A = \sum_{|\alpha|+\beta \leqslant l} a_{\alpha\beta}(t, x) D_t^\beta D_x^\alpha$, $a_{\alpha\beta} \in C(M, G)$, where G is the cylinder $\{(t, x) : t \in [-1, 1], \; x \in \mathbb{R}^{n-1}, \; |x| \leqslant 1\}$ and $a_{0l}(t, x) \neq 0$ in G;

(b) *the class* $C(M, G)$ *is quasi-analytic, i.e.,*

$$\int_1^\infty \frac{\log T(r)}{r^2} \, dr = \infty, \quad T(r) \stackrel{\text{def}}{=} \sup_k \frac{r^k}{M_k}.$$

If $u \in \mathcal{A}_\infty(M, G)$ *and* $\mathcal{D}_t^p A^j(0, x) = 0$ $(0 \leqslant p \leqslant l-1, \; j = 0, 1, \dots, |x| \leqslant 1)$ *then* $u \equiv 0$ *in a neighbourhood of the origin.*

Fix now a multi-index λ with decreasing natural coordinates. For every multi-index α put $|\alpha : \lambda| = \sum_1^n \alpha_k/\lambda_k$. Suppose that

$$A = \sum_{|\alpha:\lambda| \leqslant 1} a_\alpha(x) \mathcal{D}^\alpha,$$

where $a_\alpha \in C(M, K, d)$ on every compact K, $K \subset \Omega$, and $d_j \overset{\text{def}}{=} \lambda_1/\lambda_j$ $(j = 1, \dots, n)$. Assume moreover that

$$\sum_{|\alpha:l| \leqslant 1} a_\alpha(x)\xi^\alpha \neq 0 \quad ((x,\xi) \in \Omega \times (\mathbb{R}^n \setminus \{0\})) \ .$$

THEOREM 2 (M. M. Malamud). *Under the above conditions*

1. $\mathcal{A}_p(M) \subset C(M, d, L_p(\Omega))$ $(1 < p < \infty)$ *(the class* $C(M, d, L_p(\Omega))$ *is defined like* $C(M, d, K)$ *above with* $\| \cdot \|_{L^P(\Omega)}$ *instead of* $\| \cdot \|_{C(K)}$*).*

2. *Suppose* $\lambda_1 = \cdots = \lambda_m > \lambda_{m+1} \geqslant \cdots > \lambda_n$ *and* $E = \operatorname{clos} E \subset \Omega$. *If* $P_m(E) = P_m(\Omega)$ *(P_m being the projection* $(x_1, \dots, x_n) - (0, \dots, 0, x_{m+1}, \dots, x_n))$*, then*

$$(*) \Longleftrightarrow \left\{ u \in \mathcal{A}_p(M, \Omega) : \mathcal{D}_{x_1}^{\alpha_1} \cdots \mathcal{D}_{x_m}^{\alpha_m} u | E = 0 \quad 0 \leqslant |\alpha| < \infty \right\} = \{0\}.$$

In 1., the sequence M is not supposed to satisfy $(*)$. The condition concerning $P_m(E)$ cannot be dropped from 2. If A is elliptic (in that case $l_1 = \cdots = l_n$), then every singleton $\{x_0\}$ can serve as the set E in 2., and $\operatorname{Ker} A$ and $\mathcal{A}_p(M, \Omega)$ are both quasianalytic if $C(M, L_p(\Omega))$ is (this result is due to M. M. Malamud and A. E. Shishkov, see also [3]).

V. P. Palamodov has pointed out (a private communication to the editors) that one can confirm the Conjecture in the case of an ultra-hyperbolic operator using methods of §17, Ch. VI of [4].

We conclude by a list of works connected with our theme.

REFERENCES

3. Lions J.-L., Magenes P., *Problèmes aux limites nonhomogènes et applications*, Vol. 3, Dunod, Paris, 1970.
4. Courant R., *Partial differential equations*, N.Y., London, 1962.
5. Ljubich Ju. I., Tkachenko V. A., *The abstract quasi-analyticity problem*, Teor. Funk. Funktz. Anal. i Prilozhen. **16** (1972), 18–29. (Russian)
6. Chernyavskii A. G., *Quasi-analytic classes generated by hyperbolic operators with fixed coefficients in* $\mathbb{R}^2$, Teor. Funk. Funktz. Anal. i Prilozhen. **37** (1982), 122–127. (Russian)
7. Chernyavskii A. G., *On a generalization of Holmgren's uniqueness theorem*, Sibirsk. Mat. Zh. **22** (1981), 212–215. (Russian)
8. Kotaké T., Narasimhan M. S., *Regularity theorems for fractional powers of a linear elliptic operator*, Bull. Soc. Math. France **90** (1962), 449–471.

TWO PROBLEMS INVOLVING QUASIANALYTIC FUNCTIONS

JEAN ECALLE

0. Reminder about quasianalytic and cohesive functions.

For any positive sequence M_n we say that a C^∞ function φ is of Carleman class $C(M_n)$ on $I = [a, b]$ if its derivatives are uniformly bounded by $c_0 c_1^n M_n$ on I for some constants $c_0, c_1 < +\infty$. The class $C(M_n)$ is said to be quasianalytic if any φ in it vanishes on the whole interval I as soon as all its derivatives vanish at a given point $x_0 \in I$. A necessary and sufficient condition for quasianalyticity is:

$$(1) \qquad \sum \frac{1}{\beta_n} = +\infty \text{ with } \beta_n = \inf_{m \geqslant n} \left(M_m\right)^{1/m}.$$

For any finite or transfinite ordinal $\alpha < \omega^\omega$ of the form:

$$(2) \qquad \alpha = \omega^r n_r + \ldots \omega^2 n_2 + \omega n_1 + n_0 \qquad \left(r \in \mathbb{N}; n_i \in \mathbb{N}\right)$$

we define a transfinite iterate of $L = \log$ of order α by:

$$(3) \qquad L_\alpha = (L)^{\circ n_0} \circ \left(L^*\right)^{\circ n_1} \circ \left(L^{**}\right)^{\circ n_2} \circ \ldots \left(L^{**\ldots *}\right)^{\circ n_r}$$

where $L^*, L^{**}, \ldots$ are *any* C^∞, increasing functions on $\mathbb{R}^+$ satisfying:

$$(4) \qquad L^* \circ L = -1 + L^*; \quad L^{**} \circ L^* = -1 + L^{**}; \quad L^{***} \circ L^{**} = -1 + L^{***}; \text{ etc...}$$

With the ordinal α we also associate the Denjoy class $^\alpha D$ defined by:

$$(5) \qquad ^\alpha D = C(^\alpha M_n) \text{ with } ^\alpha M_n = \left(1/L'_{1+\alpha}(n)\right)^n.$$

For α finite, $^\alpha D$ coincides with the usual Denjoy class. For $\omega \leqslant \alpha < \omega^\omega$, the sequence $^\alpha M_n$ depends on the actual choice of L^*, L^{**}, etc, but the class $^\alpha D$ is independent of that choice. Each class $^\alpha D$ is quasianalytic, and so is the union:

$$(6) \qquad \text{COHES} = \cup_\alpha {}^\alpha D \qquad (\alpha < \omega^\omega)$$

which is known as the class of *cohesive functions*.

1. Problem 1. Abel's lemma for transseries.

Let $\mathbb{R}[[x^{-1}]]$ denote the ring of formal power series in x^{-1} with real coefficients and let $\mathbb{R}[[[x]]]$ be its natural closure under the operations $+$, $\times$, ∂, 0 and their inverses (∂ = derivation; 0 = composition). The elements of $\mathbb{R}[[[x]]]$ are known as formal *transseries*. Each transséries $\tilde{\varphi}(x)$ can be expressed as a well ordered sum

$$(7) \qquad \tilde{\varphi}(x) = \sum a_\sigma \tilde{A}_\sigma(x) \qquad (a_\sigma \in \mathbb{R}; \, \sigma < \alpha < \omega^\omega, \, A_{\sigma_1} \ll A_{\sigma_1} \text{ if } \sigma_1 < \sigma_2)$$

of *transmonomials* $\tilde{A}_\sigma(x)$ which are *irreductible* concatenations of symbols $+$, $\times$, ∂, 0, E, L (with $E = \exp$, $L = \log$) and $x \gg 1$. *Irreductible* here means that all (formally) infinitesimal expressions have been expelled from the exponentials and logarithms according to the obvious rules:

$$(8) \qquad \exp(\tilde{A} + \tilde{B}) = \left(1 + \sum \frac{(\tilde{B})^n}{n!}\right) \cdot \exp(\tilde{A})$$

$$(9) \qquad \log(\tilde{A}(1 + \tilde{B})) = \log \tilde{A} \sum (-1)^{n-1} \frac{(\tilde{B})}{n}$$

($\tilde{A}(x)$ infinitely large; $\tilde{B}(x)$ infinitely small; $x \gg 1$). For a precise definition and set of axioms, see [2] or [3]. For the very special transseries of the form (7) with

$$(10) \qquad a_\sigma > 0 \text{ and } \tilde{A}_\sigma(x) = 1/\tilde{B}_\sigma(x)$$

with *infinitely large* transmonials $\tilde{B}_\sigma$ expressible as *finite* concatenations of symbols $+$, $\times$, ∂, 0, E, $x \gg 1$ (no logarithms!) and *positive* coefficients, we have a lemma reminiscent of Abel's lemma for ordinary power series, namely:

RESTRICTED ABEL'S LEMMA FOR TRANSSERIES. *Whenever a transseries $\tilde{\varphi}(x)$ of the form (7) + (10) converges absolutely for some finite value x_0 of the variable, it converges for all $x \geqslant x_0$ to a sum $\varphi(x)$ which is* **cohesive.**

This applies for instance for

(11) $B_\sigma(x) = \exp \exp \exp(nx)$ $\qquad$ $(\sigma = n)$

(12) $B_\sigma(x) = \exp(n_0 \exp(n_1 \exp(n_2 x + n_3 x^2)))$ $\quad$ $(\sigma = \omega^3 n_3 + \omega^2 n_2 + \omega n_1 + n_0)$.

The condition (10) can obviously be relaxed in many ways. Hence the

PROBLEM 1. *Find necessary and sufficient conditions on the form of a transseries $\tilde{\varphi}$ for its sum to be* **cohesive.**

2. Problem 2. Singular differential equations with quasianalytic coefficients.

Let ζ, ζ', ε be > 0 and let $K_\varepsilon(\zeta, \zeta')$ be a family of measures (in ζ') on $\mathbb{R}^+$ which for $\varepsilon \to 0$ tend to the Dirac measure $\delta(\zeta' - \zeta)$. A function $\hat{\varphi}$ continuous on $\mathbb{R}^+$ is said to be K_ε-summable if the functions $\hat{\varphi}_\varepsilon$:

$$(13) \qquad \hat{\varphi}_\varepsilon(\zeta) = \int_0^{+\infty} K_\varepsilon(\zeta, \zeta') \hat{\varphi}(\zeta') \, d\zeta' \qquad (\zeta \in \mathbb{R}^+)$$

have exponential growth at $+\infty$ and if their Laplace transforms:

$$(14) \qquad \varphi_\varepsilon(z) = \int_0^{+\infty} \hat{\varphi}_\varepsilon(\zeta) \cdot e^{-z\zeta} \, d\zeta \qquad (z > 0)$$

tend to a limit $\varphi(z)$ uniformly on a segment $[\ldots, +\infty]$.

PROBLEM 2. *Find a family K_ε of measures such that*

(15) $\qquad \hat{\varphi}_1 * \hat{\varphi}_2$ *be K_ε-summable whenever $\hat{\varphi}_1, \hat{\varphi}_2$ are K_ε-summable*

(16) $\qquad \hat{\varphi}_1 \cdot \hat{\varphi}_2$ *be K_ε-summable whenever $\hat{\varphi}_1, \hat{\varphi}_2$ are K_ε-summable*

$\qquad \hat{\varphi}$ *be K_ε-summable to φ whenever $\hat{\varphi}$ is the Borel transform*

(17) $\qquad$ *of a formal power series $\hat{\varphi}(z)$ associated with a functions $\hat{\varphi}(z)$*

$\qquad$ *cohesive in $x = z^{-1} \in [0, \dots]$.*

Here $*$ denotes the finite convolution:

$$(18) \qquad \hat{\varphi}_1 * \hat{\varphi}_2(\zeta) = \int_0^\zeta \hat{\varphi}_1(\zeta_1) \hat{\varphi}_2(\zeta - \zeta_1) \, d\zeta_1$$

and in (17) $\hat{\varphi}$ denotes the term-wise Borel transform of $\tilde{\varphi}$:

$$(19) \qquad \tilde{\varphi}(z) = \sum a_n z^{-n} = \sum a_n x^n \mapsto \hat{\varphi}(\zeta) = \sum a_n \zeta^{n-1} / \Gamma(n).$$

Motivation: Formal solutions $\tilde{\varphi}$ of singular differential equations $E(\varphi) = 0$ with *analytic coefficients* can be summed by the procedure $\tilde{\varphi} \mapsto \hat{\varphi} \mapsto \varphi$ (or some modified form of it) and a K_ε-summation satisfying (15), (16), (17) would enable one to do the same for singular differential equations $E(\varphi) = 0$ with *cohesive coefficients*.

REFERENCES

1. Mandelbrojt S., *Séries adhérentes, régularisation des suites et applications*, Gauthier–Villars, Paris, 1952.
2. Ecalle J., *Finitude des cycles-limites et accéléro-sommation de l'application de retour*, Bifurcation of Planar Vector Fields, Proceedings, Luminy 1989, Springer-Verlag, pp. 74–160, Lect. Notes in Math **1455**.
3. Ecalle J., *Introduction aux fonctions analysables et preuve constructive de la conjecture de Dulac*, Actualités mathématiques (1991), Hermann (to appear).

UNIVERSITÉ PARIS-XI
91405 ORSAY
FRANCE

LOCAL OPERATORS ON FOURIER TRANSFORMS

L. DE BRANGES

If f is a square integrable function of a real variable, let $\hat{f}$ denote its Fourier transform. If k is a measurable function of a real variable, the notation $\hat{k}$ is also used to denote a partially defined operator taking $\hat{f}$ into $\hat{g} = \hat{k} * \hat{f}$ whenever f and $g = kf$ are square integrable. The operator $\hat{k}$ is said to be *local* if, whenever a function $\hat{f}$ in its domain vanishes almost everywhere in a set of positive measure, the function $\hat{g} = \hat{k} * \hat{f}$ also vanishes almost everywhere in the same set.

The operator $\hat{k}$ is CONJECTURED *to be local if k is the restriction to the real axis of an entire function of minimal exponential type.*

If the operator $\hat{k}$ is local and if it has in its domain a non-zero function which vanishes almost everywhere in a set of positive measure, then it is CONJECTURED that k *agrees almost everywhere with the restriction to the real axis of an entire function of minimal exponential type which satisfies the convergence condition*

$$(*) \qquad \int (1+t^2)^{-1} \log^+ |k(t)|\, dt < \infty .$$

If k is a function which satisfies the convergence condition, if $k \geqslant 1$, and if $\log k$ is uniformly continuous, then it is CONJECTURED that *for every positive number ε a non-zero function $\hat{f}$ in the domain of $\hat{k}$ exists which vanishes almost everywhere outside of the interval $(-\varepsilon, \varepsilon)$.*

If k is a function which does not satisfy the convergence condition, if $k \geqslant 1$, and if $\log k$ is uniformly continuous, then it is CONJECTURED that *no non-zero function exists in the domain of $\hat{k}$ which vanishes in a set of positive measure.*

PURDUE UNIVERSITY
DEPARTMENT OF MATHEMATICS
LAFAYETTE, IN 47907
USA

FROM THE AUTHOR'S SUPPLEMENT, 1983

The problem originates in a theorem on quasi-analyticity due to Levinson [2]. This theorem states that $\hat{f}$ cannot vanish in an interval without vanishing identically if it is in the domain of $\hat{k}$ where k is sufficiently large and smooth. *Large* means that the integral in $(*)$ is infinite, *small* that it is finite. The smoothness condition assumed by Levinson was that $\log^+ |k|$ is non-decreasing, but it is more natural [3] to assume that $\log^+ |k|$ is uniformly continuous (or satisfies the Lipschitz condition).

A stronger conclusion was obtained by Beurling [4] under the Levinson hypothesis. A function $\hat{f}$ in the domain of $\hat{k}$ cannot vanish in a set of positive measure unless it vanishes identically. The Beurling argument pursues a construction of Levinson and Carleman which is distinct from the methods based on the operational calculus concerned with the concept of a local operator. The Beurling theorem can be read as the assertion that certain operators are local with a trivial domain. It would be interesting to obtain the Beurling theorem as a corollary of properties of local operators with nontrivial domain.

The author thanks Professor Sergei Khrushchev for informing him that a counterexample to our locality conjecture has been obtained by Kargaev.

COMMENTARY

P. P. Kargaev [5] has *disproved* the FIRST and the LAST CONJECTURES. As to the first, he has constructed an entire function k not only of minimal exponential type, but of *zero order* such that $\hat{k}$ is not local. Moreover, the following is true:

THEOREM (Kargaev). *Let φ be a positive function decreasing to zero on $[0, +\infty)$. Then there exists a function $d\colon \mathbb{C} \to \mathbb{Z}_+$ ("a divisor"), a set $e \subset \mathbb{R}$, and a function $f \in L^2(\mathbb{R})$ with the following properties:*

$$\sum_{\lambda \in \mathbb{C}} d(\lambda)\varphi\left(|\lambda|\right) < +\infty, \qquad \sum_{\lambda \in \mathbb{C}} \frac{d(\lambda)}{|\lambda|} < +\infty,$$

$$\operatorname{mes} e > 0, \quad \hat{f} \in L^1(\mathbb{R}) \cap L^\infty(\mathbb{R}), \ k_d \hat{f} \in L^2(\mathbb{R}), \ f|e = 0,$$

$\hat{k}_d * f$ *is bounded away from zero on* e, *where*

$$k_d \overset{\text{def}}{=} \prod_{\lambda \in \mathbb{C}} \left(1 - \frac{z}{\lambda}\right)^{d(\lambda)}.$$

Choosing $\varphi(x) = [\log(2 + x)]^{-1}$ we see k_d is of zero order, but $\hat{k}_d$ is not local.

The LAST CONJECTURE is disproved by the fact (also found by Kargaev) that *there exist real finite Borel measures μ on $\mathbb{R}$ with very large lacunae in* $\operatorname{supp}\mu$ *(i.e. there is a sequence $\{(a_n, b_n)\}_{n=1}^\infty$ of intervals free of $|\mu|$, $b_n < a_{n+1}, n = 1, 2, \ldots, \frac{b_n - a_n}{a_n}$ tending to infinity as rapidly as we please) and with $\hat{\mu}$ vanishing on a set of positive length.* Take $h = \mu * \varphi$ where φ is a suitable mollifier and $k = \exp(\operatorname{dist}(x, \operatorname{supp} h))$. Then $\log k$ is a Lipschitz function and $\int \frac{\log k(x)}{1+x^2} dx = +\infty$, if $\frac{b_n - a_n}{a_n}$ grows rapidly enough. Then the inverse Fourier transform vanishes on a set of positive length and belongs to the domain of $\hat{k}$.

The THIRD CONJECTURE is true and follows from the Beurling-Malliavin multiplier theorem (this fact was overlooked both by the author and by the editors). Here is the PROOF: There exists an entire function f of exponential type ε, $f \neq 0$, satisfying $|fk| \leqslant 1$ on $\mathbb{R}$. Then $\mathcal{F}^{-1}\left(f \frac{\sin \varepsilon x}{x}\right)$ is in the domain of $\hat{k}$.

REFERENCES

1. de Branges L., *Espaces Hilbertiens de Fonctions Entières*, Masson, Paris, 1972.
2. Levinson N., *Gap and Density theorems*, Amer. Math. Soc., Providence, 1940.
3. de Branges L., *Local operators on Fourier transforms*, Duke Math. J. **25** (1958), 143–153.
4. Beurling A., *Quasianalyticity and generalized distributions*, unpublished manuscript (1961).
5. Kargaev P. P., Mat. Sb. **117** (1982), 397–411 (Russian); English transl. in Math. USSR Sbornik **45** (1983), 397–410.

NON-SPANNING SEQUENCES OF
EXPONENTIALS ON RECTIFIABLE PLANE ARCS

J. A. SIDDIQI

Let $\Lambda = (\lambda_n)$ be an increasing sequence of positive numbers with a finite density and let γ be a rectifiable arc in $\mathbb{C}$. Let $C(\gamma)$ denote the Banach space of continuous functions on γ with the usual sup-norm. If the relation of order on γ is denoted by $<$ and if z_0 and z_1 are two points on γ such that $z_0 < z_1$, we set

$$\gamma_{z_0, z_1} = \{z \in \gamma \mid z_0 < z < z_1\}.$$

The following theorem due to P. Malliavin and J. A. Siddiqi [7] gives a necessary condition in order that the sequence $(e^{\lambda z})_{\lambda \in \Lambda}$ be non-spanning in $C(\gamma)$.

THEOREM. *If the class*

$$C_{00}(M_{n-2}^{\Lambda}, \gamma_{z_0, z_1}) = \left\{ f \in C^{\infty}(\gamma_{z_0, z_1}) : \|f^{(n)}\|_{\infty} \leqslant A M_{n-2}^{\Lambda} \quad f^{(n)}(z) \to 0 \text{ as } z \to z_i, i = 0, 1, \forall_n \right\}$$

is non-empty for some $z_0, z_1 \in \gamma$, *where*

$$M_n^{\Lambda} = \sup_{r \geqslant 0} \frac{r^n}{G_{\Lambda}(ir)} \text{ and } G_{\Lambda}(z) = \prod_{\lambda \in \Lambda} \left(1 - \frac{z^2}{\lambda^2}\right) \qquad (z = re^{i\theta}),$$

then $(e^{\lambda z})_{\lambda \in \Lambda}$ *is non-spanning in* $C(\gamma)$.

It had been proved earlier by P. Malliavin and J. A. Siddiqi [6] that if γ is a piecewise analytic arc then the hypothesis of the above theorem is equivalent to the Müntz condition $\sum \lambda_n^{-1} < \infty$. In connection with the above theorem the following problem remains open.

PROBLEM 1. *Given any non-quasi-analytic class of functions on* γ *in the sense of Denjoy-Carleman, to find a non-zero function belonging to that class and having zeros of infinite order at two points of* γ.

With certain restrictions on the growth of the sequence $\{\lambda_n\}$, partial solutions of the above problem were obtained by T. Erkamma [3] and subsequently by R. Couture [2], J. Korevaar and M. Dixon [4] and M. Lundin [5].

Under the hypothesis of the above theorem, A. Baillette and J. A. Siddiqi [1] proved that $(e^{\lambda z})_{\lambda \in \Lambda}$ is not only non-spanning but also topologically linearly independent by effectively constructing the associated biorthogonal sequence. In this connection the following problem similar to one solved by L. Schwartz [8] in the case of linear segments remains open.

PROBLEM 2. *To characterize the closed linear span of* $(e^{\lambda z})_{\lambda \in \Lambda}$ *in* $C(\gamma)$ *when it is non-spanning.*

References

1. Baillette A., Siddiqi J., *Approximation de fonctions par des sommes d'exponentielles sur un arc rectifiable*, J. d'Analyse Math. **40** (1991), 263–268.
2. Couture R., *Un théorème de Denjoy-Carleman sur une courbe du plan complexe*, Proc. Amer. Math. Soc. **85** (1982), 401–406.
3. Erkamma T., *Classes non-quasi-analytiques et le théorème d'approximation de Müntz*, C. R. Acad. Sc. Paris **283** (1976), 595–597.
4. Korevaar J., Dixon M., *Non-spanning sets of exponentials on curves*, Acta Math. Acad. Sci. Hungar. **33** (1979), 89–100.
5. Lundin M., *A new proof of a Müntz-type Theorem of Korevaar and Dixon*, preprint no. 1979-7, Chalmers University of Technology and The University of Göteborg.
6. Malliavin P., Siddiqi J. A., *Approximation polynômiale sur un arc analytique dans le plan complexe*, C. R. Acad. Sc. Paris **273** (1971), 105–108.
7. Malliavin P., Siddiqi J. A., *Classes de fonctions monogènes et approximation par des sommes d'exponentielles sur un arc rectifiable de* ℂ, C. R. Acad. Sc. Paris **282** (1976), 1091–1094.
8. Schwartz L., *Études des sommes d'exponentielles*, Hermann, Paris, 1958.

Department de Mathématiques
Université Laval
Quebec, Canada, GIK 7P4

AN ALTERNATIVE FOR ANALYTIC CARLEMAN CLASSES

Hruščev S. V.

Given a sequence of positive numbers $\{M_n\}_{n\geqslant 0}$ let $C\{M_n\}$ be the Carleman class of infinitely differentiable functions on the unit circle $\mathbb{T}$ satisfying

$$\sup_{z\in\mathbb{T}}\left|f^{(n)}(z)\right| \leqslant C_f\,Q_f^n\,M_n$$

for $n = 0, 1, 2, \ldots$ and some positive constants C_f, Q_f.

A class of functions defined on $\mathbb{T}$ is called quasi-analytic if it does not contain any function with $f^{(n)}(\zeta) = 0$ for some ζ in $\mathbb{T}$ and every $n = 0, 1, \ldots$, besides $f \equiv 0$. Otherwise, the class is called non-quasianalytic. Clearly, each non-quasianalytic Carleman class contains a non-zero function vanishing on any given proper sub-arc of $\mathbb{T}$. The well-known test of Carleman [1] provides a convenient criterion in terms of $\{M_n\}$ to determine whether $C\{M_n\}$ is quasi-analytic or not.

The analytic Carleman classes $C_A\{M_n\} \stackrel{\text{def}}{=} \{f \in C\{M_n\} : \int_{\mathbb{T}} f\bar{z}^k\,dm = 0,\ k = -1, -2, \ldots\}$ can also be split into quasianalytic and non-quasianalytic ones. There exists an analogue of Carleman's test for such classes [2], but in contradistinction to the classical Carleman classes, a non-zero function in $C_A\{M_n\}$, being the boundary values of a bounded holomorphic function in the unit disc, cannot vanish on any subset of $\mathbb{T}$ having positive Lebesgue measure. Nevertheless, for some non-quasianalytic classes $C_A\{M_n\}$ zero-sets of functions can be rather thick. This is the case, for example, if $M_n = (n!)^{1+1/\alpha}$, $n = 0, 1, \ldots$, where $0 < \alpha < 1$ (see [3]). Therefore it looks reasonable to formulate as a conjecture the following alternative.

CONJECTURE 1. *For every positive sequence* $\{M_n\}_{n\geqslant 0}$, *either the analytic Carleman class* $C_A\{M_n\}$ *is quasi-analytic or there exists a non-empty perfect subset* E *of* $\mathbb{T}$ *and a non-zero function* f *in* $C_A\{M_n\}$ *such that* $f|E \equiv 0$.

The alternative, if true, would have a nice application to dissipative Schrödinger operators. Consider the class B of all bounded measurable real functions V on $[0, +\infty)$ satisfying

$$\int_0^\infty r\,|V(r)|\,dr < +\infty\,.$$

Given $h \in \mathbb{C}$, let l_h be the Schrödinger operator in $L^2(\mathbb{C}, +\infty)$ defined by

$$l_h y = -y'' + Vy, \quad -y'(0) - hy(0) = 0\,.$$

The operator l_h is self-adjoint for real V and real h and it can have only a finite number of bound states, i.e., eigenvalues, if $V \in B$. For complex h the situation changes considerably. Now the number of bound states is finite if

$$|V(r)| \leqslant \exp\{-Cr^{1/2}\}$$

and on the other hand for each α in $(0,1)$ there exist a real-valued potential V satisfying

$$(1) \qquad\qquad |V(r)| \leqslant \exp\left\{-Cr^{\frac{\alpha}{1+\alpha}}\right\}$$

and $h \in \mathbb{C}$, $\operatorname{Im} h > 0$, such that l_h has infinitely many bound states (see [4]). It can be even shown that the family of all closed subsets of $\mathbb{R}$, which may serve as derived sets of the point spectrum of l_h with the potential V satisfying (1), coincides with the family of compact non-uniqueness sets in $(0, +\infty)$ for the Gevrey class $G_\alpha \overset{\text{def}}{=} C_A\{M_n\}$, $M_n = (n!)^{1+1/\alpha}$ (see [5]). The above considerations make plausible the following conjecture.

CONJECTURE 2. *Let T be a positive function on $[0, +\infty)$ such that $t \to \log T(e^t)$ is convex. Then either every Shrödinger operator l_h with the potential V satisfying*

$$(2) \qquad\qquad |V(r)| \leqslant \frac{const}{T(r)}, \qquad r \geqslant 0$$

has only finite number of bound states or there exist V satisfying (2) and $h \in \mathbb{C}$, $\operatorname{Im} h > 0$, such that the derived set of the point spectrum of l_h is non-empty and perfect.

REFERENCES

1. Mandelbrojt S., *Séries adhérentes, Régularisation des suites, Application*, Paris, 1952.
2. R.-Salinas B., *Functions with null moments*, Rev. Acad. Ci. Madrid **49** (1955), 331–368.
3. Hruščev S. V., *Sets of uniqueness for the Gevrey classes*, Ark. Mat. **15** (1977), 253–304.
4. Pavlov B. S., *The non-self-adjoint Schrödinger operator* I, II, III, Topics in Math. Physics, Consultants Bureau, N.Y., 1967, 1968, 1969, pp. 87–114, 111–134, 53–71.
5. Hruščev S. V., *Spectral singularities of dissipative Schrödinger operators with rapidly decreasing potential*, Indiana Univ. Math. J. **33** (1984), 613–638.

EULER INTERNATIONAL MATHEMATICAL INSTITUTE
PESOCHNAYA NAB., 10
ST. PETERSBURG, 197022
RUSSIA

ON A UNIQUENESS THEOREM

V. V. Napalkov

The symbol $H(\mathcal{D})$, $\mathcal{D}$ being an open set in $\mathbb{C}^n$, denotes the set of all functions analytic in $\mathcal{D}$. Let Ω, $\Omega \subset \mathbb{R}^n$ ($n > 1$) be an arbitrary domain, $\Sigma^{(n)} \overset{\text{def}}{=} \{\sigma = (\sigma_1, \sigma_2, \ldots, \sigma_n) : \sigma_j = \pm 1\}$ and let $c = (c_1, c_2, \ldots, c_n) \in \mathbb{R}^n_+$. Define the following sets:

$$\mathcal{D}_\sigma(\Omega, c) = \{z = x + iy \in \mathbb{C}^n : x \in \Omega,\ y_j \sigma_j > c_j,\ j = 1, \ldots, n\},$$

$$\widetilde{\mathcal{D}}_\sigma(\Omega, c) = \operatorname{clos} \mathcal{D}_\sigma(\Omega, c),$$

$$\mathcal{D}^k_\sigma(\Omega, c) = \{z \in \mathbb{C}^n : x \in \Omega,\ y_j \sigma_j > c_j,\ j \neq k\}, \quad k = 1, \ldots, n,$$

$$\mathcal{D}(\Omega, c) = \cup_{\sigma \in \Sigma^{(n)}} \mathcal{D}_\sigma(\Omega, c), \qquad \mathcal{D}^k(\Omega, c) = \cup_{\sigma \in \Sigma^{(n)}} \mathcal{D}^k_\sigma(\Omega, c).$$

Suppose that for a function f, $f \in H(\mathcal{D}(\Omega, 0))$, the restriction $f|\mathcal{D}_\sigma(\Omega, 0) \overset{\text{def}}{=} f_\sigma$ is continuous on the set $\widetilde{\mathcal{D}}_\sigma(\Omega, 0)$. Then the function g,

$$g(x) = \sum_{\sigma \in \Sigma^{(n)}} \operatorname{sign} \sigma \cdot f_\sigma(x), \qquad x \in \Omega, \quad \operatorname{sign} \sigma = \sigma_1 \cdot \sigma_2 \cdot \ldots \cdot \sigma_n,$$

is well-defined in Ω. The following uniqueness theorem has been proved in [1].

If there exist $c = (c_1, c_2, \ldots, c_n) \in \mathbb{R}^n_+$ and functions h^k, $h^k \in H(\mathcal{D}^k(\Omega, c))$, $k = 1, \ldots, n$ such that $f(z) = \sum_{k=1}^n h^k(z)$, $z \in \mathcal{D}(\Omega, c)$ then $g \equiv 0$ on

$$(1) \qquad \Omega^c = \left\{ x \in \Omega : \operatorname{dist}(x, \partial\Omega) > \|c\| = \sqrt{c_1^2 + c_2^2 + \cdots + c_n^2} \right\}.$$

Note that the theorem is important for studying homogeneous convolution equations in domains of real ($\mathbb{R}^n$) or complex ($\mathbb{C}^n$) spaces (see [1], [2], [3]). One might think that $g \equiv 0$ on Ω, as it occurs in the one-dimensional case. However, there exists an example (see [1]), where all conditions of the uniqueness theorem are satisfied, but $g \not\equiv 0$ in Ω (for sufficiently large $\|c\|$). Hence the appearance of the set Ω^c is therefore inevitable, although Ω^c does not seem to be the largest set where $g \equiv 0$.

PROBLEM. *Find the maximal open subset of the domain Ω where $g \equiv 0$.*

References

1. Napalkov V. V., *On a uniqueness theorem in the theory of functions of several complex variables and homogeneous equations of convolution type in tube domains of $\mathbb{C}^2$*, Izv. Akad. Nauk SSSR **40** (1976), 115–132 (Russian); English transl. in Math. USSR Izvestia **10** (1976), 111–126.
2. Napalkov V. V., *Homogeneous systems of equations of convolution type on convex regions in $\mathbb{R}^n$*, Dokl. Akad. Nauk SSSR **219** (1974), 804–807 (Russian); English transl. in Soviet Math. Dokl. **15** (1974), 1672–1675.
3. Napalkov V. V., *On solutions of equations of infinite order in the real domain*, Mat. Sb. **102** (1977), 449–510 (Russian); English transl. in Math. USSR Sbornik **31** (1977), 445–455.

MATHEMATICAL INSTITUTE URO RAN
CHERNYSHEVSKOGO 112
UFA 450025
RUSSIA

HOW FEW CUMULANTS SPECIFY A DISTRIBUTION?

Geoffrey Grimmett

Let F be a probability distribution function and let $c_j(F)$ denote the j-th cumulant (or semi-invariant) of F. It is a famous theorem of Marcinkiewicz [8] that, if there exists a positive integer J such that $c_j(F) = 0$ for $j \geqslant J$, then F is a normal distribution, which is to say that $F = \Phi_{\mu,\sigma}$ for some μ and $\sigma(>0)$, where $\Phi_{\mu,\sigma}(x) = \Phi((x-\mu)/\sigma)$,

$$\Phi(u) = \int_{-\infty}^{u} \frac{1}{\sqrt{2\pi}} e^{-\frac{1}{2}v^2} \, dv.$$

That is to say, the normal distribution is specified by its large cumulants. The challenge of this note is to decide which distributions share this property, in the sense that (modulo a normal factor) they are specified by their large cumulants.

We write φ_F for the characteristic function of F, $\varphi_F(t) = \int_{-\infty}^{\infty} e^{itx} \, dF(x)$. If the j-th moment of F, $m_j(F) = \int_{-\infty}^{\infty} x^j \, dF(x)$, is finite for some integer $j \geqslant 1$, then φ_F is j times continuously differentiable on some neighbourhood of 0. In this case, the j-th cumulant is defined to be

$$c_j(F) = (-i)^j \frac{d^j}{dt^j} \log \varphi_F(t) \Big|_{t=0}$$

It is straightforward to see that $c_1(F) = m_1(F)$, $c_2(F) = m_2(F) - m_1(F)^2$, and $c_j(F)$ is a polynomial function of $m_1(F), m_2(F), ..., m_j(F)$. (See Cramér [3]). We say that F *is specified by its cumulants* if $c_j(G) = c_j(F)$ for all j implies that $G = F$.

It is convenient to think of the cumulants as the coefficients in the Taylor expansion of $\log \varphi_F$: if $\sum_1^\infty |c_j(F)| T^j/j! < \infty$ for some $T > 0$ then φ_F may be expressed as

$$\varphi_F(t) = \exp\left(\sum_{j=1}^{\infty} \frac{(it)^j}{j!} c_j(F) \right), \qquad |t| < T.$$

If F is the normal distribution function $\Phi_{\mu,\sigma}$, then $\varphi_F(t) = \exp\left(i\mu t - \frac{1}{2}\sigma^2 t^2\right)$, and the cumulants are $c_1(F) = \mu$, $c_2(F) = \sigma^2$, $c_j(F) = 0$ for $j \geqslant 3$. Marcinkiewicz's theorem may be restated as saying that, if $\varphi_F(t) = e^{p(t)}$ for some polynomial $p(t)$, then the degree of $p(t)$ is two or less.

Let G and H be probability distribution functions, and write $G*H$ for the distribution defined by the convolution $K(x) = \int_{-\infty}^{\infty} G(x-y) \, dH(y)$, and the note that $\varphi_{G*H} = \varphi_G \varphi_H$. It is immediate that $c_j(G * H) = c_j(G) + c_j(H)$ whenever these quantities are defined. In particular, taking H to be a normal distribution, we find that there exist distinct distributions $F(= G * H)$ and G such that $c_j(F) = c_j(G)$ if $j \neq 1, 2$.

We say *the distribution function F is specified by its large cumulants* if the statement "there exists a positive integer J such that $c_j(F) = c_j(G)$ for all $j \geqslant J$" implies that F and G differ only by a normal factor, in the sense that $\varphi_F(t)/\varphi_G(t) = e^{p(t)}$ for some polynomial of degree two or less. A normal distribution function is specified by its large cumulants.

QUESTION. *Which distributions are specified by their large cumulants?*

Gol'dberg [5] has given an example of a probability distribution function F which is not specified by its large cumulants. More specifically, he has exhibited a distribution function F with the following property: for any even polynomial p with real coefficients and satisfying $p(0) = 0$, there exist a positive ε and a distribution function G with characteristic function $\varphi_G(t) = \varphi_F(t)e^{p(\varepsilon t)}$. It is striking that the tail behaviour of Gol'dberg's F is the same as that of the log-normal distribution (Feller [4] p. 227), a distribution which is not specified by its cumulants; indeed, it may be shown by use of the Krein condition (see Stoyanov [9]) that F has this property also. It would be interesting to find such an example in which the function F has the additional property of being specified by its cumulants. Such an example may not exist, and this would require that F is specified by its *large* cumulants whenever it is specified by its cumulants. Since the cumulants of F find a natural expression in terms of $\log \varphi_F$, it is tempting to conjecture the following.

CONJECTURE. *If F has a finite moment generating function, in that $|\varphi_F(it)| < \infty$ for real t lying in some neighbourhood of the origin, then F is specified by its large cumulants.*

In the positive direction, while generalizing a result of Lukacs [7] p. 223), Christensen [2] has found a category of distributions which are specified by their large cumulants. This category is rather limited in extent, containing distributions of certain Poisson-distributed sums of lattice variables.

Apart from the intrinsic interest of the matter, there is an important potential application to the question of convergence of probability measures. It is basic that, if $c_j(F_n) \to c_j(F)$ as $n \to \infty$ for all j, and in addition F is specified by its cumulants, then $F_n \Rightarrow F$ as $n \to \infty$ (in the sense of weak convergence, see Billingsley [1]). Janson [6] has extended this via Marcinkiewicz's theorem in the case of normal limits, to obtain that $F_n \Rightarrow F$, in the case when F is normal with mean μ and variance σ^2 if there exists a positive integer $J(\geqslant 3)$ such that

$$c_j(F_n) \to c_j(F) \text{ for } j = 1, 2, \text{ and } j \geqslant J.$$

It may be shown without great difficulty that the corresponding statement is valid for any limit F which is specified by its large cumulants.

REFERENCES

1. Billingsley P., *Convergence of probability measures*, John Wiley, New York, 1968.
2. Christensen I. F., *Some further extensions of a theorem of Marcinkiewicz*, Pacific Journal of Mathematics **12** (1962), 59–67.
3. Cramér H., *Mathematical methods of statistics*, Almqvist Wiksell, Uppsala, 1945.

4. Feller W., *An introduction to probability theory and its applications*, Vol. 2, John Wiley, New York, 1971.
5. Gol'dberg A. A., *On a problem of Yu. V. Linnik*, Dokl. Akad. Nauk SSSR **211** (1973), 31–34 (Russian); English transl in Soviet Math. Dokl. **14** (1973), 950–953.
6. Janson S., *Normal convergence by higher semi-invariants with applications to sums of dependent random variables and random graphs*, Annals of Probability **16** (1988), 305–312.
7. Lukacs E., *Characteristic functions*, 2nd ed., Griffin, London, 1970.
8. Marcinkiewicz J., *Sur une propriété de large loi de Gauss*, Mathematische Zeitschrift **44** (1939), 612–618.
9. Stoyanov J., *Counterexamples in probability*, John Wiley, Chichester, 1987.

SCHOOL OF MATHEMATICS
UNIVERSITY WALK
BRISTOL BS8 1TW
UNITED KINGDOM

BOUNDARY VALUES OF QUASIREGULAR MAPPINGS

J. Manfredi,   E. Villamor

The basic problem concerning the boundary values of quasiregular mappings is this:

Problem 1. *Let $f : \mathbb{B}^n \to \mathbb{R}^n$, $n \geqslant 3$, be a bounded quasiregular mapping. Consider the set*

$$E = \left\{ \omega \in \partial \mathbb{B}^n \text{ such that } \lim_{t \to 1^-} f(t\omega) \text{ exists} \right\}.$$

Show that E is non-empty.

Indeed we expect that $\partial \mathbb{B}^n \setminus E$ is small in some sense. We could ask the same question with radial limits replaced by nontangential limits, also called angular limits. It turns out that this is not really relevant since the existence of a radial limit implies the existence of the corresponding nontangential limit whenever the Harnack inequality is satisfied.

It seems that the current technology to prove Fatou theorems using PDE methods cannot deal with nonsmooth coefficients in the nonlinear case, which is precisely what would be needed to treat the quasiregular case in higher dimensions. See [5] and [3].

It is therefore natural to turn to more geometric methods of studying boundary values in the hope of advancing our understanding of Problem 1. A simpler situation yet is obtained requiring that f satisfies, in addition, the condition

$$(1) \qquad \qquad \int_{\mathbb{B}^n} |Df|^p \, dx < \infty,$$

for some p, $1 < p \leqslant n$. The case $p = n$ corresponds to considering mappings in the Dirichlet class. In the two dimensional case $n = 2$ a classical theorem of Beurling [1] states that a harmonic function u on the disk $\mathbb{D}$ with finite Dirichlet integral has radial limits

$$\lim_{r \to 1^-} u(rx)$$

for all $x \in \partial \mathbb{D} \setminus E$, where E is a set of (logarithmic) capacity zero. This theorem has been generalized in many directions. The function u does not have to be harmonic. It is enough for u to be minimally regular (in the class ACL^2). See Section V in Carleson's book [2] for a detailed discussion and references.

A generalization to higher dimensions was given by Miklyukov in [4] where it is proven that a bounded quasiregular mapping $f : \mathbb{B}^n \to \mathbb{R}^n$ satisfying (1) has angular boundary values everywhere on $\partial \mathbb{B}^n$ with the possible exception of a set of p-capacity zero. In this setting the boundedness hypothesis is not natural. It is used by Miklyukov to show that f is normal, which is always true for $p = n$. See Vuorinen's book [6] for the definition of normality and the proof of this fact (p. 189).

Chapter 14. UNIQUENESS, MOMENTS, NORMALITY

PROBLEM 2. Let $f: \mathbb{B}^n \to \mathbb{R}^n$ be a quasiregular mapping satisfying (1). For what value of p is f normal?

PROBLEM 3. Does Miklyukov's theorem hold without the boundedness hypothesis?

We have recently found geometric proofs, using moduli of curve families, of the Beurling and Miklyukov theorems that give a positive answer to Problem 3 for $p = n$. We also conjecture that both Problems (2 and 3) have affirmative answers for $n-1 < p \leqslant n$.

1. Beurling A., *Ensembles exceptionelles*, Acta Math. **72** (1940).
2. Carleson L., *Selected Problems in Exceptional Sets*, Van Nostrand Mathematical Studies #13.
3. Fabes E., Garofalo N., Marin-Malave S., Salsa S., *Potential Theory and Fatou theorems for some nonlinear elliptic operators*, Rev. Mat. Iberoamericana (to appear).
4. Miklyukov V., *Boundary property of n-dimensional mappings with bounded distortion*, Mat. Zametki **11** (1972), no. 2, 159–164 (Russian); English transl in Math. Notes **11** (1972), 102–105.
5. Manfredi J., Weitsman A., *On the Fatou theorem for p-harmonic functions*, Comm. Partial Differential Equations **13** (1988), no. 6, 651–688.
6. Vuorinen M., *Conformal Geometry and Quasiregular Mappings*, Lecture Notes 1319, Springer–Verlag, 1988.

DEPARTMENT OF MATHEMATICS
UNIVERSITY OF PITTSBURGH
PITTSBURGH, PA 15260
USA

DEPARTMENT OF MATHEMATIICS
FLORIDA INTERNATIONAL UNIVERSITY
MIAMI, FL 33139
USA

PATHS FOR SUBHARMONIC FUNCTIONS

John L. Lewis

Let $x = (x_1, \ldots, x_n)$ denote a point in Euclidean n space ($\mathbb{R}^n$), and put $B(x, r) = \{y : |y - x| < r\}$, $r > 0$, $x \in \mathbb{R}^n$. Let $\bar{E}, \partial E$, denote the closure and boundary of E. Let u, $0 \leqslant u \leqslant 1$, be subharmonic in $B(0, 1)$ with $u(0) = \varepsilon > 0$. In this note we consider problems of the following type :

(a) Estimate the length of the *shortest* path γ joining 0 to $\partial B(0, 1)$ with $u > 0$ on $\gamma \cap B(0, 1)$.

(b) Estimate $r(\varepsilon)$, $A(\varepsilon)$ with the following property: If u is as above, $|y| \leqslant r(\varepsilon)$, and $u(y) \geqslant \varepsilon$, then there exists a path γ connecting 0 to y with $u > 0$ on γ and $|\gamma| \leqslant A(\varepsilon)|y|$.

In (b), $|\gamma|$ denotes the length of γ. Estimates of the above type were obtained by Burgess Davis and the author in [2, Theorems 1 and 3] when $n = 2, 3$. After [2] appeared, Alano Ancona pointed out to the author that a theorem of Brelot and Choquet on Green lines could be used to obtain estimates in (a) and (b) whenever $n \geqslant 2$. Since his proof is short, very elegant, and to our knowledge has not appeared elsewhere, we outline this proof for $n \geqslant 3$.

Let $\mathcal{O}$ be a bounded open set with diameter ρ and let

$$(1) \qquad G(x) = |x - x_0|^{2-n} - \int |x - y|^{2-n} \, d\mu(y),$$

be Green's function for $\mathcal{O}$ with pole at $x_0 \in \mathcal{O}$. Here, μ is harmonic measure for $\mathcal{O}$ relative to x_0. Recall that a Green line relative to x_0 is a curve beginning at x_0 and ending at a point on $\partial \mathcal{O}$ with the property that at each interior point of this curve, ∇G is nonzero and tangent to the curve. Then Brelot and Choquet proved (see [1, IV]) that

$$(2) \qquad \int_{\mathcal{O}} |\nabla G| \, dx = \int_{\partial \mathcal{O}} l(y) \, d\mu(y),$$

where $l(y)$ denotes the length of the Green line connecting 0 to $y \in \partial \mathcal{O}$, whenever this curve exists uniquely. Now from (1), (2), and the Tonelli theorem we deduce

$$\alpha \mu(\{y \in \partial \mathcal{O} : l(y) \geqslant \alpha\}) \leqslant \int_{\partial \mathcal{O}} l(y) \, d\mu(y) = \int_{\mathcal{O}} |\nabla G| \, dx$$

$$(3) \qquad \leqslant (n - 2) \int_{\mathcal{O}} |x - x_0|^{1-n} \, dx + (n - 2) \int_{\partial \mathcal{O}} \left(\int_{\mathcal{O}} |x - y|^{1-n} \, dx \right) d\mu(y)$$

$$\leqslant 2(n - 2)\rho \beta_n,$$

where β_n denotes the surface area of the unit sphere in $\mathbb{R}^n$. To obtain estimates for (a) we argue as in [3], [4]. First suppose u is continuous in $B(0, 1)$ and put $\mathcal{O} =$

$\{x : 0 < u(x) < 2\varepsilon\}$, $x_0 = 0$. Then from the maximum principle for harmonic functions we see that

(4) $$\frac{1}{2} = (2\varepsilon)^{-1}u(0) \leqslant \mu\big(\partial\mathcal{O} - \{x : u(x) = 0\}\big).$$

Using (3) with $\alpha = 8(n-1)\beta_n$, $\rho = 2$, and (4) we see there exists a curve γ_1 connecting 0 to a point x_1 with either $u(x_1) = 2\varepsilon$ or $|x_1| = 1$ and $u > 0$ on γ_1. Also, $|\gamma_1| \leqslant 8(n-1)\beta_n$. If $|x_1| < 1$ we put $\mathcal{O} = \{x : 0 < u(x) < 4\varepsilon\}$, $x_1 = x_0$. Arguing as previously we obtain γ_2, connecting x_1 to a point x_2 with either $u(x_2) = 4\varepsilon$ or $|x_2| = 1$. Moreover $u > 0$ on γ_2 and $|\gamma_2| \leqslant 8(n-1)\beta_n$. Then $\gamma_1 \cup \gamma_2$ connects 0 to x_2. Continuing in this way we obtain after at most m times , where m is the least positive integer such that $2^m \varepsilon \geqslant 1$, a curve γ connecting 0 to $\partial B(0,1)$ with $u > 0$ on $\gamma \cap B(0,1)$ and

(5) $$|\gamma| \leqslant 8m(n-1)\beta_n \leqslant 16(n-1)\beta_n \log\left(\frac{2}{\varepsilon}\right).$$

To remove the assumption that u is continuous let $\bar{u} = \max\left[u - \frac{\varepsilon}{2}, 0\right]$, $r_j = 1 - 2^{-j}$, $j = 1, 2, \ldots$. Let $0 \leqslant \varphi \in C_0^\infty([0,1])$ with $\int \varphi\, dx = 1$. Put $\varphi_j(x) = 2^{nj}\varphi(2^j|x|)$, and $u_j = \bar{u} * \varphi_j$ in $B(0, r_j)$, $j = 1, 2, \ldots$, where $*$ denotes convolution. Finally set $\bar{u}_j(x) = u_j(r_j x)$ when $x \in B(0,1)$. Then $\bar{u}_j(0) \geqslant \bar{u}(0) = \frac{\varepsilon}{2}$, and $0 \leqslant \bar{u}_j \leqslant 1$. Furthermore, $\bar{u}_j$ is subharmonic and continuous in $B(0,1)$ for $j = 1, 2, \ldots$. Hence there exists Γ_j, $j = 1, 2, \ldots$, connecting 0 to $\partial B(0,1)$ with $\bar{u}_j > 0$ on $\Gamma_j \cap B(0,1)$. Moreover (5) holds with γ replaced by Γ_j, $j = 1, 2, \ldots$, and ε by $\frac{\varepsilon}{2}$. From (5) it follows that we can parametrize Γ_j, $j = 1, 2, \ldots$, so that $\{\Gamma_j\}$ is uniformly Lipschitz on $[0,1]$. From Ascoli's theorem and (5) we deduce that a subsequence of this sequence converges uniformly on $[0,1]$ to a Lipschitz curve Γ connecting 0 to $\partial B(0,1)$ with

(6) $$|\Gamma| \leqslant 16(n-1)\beta_n \log\left(\frac{4}{\varepsilon}\right).$$

We claim that $u \geqslant \frac{\varepsilon}{2}$ on $\Gamma \cap B(0,1)$. Otherwise, it would follow from the upper semi-continuity of u that there exists z in $\Gamma \cap B(0,1)$ and $\delta > 0$ with $u < \varepsilon/2$ in $B(z, 2\delta)$. Then $\bar{u}_j = 0$ in $B(z, \delta)$, $j \geqslant N$, provided N is large enough. From uniform convergence of a subsequence of $\{\Gamma_j\}$ to Γ it now follows for some j that $\bar{u}_j = 0$ on an arc of Γ_j, which is a contradiction. Hence, (6) gives an estimate for (a).

PROBLEM 1. *Are the functions of ε, n in (6) asymptotically sharp as $\varepsilon \to 0$ and/or as $n \to \infty$?*

Here is an example which shows that the *shortest curve* may have length at least, $c \log\log(1/\varepsilon)$. Let S_j, $j = 1, 3, \ldots$, be the spherical cap on $\partial B(0, r_j)$ centered at $(1, 0, \ldots, 0)$, whose surface area is 3/4's of the surface area of $\partial B(0, r_j)$. Let S_j, $j = 2, 4, \ldots$, be the spherical cap on $\partial B(0, r_j)$ centered at $(-1, 0, \ldots, 0)$, whose surface area is 3/4's the surface area of $\partial B(0, r_j)$. Let $D = B(0,1) \setminus \cup_{j=1}^m S_j$, where again m is the least positive integer such that $2^{-m} < \varepsilon$. Define u in $\bar{B}(0,1)$ so that u is continuous in $\bar{B}(0,1)$, harmonic in D, and

$$\lim_{x \to y} u(x) = 1, \ y \in \partial B(0,1),$$

$$\lim_{x \to y} u(x) = 0, \ y \in \partial D \cap B(0,1).$$

Clearly, u is subharmonic in $B(0,1)$, $0 \leqslant u \leqslant 1$, and the length of any path connecting 0 to $\partial B(0,1)$ on which $u > 0$, is at least cm. Using Harnack's inequality it is also easily seen that

$$\log \log(1/u(0)) \leqslant cm.$$

To get estimates for (b), we put

$$(7) \qquad\qquad r = r(\varepsilon) = \left(\frac{\varepsilon}{2}\right)^{c/\varepsilon}$$

and suppose that $|y| \leqslant r$. If $u(y) \geqslant \varepsilon$, we argue as in [2, Lemma 3] to deduce for c large enough (depending only on n), the existence of $s \geqslant 2|y|$, and a closed set $E \subset \bar{B}(0,s) \cap \{x : u(x) \geqslant \frac{\varepsilon}{2}\}$ with the properties that:

(i) $s \leqslant |y|r^{-1}$,

(ii) any two points in E can be connected by a curve of length at most $4s$,

(iii) $\partial B(0,s) - E$ has surface area at most, $\left(\frac{1}{2}\right)^{n+2} \beta_n \varepsilon s^{n-1}$.

Now if u is continuous in $B(0,1)$, $u^*(x) = \bar{u}(sx)$, $x \in B(0,1)$ ($\bar{u}$ as above), and $\mathcal{O} = \{x : u^*(x) > 0\}$, then as in (3), (4), we see there are closed sets $F_1, F_2 \subset \partial B(0,1) \cap \partial \mathcal{O}$ of harmonic measure (relative to $\mathcal{O}, y/s$, respectively) at least $\frac{\varepsilon}{4}$ such that every point in F_1, F_2 can be joined to $0, y/s$, by a curve of length at most $16(n-2)\beta_n \varepsilon^{-1}$. Moreover $u^* > 0$ on each curve. Using the Poisson integral formula and comparing harmonic measures of F_1, F_2, relative to $\mathcal{O}$, $B(0,1)$, we find that F_1, F_2, each has surface area at least, $\left(\frac{1}{2}\right)^{(n+1)} \beta_n \varepsilon$. Dilating by s and using (iii) we see that $0, y$ can be joined to points z_1, z_2 in $F_1 \cap E$, $F_2 \cap E$, respectively, by curves σ_1, σ_2. Also, $|\sigma_i| \leqslant 13(n-2)\beta_n \varepsilon^{-1}s$ for $i = 1, 2$. From (ii) we can then join z_1, z_2, to each other by a curve σ_3 of length at most $4s$. If $\gamma = \cup_{i=1}^3 \sigma_i$, then it follows that γ connects 0 to y, $u > 0$ on γ, and

$$|\gamma| \leqslant 40s(n-2)\beta_n \varepsilon^{-1} \leqslant 40(\varepsilon r)^{-1}(n-2)\beta_n |y|,$$

where (i) was used in the last inequality. The assumption that u is continuous can be removed by smoothing as in (6). Thus (7) and (8) provide estimates for (b).

PROBLEM 2. *Can the functions of ε in (7), (8), be replaced by ε^a, ε^{-a} for some positive a, depending only on n? If not, how about estimates for c in (7)?*

REFERENCES

1. Brelot M., Choquet C., *Espaces et lignes de Green*, Ann. Inst. Fourier **3** (1951), 199–263.
2. Davis B., Lewis J., *Paths for subharmonic functions*, Proc London Math. Soc. **48** (1984), no. 3, 401–427.
3. Lewis J., Rossi J., Weitsman A., *On the growth of subharmonic functions along paths*, Ark. Mat. **22** (1984), no. 1, 109–119.
4. Wu J. M., *Lengths of paths for subharmonic functions*, J. London Math. Soc. **32** (1985), no. 2, 497–505.

DEPARTMENT OF MATHEMATICS
UNIVERSITY OF KENTUCKY
LEXINGTON, KY 40506-0027
USA

14.17

WHEN IS A PSEUDOSPHERE A QUASISPHERE?

John Lewis

In [2] Harold Shapiro defined a pseudosphere as follows: let D be a bounded domain in R^n. Then D is said to be a pseudosphere if

(a) ∂D is homeomorphic to the unit sphere in R^n,

(b) $g(0) = a \int_{\partial D} g \, dH^{n-1}$, whenever g is harmonic in D and continuous in the closure of D.

In (b), a denotes a constant and H^{n-1} denotes $(n-1)$-dimensional Hausdorff measure in R^n. Lewis and Vogel [1], have established the existence of pseudospheres which are not spheres in R^n, $n \geqslant 3$. The homeomorphism f which they construct is in fact a homeomorphism from R^n to R^n and f, f^{-1} are Hölder α whenever $0 < \alpha < 1$. *Do there exist pseudospheres (other than spheres) in R^n which are also quasispheres? That is, can the homeomorphism in (a) be chosen to be a quasiconformal homeomorphism?*

The answer to this question in R^2 is yes as can be deduced from some examples of Keldysh and Lavrentiev (see [1] for references).

References

1. Lewis J., Vogel A., *On pseudospheres*, Rev. Math. Iberoamericana **7** (1991), 25–54.
2. Shapiro H. S., *Remarks concerning domains of Smirnov type*, Michigan Math. J. **13** (1966), 341–348.

Department of Mathematics
University of Kentucky
Lexington, KY 40506–0027
USA

UNIQUE CONTINUATION WITH L^p LOWER ORDER TERMS

T. WOLFF

PROBLEM. *Prove, or more likely disprove, the following statement: if u is a smooth function on a domain in $\mathbb{R}^d$, $d \geqslant 2$, satisfying a differential inequality*

$$|\Delta u| \leqslant A|u| + B|\nabla u|$$

with $A \in L^p$, $B \in L^q$, $p = q = 1$, and if u vanishes on an open set, then it vanishes identically.

The key point here is that A and B are only assumed in L^1. What is known as a consequence of the classical Carleman method and arguments from harmonic analysis is that the statement is true when $p = \frac{d}{2}$, $q = d$, $d \geqslant 3$. This was proved in [5]; an earlier related paper [1] should also be mentioned. The question is also open for any pair (p, q) such that $p \geqslant 1$, $q \geqslant 1$ and either $p < \frac{d}{2}$ or $q < d$. If p or q is < 1 then there are easy counterexamples obtained by starting from one-dimensional counterexamples and adjoining dummy variables.

There is a long history of complicated counterexamples in this sort of problem. We note the results of A. Plis [3] and K. Miller [2]. In the latter paper an example is given in $\mathbb{R}^d$, $d \geqslant 3$ of an equation

$$\operatorname{div}(A\nabla u) = 0$$

with A Hölder continuous and elliptic, and having a nontrivial smooth solution vanishing on an open set. As far as we have been able to find out, it is unknown whether such examples or similar examples with A just bounded and uniformly elliptic can be constructed in $\mathbb{R}^2$. This question, of course, reduces to the case where A is a scalar multiple of the identity via the Beltrami equation.

A related QUESTION is this: *What happens when u vanishes to infinite order at a single point instead of on an open set?* We assume for simplicity that $d \geqslant 3$ here. For the inequality

$$|\Delta u| \leqslant A|u|, \quad A \in L^p$$

the question is answered in [1]: if $p = \frac{d}{2}$ then u must vanish identically, and if $p < \frac{d}{2}$ there are easy counterexamples. For the first-order inequality

$$|\Delta u| \leqslant B|\nabla u|, \quad B \in L^q$$

the question is open if $d \geqslant 5$ and $d < q < \frac{3d-4}{2}$. There are easy counterexamples in any dimension if $q < d$ and more complicated counterexamples due to the author if $d \geqslant 5$, $q = d$ (to appear). If $d \leqslant 4$ and $q \geqslant d$, or $d \geqslant 5$ and $q \geqslant \frac{3d-4}{2}$ there is a positive result (cf. [4], where references to relevant earlier work are also given).

REFERENCES

1. Jerison D., Kenig C., *Unique continuation and absence of positive eigenvalues for Schrödinger operators*, Ann. Math. **121** (1985), 463–494.
2. Miller K., *Nonunique continuation for certain ODE's in Hilbert space and for uniformly parabolic and elliptic equations in self-adjoint divergence form*, Symposium on non-well-posed problems and logarithmic convexity (R.J. Knops, ed.), Lecture Notes in Math., vol. 316, Springer-Verlag, 1973, pp. 85–101.
3. Plis A., *On nonuniqueness in Cauchy problems for an elliptic second-order differential operator*, Bull. Acad. Polon. Sci. **11** (1963), 95–100.
4. Wolff T., *Unique continuation for $|\Delta u| \leqslant V|\nabla u|$ and related problems*, Rev. Mat. Iberoamericana **6** (1990), 155–200.
5. Wolff T., *A property of measures in $\mathbb{R}^n$ and an application to unique continuation*, Geom. Funct. Anal. **2** (1992), 225–284.

DEPARTMENT OF MATHEMATICS
UNIVERSITY OF CALIFORNIA
BERKELEY, CA 94720
USA

WHEN IS $\int_{\mathbb{T}} \log |f| \, dm > -\infty$?

A. L. VOLBERG

It is a well-known fact of Nevanlinna theory that the inequality in the title holds for boundary values of non-zero holomorphic functions which belong to the Nevanlinna class in the unit disc. But what can be said about summable functions f with non-zero Riesz projection $\mathbb{P}_- f \neq 0$? Here $\mathbb{P}_- f \overset{\text{def}}{=} \sum_{n>0} \hat{f}(-n) z^n$, $|z| \leqslant 1$.

Given a positive sequence $\{M_n\}_{n \geqslant 0}$ define

$$C\{M_n\} = \left\{ f \in C^\infty(\mathbb{T}) : \|f^{(n)}\|_\infty \leqslant C_f \, Q_f^n \, M_n \right\}.$$

It is assumed that (a) $M_n^2 \leqslant M_{n-1} M_{n+1}$, $n = 1, 2, \ldots$; (b) $\lim_{n \to \infty} \frac{x^n}{M_n} = 0$, $x \geqslant 0$. This does not restrict the generality because every Carleman class coincides with one defined by a sequence satisfying (a) and (b). Let $T(x) = \sup \frac{x^n}{M_n}$, $x > 0$ Then $C\{M_n\}$ is a quasianalytic class iff $\int^\infty \frac{\log T(x)}{x^2} dx = +\infty$. In case $C\{M_n\}$ is non-quasianalytic, there are, of course, functions f in $C\{M_n\}$ with $\int_{\mathbb{T}} \log |f| dm = -\infty$. In fact there exists an f in $C\{M_n\}$ equal to zero on an open subset of $\mathbb{T}$.

QUESTION. *Suppose* $C\{M_n\}$ *is a quasianalytic class and let* $f \in L^1(\mathbb{T})$ *with* $P_- f \in C\{M_n\}$. *Is it true that*

$$\int_{\mathbb{T}} \log |f| \, dm > -\infty ?$$

Under some additional assumptions on regularity of $\{M_n\}$ the answer is yes [1].

REFERENCE

1. Volberg A. L., *The logarithm of an almost analytic function is summable*, Dokl. Akad. Nauk SSSR **265** (1983), 1317–1323 (Russian); English transl. in Soviet Math. Dokl. 26 (1982), 238–243.

DEPARTEMENT OF MATHEMATICS
MICHIGAN STATE UNIVERSITY
EAST LANSING, MI 48824   AND
USA

STEKLOV MATHEMATICAL INSTITUTE
ST. PETERSBURG BRANCH
FONTANKA 27
ST. PETERSBURG, 191011
RUSSIA

COMMENTARY BY THE AUTHOR

The question that has been raised here has a negative answer. A counterexample can be based on the following result:

THEOREM (see [2]). *There exists a bounded monotone function w on $[0,1]$ and a function $f \in C^1(\mathbb{D}) \cap L^\infty(\mathbb{D})$ such that*

(1)
$$|\bar{\partial} f(z)| \leqslant w(1 - |z|)$$

(2)
$$\int_0^1 \log \log \frac{1}{w(x)} \, dx = +\infty$$

(3)
$$\int_0^{2\pi} \log |f(e^{i\theta})| \, d\theta = -\infty$$

but $f(e^{i\theta}) \not\equiv 0$ on $\mathbb{T}$.

Here $f(e^{i\theta})$ is to be understood as the radial (or non-tangential) boundary values of f, which exist by virtue of property (1) and the bound on f. We may also assume without loss of generality that w is logarithmically convex (i.e. $\log \frac{1}{w(e^{-x})}$ is convex in x).

The proof of the theorem is based on a deep result of Widom [3]. Using the stated theorem we can construct a quasianalytic class $C\{M_n\}$ and a function $f \not\equiv 0$ on $\mathbb{T}$ so that $\mathbb{P}_- f \in C\{M_n\}$ and property (3) holds. In this way a negative answer to our original question is easily obtained.

We begin with the function f whose existence is guaranteed by our theorem. It remains to construct a quasianalytic sequence $\{M_n\}$ for which the Cauchy integral

$$F(\zeta) = -\frac{1}{\pi} \int_{\mathbb{D}} \frac{\bar{\partial} f(z)}{z - \zeta} \, dx \, dy, \quad z = x + iy$$

belongs to $C\{M_n\}$ on $\mathbb{T}$. It is a straightforward consequence of Green's theorem that $\mathbb{P}_- f = F$ on $\mathbb{T}$. And, to accomplish our objective it is sufficient to show that

(4)
$$\mathcal{D}_w \subset C\{M_n\},$$

where, by definition, $\mathcal{D}_w = \{ H \in L^\infty(2\mathbb{D}) : |\bar{\partial} H(z)| \leqslant w(|1 - |z||) \}$. Consider, therefore, the moment sequence of w:

(5)
$$\Lambda_n = \sup_{0 < x < 1} \frac{w(x)}{x^n}.$$

The sequence $\{\Lambda_n\}$ is logarithmically convex (i.e. $\Lambda_n^2 \leqslant \Lambda_{n-1}\Lambda_n$) and so consequently is the sequence $\{M_n\}$ obtained by setting $M_n = n! \, \Lambda_n$. Because $F \in \mathcal{D}_w$ its boundary function $F(e^{i\theta})$ can be represented in the form

$$F(e^{i\theta}) = -\frac{1}{\pi} \int_{|\zeta| < 2} \frac{\bar{\partial} F(z)}{\zeta - e^{i\theta}} \, d\xi \, d\eta + A(e^{i\theta}),$$

where A is defined and holomorphic in $|z| < 2$ and $\zeta = \xi + i\eta$. It follows that

$$|F^{(n)}(e^{i\theta})| \leqslant \frac{n!}{\pi} \int_{|\zeta| < 2} \frac{w(|1 - |z||)}{|1 - |\zeta||^n |\zeta - e^{i\theta}|} \, d\xi \, d\eta + C^n n!$$
$$\leqslant B n! \, \Lambda_n = B M_n.$$

Thus, we have only to check that $C\{M_n\}$ is a quasianalytic class.

CLAIM. *If w is logarithmic convex and if (2) holds then $C\{M_n\}$ is a quasianalytic class.*

Note that logarithmic convexity is essential here. It is possible to construct a monotonic w for which (2) is satisfied, but nevertheless $\mathcal{D}_w$ is not contained in any quasianalytic Carleman class. This fact seems to be generally known, although we have not been able to find a reference to it in the literature. The construction, however, can be based on an idea of B. M. Makarov [4].

Proof of the claim. Let

$$a = a_w(t) = \inf_{0 < x < 1} \frac{1}{w(x)(1-x)^t}$$

and set $p = p_w(t) = \log a_w(t)$. By the Denjoy–Carleman theorem $C\{M_n\}$ is a quasianalytic class if and only if

(6)
$$\int^\infty \frac{\log T_0(x)}{1 + x^2}\, dx = +\infty,$$

where by definition $T_0(x) = \sup_{n>0} \frac{x^n}{M_n}$. On the other hand, we know that $\int_0 \log\log \frac{1}{w(x)}\, dx = +\infty$ if and only if

(7)
$$\int^\infty \frac{\log a(t)}{1 + t^2}\, dt = +\infty.$$

Unfortunately, we cannot use a to directly estimate T_0 from below. But, we can obtain the required estimate in terms of a closely related function a_0 defined by:

(i) $a_0(t) = \sup\limits_{n>0} \dfrac{t^n}{A_n}$,

(ii) $A_n = \sup\limits_{0 < t < \infty} \dfrac{t^n}{a(t)}$.

In fact, it can be shown that

(8)
$$T_0((1+\varepsilon)t) > a_0(t), \quad t \geqslant t_\varepsilon$$

by simply verifying that $A_n \geqslant \dfrac{n!\, A_n}{(1+\varepsilon)^n}$. The latter is in turn a consequence of the simple calculation

$$A_n = \sup_t \frac{t^n}{a(t)} = \sup_{0 < x < 1}\ \sup_t w(x)\, t^n (1-x)^t$$

$$= \sup\left\{ \frac{w(x)}{x^n} \left(\frac{n}{e}\right)^n \left(\frac{1}{x} \log \frac{1}{1-x}\right)^{-n} \right\}.$$

Because $\frac{1}{x} \log \frac{1}{(1-x)} > 1$, we conclude that $A_n \geqslant \dfrac{n!\, A_n}{(1+\varepsilon)^n}$ and therefore (8). Finally, it is well-known that

(iii) w logarithmically convex $\Rightarrow a = a_w$ is also logarithmically convex.

(iv) if a is logarithmically convex then

(9)
$$\int^\infty \frac{\log a(t)}{1 + t^2}\, dt = +\infty \quad \Longleftrightarrow \quad \int^\infty \frac{\log a_0(t)}{1 + t^2}\, dt = +\infty.$$

For this assertion see [5].

Combining (7), (8) and (9) we have the claim. A counterexample to our original problem now follows from the claim and our earlier remarks.

Chapter 14. UNIQUENESS, MOMENTS, NORMALITY

REFERENCES

2. Borichev A. A., Vol'berg A., *Uniqueness theorems for almost analytic functions*, Algebra i Analiz **1** (1989), 146–177 (Russian); English transl. in Leningrad Math. J. **1** (1990), 157–191.
3. Widom H., H^p *sections of vector bundles over Riemann surfaces*, Ann. of Math. **94** (1971), 304–324.
4. Makarov B. M., *The problem of moments in some functional spaces*, Dokl. Akad. Nauk SSSR **127** (1959), 957–960. (Russian)
5. Hruščёv S. V., *The problem of simultaneous approximation and removal of singularities of Cauchy-type integrals*, Trudy Mat. Inst. Steklov Akad. Nauk SSSR **130** (1978), 124–195 (Russian); English transl. in Proc. Steklov Inst. Math. **130** (1979), 133–203.

S.14.20
old

MOMENT PROBLEM QUESTIONS

ALLEN DEVINATZ

Let N_0 be the non-negative integers and N_0^n the set of all multi-indices $\alpha = (\alpha_1, \ldots, \alpha_n)$ with each $\alpha_j \in N_0$. For any $x = (x_1, \ldots, x_n) \in \mathbb{R}^n$ write $x^\alpha = x_1^{\alpha_1} \ldots x_n^{\alpha_n}$, where $x_j^0 = 1$. Denote by $\mathcal{P}_0$ the complex vector space of all polynomials, $p(x) = \sum a_\alpha x^\alpha$, considered as functions from $\mathbb{R}^n \to \mathbb{C}$. A multi-sequence $\mu_\alpha \colon N_0^n \to \mathbb{R}$ is called a moment sequence if there exists a bounded non-negative Borel measure $d\mu(x)$ so that

$$\mu_\alpha = \int_{\mathbb{R}^n} x^\alpha \, d\mu(x).$$

The moment sequence is said to be determined if there exists a unique representing measure. We refer the reader to the recent expository article by B. Fuglede [2] for a discussion of this problem together with an up-to-date set of references.

If $n = 1$, it is well known that if the moment sequence is determined, then $\mathcal{P}_0$ is dense in $L^2(d\mu)$. Indeed, more is known. If $\mu_0 = 1$, and $d\mu$ is an extreme point of the convex set of representing measures for μ_α, then $\mathcal{P}_0$ is dense in $L(d\mu)$.

In 1978, the theoretical physicist, Professor John Challifour of Indiana University proposed (in private conversation with the author) the following question:

QUESTION 1. *For $n > 1$, is it still true that if a multi-parameter moment sequence μ_α has a unique representing measure $d\mu$, then $\mathcal{P}_0$ is dense in $L^2(d\mu)$?*

To turn to a second question, suppose μ_α and ν_α are moment sequences from $N_0 \to \mathbb{R}$, $(\mu * \nu)_\alpha$ is the moment sequence formed from the convolution measure $d(\mu * \nu)$, then it was shown in [1] that if $(\mu * \nu)_\alpha$ is a determined moment sequence, then so are the individual moment sequences μ_α and ν_α. Very recently, the statistician, Persi Diaconis of Stanford University proposed (again in private conversation with the author) the following question:

QUESTION 2. *If μ_α and ν_α are determined moment sequences, is it true that $(\mu * \nu)_\alpha$ is a determined moment sequence?*

REFERENCES

1. Devinatz A., *On a theorem of Levy–Raikov*, Ann. of Math. Statistics **30** (1959), 538–586.
2. Fuglede B., *The multidimensional moment problem*, Expo. Math. **1** (1983), 47–65.

NORTHWESTERN UNIVERSITY
DEPARTMENT OF MATHEMATICS
EVANSTON, IL 606201
USA

Chapter 14. UNIQUENESS, MOMENTS, NORMALITY

COMMENTARY BY CHRISTIAN BERG

ANSWER TO QUESTION 1.(Challifour, Devinatz).

Let μ be a measure of R^n with moments of every order, and assume that μ is *determinate*, i.e. there is no other measure having the same moments as μ. By a classical theorem of M. Riesz the polynomials are dense in $L^2(\mu)$ in dimension $n = 1$, and the question is whether this result remains true in dimension $n > 1$. The question was also raised by Fuglede [6], and he was led to introduce the notion of strong determinacy. The question has been answered in the negative in a paper by Marco Thill and the author [1]. We prove the following:

There exist rotation invariant measures μ on R^n, $n > 1$, which are determinate but for which the polynomials are not dense in $L^2(\mu)$. Such measures μ are not strongly determinate, and they are necessarily of the following very special form

$$\mu = \sum_{i=0}^{\infty} \alpha_i \omega_{r_i},$$

where $\alpha_i > 0$, ω_r is the normalized uniform distribution on the sphere $\|x\| = r$ and $0 \leqslant r_0 < r_1 < \cdots \to \infty$ are the zeros of an entire function of order $\leqslant 2$.

The above result is obtained by studying the relation between a rotation invariant measure μ and its image σ under the mapping $x \mapsto \|x\|^2$ of R^n onto $[0, \infty]$. A major step is the construction and characterization of measures σ on $[0, \infty]$ which are determinate in the sense of Stieltjes, but for which the polynomials are not dense in $L^2(t^k \, d\sigma(t))$ for some integer k (necessarily $\geqslant 3$). This theme is expanded in the joint paper [2].

ANSWER TO QUESTION 2.(Diaconis [5], Devinatz).

Let μ and ν be determinate measures on R. The question is whether $\mu * \nu$ is again determinate. This was answered negatively in [3] and [4]. In [3] a determinate measure μ is constructed such that $\mu * \mu$ is indeterminate. In [4] the following result is given: For any integer $m \geqslant 0$ there exists a determinate measure μ and a measure ν of finite support such that $\mu * \nu$ is indeterminate and m-canonical.

REFERENCES

1. Berg C., Thill M., *Rotation invariant moment problems*, Acta. Math. **167** (1991), 207–227.
2. Berg C., Thill M., *A density index for the Stieltjes moment problem*, IMACS Annals on computing and applied mathematics **9** (1991), 185–188;Claude Brezinski, Laura Gori and André Ronveaux (eds.), *Orthogonal polynomials and their applications*, Proceedings from the Third International Symposium on Orthogonal Polynomials, Erice, 1990.
3. Berg C., *On the preservation of determinacy under convolution*, Proc. Amer. Math. Soc. **93** (1985), 351–357.
4. Buchwalter H., Cassier G., *Mesures canoniques dans le problème classique des moments*, Ann. Inst. Fourier (Grenoble) **34** (1984), no. 2, 45–52.
5. Diaconis P., Ylvisaker D., *Quantifying prior opinion*, Baysian Statistics 2 (J. Bernardo et al., eds.), North Holland, Amsterdam, 1985, pp. 133–156.
6. Fuglede B., *The multidimensional moment problem*, Expo. Math. **1** (1983), 47–65.

MATEMATISK INSTITUT
UNIVERSITETSPARKEN 5
DK-2100 COPENHAGEN Ø
DENMARK

Chapter 15

INTERPOLATION, BASES, MULTIPLIERS

Edited by

N. K. Nikolski

Université Bordeaux-I
UFR Mathématiques
351, cours de la Libération   and
33405 Talence CEDEX
France

Steklov Mathematical Institute
St. Petersburg Branch
Fontanka 27
St. Petersburg, 191011
Russia

INTRODUCTION

The chapter succeeds Chapter 10 of the previous edition (LNM, 1043) and we reproduce (slightly updated) Preface to this chapter.

We discuss in this introduction only one of various aspects of interpolation, namely the **free** (or Carleson) interpolation by analytic functions.

Let X be a class of functions analytic in the open unit disc $\mathbb{D}$. We say that the interpolation by elements of X on a set $E \subset \mathbb{D}$ is free if the set $X|E$ (of all restrictions $f|E$, $f \in X$) can be described in terms not involving the complex structure inherited from $\mathbb{D}$. So, for example, if E satisfies the well-known Carleson condition (see formula (C) in Problems 15.5, 15.7 below), the interpolation by elements of H^∞ on E is free in the following sense: any function, bounded on E, belongs to $H^\infty|E$. The freedom of interpolation for many other classes X means (as in the above example) that the space $X|E$ is ideal (i.e. $\varphi \in X|E$, $|\psi| \leqslant |\varphi|$ on $E \Longrightarrow \psi \in X|E$). Sometimes the freedom means something else, as it is the case with classes X of analytic functions enjoying certain smoothness at the boundary (see Problem 15.5), or with the Hermite interpolation with unbounded multiplicities of knots (this theme is treated in [1], [2]).

Problems 15.1, 15.3, 15.5, 15.6 below deal with free interpolation which is also the theme (main or peripheral) of Problems 9.5, 10.14, 10.21, 11.7, S.14.19, S.16.22. But the information, contained in the volume, does not exhaust the subject, and we recommend survey [3], book [4] and doctoral thesis [5].

There exists a simple but important connection of interpolation (or, in other words, of the moment problem) with the study and classification of biorthogonal expansions (bases). This fact was (at last) widely realized during the past 20–25 years, though it was explicitly used already by S. Banach and T. Carleman. Namely, every pair of biorthogonal families $\mathcal{F} = \left\{ f_\lambda \right\}_{\lambda \in \sigma}$, $\mathcal{F}' = \left\{ f'_\lambda \right\}_{\lambda \in \sigma}$ (f_λ are vectors in the space V, f'_λ belong to the dual space, $\langle f_\lambda, f'_\mu \rangle = \delta_{\lambda\mu}$) generates the following interpolation problem: to describe the coefficient space $\mathcal{J}V \left(\mathcal{J}f \stackrel{\text{def}}{=} \left\{ \langle f, f'_\lambda \rangle \right\}_{\lambda \in \sigma} \right)$ of formal Fourier expansions $f \sim \sum_\lambda \langle f, f'_\lambda \rangle f_\lambda$. There are also continuous analogues of this connection which are of importance for the spectral theory. "Freedom" of this kind of interpolation (or, to be more precise, the ideal character of the space $\mathcal{J}V$) means that $\mathcal{F}$ is an unconditional basis in its closed linear hull. This observation plays now a significant role in the intersection of interpolation methods with the spectral theory and convolution equations, the latters being the principal suppliers of concrete biorthogonal families. These families usually consist of eigen- or root-vectors of an operator T (in Function Theory T is often differentiation or the backward shift, the two being isomorphic): $Tf_\lambda = \lambda f_\lambda$, $\lambda \in \sigma$. Thus the properties of the equation $\mathcal{J}f = g$ (g is the given function defined on σ) depend on the amount of *multipliers* of $\mathcal{F}$, i.e. of operators $V \mapsto V$ sending f_λ to $\mu(\lambda)f_\lambda$, where μ denotes a function $\sigma \to \mathbb{C}$ or the multiplier itself. These multipliers μ may turn out to be functions of T ($\mu = \varphi(T)$) and then we come to another interpolation problem (given μ, find φ). The solution of this "multiplier" interpolation problem often leads to the solution of the initial problem $\mathcal{J}f = g$. We cannot enter here into more details or enlist the literature and refer the reader to the mentioned book [1] and to papers [6], [7].

Interpolation, bases and multipliers are related approximately in the mentioned way in Problems 15.1, 15.3, 15.9, whereas Problem 15.10 deals with Fourier multipliers in their own right. These occur, as is well-known, in numerous problems of Analysis, but in the present context the amount of multipliers determines the convergence (summability) properties of standard Fourier expansions in the given function space. (By the way, the word "interpolation" in the title of Problem 15.10 has almost nothing to do with the same term in the Chapter title, and means the interpolation of operators. We say "almost" because the latter is often and successfully used in free interpolation).

Problem 15.2 concerns biorthogonal expansions of analytic functions. The theme of bases is discussed also in 15.3, 15.4 and in 1.3, 1.15, 1.17.

By the way, the last 10–15 years one could mention a visible success of the free interpolation—unconditional bases tandem in applications to convolution equations (and, in particular, to partial differential equations). Namely, the point is in a slight specification of (not necessary free) interpolation problems, that is in the existence of a *linear* interpolation operator $I \colon X|E \to X$ (i.e. such that $RI = id$ where R stands for the restriction $Rf = f|E$, $f \in X$). Usually, the existence of I, being equivalent to the fact that the "ideal" $X_E = \{ f \in X : Rf = 0 \}$ is a complemented subspace of X, does not imply any "freedom" of the interpolation problem, and, vice versa, the ideal character of interpolation data space $X|E$ does not guarantee the complementation of X_E. But it help to build an interpolation operator: let $f'_\lambda(g) = g(\lambda)$, $g \in X$ be point evaluation functionals on X, and $\{ \dot{f}_\lambda \}$ their bi-orthogonal companion in the quotient space X/X_E; if the coefficient space $\mathcal{J}(X / X_E) = RX$ is an ideal one it is much easier to choose representatives $f_\lambda \in \dot{f}_\lambda = \dot{f}_\lambda + X_E$ in such a way that interpolation (Lagrange) series $If = \sum_\lambda \langle f, f'_\lambda \rangle f_\lambda$ to be convergent (or summable) in X for each f, $f \in X$. On the other hand, the complementation is of interest for convolution equations, in particular due to its links to the surjectivity problem $C_\mu Y = Y$ where $C_\mu f = f * \mu$; the matter is then on the splitting of Y into the sum of the kernel $\operatorname{Ker} C_\mu$ and its complement; this is equivalent to the linear interpolation for the adjoint space of Y on the zeros set $E = \{ \lambda : \hat{\mu}(\lambda) = 0 \}$ where $\hat{\mu}$ stands for Fourier–Laplace transform. We can not enter here in any detail and refer the interested reader to items [6]–[10] from the reference list of 15.1 (this problem is related to the philosophy described above; for more information see also Chapter 11).

Problem 15.8 represents an interesting and vast aspect of interpolation, namely, its "real" aspect. We mean here extension theorems à la Whitney tending to the constructive description of traces of function classes determined by global conditions.

Free interpolation by analytic functions in $\mathbb{C}^n$ (and by harmonic functions in $\mathbb{R}^n$) is a fascinating area (see, e.g., Preface to Garnett's book). It is almost unexplored not counting classical results on extensions from complex submanifolds and their refinements. Free interpolation in $\mathbb{C}^n$ is discussed in Problems 15.6, 15.7.

And, at last, problems 15.11–15.13 have quite a different flavour: they are related to the classical extension problems for positive defined functions, a kind of interpolation but—if thinking in an analytic functions context—rather a Carathéodory type interpolation than the free one. See also problems 14.1, 14.2, S.14.20 where some aspects of various moment problems are discussed.

References

1. Nikol'skii N., *Treatise on the shift operator*, Springer-Verlag, Heidelberg etc., 1986.
2. Vinogradov S. A., Rukshin S. E., *On free interpolation of germs of analytic functions in the Hardy spaces*, Zapiski Nauchn. Semin. LOMI **107** (1981), 36–45 (Russian); English transl. in J. Soviet Math. **36** (1987), no. 3.
3. Vinogradov S. A., Havin V. P., *The free interpolation in H^∞ and in some other function classes*, Zapiski Nauchn. Semin. LOMI **47** (1974), 15–54; **56** (1976), 12–58 (Russian); English transl. in J. Soviet Math. **9** (1978), no. 2; **14** (1980), no. 2.
4. Garnett J., *Bounded analytic functions*, Academic Press, NY, 1981.
5. Vinogradov S. A., *Free interpolation in spaces of analytic functions*, Leningrad State University (2 nd doctoral thesis), 1982.
6. Hruščëv S. V., Nikolski N. K., Pavlov B. S., *Unconditional bases of exponentials and reproducing kernels*, Lect. Notes Math. **864** (1981), 214–335.
7. Nikolski N., Hruščëv S., *A functional model and some problems in the spectral function theory*, Trudy Math. Inst. Steklov **176** (1987), 97–201 (Russian); English transl. in Proc. Steklov Inst. of Math. (1988), no. 3, 111–214, AMS series of publications.

NECESSARY CONDITIONS FOR INTERPOLATION
BY ENTIRE FUNCTIONS

B. A. TAYLOR

Let ρ be a subharmonic function on $\mathbb{C}$ such that $\log(1 + |z|) = O(\rho(z))$ and let A_ρ denote the algebra of entire functions f such that $|f(z)| \leqslant A \exp(B\rho(z))$ for some $A, B > 0$. Let V denote a discrete sequence of points $\{a_n\}$ of $\mathbb{C}$ together with a sequence of positive integers $\{p_n\}$ (the multiplicities of $\{a_n\}$). If $f \in A_\rho$, $f \not\equiv 0$, then $V(f)$ denotes the sequence $\{a_n\}$ of zeros of f and p_n is the order of zero of f at a_n.

In this situation, there are THREE NATURAL PROBLEMS to study.

I. ZERO SET PROBLEM. *Given ρ, describe the sets $V(f)$, $f \in A_\rho$.*

II. INTERPOLATION PROBLEM. *If $\{a_n, p_n\} = V \subset V(f)$ for some f, $f \in A_\rho$, describe all sequences $\{\lambda_{n,k}\}$ which are of the form*

$$(1) \qquad \lambda_{n,k} = \frac{g^{(k)}(a_n)}{k!}, \quad 0 \leqslant k < p_n, \ n = 1, 2, \ldots \quad \text{for some } g, \ g \in A_\rho.$$

III. UNIVERSAL INTERPOLATION PROBLEM. *If $V \subset V(f)$ for some f, $f \in A_\rho$, under what conditions on V is it true that for every sequence $\{\lambda_{n,k}\}$ such that $|\lambda_{n,k}| \leqslant A \exp(B\rho(a_n))$ there exists g, $g \in A_\rho$, satisfying (1).*

In case $\rho(z) = \rho(|z|)$ (and satisfies some mild, technical conditions), quite good solutions to problems I–III are known. This work has been carried out by A. F. Leont'ev and others (see e.g. [1] for a survey). However, when ρ is not a function of $|z|$, the general solutions are not known.

The purpose of this note is to call attention to an interesting special case of III. Consider the case $\rho(z) = |\operatorname{Im} z| + \log(1 + |z|^2)$. Then $A_\rho = \widehat{\mathcal{E}'}$, the space of all entire functions of exponential type with polynomial growth on the real axis. The space $\widehat{\mathcal{E}'}$ is of special interest because, by the Paley–Wiener–Schwartz Theorem, it is the space of Fourier transforms of distributions on $\mathbb{R}$ with compact support. The problems I–III are then dual to some problems about convolution operators on the space $\mathcal{E} = C^\infty(\mathbb{R})$ (see e.g. [1], [2], or [3]).

Specifically, suppose that for some $\varepsilon > 0$, $c > 0$, $f \in \widehat{\mathcal{E}'}$, we have

$$(2) \qquad V = \{a_n, p_n\} \subset V(f), \quad \text{where } \frac{|f^{(p_n)}(a_n)|}{p_n!} \geqslant \varepsilon \exp\left(-\frac{\rho(a_n)}{p_n}\right)$$

$(\rho(z) = |\operatorname{Im} z| + \log(1 + |z|^2))$. Then it is not hard to show that (2) is a sufficient condition that V has the universal interpolation property III. We wish to pose the converse problem.

PROBLEM. *Suppose that $V \subset V(F)$ for some F, $F \in \widehat{\mathcal{E}}'$, and that V is a universal interpolating sequence: i.e. III holds. Is it true that (2) must hold for some f, $f \in \widehat{\mathcal{E}}'(\mathbb{R})$?*

In all the cases known to the author where the Problem has answer yes, it is also true that the range of the multiplication operator $M_F \colon A_\rho \to A_\rho$ given by $M_F(f) = Ff$ is closed. Is the fact that M_F has closed range necessary for a "yes" answer? (In the case $A_\rho = \widehat{\mathcal{E}}'$, if M_F has closed range, then the Problem has answer yes, as can be shown by the techniques of [4]). However, the main interest in the Problem is to find if (2) must hold with no additional assumptions on F.

REFERENCES

1. Leont'ev A. F., *On properties of sequences of linear aggregates that converge in a region in which the system of functions generating the linear aggregates is not complete*, Uspekhi Matem. Nauk **11** (1956), no. 5, 26–37. (Russian)
2. Ehrenpreis L., *Fourier Analysis in Several Complex Variables*, Wiley–Interscience, New York, 1970.
3. Palamodov V. P., *Linear differential operators with constant coefficients*, Nauka, Moscow, 1967. (Russian)
4. Ehrenpreis L., Malliavin P., *Invertible operators and interpolation in AU spaces*, J. Math. Pure Appl. **13** (1974), 165–182.
5. Borisevich A. I., Lapin G. P., *On interpolation of entire functions*, Sib. Mat. Zh. 9 (1968), no. 3, 522–529. (Russian)

DEPARTMENT OF MATHEMATICS
UNIVERSITY OF MICHIGAN
ANN ARBOR, MI 48109
USA

COMMENTARY

Papers [6], [7] contain useful information concerning the Problem (see, for instance, Theorem 4 of [6]). Taking into account the duality between interpolation and bases (as described in the introduction to the chapter) see also [8–10].

REFERENCES

6. Berenstein C. A., Taylor B. A., *A new look at interpolation theory for entire functions of one variable*, Adv. Math. **33** (1979), no. 2, 109–143.
7. Squires W. A., *Necessary conditions for universal interpolation in $\widehat{\mathcal{E}}'$*, Canad. J. Math. **33** (1981), no. 6, 1356–1364 (MR 83g: 30040).
8. Meise R., *Sequence space representations for (DFN)-algebras of entire functions modulo closed ideals*, J. Reine Angew. Math. **363** (1985), 59–95.
9. Braun R. W., Meise R., *Generalized Fourier expansion for zero-solutions of surjective convolution operators on $\mathcal{D}_{\{\omega\}}(\mathbb{R})'$*, Arch. Math..
10. Napalkov V. V., Komarov A. V., *On expansions of analytic functions in series of elementary solutions of aconvolution equation*, Matem. Sbornik **181** (1990), no. 4, 556-563. (Russian)

REPRESENTATIONS OF FUNCTIONS BY EXPONENTIAL SERIES

A. F. LEONTIEV

1. Let L be an entire function of exponential type with zero divisor $k = k_L$ ($k(\lambda)$ is the zero multiplicity of L at the point λ, $\lambda \in \mathbb{C}$), and let γ the Borel transform of L, namely

$$L(\lambda) = \frac{1}{2\pi i} \int_C \gamma(t) e^{\lambda t} dt \qquad (\lambda \in \mathbb{C})$$

where the closed contour C embraces a closed set $\mathcal{D}$ containing all singularities of γ. There exists a family $\{\psi_{k,\lambda} : 0 \leqslant k < k(\lambda)\}$ of functions analytic in $\mathbb{C} \backslash \mathcal{D}$ and biorthogonal to the family $\{z^s e^{\lambda z} : 0 \leqslant s < k(\lambda)\}$, so that

$$\frac{1}{2\pi i} \int_C z^s e^{\lambda z} \psi_{k,\mu}(z)\, dz = \delta_{\lambda \mu} \cdot \delta_{sk},$$

where $\delta_{\alpha\beta}$ is the Kronecker delta (see the construction of $\psi_{k,\lambda}$ in [1], p. 228). Any analytic function on $\mathcal{D}$ can be expanded in Fourier series

$$(1) \qquad f \sim \sum_{\lambda, k(\lambda) > 0} \sum_{k=0}^{k(\lambda)-1} a_{k,\lambda} z^k e^{\lambda z}; \qquad a_{k,\lambda} = \frac{1}{2\pi i} \int_C f(z) \psi_{k,\lambda}(z)\, dz.$$

The following uniqueness theorem is known ([1], p. 255): if L has infinitely many zeros and $\mathcal{D}$ is a convex set then $a_{k,\lambda} \equiv 0 \implies f \equiv 0$. The proof uses in an essential way the convexity of $\mathcal{D}$.

PROBLEM 1. *Does the uniqueness theorem hold without the convexity assumption?*

2. Let $\mathcal{D}$ be the closed convex envelope of the set of singular points of γ and suppose that L has simple zeros only (i.e. $k(\lambda) \leqslant 1$, $\lambda \in \mathbb{C}$). The necessary and sufficient conditions for series (1) to converge to f in the interior of $\mathcal{D}$ for any f analytic in $\mathcal{D}$ are the following:

   a) $|L'(\lambda)| > e^{[h(\varphi)-\varepsilon]|\lambda|}$, $\varphi \equiv \arg \lambda$, $k(\lambda) > 0$, $|\lambda| > r(\varepsilon)$ for any $\varepsilon > 0$ (h stands for the growth indicator of L, cf. [1]);

   b) there exist numbers $p > 0$ and r_k, $0 < r_k \nearrow \infty$ such that $|L(\lambda)| > e^{p|\lambda|}$, $|\lambda| = r_k$, $k \geqslant 1$.

Condition a) ensures the convergence of (1) in int $\mathcal{D}$ and b) implies that the obtained sum equals f.

PROBLEMS 2. *Is b) implied by a)?*

The negative answer would mean that series (1) generated by f may converge to a function different from f.

3. Suppose that int $\mathcal{D}$ is an unbounded convex domain containing $(-\infty, 0)$. Suppose further that φ ranges over the interval $(-\varphi_0, \varphi_0)$, $0 < \varphi_0 \leqslant \pi/2$ when $\ell_\varphi \overset{\text{def}}{=} \{\, x + iy :$ $x \cos\varphi + y \sin\varphi - k(\varphi) = 0 \,\}$ ranges through the set of all supporting lines of int $\mathcal{D}$. The possibility for ℓ_{φ_0} to be supporting lines is not excluded but in such a case, evidently, the boundary $\partial\mathcal{D}$ eventually coincides with $\ell_{\pm\varphi_0}$.

Let $h(\varphi) = k(-\varphi)$ and let

$$(2) \qquad \left| L\left(re^{i\varphi}\right) \right| < \frac{C}{1 + r^\alpha} e^{rh(\varphi)}, \qquad \alpha > 1, \quad |\varphi| < \varphi_0$$

(may be with $\varphi = \pm\varphi_0$ in the above mentioned case). All zeros of L are assumed to be simple. Let $\{\psi_\lambda : k(\lambda) > 0\}$ be the biorthogonal family to $\{e^{\lambda z} : k(\lambda) > 0\}$,

$$\psi_\lambda(z) \overset{\text{def}}{=} \frac{1}{L'(\lambda)} \int_0^{\infty e^{i\varphi}} \frac{L(t)}{t - \lambda} e^{-tz} \, dt, \qquad |\varphi| < \varphi_0, \quad k(\lambda) > 0.$$

Condition (2) implies that ψ_λ are bounded (by the constants, which may depend on λ).

Let $B(\mathcal{D})$ be the class of all functions f analytic in int $\mathcal{D}$, continuous in $\mathcal{D}$ and such that

$$f(z) = O\left(\frac{1}{|z|^\mu}\right), \qquad \mu > 1, \quad z \in \mathcal{D}, \quad z \to \infty.$$

Putting $C = \partial\mathcal{D}$, $\psi_{k,\lambda} = \psi_\lambda$, $k(\lambda) > 0$ in (1), associate with every function $f \in B(\mathcal{D})$ its Fourier series. In this section it is convenient to enumerate the zeros of L: $\{\lambda_\nu\}_{\nu \geqslant 1}$.

We shall be concerned with the convergence of (1) to f in int $\mathcal{D}$. Suppose that L satisfies the following additional requirement. There is a family of closed contours Γ_k $(k \geqslant 1)$ and a family of curvilinear annuli containing these contours

$$\mathcal{P}_k = \bigcup_{t \in \Gamma_k} \left\{ z : |z - t| \leqslant e^{-\varepsilon_k r_k} \right\}, \quad \varepsilon_k > 0, \quad \lim_k \varepsilon_k = 0, \quad \lim_k r_k = \infty \quad \left(r_k \overset{\text{def}}{=} \min_{t \in \Gamma_k} |t|\right)$$

satisfying

 a) for all γ, $\gamma > 0$, and ε, $\varepsilon > 0$,

$$\lim_{k \to \infty} \min_{\lambda \in \Gamma_k'} H(\lambda) = \infty, \quad \text{where } H(\lambda) = \frac{\log |L(\lambda)|}{|\lambda|};$$

for $k > k(\gamma, \varepsilon)$ the function H is greater than $h(\varphi) - \varepsilon$ on $\mathcal{P}_k'$, greater than $h(\varphi_0) - \varepsilon$ on $\mathcal{P}_k''$, greater than $h(-\varphi_0 + \gamma) - \varepsilon$ on $\mathcal{P}_k'''$ where Γ_k' is the part of Γ_k lying in the complement of the angle $|\varphi| < \varphi_0 + \gamma$, $\mathcal{P}_k'$ is the part of $\mathcal{P}_k$ lying in the angle $|\varphi| < \varphi_0 - \gamma$, $\mathcal{P}_k''$ and $\mathcal{P}_k'''$ are the parts of $\mathcal{P}_k$ lying in the small angles $|\varphi - \varphi_0| \leqslant \gamma$, $|\varphi + \varphi_0| \leqslant \gamma$, correspondingly;

 b) if the boundary of the curvilinear half-annulus $\mathcal{P}_k' \cup \mathcal{P}_k'' \cup \mathcal{P}_k'''$ is divided into the parts C_k' and C_k'' by Γ_k (moreover let C_k'' be inside Γ), then the lengths of the curves Γ_k', C_k', C_k'' are $\exp[O(1)r_k]$ when $k \to \infty$;

 c) L has only one zero in the annulus between Γ_k and Γ_{k+1}, namely λ_k.

Under these conditions it has been proved in [2] that if $f \in B(\mathcal{D})$ and E is a compact subset of int $\mathcal{D}$, then

$$\left| f(z) - \sum_{\nu=1}^{n} a_{\lambda_\nu} e^{\lambda_\nu z} \right| < e^{-\delta_0 r_n}, \qquad \delta_0 = \delta_0(E) > 0, \qquad n > n_0 \qquad (z \in E),$$

so

$$f = \sum_{\nu \geqslant 1} a_{\lambda_\nu} e^{\lambda_\nu z}, \qquad z \in \mathcal{D}.$$

It was shown in [2] and [3] how the general case (i.e. the case of an arbitrary f analytic in int $\mathcal{D}$) can be reduced to the case $f \in B(\mathcal{D})$.

PROBLEM 3. *Show that for any domain* int $\mathcal{D}$ *there exists a function* L *with the properties* (1), a), b), c).

REFERENCES

1. Leontiev A. F., *Series of exponents*, Nauka, Moscow, 1976. (Russian)
2. Leontiev A. F., *On the problem of representation of analytic functions in an infinite convex domain by Dirichlet series*, Doklady Akad. Nauk SSSR **225** (1975), no. 5, 1013–1015 (Russian); English transl. in Soviet Math.–Doklady.
3. Leontiev A. F., *On a representation of an analytic function in an infinite convex domain*, Anal. Math. **2** (1976), 125–148.

BASES OF REPRODUCING KERNELS
AND EXPONENTIALS

N. K. Nikolski

1. Bases of exponential polynomials. For a non-negative integer-valued function k (a divisor) in the complex plane $\mathbb{C}$ let us denote by $\mathcal{E}(k)$ the family $\{\mathcal{E}_\lambda : \lambda \in \mathbb{C}\}$ of exponential polynomial subspaces $\mathcal{E}_\lambda = \{pe^{i\lambda x} : p$ is a polynomial, $\deg p < k(\lambda)\}$.

QUESTION 1. *For what divisors k does the family $\mathcal{E}(k)$ form an unconditional basis in the space $L^2(0,a)$, $a > 0$?*

"Unconditional basis" is used in the usual sense and means the existence, uniqueness and unconditional convergence of the expansion

$$(*) \qquad\qquad f = \sum_{\lambda \in \mathbb{C}} p_\lambda e^{i\lambda x}; \qquad \deg p_\lambda < p(\lambda), \quad \lambda \in \mathbb{C}$$

for every function $f \in L^2(0,a)$. It is clear that in this case $k \equiv 0$ off a countable discrete set $\sigma = \operatorname{supp} k$ and the starred expansion turns out to be a generalized Fourier series with respect to the minimal family of subspaces $\mathcal{E}(k)$.

The most interesting problem arises for $k \leqslant 1$ (i.e. $k = \chi_\sigma$, the characteristic function of σ) in which case the reader deals, as a matter of fact, with the well-known problem on exponential bases on intervals of the real axis.

Here is a bit of known information:

1) for $k = \chi_\sigma$, $\sigma \subset \mathbb{C}_+ \overset{\text{def}}{=} \{\lambda : \operatorname{Im} \lambda > 0\}$ the Question has been answered in [1]. Namely, σ must be a (Carleson) interpolation subset of $\mathbb{C}_+$ and the function $x \to \arg \Theta \overline{B}_\sigma(x)$ must satisfy the Devinatz–Widom condition, where B_σ is the Blaschke product with zero set σ and $\Theta = \exp(iaz)$. In case $\sup \operatorname{Im} \sigma < +\infty$ the answer can be reformulated in terms of density of σ. Paper [1] contains also exhausting historical remarks.

2) In the limit case $a = \infty$ (which implies $\operatorname{supp} k \subset \mathbb{C}_+$) no $\mathcal{E}(k)$ forms a basis in $L^2(0, +\infty)$. The right analogue of the problem in such a situation is to describe all divisors k for which $\mathcal{E}(k)$ is an unconditional basis in the closed linear span of $\mathcal{E}(k)$ ($\overset{\text{def}}{=} \operatorname{span} \mathcal{E}(k)$). This problem has been solved in its complete generality in [2] in terms if the generalized (multiple) Carleson condition.

3) It is not hard to see that for $\operatorname{supp} k \subset \mathbb{C}_+$ the Question is equivalent to a kind of multiple free interpolation problem for entire functions of exponential type $a/2$ (see [1], [3] for details).

2. Bases extending a given basis. Exponential (or exponential-polynomial) bases problem is a special case ($\Theta = \exp(iaz)$) of the problem on reproducing bases in the model space

$$K_\Theta = H_+^2 \ominus \Theta H_+^2,$$

where H_+^2 stands for the usual Hardy space in $\mathbb{C}_+$ and Θ is an inner function. Denote by

$$k_\Theta(z, \lambda) = \frac{1 - \overline{\Theta(\lambda)}\Theta}{z - \bar\lambda}, \qquad \operatorname{Im}\lambda > 0$$

the reproducing kernel for K_Θ and put $\mathcal{K}_\Theta(\sigma) = \{\, k_\Theta(\cdot, \lambda) : \lambda \in \sigma \,\}$ for $\sigma \subset \mathbb{C}_+$.

QUESTION 2. *Let $\mathcal{K}_\Theta(\sigma)$ be an unconditional basis in* $\operatorname{span}\mathcal{K}_\Theta(\sigma)$. *Is it true that there exist unconditional bases $\mathcal{K}_\Theta(\sigma')$ in the whole space K_Θ containing $\mathcal{K}_\Theta(\sigma)$ (i.e. such that $\sigma \subset \sigma' \subset \mathbb{C}_+$)?*

QUESTION 2′. *Let $a > 0$ and let $\mathcal{E}(k)$ be an unconditional basis in* $\operatorname{span}\mathcal{E}(k)$. *Is it true that there exist unconditional bases $\mathcal{E}(k')$ in the whole space $L^2(0, a)$ containing $\mathcal{E}(k)$ (i.e. such that $k \leqslant k'$)? Is it possible to choose such a k' multiplicity-free (i.e. $k' = \chi_{\sigma'}$) provided $k = \chi_\sigma$?*

The second part of Question 2′ is a special case of Question 2 ($\Theta = \exp(iaz)$). The answer to this part of Question is known to be positive (V. I. Vasyunin, S. A. Vinogradov) under some additional assumptions (i.e. a quantitative relation between $\inf\operatorname{Im}\sigma$, $\sigma = \operatorname{supp} k$ and the interpolating constant of σ, see [1]).

3. Existence of a basis.

QUESTION 3. *In which model space K_Θ does there exist an unconditional basis of the form $\mathcal{K}_\Theta(\sigma)$?*

Each of the following two QUESTIONS 3′ and 3″ is equivalent to Question 3 (see [1], [3] for the proofs). *For which inner functions Θ does there exist an interpolating Blaschke product B such that*

3′) $\operatorname{dist}(\Theta, BH^\infty) < 1$, $\operatorname{dist}(B, \Theta H^\infty) < 1$?

or

3″) *the Toeplitz operator $T_{\Theta\bar B}f \overset{\text{def}}{=} P_+\Theta\bar Bf$, $f \in H^2$ is invertible in H^2?*

It is proved in [4] that, Θ being an inner function, there exist interpolating Blaschke products B, B' such that $\|\Theta B - B'\|_\infty < 1$. It follows that the space K_Θ can be "complemented" by the space K_B with an unconditional basis of reproducing kernels ($\equiv$ of rational fractions in this case) to the space $K_{\Theta B} = \operatorname{clos}(K_\Theta + K_B)$, $K_\Theta \cap K_B = \{0\}$ in such a way that $K_{\Theta B}$ has also an unconditional basis of the form $\mathcal{K}_{\Theta B}(\sigma)$.

A limit case of the problem (the existence of *orthogonal* bases of the form $k_\Theta(\cdot, \lambda)$, $|\lambda| = 1$) is considered in [5].

REFERENCES

1. Hruščëv S. V., Nikol'skii N. K., Pavlov B. S., *Unconditional bases of exponentials and of reproducing kernels*, Lect. Notes in Math. **864** (1981), 214–335.
2. Vasyunin V. I., *Unconditionally convergent spectral decompositions and interpolation problems*, Trudy Matem. Inst. Steklova AN SSSR **130** (1978), 5–49 (Russian); English transl. in Proc. Steklov Inst. Math. **4** (1979), 1–53.
3. Nikol'skii N. K., *Treatise on the Shift Operator*, Springer-Verlag, 1986 (transl. from Russian).
4. Jones P. W., *Ratios of interpolating Blaschke products*, Pacific J. Math. **95** (1981), no. 2, 311–321.
5. Clark D. N., *On interpolating sequences and the theory of Hankel and Toeplitz matrices*, J. Funct. Anal. **5** (1970), no. 2, 247–258.

Chapter 15. INTERPOLATION, BASES, MULTIPLIERS

STEKLOV MATHEMATICAL INSTITUTE
ST. PETERSBURG BRANCH
FONTANKA 27   AND
ST. PETERSBURG, 191011
RUSSIA

UNIVERSITÉ BORDEAUX-I
UFR DE MATHÉMATIQUES
351, COURS DE LA LIBÉRATION
33405 TALENCE CEDEX
FRANCE

COMMENTARY BY K. M. DYAKONOV

In connection with Question 3 (or rather its restatement 3′) I cite my recent result proved in [6].

THEOREM. *There exists an absolute (numerical) constant N, $N \in \mathbb{N}$, with the following property. For any inner function θ one can find an interpolating Blaschke product B such that*

$$\operatorname{dist}(B, \theta H^\infty) < 1 \qquad \operatorname{dist}(\theta^N, BH^\infty) < 1$$

In other words, one can always get a subspace of the form K_B (B being a suitable interpolating Blaschke product) "squeezed in" between K_θ and K_{θ^N} in the sense that both Toeplitz operators $T_{\bar\theta B}$ and $T_{\theta^N \bar B}$ become left-invertible.

It would be somewhat surprising if the least possible value of N turned out to be > 1.

PR. KHUDOZHNIKOV, 24-1-412
ST. PETERSBURG, 194295
RUSSIA

CHAPTER EDITOR'S NOTE

Question 1 seems to be answered for the case of *simple* exponentials ($k = \chi_\sigma$, $\sigma \subset \mathbb{C}$) by A. M. Minkin [7]. The answer claimed in [7] says $\mathcal{E}(\chi_\sigma)$ forms an unconditional basis of $L^2(-a, a)$ iff

(1) $\inf_{n \neq k} |\lambda_n - \lambda_k| > 0$;

(2) $\sigma_\pm \overset{\text{def}}{=} \sigma \cap \mathbb{C}_\pm \in (C)$;

(3) $\operatorname{dist}(e^{2iaz}, B_y H_+^\infty) < 1$, $\operatorname{dist}(B_y, e^{2iaz} H_+^\infty) < 1$,

where B_y is the Blaschke product for the set $y + (\sigma_+ \cup \overline{\sigma_-})$, $y > 0$ (the bar stands for the complex conjugation).

REFERENCES

6. Dyakonov K. M., *Interpolating functions of minimal norm, star-invariant subspaces, and kernels of Toeplitz operators*, Proc. Amer. Math. Soc. (to appear).
7. Minkin A. M., *Reflection of frequencies and unconditional bases of exponentials*, Algebra i analiz **3** (1991), no. 5, 109–134 (Russian); English transl. in Petersburg Math. J. **3** (1992), no. 5.

UNCONDITIONAL BASES
GENERATED BY MUCKENHOUPT WEIGHTS

G. M. Gubreev

Denote by w^2 an arbitrary Muckenhoupt weight [1] on $\mathbb{R}$. Let $|w_-|^2$ be its standard extension into the lower half-plane defined by means of the outer function w_- which can be represented in the following form [2]:

$$w_-(z) = z \int_{\mathbb{R}_+} e^{-izt} y_w(t)\, dt, \qquad \operatorname{Im} z < 0$$

where $y_w \in L^2_{\mathrm{loc}}(\mathbb{R}_+)$. Put

$$y_w(\lambda, t) = \frac{d}{dt} \int_0^t y_w(t-s)e^{i\lambda s}\, ds, \qquad \lambda \in \mathbb{C}, \quad t \in \mathbb{R}_+$$

Unconditional bases of the space $L^2(0, \sigma)$ of the form

$$(1) \qquad \{\, y_w(\lambda_k, t) : \lambda_k \in \Lambda \,\}, \qquad \inf_{\lambda_k \in \Lambda} \operatorname{Im} \lambda_k > 0$$

were studied in [2]–[5]. Let us state a result generalizing B. S. Pavlov's theorem on unconditional bases of exponentials.

THEOREM [2]. *Family* (1) *is an unconditional basis of* $L^2(0, \sigma)$ *if and only if* Λ *is the set of (simple) zeros of an entire function* φ *of finite degree satisfying the following conditions:*

1) $\varlimsup_{r \to \infty} r^{-1} \log|\varphi(re^{i\pi R})| = 0, \ \varlimsup_{r \to \infty} r^{-1} \log|\varphi(re^{-i\pi R})| = \sigma;$
2) $w^{-2}|\varphi|^2$ *satisfies the* A_2*–condition on* $\mathbb{R}$;
3) $\inf_k \prod_{j \neq k} |\lambda_k - \lambda_j| \, |\lambda_k - \bar\lambda_j|^{-1} > 0.$

It seems plausible that bases of this kind are "universal" (in a sense). Let $\mathcal{H}$ be a separable Hilbert space, B a complete non-selfadjoint Volterra dissipative operator ($\operatorname{Im} B \geqslant 0$) in $\mathcal{H}$. Consider the entire vector function

$$y(\lambda) = (I - \lambda B)^{-1} y, \qquad y \in \mathcal{H}.$$

QUESTION 1. *Suppose* $(I - \lambda B)^{-1}$ *is of finite degree* c *(as a function of* λ*). Suppose*

$$(2) \qquad \{\, y(\lambda_k) : \lambda_k \in \Lambda \,\} \quad \text{where} \quad \inf_{\lambda_k \in \Lambda} \operatorname{Im} \lambda_k > 0$$

is an unconditional basis of $\mathcal{H}$*. Is there an* A_2*–weight* w^2 *on* $\mathbb{R}$ *such that this basis is isomorphic to a basis* (1) *in* $L^2(0, \sigma)$ *generated by the same sequence* Λ *and by* w^2*?*

Recall that bases are isomorphic if there is a continuous and continuously invertible operator $S \colon \mathcal{H} \to L^2(0, \sigma)$ such that $Sy(\lambda_k) = y_w(\lambda_k, t)$, $\lambda_k \in \Lambda$.

The affirmative answer to Question 1 would follow if the following question could be answered in the affirmative.

QUESTION 2. *Suppose all conditions of Question 1 are satisfied. Is it true that the weight*

$$W^2(x) = \left\|(I - xB)^{-1}y\right\|, \qquad x \in \mathbb{R}$$

satisfies the condition (A_2) *on* $\mathbb{R}$?

Let us mention in this connection the following result of the author: if (2) is an unconditional basis, then $\int_{\mathbb{R}} W^{\pm 2}(x) \cdot (1 + x^2)^{-1}\, dx < \infty$. Thus, whenever $W^2(x)$ behaves like a power, i.e. if

$$\left\|(I - xB)^{-1}y\right\|^2 \asymp |x|^{\alpha_1}, \qquad x \to +\infty,$$

$$\left\|(I - xB)^{-1}y\right\|^2 \asymp |x|^{\alpha_2}, \qquad x \to -\infty, \qquad \alpha_1, \alpha_2 \in \mathbb{R}$$

then the above questions can be answered in the affirmative. But in general the answers are unknown, even in the simplest situation:

$$\mathcal{H} = L^2(0, \sigma), \qquad (Bh)(t) = i \int_0^t h(s)\, ds.$$

REFERENCES

1. Muckenhoupt B., *Weighted norm inequalities for the Hardy maximal function*, Trans. Amer. Math. Soc. **165** (1972), 207–226.
2. Gubreev G. M., *Spectral analysis of biorthogonal expansions generated by Muckenhoupt weights*, Zapiski Nauchn. Semin. LOMI **190** (1991), 34–80 (Russian); English transl. in J. Soviet Math.
3. Hruščev S. V., *Unconditional bases in $L^2(0, a)$*, Preprint Inst. D'Estudios Catalans, Barselona, 1985.
4. Hruščev S. V., *Unconditional bases in $L^2(0, a)$*, Proc. Amer. Math. Soc. **99** (1987), no. 4, 651–656.
5. Gubreev G. M., *Generalized Dzhrbashyan transform and some of their applications*, Izvestiya AN Arm. SSR, Matematika **21** (1986), no. 3, 306–310. (Russian)
6. Pavlov B. S., *Bases of exponentials and the Muckenhoupt condition*, Dokl. AN SSSR **247** (1979), no. 1, 37–40. (Russian)
7. Hruščev S. V., Nikloski N. K., Pavlov B. S., *Unconditional bases of exponentials and of reproducing kernels*, Lecture Notes in Math. vol. 864, 1981.

ODESSA PEDAGOGICAL INSTITUTE
STAROPORTOFRANKOVSKAYA 26
ODESSA 270020
UKRAINE

CHAPTER EDITOR'S NOTE

See also a related theorem by S. Hruščev in [7] from the Preface reference list (namely, Theorem 114 of [7]).

FREE INTERPOLATION IN REGULAR CLASSES

J. Bruna

Let $\mathbb{D}$ denote the open unit disc in $\mathbb{C}$ and let X be a closed subset of $\overline{\mathbb{D}}$. For $0 < \alpha < 1$, let Λ_α denote the algebra of holomorphic function in $\mathbb{D}$ satisfying a Lipschitz condition of order α. The set X is called an interpolation set for Λ_α if the restriction map

$$\Lambda_\alpha \to \mathrm{Lip}(\alpha, X)$$
$$f \mapsto f|X$$

is onto. The interpolation sets for Λ_α, $0 < \alpha < 1$ (and also of other classes of functions) were characterized by Dyn'kin in [3] as those for which the following conditions hold:

The condition (K): if $d(z) = \inf\{ |z - w| : w \in X \}$, then for all arcs $I \subset T$,

$$\sup_{z \in I} d(z) \geqslant c|I|$$

where $|I|$ denotes the length of I.

The Carleson condition (C): $X \cap \mathbb{D}$ must be a sequence (z_n) such that

(C) $$\inf_n \prod_{m \neq n} \frac{|z_m - z_n|}{|1 - \bar{z}_n z_m|} > 0.$$

In the limit case $\alpha = 1$ there are (at least) three different ways of posing the problem:

1. We can simply ask when the restriction map $\Lambda_1 \to \mathrm{Lip}(1, X)$ is onto, Λ_1 being the class of holomorphic functions in $\mathbb{D}$ satisfying a Lipschitz condition of order 1.

2. We can also consider the class

$$A^1 = H(\mathbb{D}) \cap C^1(\overline{\mathbb{D}})$$

and call X an interpolation set for A^1 if for all $\varphi \in C^1(X)$ (the space of Whitney jets) such that $\bar{\partial}\varphi = 0$ there exists f in A^1 with $f = \varphi$, $f' = \partial f$ on X.

3. Finally one can consider the Zygmund class version of the problem. Let Λ_* denote the class of holomorphic function in $\mathbb{D}$ having continuous boundary values belonging to the Zygmund class of T. We say that X is an interpolation set for Λ_* if for any φ in the Zygmund class of $\mathbb{C}$ there exists f in Λ_* such that $f = \varphi$ on X.

In [1] and [2], it has been shown that Dyn'kin's theorem also holds for A^1-interpolation sets. For Λ_1 interpolation sets the Carleson condition must be replaced by

(2C): $X \cap \mathbb{D}$ is a union of two Carleson sequences.

Chapter 15. INTERPOLATION, BASES, MULTIPLIERS

Our PROBLEM is the following: *which are the interpolation sets for the Zygmund class?*

Considering the special nature of the Zygmund class, I am not sure whether the condition describing the interpolation sets for Λ_* (one can simply think about the boundary interpolation, i.e. $X \subset \mathbb{T}$) should be different or not from condition (K). Recently I became aware of the paper [4], where a description of the trace of Zygmund class (of $\mathbb{R}^n$) on any compact set and a theorem of Whitney type are given. These are two important technical steps in the proofs of the results quoted above and so it seems possible to apply the same techniques.

REFERENCES

1. Bruna J., *Boundary interpolation sets for holomorphic functions smooth to the boundary and BMO*, Trans. Amer. Math. Soc. **264** (1981), no. 2, 393–409.
2. Bruna J., Tugores F., *Free interpolation for holomorphic functions regular up to the boundary*, Pacific J. Math. **108** (1983), 31–49.
3. Dyn'kin E. M., *Free interpolation sets for Hölder classes*, Mat. Sb. **109(151)** (1979), no. 1, 107-128 (Russian); English transl. in Math. USSR Sbornik **37** (1980), 97–117.
4. Jonsson A., Wallin H., *The trace to closed sets of functions in $\mathbb{R}^n$ with second difference of order $O(h)$*, J. Approx. Theory **26** (1979), 159–184.

DEPARTAMENT DE MATEMÀTIQUES
UNIVERSITAT AUTÒNOMA DE BARCELONA
08193 BELLATERRA (BARCELONA)
ESPAÑA

CHAPTER EDITOR'S NOTE

See also recent paper [5] dealing with the free interpolation in Hölder, Sobolev and Gevrey spaces.

REFERENCE

5. Boricheva I., Dyn'kin E., *A non-classical free interpolation problem*, Algebra i Analiz 4 (1992), no. 5, 45–90 (Russian); English transl. in St. Petersburg Math. J. **4** (1993), no. 5.

TRACES OF $H^\infty(\mathbb{B}^N)$-FUNCTIONS ON HYPERPLANES

N. A. SHIROKOV

Let $\mathbb{B}^N$ be the unit ball of $\mathbb{C}^N$ ($N > 1$) and denote by $H^\infty(\mathbb{B}^N)$ the algebra of all bounded holomorphic functions in $\mathbb{B}^N$. An analytic subset E of $\mathbb{B}^N$ is said to be *a zero-set for $H^\infty(\mathbb{B}^N)$* (in symbols: $E \in ZH^\infty(\mathbb{B}^N)$) if there exists a non-zero function f in $H^\infty(\mathbb{B}^N)$ with $E = f^{-1}(0)$; E is said to be *an interpolation set for $H^\infty(\mathbb{B}^N)$* (in symbols: $E \in IH^\infty(\mathbb{B}^N)$) if for any bounded holomorphic function φ on E there exists a function f in $H^\infty(\mathbb{B}^N)$ with $f|E = \varphi$. The problem to describe the sets of classes $ZH^\infty(\mathbb{B}^N)$ and $IH^\infty(\mathbb{B}^N)$ proves now to be very difficult. I would like to propose some partial questions concerning this problem; the answers could probably suggest conjectures in the general case.

Let A be a countable subset of $\mathbb{B}^N$. Set $T_A = \{\, z \in \mathbb{B}^N : (z, a) = |a|^2 \,\}$, $T_A = \cup_{a \in A} T_a$.

PROBLEM 1. *What are the sets A such that $T_A \in ZF^\infty(\mathbb{B}^N)$?*

PROBLEM 2. *What are the sets A such that $T_A \in IH^\infty(\mathbb{B}^N)$?*

It follows easily from results of G. M. Henkin [1] and classical results concerning the unit disc that the following two conditions are necessary for $T_A \in Z\bar{H}^\infty(\mathbb{B}^N)$:

(1)
$$\sum_{a \in A} \max\left(1 - \frac{|a|^4}{|(a, \theta)|^2}, 0\right) < \infty, \qquad \theta \in \mathbb{C}^n, |\theta| = 1$$

(2)
$$\sum_{a \in A}(1 - |a|^2)^N \leqslant \infty$$

These conditions do not seem however to be sufficient.

A necessary (insufficient) condition for $T_A \in IH^\infty(\mathbb{B}^N)$ can also be indicated. Namely, $T_A \in IH^\infty(\mathbb{B}^N)$ implies that there exists $\delta_A > 0$ such that for every $a \in A$ the set T_a intersects no ellipsoid $Q_{\delta_A}(a')$ with $a' \neq a$, where

(3) $\quad Q_\delta(a) \overset{\text{def}}{=} \left\{ z \in \mathbb{C}^N : \left| \frac{1 - |a|^2 \delta^2}{\delta(1 - |a|^2)} \cdot \frac{(z, a)}{|a|} - \frac{a}{\delta} \frac{1 - \delta^2}{1 - |a|^2} \right| + \right.$

$$\left. \frac{1 - |a|^2 \delta^2}{(1 - |a|^2)}\left(|z|^2 - \frac{|(z, a)|^2}{|a|^2}\right) < 1 \right\}$$

If A lies in a "sufficiently compact" subset of $\mathbb{B}^N$, it is possible to give complete solutions to Problem 1 and 2. Let $\rho, q \in (0, 1)$, $c > 0$; the (ρ, c, q)–wedge with the top at a point $e_0 \in \partial \mathbb{B}^N$ is, by definition, the union of the ball $\left\{ z \in \mathbb{B}^N : |z| < \rho \right\}$ and the set

$$E(e_0) \stackrel{\text{def}}{=} \left\{ z \in \mathbb{B}^N : \left| \operatorname{Im}\left(1 - (z, e_0)\right) \right| \leqslant c \operatorname{Re}\left(1 - (z, e_0)\right); \right.$$
$$\left. |z|^2 - \left|(z, e_0)\right|^2 \leqslant q\left(1 - \left|(z, e_0)\right|^2\right); \quad \operatorname{Re}\left(1 - (z, e_0)\right) \leqslant \frac{1}{c^2 + 1} \right\}.$$

The scale of all (ρ, c, q)–wedges in equivalent (in a sense) to that of Fatou–Korányi–Stein wedges [1]. The following theorem holds.

THEOREM. *Let A be a subset of a finite union of (ρ, c, q)–wedges with $q < 1/2$. Then $T_A \in ZH^\infty(\mathbb{B}^N)$ if and only if $\sum_{a \in A}(1 - |a|^2) < \infty$; $T_A \in IH^\infty(\mathbb{B}^N)$ if and only if there exists $\delta > 0$ such that T_A intersects no one of the sets $Q_\delta(a')$ with $a' \neq a$, $Q_\delta(a)$ being defined by (3).*

In view of this theorem the following specializations of Problems 1 and 2 are of interest. Let A be a subset of a (ρ, c, q)–wedge, but in contrast to the theorem q can be an arbitrary number from $(0, 1)$.

PROBLEM 1′. *Is it true that $T_A \in ZH^\infty(\mathbb{B}^N)$ implies $\sum_{a \in A}(1 - |a|^2) < \infty$?*

PROBLEM 2′. *Is it true that $T_a \cap Q_\delta(a') \neq \emptyset$ for all $a' \neq a$ implies $T_A \in IH^\infty(\mathbb{B}^N)$?*

REFERENCE

1. Henkin G. M., *G. Lewi equation and analysis on a pseudoconvex manifold*, Mat. Sb. **102** (1977), no. 1, 71–108. (Russian)

TOVARISCHESKIĬ PR.
9, AP. 21
ST. PETERSBURG, 193231
RUSSIA

TRACES OF PLURIHARMONIC FUNCTIONS ON CURVES

JOAQUIM BRUNA

Let B^n denote the unit ball in $\mathbb{C}^n$, S its boundary. In a paper with the same title ([1]) the following theorem is proved.

THEOREM. *If γ is a simple curve on S of class C^2, the trace on γ of the space of pluriharmonic functions in B^n, which are continuous on $\bar{B}^n$, is a closed subspace of finite codimension in the space $C(\gamma)$ of all continuous functions on γ.*

The proof in [1] consists in constructing an approximate integral extension operator L associating to every $\varphi \in C(\gamma)$ a pluriharmonic function $L(\varphi)$, continuous up to S, such that $L(\varphi) = \varphi + E(\varphi)$ on γ, where the error term $E(\varphi)$ defines a compact operator, in the same way that Dirichlet problem is solved by using layer potentials. The operator L is given by the kernel

$$K(t,z) = \frac{1}{2\pi} \operatorname{Im} \frac{(1 + i\dot{T}(t))z \cdot \overline{\dot{\gamma}(t)} + (1 - i\dot{T}(t))\gamma(t) \cdot \overline{\dot{\gamma}(t)}}{z\overline{\gamma(t)} - 1}.$$

where $\gamma(t)$ is arc-length parametrization and $iT(t) = \gamma(t) \cdot \overline{\dot{\gamma}(t)}$, and thus involves second-order derivatives of $\gamma(t)$.

PROBLEM 1. *Is the theorem true for a less smooth curve γ, say a rectifiable curve?*

For instance, when the curve is transverse at all its points (i.e. $T(t) \neq 0 \ \forall t$), C^1 is enough, for one can take a parametrization $\gamma(s)$ for which $T(s) \equiv 1$.

The codimension mentioned in the theorem is zero if the curve is polynomially convex (since then holomorphic polynomials are dense in $C(\gamma)$), and this is the case when γ is an arc or has one complex-tangential point $\gamma(t_0)$ (i.e. $T(t_0) = 0$ for one t_0). This latter fact can be found in [3] and is also a consequence of a result of Forstnerič [2], which says that if γ has a non-trivial hull, then this hull must be an analytic variety with at most a finite number of singularities intersecting S transversally (the theorem shows that $P(\gamma)$, the closure on γ of holomorphic polynomials, is a Hypodirichlet algebra, and Forstnerič' result is related with a general structure theorem of the spectrum of such an algebra).

Denoting by $\hat{\gamma}$ the hull of γ it is natural to pose:

PROBLEM 2. *Compute the codimension in terms of $\hat{\gamma} \backslash \gamma$ (possibly in terms of the number and type of the singularities).*

Some results on Problem 2 have been obtained in [4].

PROBLEM 3. *Is it possible to compute the codimension in terms of γ alone?*

REFERENCES

1. Berndtsson B., Bruna J., *Traces of pluriharmonic functions on curves*, Ark. Mat. **28** (1990), no. 2, 221–230.
2. Forstnerič F., *Regularity of varieties in strictly pseudoconvex domains*, Publicacions Matemàtiques **32** (1988), 145–150.
3. Rosay J.–P., *A remark on a theorem by B. Forstnerič*, Ark. Mat. **28** (1990), 311–314.
4. Stout E. L., written communication.

DEPARTAMENT DE MATEMÀTIQUES
UNIVERSITAT AUTÒNOMA DE BARCELONA
08193 BELLATERRA (BARCELONA)
SPAIN

THE TRACES OF DIFFERENTIABLE FUNCTIONS
TO SUBSETS OF $\mathbb{R}^n$

Yu. Brudny, P. Shvartsman

1. The text below is an essentially revised and enlarged variant of the note [1], problem 10.7. Taking into account recent results some new problems and corresponding conjectures are formulated here.

We begin with the account of problems arising from the classical work [2] of Whitney, where the case of the space C^k was studied. To this end together with the space $C^k(\mathbb{R}^n)$ consisting of functions whose k-th derivatives are bounded and *uniformly continuous* in $\mathbb{R}^n$, we shall also consider the space $C^k\Lambda_\omega^s(\mathbb{R}^n)$, defined by the seminorm

$$(1) \qquad |f|_{C^k\Lambda_\omega^s} := \sup_{|\alpha|=k} \sup_{x,h\in\mathbb{R}^n} \frac{|\Delta_h^s L^\alpha f(x)|}{\omega(|h|)}.$$

here the majorant ω is monotone and $\omega(+0) = 0$.

Let us recall that in the case of a power majorant ω this space coincides with the space of functions whose higher derivatives satisfy a corresponding Hölder or Zygmund condition (i.e. belong Λ_t^s). In particular, $C^k\Lambda_{t^s}^s = W_\infty^{k+s}$.

Let S be one of the spaces considered and F be an arbitrary closed subset of $\mathbb{R}^n$.

PROBLEM 1. *Give a constructive description of the trace space $S|_F$.*

An analysis of a few particular cases resulting in a successful solution of Problem 1 permits us to formulate the finiteness hypothesis given below as a key step in the solution. To do this we need

DEFINITION 1. *Let $\{F_\alpha\}_{\alpha\in A}$ be a family of closed subsets of $\mathbb{R}^n$ and $S = C^k\Lambda_\omega^s (= C^k)$. A family of functions $\{g_\alpha \in C(F_\alpha)\}_{\alpha\in A}$ is called S-compatible if there exists a family of functions $\{f_\alpha\}_{\alpha\in A}$ bounded in $C^k\Lambda_\omega^s(\mathbb{R}^n)$ (correspondingly, precompact in $C^k(\mathbb{R}^n)$) so that $g_\alpha = f_\alpha|_{F_\alpha}$, $\alpha \in A$.*

CONJECTURE 1. *There is an integer N such that a function $f \in C(F)$ belongs to $S|_F$ if and only if the function family $\{f|_{\tilde{F}}; \text{card } \tilde{F} \leqslant N\}$ is S-compatible.*

We denote the least N with such a property by $N(S)$. Then $N(C^k(\mathbb{R})) = k+2$ (see H. Whitney [2]); $N(C^k\Lambda_\omega^s(\mathbb{R})) = k+s+1$ (see A. Jonsson [3], P. A. Shvartsman [4], V. K. Dzjadyk, I. A. Shevchuk [5], I. A. Shevchuk [6, 7]). In the multidimensional case $N(\Lambda_\omega^1(\mathbb{R}^n)) = 2$ (H. Whitney [8]), $N(\Lambda_\omega^2(\mathbb{R}^n)) = 3 \cdot 2^{n-1}$ (P. A. Shvartsman [9]), and lastly, $N(C^1(\mathbb{R}^n)) = 3 \cdot 2^{n-1}$ (Yu. A. Brudny, P. A. Shvartsman [10]).

In case Conjecture 1 is true a solution of the following problem, which has an interest in itself, would be the next step.

PROBLEM 2. *Let* card $F \leqslant N(S)$. *Construct a linear extension operator* $T : \ell^\infty(F) \to S(\mathbb{R}^n)$ *with the least norm and estimate it in terms of the "geometry" of the set* F.

In the one-dimensional case Problem 2 can be solved with the help of the Lagrange interpolation operator. The same approach is impossible in the multidimensional situation (corresponding points can be situated on the level surface of a polynomial); the only result known concerns the case $S = \Lambda_\omega^2(\mathbb{R}^2)$ (see P. A. Shvartsman [9]). One can hope that a solution of Problem 2 will lead to important generalizations of polynomial interpolation.

2. PROBLEM 3. *Is there a linear extension operator* $T : S|_F \to S(\mathbb{R}^n)$ *for any closed subset* F?

The set of such F is denoted by $\mathcal{LE}(S)$. From Whitney's work [2] it follows that the class $\mathcal{LE}(C^k(\mathbb{R}))$ coincides with the class $\mathrm{Clos}(\mathbb{R})$ of all closed subsets of $\mathbb{R}$. A similar answer is also true for the spaces $C^k \Lambda_\omega^s(\mathbb{R})$ and $\Lambda_\omega^1(\mathbb{R})$ (see the papers cited above dealing with these cases). The following result $\mathcal{LE}(\Lambda_\omega^2(\mathbb{R}^n)) = \mathrm{Clos}(\mathbb{R}^n)$ is much more difficult (see Yu. A. Brudny, P. A. Shvartsman [11]). The difficulty of the multidimensional case is emphasized by the following unexpected fact (compare with the corresponding one-dimensional result of Whitney [2]) $\mathcal{LE}(C^1(\mathbb{R}^2)) \neq \mathrm{Clos}(\mathbb{R}^2)$ (see Yu. A. Brudny, P. A. Shvartsman [10]). It evokes the following two questions.

CONJECTURE 2. $\mathcal{LE}(C^k \Lambda_\omega^s(\mathbb{R}^n)) = \mathrm{Clos}(\mathbb{R}^n)$.

PROBLEM 4 *is to describe such* F *for which there exists a linear extension operator* $C^1|_F \to C^1(\mathbb{R}^n)$.

The general case of the space C^k is interesting, but apparently hopeless.

3. Finally let us dwell on the connection between "internal" and "external" definitions of spaces of smooth functions for domains $G \subset \mathbb{R}^n$. Among possible definitions of the space $S(G)$ we have chosen a definition which gives the classical Sobolev space W_∞^k in the case $S = \Lambda_{tk}^k$. In particular, $C^k \Lambda_\omega^s(G)$ is determined by means of the seminorm (1), where now the supremum is taken at first over all x, so that $x + jh \in G_h$, $j = 0, \ldots, k$, $G_h := \{x \in G : \mathrm{dist}(x, \mathbb{R}^n \setminus G) \geqslant h\}$ and then over all $h \in \mathbb{R}^n$. We shall limit ourselves to the discussion of the scale $C^{k,\omega} := C^k \Lambda_\omega^1$, where ω is a function of the modulus of continuity type; in the case $\omega \equiv 1$ we set $C^{k,\omega} := C^k \Lambda_t^2$.

PROBLEM 5 *is to find a necessary and sufficient condition for domain* G *which ensure*

$$(2) \qquad\qquad C^{k,\omega}(G) = C^{k,\omega}(\mathbb{R}^n)_G.$$

To formulate a corresponding hypothesis define two metrics in G according to the following formula

$$d_\omega(x,y) = \inf_\gamma \int_\gamma \omega \circ \rho_G \, \frac{dl}{\rho_G}.$$

Here $\rho_G(z) := \mathrm{dist}(z, \mathbb{R}^n \setminus G)$ and the infimum is taken over all smooth curves γ connecting x and y in G. Next we set

$$\delta_\omega(x,y) := \int_{m(x,y)}^{M(x,y)} \omega(t) \, \frac{dt}{t},$$

where $m(x,y) := \min(\rho_G(x), \rho_G(y))$, $M(x,y) := |x - y| + m(x,y)$. Let us notice that $\delta_\omega(x,y) \approx |x - y|^\beta$ for $\omega(t) = t^\beta$, $0 < \beta \leqslant 1$, and $\delta_\omega(x,y) = \ln(1 + |x - y|/m(x,y))$ for $\omega \equiv 1$.

CONJECTURE 3. *If*

$$\sup_{t>0} \frac{1}{\omega(t)} \int \frac{\omega(s)}{s} \, ds < \infty$$

then the condition

$$\sup_{x,y \in G} \frac{d_\omega(x,y)}{\delta_\omega(x,y)} < \infty$$

is necessary and sufficient for the fulfillment of (2).

We note, that the sufficiency of that condition has been proved; see Yu. A. Brudny, P. A. Shvartsman [12] ($\omega(t) = t$ was known earlier; see Whitney [8], Zobin [13], Konovalov [14]). The necessity of condition (3) for the fulfilment of condition (2) is obvious in case $k = 0$ and $\omega \not\equiv 1$. Note, also that for the case of a bounded *simply connected* domain $G \subset \mathbb{R}^2$ and $\omega(t) = t$ (i.e. for W_∞^k) the necessity of condition (3) has also been established, see Zoblin [13] and Konovalov [14].

REFERENCES

1. Brudny Yu. A., *Restrictions of the Lipschitz spaces to closed sets*, Linear and Complex Analysis Problem Book, 199 Research Problems, Lect. Notes. Math., vol. 1043 Springer–Verlag, 1984, pp. 583–585.
2. Whitney H., *Differentiable functions defined in closed sets*, I, Amer. Math. Soc. Trans. **36** (1934), 369–387.
3. Jonsson A., *The trace of the Zygmund class $\Lambda_k(\mathbb{R})$ to closed sets and interpolating polynomials*, J. Approx. Theory **44** (1985), 1–13.
4. Shvartsman P. A., *The traces of functions of two variables satisfying the Zygmund condition*, Studies in the Theory of Functions of Several Real Variables, Yaroslavl State Univ., Yaroslavl, 1981, pp. 145–168. (Russian)
5. Dzjadyk V. K., Shevchuk I. A., *Continuation of functions which, on an arbitrary set of the line, are traces of functions with a given second modulus of continuity*, Izv. Akad. Nauk SSSR **47** (1983), 248–267 (Russian); English transl. in Math. USSR Izvestia **39** (1983).
6. Shevchuk I. A., *On traces of functions of class H_k^ϕ on the line*, Dokl. Akad. Nauk SSSR **273** (1983), 313–314 (Russian); English transl. in Soviet Math.– Doklady **28** (1983), 652–653.
7. Shevchuk I. A., *A constructive trace description of differentiable real variable functions*, Inst. Matem. Ukr. SSR, Kiev, 1984, preprint. (Russian)
8. Whitney H., *Analytic extensions of differentiable functions defined in closed sets*, Trans. Amer. Math. Soc. **36** (1934), 63–89.
9. Shvartsman P. A., *Traces of functions of Zygmund class*, Sibirsk. Mat. Zh. **28** (1987), 203–215 (Russian); English transl. in Siberian Math. J. **28** (1987), 853–863.
10. Brudny Yu. A., Shvartsman P. A., *The traces of differentiable functions on closed subsets of $\mathbb{R}^n$*, Proceedings of 2nd Int. Conf. on Func. Spaces, Poznan, Poland, 1991, Teubner. Zur. Mathem.
11. Brudny Yu. A., Shvartsman P. A., *A linear extension operator for a space of smooth functions defined on a closed subset of $\mathbb{R}^n$*, Dokl. Akad. Nauk SSSR **230** (1985), 268–272 (Russian); English transl. in Soviet Math. Dokl. **31** (1985), 48–51.
12. Brudny Yu. A., Shvartsman P. A., *Extensions of functions with preservation of smoothness*, Trudy Math. Inst. Steklov **180** (1987) (Russian); English transl. in Proc. Steklov Inst. Math. 3 (1989), 67–69.
13. Zobin N. M., *Investigations on the theory of nuclear spaces* Ph. D. thesis, Voronezh State Univ., Voronezh, 1975. (Russian)

14. Konovalov V. N., *A trace description for some classes of several real variables functions*, Akad. Nauk Ukr. SSR, Inst. Matem,, Kiev, 1984, preprint. (Russian)

DEPT. OF MATHEMATICS
TECHNION
HAIFA 32000
ISRAEL

DEPT. OF MATHEMATICS
TECHNION
HAIFA 32000
ISRAEL

MULTIPLICATIVE PROPERTIES OF ℓ_A^p

S. A. VINOGRADOV

Let ℓ_A^p be the Banach space of all functions $f = \sum_{k \geqslant 0} \hat{f}(k) z^k$ holomorphic in the unit disc $\mathbb{D}$ and satisfying

$$\|f\|_{\ell^p} = \left(\sum_{n \geqslant 0} |\hat{f}(n)|^p \right)^{1/p} < +\infty, \qquad 1 \leqslant p \leqslant \infty.$$

$(\|f\|_{\ell^\infty} \stackrel{\text{def}}{=} \sup_{n \geqslant 0} |\hat{f}(n)|)$. It is well–known that ℓ_A^p is not an algebra with respect to the pointwise multiplication of functions if $p \neq 1$. Therefore, when studying the multiplicative structure of ℓ_A^p, the space $M_A^p \stackrel{\text{def}}{=} \{ g \in \ell_A^p : gf \in \ell_A^p, \forall f \in \ell_A^p \}$ becomes very important. Recall that $M_A^p = M_A^q$, $\frac{1}{p} + \frac{1}{q} = 1$, $1 \leqslant p \leqslant \infty$; l_A^2 coincides with the Hardy class H^2; $M_A^1 = \ell_A^1$, $\ell_A^1 \subset M_A^p \subset H^\infty$, $1 \leqslant p \leqslant 2$.

The conjectures of the paper are closely connected with the theorem of L. Carleson [1] on the interpolation by bounded analytic functions. Given a subset E of $\mathbb{D}$ let R_E denote the restriction operator onto E.

THEOREM [1]. $R^E(H^\infty) = \ell^\infty(E)$ if and only if

(C) $$\delta(E) \stackrel{\text{def}}{=} \inf \left\{ \prod_{\eta \in E \setminus \{\xi\}} \left| \frac{\eta - \xi}{1 - \bar{\eta}\xi} \right| : \xi \in E \right\} > 0.$$

Note that (C), being necessary for $R^E(M_A^p) = \ell^\infty(E)$, $1 < p < 2$, is not sufficient.[1] On the other hand it turns out to be sufficient for $1 < p \leqslant 2$ if E satisfies the Stolz condition (i.e. E is constrained in a finite union of domains $S_\lambda(\zeta) \stackrel{\text{def}}{=} \{ \eta \in \mathbb{D} : |\zeta - \eta| \leqslant \lambda(1 - |\eta|) \}$, where $1 < \lambda < \infty$, $\zeta \in \mathbb{T}$), cf. [2].

Suppose that E satisfies the Stolz condition. Then it is easy to check (see [4]) that

$$\delta(E) > 0 \iff \sigma(E) \stackrel{\text{def}}{=} \inf \left\{ \left| \frac{\xi - \eta}{1 - \bar{\eta}\xi} \right| : \xi, \eta \in E, \xi \neq \eta \right\} > 0 \iff$$

$$\sigma(E) > 0 \; \& \; \gamma(E) \stackrel{\text{def}}{=} \sup \left\{ \sum_{\substack{\eta \in E \\ |\eta| \geqslant |\xi|}} \frac{1 - |\eta|}{1 - |\xi|} : \xi \in E \right\} < +\infty.$$

The conditions $\sigma(E) > 0$ and $\gamma(E) < +\infty$ are important for the problems of interpolation theory in ℓ_A^p as well as in other spaces [3]. Everything said above makes plausible the following conjecture.

---

[1] That (C) does not imply $R^E(M_A^p) = \ell^\infty(E)$, $1 < p < 2$ can be proved with help of [3]

CONJECTURE 1. $\sigma(E) > 0$, $\gamma(E) < +\infty \implies R^E(M_A^p) = \ell^\infty(E)$.

Conjecture 1 is related to

CONJECTURE 2. $\gamma(E) < +\infty \implies B^E \in \cap_{1 < p \leqslant 2} M_A^p$, where $B^E \overset{\text{def}}{=} \prod_{\eta \in E} \frac{|\eta|}{\eta} \frac{\eta - z}{1 - \bar{\eta}z}$ stands for the Blaschke product generated by E.

Conjecture 1 follows from Conjecture 2. To see this it is sufficient to apply the Earl theorem [5] about the interpolation by Blaschke products. It is not hard to show that the zero set E of the corresponding Blaschke product can be chosen in this case satisfying $\sigma(E) > 0$, $\gamma(E) < +\infty$ (see [6], §4 for details).

It follows from $M_A^p = M_A^q$ ($\frac{1}{p} + \frac{1}{q} = 1$, $1 \leqslant p \leqslant \infty$) that every inner function $I \in M_A^p$ satisfies

$$(1) \qquad \left(I \cdot F \in \ell_A^p, \ F \in H^1 \right) \implies F \in \ell_A^p$$

for $1 < p \leqslant 2$.
Therefore the proof of Conjecture 2 would give new non–trivial examples of Blaschke products I with property (1). In this direction at present, apparently, only the following is known.

1. $I = \exp \frac{z+1}{z-1}$ does not satisfy (1) for $1 \leqslant p \leqslant 4/3$ (see [7], [8]).
2. For $1 < p < 2$ B^E satisfies (1) provided $\gamma(E) < \infty$ and E satisfies the Stolz condition (see [2]).

CONJECTURE 3.

(a) Suppose $\gamma(E) < \infty$. Then B^E satisfies (1), $1 < p < 2$.
(b) If $\gamma(E) < +\infty$ and E satisfies the Stolz condition then B^E satisfies (1) with $p = 1$.

Analogous conjectures can be formulated for multipliers of

$$\left\{ \int_0^\infty f(t)e^{itz}\, dt : f \in L^p(0, +\infty) \right\}.$$

REFERENCES

1. Carleson L., *An interpolation problem for bounded analytic functions*, Amer. J. Math **80** (1958), no. 4, 921–930.
2. Vinogradov S. A., *Multipliers of power series with sequence of coefficients from l^p*, Zapiski nauchn. sem. LOMI **39** (1974), 30–40 (Russian); English transl. in J. Soviet Math. **8** (1977), no. 1.
3. Vinogradov S. A., *Exponential bases and free interpolation in Banach spaces with the L^p-norm*, Zapiski nauchn. sem. LOMI **65** (1976), 17–68 (Russian); English transl. in J. Soviet Math. **16** (1981), no. 3, 1060–1065.
4. Vinogradov S. A., Havin V. P., *Free interpolation in H^∞ and some other classes of functions*, Zapiski nauchn. sem. LOMI **47** (1974), 15–54 (Russian); English transl. in J. Soviet Math. **9** (1978), no. 2.
5. Earl J. P., *On the interpolation of bounded sequences by bounded analytic functions*, J. London Math. Soc. **2** (1970), no. 2, 544–548.
6. Vinogradov S. A., Havin V. P., *Free interpolation in H^∞ and some other classes of functions*, Zapiski nauchn. sem. LOMI **56** (1976), 12–58 (Russian); English transl. in J. Soviet Math. **14** (1980), no. 2, 1027–1065.

7. Gurarii V. P., *The factorization of absolutely convergent Taylor series and Fourier integrals*, Zapiski nauchn. sem. LOMI **30** (1972), 15–32 (Russian); English transl. in J. Soviet Math. **4** (1975), no. 4.
8. Shirokov N. A., *Some properties of primary ideals of absolutely convergent Taylor series and Fourier integrals*, Zapiski nauchn. sem. LOMI **39** (1974), 149–161 (Russian); English transl. in J. Soviet Math. **8** (1977), no. 1.

COMMENTARY BY THE AUTHOR

Conjecture 2 and 3 are disproved in [9] (see Corollary 1 in [9] disproving Conjecture 2 and Corollary 2 and Theorem 5 in [9] disproving Conjecture 3). The results of [8], [9] and [10] lead to the following question.

QUESTION. *Is there a singular inner function in* $\bigcup_{1 \leqslant p < 2} M_A^p$?

REFERENCES

9. Vinogradov S. A., *Multiplicative properties of power series with a sequence of coefficients from l^p*, Doklady Akad. Nauk SSSR **254** (1980), no. 6, 1301–1306 (Russian); English transl. in Soviet Math. Dokl. **22** (1980), no. 2, 560–565.
10. Verbitskii I. E., *Multipliers of spaces l_A^p*, Funktsional. Anal. i Prilozhen. **14** (1980), no. 3, 67–68 (Russian); English transl. in Funct. Anal. Appl. **14** (1980), no. 3, 219–220.

DEPARTMENT OF MATHEMATICS AND MECHANICS
ST. PETERSBURG STATE UNIVERSITY
STARYI PETERHOF
ST. PETERSBURG, 198904
RUSSIA

MULTIPLIERS, INTERPOLATION, AND $\Lambda(p)$ SETS

Misha Zafran

Let G be a locally compact Abelian group, with dual group Γ. An operator

$$T\colon L_p(G) \to L_p(G)$$

will be called a multiplier provided there exists a function $\widehat{T} \in L_\infty(\Gamma)$ so that $T(f)\widehat{}= \widehat{T}\widehat{f}$, for all integrable simple functions f. The space of multipliers on $L_p(G)$ is denoted by $M_p(G)$. Let $CM_p(G) = \{T \in M_p(G) : \widehat{T} \in C(\Gamma)\}$. In response to a question of J. Peetre, the author has recently shown that for the classical groups, $CM_p(G)$ is not an interpolation space between $M_1(G) = M(G)$ and $CM_2(G) = L_\infty(\Gamma) \cap C(\Gamma)$. More specifically, we obtained the following theorem (see [2]).

THEOREM 1. *Let G denote one of the groups $\mathbb{T}^n$, $\mathbb{R}^n$, or $\mathbb{Z}^n$. Then there exists an operator T so that*

(a) *T is a bounded operator on $L_\infty(\Gamma) \cap C(\Gamma)$.*
(b) *$T|M(G)$ is a bounded operator on $M(G)$.*
(c) *$T|CM_p(G)$ is **not** a bounded operator on $CM_p(G)$, $p \in (1,2)$.*

Observe that T is *independent* of p, $1 < p < 2$. Our method of construction makes essential use of certain results concerning $\Lambda(q)$ sets. Recall that a set $E \subseteq \mathbb{Z}$ is said to be of type $\Lambda(q)$ $(1 < q < \infty)$ if whenever $f \in L_1(\mathbb{T})$ and $\widehat{f}(n) = 0$ for all $n \notin E$, we have $f \in L_q(\mathbb{T})$. We used the following elegant result of W. Rudin [1].

THEOREM 2. *Let $s \geqslant 2$ be an integer, let N be a prime with $N > s$, and let $M = s^{s-1}N^s - 1$. Then there exists a set $F \subseteq \{0,1,2,\ldots,M\}$ so that*

(a) *F contains exactly N points, and*
(b) *$\|f\|_{2s} \leqslant C\|f\|_2$, for every trigonometric polynomial f, with $\widehat{f}(n) = 0$ for $n \notin F$. (Such f are called F-polynomials).*

(Here C is *independent* of N).

As a consequence of Theorem 2, Rudin showed that there exist sets of type $\Lambda(2s)$ which are not of type $\Lambda(2s + \varepsilon)$, for all $\varepsilon > 0$ (see [1]).

An obvious conjecture arising from Theorem 1 is the following:

CONJECTURE 1. *Let $1 \leqslant p_1 < p_2 \leqslant 2$. Then there exists an operator T so that*

(a) *T is a bounded operator on CM_{p_2}.*
(b) *$T|CM_{p_1}$ is a bounded operator on CM_{p_1}.*
(c) *$T|CM_p$ is **not** a bounded operator on CM_p, for all $p \in (p_1, p_2)$.*

It is natural to attempt to analyze this conjecture by means of the techniques used to obtain Theorem 1. But it soon becomes evident that such an analysis requires a deep extension of Rudin's theorem. Specifically, we require a result of the following form:

CONJECTURE 2. (The $\Lambda(p)$ Problem). Let $2 < p < \infty$. Then there exists a set of type $\Lambda(p)$ which is not of type $\Lambda(p + \varepsilon)$, for all $\varepsilon > 0$.

This conjecture (which was essentially posed by Rudin) has remained unresolved for nearly a quarter of a century, and is one of the fundamental open questions in harmonic analysis. Its solution will undoubtedly require very subtle new ideas involving estimation in L_p. Conjecture 1 may be just one of the manifold consequences of the $\Lambda(p)$ problem.

Let us attempt to briefly outline one possible approach to the study of Conjecture 2. Let $p = 2s/k$ where $s > 1$ and $k \geqslant 1$ are integers, and $s > k$. Let $F = F_N$ be the set of Theorem 2, and let $F_{k,N}$ denote the k-fold sum $F + \cdots + F$. In essence, the "piecing together" of the $F_{k,N}$ (for an infinity of N's) provides an example of a set which is not of type $\Lambda(p + \varepsilon)$, for all $\varepsilon > 0$. The difficulty is in proving that $F_{k,N}$ is of type $\Lambda(p)$ (with all constants uniform in N). One may seek to accomplish this by writing an $F_{k,N}$-polynomial f in a judicious way as a sum of products of F-polynomials, and carefully examining the resultant representation of f. However, new estimation techniques for L_p norms would still be very much a necessity in order to carry out this program.

REFERENCES

1. Rudin W., *Trigonometric series with gaps*, J. Math. Mech. 9 (1960), 203–227.
2. Zafran M., *Interpolation of multiplier spaces*, Amer. J. Math. 105 (1983), 1405–1416.

DEPARTMENT OF MATHEMATICS
UNIVERSITY OF WASHINGTON
SEATTLE, WA 98195
USA

BANACH ALGEBRAS OF FUNCTIONS
GENERATED BY THE SET
OF ALL ALMOST PERIODIC POLYNOMIALS
WHOSE EXPONENTS BELONG TO A GIVEN INTERVAL

M. G. KREIN

1. For any $A \in (0, \infty]$ let $\mathbb{P}_A$ denote the linear set of all almost periodic polynomials

$$\mathcal{P}(t) = \sum_k c_k e^{i\lambda_k t}$$

with exponents $\lambda_k \in (-A, A)$, endowed with the sup-norm $\|\mathcal{P}\| = \sup\{ |\mathcal{P}(t)| : -\infty < t < \infty \}$.

Evidently every linear functional Φ on $\mathbb{P}_A$ is completely defined by its χ-function (characteristic function): $f(\lambda) = \Phi(e_\lambda)$ $(-A < \lambda < A)$, where $e_\lambda(t) = \exp(i\lambda t)$ $(-\infty < t < \infty)$. The χ-function may be an arbitrary function from $(-A, A)$ to $\mathbb{C}$.

Let us denote by $\mathcal{R}_A$ the set of all $f : (-A, A) \to \mathbb{C}$ generating linear continuous functionals Φ_f on $\mathbb{P}_A$ $(\Phi_f \in \mathbb{P}_A^*)$. For $f \in \mathcal{R}_A$ we put $\|f\|_A = \|\Phi_f\|$. It is easy to see that $\mathcal{R}_A$ is a Banach algebra of functions. By $\mathcal{R}_A^c$ $(\mathcal{R}_A^m)$ we denote the Banach subalgebras consisting of all continuous (measurable) $f \in \mathcal{R}_A$.

The following proposition was proved in [1] for $A = \infty$ and in [2] for $(0 <) A < \infty$.

THEOREM 1. *Let f be a function from $(-A, A)$ to $\mathbb{C}$. Then $f \in \mathcal{R}_A^c$ iff it admits the representation*

$$(1) \qquad f(\lambda) = \int_{-\infty}^{\infty} e^{i\lambda t} \, d\sigma(t) \qquad \forall \lambda \in (-A, A),$$

where σ is a complex measure on $\mathbb{R}$ of bounded variation $\text{Var} \, \sigma < \infty$. Moreover, every $f \in \mathcal{R}_A^c$ admits representation (1) with $\text{Var} \, \sigma = \|f\|_A$.

Clearly it follows from this theorem, that every function $f \in \mathcal{R}_A^c$ $(0 < A < \infty)$ admits an extension $\tilde{f} \in \mathcal{R}_\infty^c$ with $\|\tilde{f}\|_\infty = \|f\|_A$. On the other hand every $f \in \mathcal{R}_A$ $(0 < A < \infty)$ admits an extension $\tilde{f} \in \mathcal{R}_A$ with $\|\tilde{f}\|_\infty = \|f\|_A$ by the Hahn–Banach theorem.

QUESTION I. *Does every $f \in \mathcal{R}_A^m$ $(0 < A < \infty)$ admit an extension $\tilde{f} \in \mathcal{R}_\infty^m$?*

QUESTION II. *If every $f \in \mathcal{R}_A^m$ has an extension $\tilde{f} \in \mathcal{R}_\infty^m$, does there exist for this f such an extension $\tilde{f}_0$ that $\|\tilde{f}_0\|_\infty = \|f\|_A$?*

QUESTION III. *Does every $f \in \mathcal{R}_A^m$ $(0 < A \leqslant \infty)$ admit a decomposition $f = f_c + f_m$, where $f_c \in \mathcal{R}_A^c$, $f_m \in \mathcal{R}_A^m$ and f_m equals zero a.e.?*

2. A functional $\Phi \in \mathbb{P}_A^*$ is said to be real if it takes real values on real $\mathcal{P}$'s, $\mathcal{P} \in \mathbb{P}_A$. A $\Phi \in \mathbb{P}_A^*$ is real iff its χ-function f $(\Phi = \Phi_f)$ is Hermitian: $f(-\lambda) = \overline{f(\lambda)}$ $\forall \lambda \in (-\infty, \infty)$. Every functional $\Phi \in \mathbb{P}_A^*$ admits a unique decomposition $\Phi = \Phi_\mathcal{R} + i\Phi_\mathcal{J}$, where $\Phi_\mathcal{R}, \Phi_\mathcal{J} (\in \mathbb{P}_A^*)$ are real. Therefore, it is easy to see, that in Questions I, II, III we may restrict ourselves to the case of Hermitian f only.

Denote by $\mathbb{K}_A^*$ $(0 < A \leqslant \infty)$ the cone of all non-negative $\Phi \in \mathbb{P}_A^*$. Naturally, a $\Phi (\in \mathbb{P}_A^*)$ is said to be non-negative if $\Phi(\mathcal{P}) \geqslant 0$ as soon as $\mathcal{P}(t) \geqslant 0$ $\forall t \in (-\infty, \infty)$ $(\mathcal{P} \in \mathbb{P}_A)$.

Denote by $\mathfrak{P}_A$ $(0 < A \leqslant \infty)$ the cone of all χ-functions corresponding to the elements of $\mathbb{K}_A^*$. The subcone of all continuous (measurable) $f \in \mathfrak{P}_A$ will be denoted by $\mathfrak{P}_A^c (\mathcal{K}_A^m)$.

It is easy to see that for every $f \in \mathfrak{P}_A$ we have: $f(0) = \|f\|_A$. For any Hermitian $f \in \mathcal{R}_A$ there exists a decomposition $f = f_+ - f_-$, where $f_\pm \in \mathfrak{P}_A$ and $\|f\|_A = f_+(0) + f_-(0)$.

To establish the last assertion it is sufficient to do this for $A = \infty$. In this case $\mathbb{P}_\infty$ forms a linear dense set in the Banach space $\mathbb{B}$ of all almost periodic Bohr functions with sup-norm. The cone $\mathbb{K}_\infty^*$ is dual to the cone $\mathbb{K}$ of all non-negative functions in $\mathbb{B}$. As $\mathbb{B}$ may be identified with a Banach space of all continuous functions on a compact space, the existence of the required decomposition for Hermitian $f \in \mathfrak{P}_\infty$ follows. Moreover, this decomposition is unique and minimal in this case $(A = \infty)$.

If $f \in \mathcal{R}_A^c$ is Hermitian, then there exists a decomposition $f = f_+ - f_-$ with $f_\pm \in \mathfrak{P}_A^c$ and $\|f\|_A = f_+(0) + f_-(0)$. Indeed, for Hermitian $f \in \mathcal{R}_A^c$ we can obtain (1) with a real measure σ, $\operatorname{Var}\sigma = \|f\|_A$, admitting a unique decomposition $\sigma = \sigma_- - \sigma_-$ with non-negative measures $\sigma_\pm$ such that $\operatorname{Var}\sigma = \operatorname{Var}\sigma_+ + \operatorname{Var}\sigma_-$. This decomposition yields (via (1)) the required decomposition for f.

QUESTION IV. *Does every Hermitian $f \in \mathcal{R}_A^m$ admit a decomposition $f = f_+ - f_-$ with $f_\pm \in \mathfrak{P}_A^m$?*

It turns out, that the affirmative answer to this question implies the same for Question III. This connection is due to a theorem of [4] according to which every function $g \in \mathfrak{P}_A^m$ admits a decomposition $g = g_c + g_m$, where $g_c \in \mathfrak{P}_A^c$, $g_m \in \mathfrak{P}_A^m$ and g_m equals zero a.e. This theorem has been generalized recently in [5].

It is plausible, that for any Hermitian $f \in \mathcal{R}_A^m$ in the decomposition $f = f_+ - f_-$ with $f_\pm \in \mathfrak{P}_A$, $\|f\|_A = f_+(0) + f_-(0)$ (which always exists) automatically $f_\pm \in \mathfrak{P}_A^m$.

3. For better orientation we will indicate that A. P. Artëmenko [6,7] has obtained a general proposition which contains, in particular, the following characterization of functions $f \in \mathfrak{P}_A$.

THEOREM 2. *Let $0 < A \leqslant \infty$ and let f be a function from $(-A, A)$ to $\mathbb{C}$. Then $f \in \mathfrak{P}_A$ iff for any $\lambda_j \in [0, A]$ and any $\xi_j \in \mathbb{C}$ $(j = 1, 2, \ldots, n; n = 1, 2, \ldots)$:*

$$\sum_{j,k=1}^n f(\lambda_j - \lambda_k)\xi_j \bar{\xi}_k \geqslant 0.$$

The necessity of this condition is trivial. A transparent proof of its sufficiency has been obtained by B. Ja. Levin [8].

For a continuous function f from $(-A, A)$ to $\mathbb{C}$ the assertion of Theorem 2 (for $A = \infty$) is contained in the well-known theorem of Bochner and for $(0 <) A < \infty$ in the author's corresponding theorem [2].

A series of unsolved problems concerning extensions of functions $f \in \mathfrak{P}_A^m$ $(0 < A < \infty)$ is formulated in [4]. In this connection we also mention [9].

REFERENCES

1. Bochner S., *A theorem on Fourier – Stieltjes integrals*, Bull. Amer. Math. Soc. **40** (1934), no. 4, 271–276.
2. Krein M. G., *On a problem of continuation of Hermitian-positive continuous functions*, Doklady Akad. Nauk SSSR **26** (1940), no. 1, 17–21. (Russian)
3. Krein M. G., *On representation of functions by Fourier-Stieltjes integrals*, Učen. Zap. Kuibyš. GPI (1943), no. 7, 123–147. (Russian)
4. Krein M. G., *Measurable Hermitian-positive functions*, Mat. Zametki **23** (1978), no. 1, 79–89 (Russian); English transl. in Math. Notes **23** (1978), no. 1, 45–50.
5. Langer H., *On measurable Hermitian indefinite functions with a finite number of negative squares*, Acta Sci. Math. Szeged **45** (1984), 281–292.
6. Artemenko A. P., *On positive linear functionals in the space of almost periodic functions of H. Bohr*, Comm. Inst. Sci. Math. Mec. Univ. Kharkov **(4) 16** (1940), 111–119. (Russian)
7. Artemenko A. P., *Hermitian-positive functions and positive functionals. I*, Teor. Funkts., Funktsion. Anal. i Prilozhen. (1984), no. 41, 3–16. (Russian)
8. Levin B. Ya., *On a generalization of the Fejer-Riesz theorem*, Dokl. Akad. Nauk SSSR **52** (1946), 291–294. (Russian)
9. Crum M. M., *On positive definite functions*, Proc. London Math. Soc. **6** (1956), no. 3, 548–560.

SOME QUESTIONS CONNECTED WITH POSITIVE
DEFINITE KERNELS OF TWO VARIABLES

I. E. OVCHARENKO

1. Intrinsic positive functions; Π-lifting. Let $B(E)$ be a Hilbert space of entire functions corresponding to the indeterminate case of the problem of moments (a particular case of *de Branges* general Hilbert spaces of entire functions (HSEF), see [1]. Let us call a function $f \in B(E)$ intrinstic positive in $B(E)$ if for arbitrary polynomials $P(\lambda) \geqslant 0$, $-\infty < \lambda < +\infty$ one has $\langle P, f \rangle \geqslant 0$. A typical example of such a function is $K(z,t)$, $-\infty < \lambda < +\infty$, where $K(z,t)$ is the reproducing kernel of $B(E)$. Let $\rho(d\lambda)$ be any measure representing the scalar product on $B(E)$. Let us call a function $\tilde{f}(\lambda) \geqslant 0 \in L^2_\rho$ a lifting of the function f if $\mathrm{Pr}_{B(E)} \tilde{f} = f$.

PROBLEM. *To give an adequate description of the lifting of a given i. p. function in space L^2_ρ.*

At the physical level of accuracy any i. p. function is

$$f(\lambda) = \int_{-\infty}^{+\infty} K(\lambda, t)\mu(dt), \qquad \mu(dt) \geqslant 0$$

This equation requires precise description. I believe that similar questions are especially interesting in the context of *de Branges* spaces [1].

2. Intrinsic p. d. sequences. Let $s_0(\lambda)(\equiv 1), s_1(\lambda), \ldots, s_n(\lambda) \in B(E)$; let us call this sequence as i. p. d. in $B(E)$ if $\sum_{j,k} s_{j+k}\xi_j\xi_k$ is i. p. in $B(E)$ for every ξ_j; let $\rho(\lambda) \geqslant 0$ be any measure representing $\langle \ , \ \rangle_{B(E)}$; the sequence $\phi_0(\lambda), \ldots, \phi_n(\lambda) \in L^2_\rho$ is called a Π_ρ-lifting of the sequence $\{s\}$ if: a) $P\phi_n = s_n$; b) $\{\phi_n\}$ is p. d. in L^2_ρ.

PROBLEM. *To give a description of the set of all Π_ρ-liftings of $\{s_n(\lambda)\}$.*

This problem is related to a traditional solution of the two dimensional power moment problem [2]. One can remark that $\phi_k \in L^{2k}_\rho$, $k = 1, 2, \ldots$

3. De Branges spaces with bounded reproducing kernel. It is important to consider particular cases of p. d. kernels of various spaces $B(E)$ for which $K(\lambda, \lambda) < M$ ($-\infty < \lambda < +\infty$).

PROBLEM. *To give a characterization of such spaces (as much constructive as possible) in various terms of de Branges theory of HSEF.*

In this direction see [1], Th. 51. (I thank Louis de Branges for the discussion of these questions).

4. Analytic p. d. sequences. Let $\phi_0(z), \phi_1(z), \phi_2(z), \ldots$ be analytic operator functions $\phi_n(z) \in [H]$, $\|z\|_H < 1$ such that the sequence $T_n(z) := \phi_n(z)$, $n \geqslant 0$, $T_n(z) = \overline{\phi^*_n(z)}$, $n < 0$, is p. d. $\forall z$.

PROBLEM. *To give a simple proof of the following property: there exists a Hilbert space* $\tilde{H} \supseteq H$ *and an analytic contraction* $B(z)$ *having meaning in* $\left[\tilde{H}\right]$ *such that* $\|z\|_{\tilde{H}} < 1$, $\|B(z)\| \leqslant 1$ *and* $T_n(z) = \mathrm{Pr}_H B^n(z)$, $n > 0$.

The extension problem for finite sequences of analytic functions (similar to those mentioned above) are also of interest. Any progress here would, in my opinion, give answers to many questions on analytic functions with non-negative real part $\mathrm{Re} f(z, w) \geqslant 0$ on the bidisc.

5. U.P.-measures. Let $\rho(d\lambda) \geqslant 0$ be a measure having all moments. Let us call the measure $\rho(d\lambda)$ a U.P.-measure if: a) the moment sequence generating by $\rho(d\lambda)$ is determined; b)

$$(f \in L_s^2) \& \left(\rho_f(n) := \int \lambda^n f(\lambda) \rho(d\lambda) \text{ is p. d.} \right) \implies f(\lambda) \geqslant 0 \quad \rho\text{-a. e.}$$

These measures was considered by M. A. Kowalski [3], but was not named. Together with the Chebyshev's recursion they are useful for a computational procedures to find the bounds of Shrödinger's equation solutions, as well as some other equations.

PROBLEM. *For a given U.P.-measure find an analogue of the well-known Riesz criterion for a sequence to be a determined moment sequence. What are the interrelations between the U.P.-property and properties of marginal projections of measure* ρ?

6. Chebyshev's recursion was discovered in 1859 and recently found important applications in a computational mathematics, quantum chemistry and physics. For instance, see [4], [5], [6] and references there. The Chebyshev's recursion gives an effective translation procedure from a system of orthogonal polynomials related to a scalar product on the space of polynomials (of one variable) to a system of orthogonal polynomials related with another scalar product.

PROBLEM. *To find suitable analogue of Chebyshev's procedure for general de Branges's spaces (continuous moment problems).*

REFERENCES

1. De Branges L., *Hilbert spaces of entire functions*, Prentice Hall, 1968.
2. Ovcharenko I. E., *On two parameters power moment sequences*, Ukrain. Math. J. **36** (1984), no. 1, 51–56. (Russian)
3. Kowalski M. A., *Representation of inner products on space of polynomials*, Acta Math. Hung. **46** (1985), no. 1–2, 101–109.
4. Gautschi W., *Orthogonal polynomials — Constructive theory and applications*, J. of Comput. Appl. Math. **12–13** (1985), 61–66.
5. Ovcharenko I. E., *Scalar products on space of polynomials and positivity*, Dokl. Akad. Nauk Ukr. SSR, ser. A **7** (1990), 17–21. (Russian)
6. Ovcharenko I. E., *Some applications of the Chebyshev's recursion*, Dokl. Akad. Nauk SSSR **319** (1991), no. 1, 12–15. (Russian)

AN UKRAINE
PL. REPINA, 1
KIEV
UKRAINE

ON NONEXTENDABLE
HERMITIAN-POSITIVE FUNCTIONS

L. A. SAKHNOVICH

Let S be a set of points in $\mathbb{R}^n$ and put $\Delta = S - S$ (Minkowski difference). A function $\varphi(x)$ is called Hermitian-positive on Δ if the inequality

$$\sum_{i,j=1}^{N} \xi_i \bar{\xi}_j \varphi(x_i - x_j) \geqslant 0$$

holds for any points $x_1, x_2, \ldots, x_N \in S$ and all numbers $\xi_1, \xi_2, \ldots, \xi_N$.

Let $n = 2$. Let $\Delta(N_1, N_2)$ be the set of all points $M(m, l)$ such that $|m| \leqslant N_1$, $|l| \leqslant N_2$. By $\mathcal{P}(N_1, N_2)$ we shall denote the class of functions which are Hermitian-positive on $\Delta(N_1, N_2)$. The following is proved in the papers [1], [2].

PROPOSITION. *There exists a function in $\mathcal{P}(3,3)$ which is not extendable to $\mathcal{P}(\infty, \infty)$.*

In [3] we gave examples of functions in $\mathcal{P}(2,2)$ which are not extendable to $\mathcal{P}(2,3)$.

PROBLEM 1. *To give a finite procedure for deciding whether a function $\varphi(x)$ in $\mathcal{P}(N_1, N_2)$ is extendable to $\mathcal{P}(\infty, \infty)$.*

PROBLEM 2. *To give a physical interpretation of the nonextendable functions in $\mathcal{P}(N_1, N_2)$.*

It is likely that the functions in question are connected with structures which have exhausted the possibilities of their growth.

It is well-known ([1], [2]) that the problem of the representation of a non-negative polynomial

(1) $$f(x,y) = \sum_{0 \leqslant k \leqslant N_1} \sum_{0 \leqslant l \leqslant N_2} a_{k,l} x^k y^l$$

in the form

(2) $$f(x,y) = \sum_{j=1}^{r} q_j^2(x,y),$$

where the $q_j(x,y)$ are real polynomials, is closely connected with the extension problem for the class $\mathcal{P}(N_1, N_2)$.

D. Hilbert [4] proved that there is a non-negative polynomial of degree 6 which can not be represented in the form (2).

PROBLEM 3. *To give a finite procedure for determining whether a non-negative polynomial of the form (1) can be represented as (2).*

REFERENCES

1. Calderón A., Pepinsky R., *On the phases of Fourier coefficients for positive real functions*, Computing methods and the phase problem in x-ray crystal analysis, Report of a conference held at the Pennsylvania State College, April 6-8, 1950 (R. Pepinsky, ed.), State College, Pa., 1952, pp. 339–348.
2. Rudin W., *The extension problem for positive definite functions*, Illinois J. Math. **3** (1963), 532–539.
3. Sakhnovich L. A., *Effective construction of nonextendable Hermitian-positive functions of several variables*, Funkts. Anal. Prilozh. **14** (1980), no. 4, 55–60 (Russian); English transl. in Funct. Anal. Appl. **14** (1980), 290–294.
4. Hilbert D., *Über die Darstellung definiter Formen als Summe von Formquadraten*, Math. Ann. **32** (1888), 342–350.

ODESSA ELECTRICAL ENGENEERING COMMUNICATION INSTITUTE
PR. DOBROVOLSKOGO 154/199
ODESSA 270111
UKRAINE

Chapter 16

ENTIRE AND SUBHARMONIC FUNCTIONS.

POTENTIAL THEORY

Edited by

A. A. Goldberg
Lvov State University
Universitetskaya 1
290602, Lvov
Ukraine

B. Ya. Levin
Dept. Math. FTINT
pr. Lenina 47
310164, Kharkov
Ukraine

I. V. Ostrovskii
Dept. Math. FTINT
pr. Lenina 47
310164, Kharkov
Ukraine

INTRODUCTION

The number of zeros of a polynomial determines its asymptotic behaviour at infinity. For a transcendental entire function the relation between zeros and asymptotics is much more complicated. In classical works by Hadamard and Borel the number of zeros of a function f in the disk $\{z : |z| \leqslant r\}$ has been compared with the maximum of the modulus of f on the same disk. However, in contrast with polynomials the asymptotic behaviour of $\log|f(re^{i\varphi})|$, $r \to \infty$ depends on φ essentially. If f is an entire function of finite order ρ then $\log|f(re^{i\varphi})|$ admits the asymptotic upper bound $h_f(\varphi)r^\rho + o(r^\rho)$, $r \to \infty$, and, outside a small exceptional set, the asymptotic lower bound $\underline{h}_f(\varphi)r^\rho + o(r^\rho)$, $r \to \infty$, where $h_f(\varphi)$ and $\underline{h}_f(\varphi)$ are the indicator and the lower indicator respectively. If $h_f(\varphi) = \underline{h}_f(\varphi)$ for every φ then the entire function f is said to be of completely regular growth (c.r.g. in short, this abbreviation will be used throughout the whole chapter). In this case there are formulae to compute $h_f(\varphi)$ in terms of the asymptotic behaviour of zeros. In the thirties B. Ya. Levin and A. Pfluger found necessary and sufficient conditions on the distribution of zeros that provide c.r.g. We suppose the reader to be familiar with notions and theorems of the theory of entire functions of c.r.g. (e.g. as presented in chapters II and III of the book [1]). Problems 16.2, 16.3, S.16.20 are connected with Levin–Pfluger theory. If zeros of an entire function do not satisfy the just mentioned conditions of c.r.g., then exact formulae for indicators do not exist but it is possible to find their estimates from above and from below. Problem 16.5 is of this kind.

General form of the indicator of an entire function of order ρ is well-known, but for various subclasses of entire functions it is possible to find some additional information. Problem 16.4 contains some open questions of this kind. Some other problems are connected with entire functions. In Problem 16.1 three questions on functions of Laguerre–Polya class are formulated (note that the answer to the first question has recently been obtained). Problem S.16.21 is also connected with this class. Problem 16.7 contains questions concerning arguments of zeros of an entire function representable by a power series with bounded gaps.

The theory of subharmonic functions is closely related with the theory of analytic functions. In particular, if f is analytic then $\log|f|$ is subharmonic. Problems 16.6, 16.8, 16.10, 16.14, S.16.22 deal with the theory of subharmonic functions.

It is difficult to classify the remaining problems. Problem 16.11 is rather a problem of approximation theory. Problem 16.12 contains 4 questions concerning transcendental entire functions and unbounded analytic functions in the unit disc. In Problem 16.13 it is asked whether an analytic solution "on the whole" of some system of linear difference equations exists. Problem 16.15 deals with properties of harmonic measure in $\mathbb{R}^n$. Problem 16.6 is related to uniqueness theorems for polyharmonic functions.

Two solved problems S.16.19, S.16.20 are connected with Nevanlinna theory.

All the problems are undoubtedly of considerable scientific interest though their diversity does not give grounds to judge what branches of the theory of entire and subharmonic functions attract the highest interest at present or where new important discoveries may be expected. It seems to the point to quote here the opinion of the

INTRODUCTION

Nobel Prize winner academician L. D. Landau [2] (addressed to physicists, of course):

"You ask me what topics one must be busy with, meaning what parts of theoretical physics are most important. I have to say that I think this question is absurd. One must be rather anecdotally immodest to think that he is destined for the "most important" scientific problems only. To my mind, every physicist must be busy with those problems which are most interesting to him, and not have vanity considerations as a starting point in his scientific work."

We thank A. E. Fryntov and M. L. Sodin for valuable assistance in preparing this chapter.

REFERENCES

1. Levin B. Ya., *Distributions of Zeros of Entire Functions*, AMS, 1980.
2. Livanova A., *L. D. Landau*, "Znaniye", Moscow, 1978, p.85. (Russian)

ENTIRE FUNCTIONS OF LAGUERRE–POLYA CLASS

B. Ya. Levin

Laguerre–Polya class $\mathcal{LP}$ plays an important role in the theory of entire functions. This class consists of functions of the form

$$f(z) = e^{-\gamma z^2 + \beta z + \alpha} \prod_1^{\omega} \left(1 - \frac{z}{a_k}\right) e^{z/a_k} \qquad (\omega \leqslant \infty)$$

where $\gamma \geqslant 0$, $\operatorname{Im}\beta = 0$, $\operatorname{Im} a_k = 0$, $\sum_1^{\omega} |a_k^{-2}| < \infty$. It is known (see [1]) that this class is the closure (in the sense of the uniform convergence on compact sets) of polynomials with real roots. It follows that $f^{(k)} \in \mathcal{LP}$, $k = 1, 2, \ldots$ for $f \in \mathcal{LP}$. In 1914 Polya proposed the following conjecture: a real entire function (i.e. an entire f with $f(\mathbb{R}) \subset \mathbb{R}$) such that f and all its derivatives have no zeros off $\mathbb{R}$ is in $\mathcal{LP}$.

There are a plenty of works devoted to this conjecture. The bibliography can be found in [3], [4]. S. Hellerstein and J. Williamson solved this problem in their works [4], [5]. They have shown that a real entire function f with all the zeros of f, f', f'' real, is in $\mathcal{LP}$.

PROBLEM 1. *Prove that a real entire function with all the zeros of f and f'' real is in $\mathcal{LP}$.*

In [3] it is shown only that $\log\log M(r, f) = O(r \log r)$ for a real entire function f such that f and f'' have only real zeros.

Consider a well-known class $\overline{HB}$ of entire function ω defined by:

a) the zeros of ω lie in the upper half-plane $\operatorname{Im} z \geqslant 0$ only;

b) if $\operatorname{Im} z < 0$ then $|\omega(z)| \geqslant |\omega(\bar{z})|$.

An arbitrary entire function ω can be represented as

$$\omega = \mathcal{P} + i\mathcal{Q},$$

where $\mathcal{P}$ and $\mathcal{Q}$ are real. It is known ([1]) that $\omega \in \overline{HB}$ if and only if for an arbitrary pair of real numbers λ, μ the function

$$\lambda \mathcal{P} + \mu \mathcal{Q}$$

has only real zeros.

Applying Hellerstein–Williamson's result we now deduce that if $\omega^{(k)} \in \overline{HB}$, $k = 0, 1, \ldots$, then $\omega \in \mathcal{P}^*$, the class $\mathcal{P}^*$ being defined by:

c) $\omega \in \overline{HB}$;

d) $\omega(z) = e^{-\gamma z^2 + \beta z + \alpha} \prod_1^{w} \left(1 - \frac{z}{a_k}\right) e^{z/a_k} \qquad (w \leqslant \infty)$, with $\gamma \geqslant 0$ and $\sum_1^w |a_k^{-2}| < \infty$.

If d) holds, c) is equivalent to the following conditions:

$$\operatorname{Im}\beta + \sum_1^w \operatorname{Im}(a_k^{-1}) \geqslant 0, \qquad \operatorname{Im} a_k \geqslant 0.$$

It is known that $\mathcal{P}^*$ is the closure of the set of polynomials having all their zeros in $\operatorname{Im} z \geqslant 0$ (see [1], for example). So we have $\omega^{(k)} \in \mathcal{P}^*$ for $\omega \in \mathcal{P}^*$.

PROBLEM 2. *Prove that* $\omega \in \mathcal{P}^*$ *if all zeros of* $\omega^{(k)}$, $k = 0, 1, \ldots$ *are in the upper half-plane* $\operatorname{Im} z \geqslant 0$.

A similar problem can be formulated for entire functions of several variables. For simplicity we assume $n = 2$. A polynomial is called an $\overline{HB}$-polynomial if it has no zeros in $\Omega = \{(z,w) : \operatorname{Im} z < 0, \operatorname{Im} w < 0\}$. The closure of the set of $\overline{HB}$-polynomials will be denoted by $\mathcal{P}_2^*$ (the information about $\overline{HB}$-polynomials and about the class $\mathcal{P}_2^*$ can be found in [1], Ch.9).

PROBLEM 3. *Prove that an entire function* ω *belongs to* $\mathcal{P}_2^*$ *if this function and all its derivatives have no zeros in* Ω.

REFERENCES

1. Levin B. Ya., *Distribution of Zeros of Entire Functions*, AMS, 1980.
2. Polya G., *Sur une question concernant les fonctions entieres* C.R. Acad. Sci. Paris **158** (1914).
3. Levin B. Ya., Ostrovskii I. V., *On a dependence of growth of an entire function on the distribution of roots of its derivatives*, Sib. Mat. Zh. **1** (1960), no. 3, 427–455. (Russian)
4. Hellerstein S., Williamson J., *Derivatives of entire functions and a question of Polya*, Trans. Amer. Math. Soc. **227** (1977), 227–249.
5. Hellerstein S., Williamson J., *Derivatives of entire functions and a question of Polya*, Bull. Amer. Math. Soc. **81** (1975), 453–455.

DEPT. MATH. FTINT,
PR. LENINA 47,
KHARKOV, 310164,
UKRAINE

CHAPTER EDITORS' NOTE

The statement of Problem 1 was proved by T. Sheil-Small [6]. See also [7].

REFERENCES

6. Sheil-Small T., *On the zeros of derivatives of real entire functions and Wiman's conjecture*, Ann. of Math. **129** (1989), 173–193.
7. Sheil-Small T., *On the zeros of $L' + L^2$ for a certain rational functions L*, Proc. Amer. Math. Soc. **107** (1989), 1013–1016.

COMPLETELY REGULAR GROWTH OF ENTIRE SOLUTIONS
OF LINEAR DIFFERENTIAL EQUATION

A. A. GOLDBERG,   I. V. OSTROVSKII

Let

(1) $$w^{(n)} + a_{n-1}(z)w^{(n-1)} + \cdots + a_1(z)w' + a_0(z)w = 0$$

be a linear differential equation whose coefficients are entire functions of completely regular growth (c.r.g.). Is it true that if a solution $w(z)$ of (1) is of finite order then it has c.r.g.?

This problem was formulated by V. P. Petrenko [1, p.132] but with the condition of c.r.g. of the coefficients $a_j(z)$, $1 \leqslant j \leqslant n - 1$ omitted. Since the book [1] was published after the death of the author (who did not read the galley–proof), it is possible that the condition of c.r.g. of coefficients missed accidentally. Without this condition the answer is negative.

THEOREM (A. A. Goldberg). *Let f be an arbitrary entire function with zeros of order at most $n-1$. Then f is a solution of some differential equation (1) with entire coefficients.*

Proof. Let n_k be the order of the zero z_k of the function f. We define the numbers $A_{s,k}$, $1 \leqslant s \leqslant p_k$ in the following way. Put $A_{p_k,k} = -f^{(n)}(z_k)/f^{(p_k)}(z_k)$. Suppose that $A_{p_k,k}, A_{p_k-1,k}, A_{p_k-j+1,k}$, $j = 1, 2, \ldots, p_k - 1$ have been defined. Then we define

$$A_{p_k-j,k} = - \left(f^{(n+j)}(z_k) + A_{p_k,k}f^{(p_k+j)}(z_k) + \cdots + A_{p_k-j+1,k}f^{(p_k+1)}(z_k) \right) \Big/ f^{(p_k)}(z_k).$$

We define entire functions $a_1, \ldots a_{n-1}$ as the solutions of the following interpolation problem:

  (i) the functions $a_{p_k+1}, a_{p_k+2}, \ldots, a_{n-1}$ have zeros of order p_k at z_k,
  (ii) $a_1(z_k) = A_{1,k}, \ldots, a_{p_k}(z_k) = A_{p_k,k}$,
  (iii) the functions $a'_1, \ldots, a'_{p_k}$ have zeros of order $p_k - 1$ at z_k.

After the functions $a_1, \ldots, a_{n-1}$ have been defined, let us define a_0 by the equation

$$a_0(z) = - \left(f^{(n)}(z) + a_{n-1}(z)f^{(n-1)}(z) + \cdots + a_1(z)f'(z) \right) \Big/ f(z).$$

It follows from the previous conditions that a_0 is an entire function.

If the coefficients of equation (1) are polynomials, then it is known [1, ch.IV, 3] that all solutions of (1) are entire functions of c.r.g.

REFERENCE

1. Petrenko V. P., *Entire Curves*, "Visha shkola", Kharkov, 1984. (Russian)

DEPT. MATH., LVOV STATE UNIVERSITY
UNIVERSITETSKAYA UL. 1
290602, LVOV
UKRAINE

DEPT. MATH., FTINT
PR. LENINA 47
310164, KHARKOV
UKRAINE

OPERATORS PRESERVING
THE COMPLETELY REGULAR GROWTH

I. V. Ostrovskii

In [1] derivatives and the integral of the function f of c.r.g. were considered and the following result was obtained:

1) f' has c.r.g. on all rays $\arg z = \theta$ except maybe for θ with $h_f(\theta) = 0$;
2) the integral

$$F(z) = \int_0^z f(t)\, dt$$

has c.r.g. in $\mathbb{C}$.

Now consider instead of the operator $D = d/dz$ a more general operator $\varphi(D)$ of infinite order, where φ is an entire function of exponential type. *Is there a result similar to that in* [1] *in this case?* For an entire function f of c.r.g. and of order r, $r < 1$, the answer is given by the following theorem [2].

THEOREM. *Let f be an entire function of c.r.g. of order ρ, $\rho < 1$, and φ be an entire function of exponential type. Then:*

1) *the function $\varphi(D)f$ has c.r.g. on all rays, maybe except for the rays with $h_f(\theta) = 0$;*
2) *every solution F of the equation $\varphi(D)F = f$ in the class of entire functions of order* * ρ *has c.r.g. in $\mathbb{C}$.*

For entire functions f of order ρ, $\rho \geqslant 1$, one can expect no analog of [1] which is as simple as in the case $\rho < 1$. This can be seen from the following example. Let $\varphi(t) = t^n - 1$, $n \geqslant 3$, and let g be an entire function of exponential type σ, $\sigma < 1/2$ which is not a c.r.g. function. Denote by G a solution of the equation $\varphi(D)G = g$ in the class of entire functions of exponential type σ (the set of solutions is nonempty by Theorem IV of [3], p.361). Set

$$f(z) = G(z) + \sum_{k=0}^{n-1} \exp\left\{ e^{\frac{2\pi k i}{n}} z \right\}.$$

Then f is an entire function of exponential type, of c.r.g., and with positive indicator, whereas $g = \varphi(D)f$ is not of c.r.g.

For functions of exponential type the following conjecture seems plausible.

---

*The set of solutions of such kind is nonempty, see A. O. Gelfond [3], p.359.

CONJECTURE. *Let f be an entire function of exponential type, of c.r.g., and φ have no zeros at the points of the boundary of the conjugate diagram of f which are common endpoints of two segments on the boundary of the diagram. Then $\varphi(D)f$ has c.r.g. In particular, if φ has no zeros on the boundary of the conjugate diagram of f, then $\varphi(D)f$ has c.r.g.*

For a function f which grows faster than functions of exponential type the answer may be still more complicated, as solutions of the equation $\varphi(D)f = 0$ in this class may be not of c.r.g.

The following QUESTION remains unsolved too. Let f be an entire function of c.r.g. and of order ρ, $\rho \geqslant 1$, φ be an entire function of exponential type. *Are there solutions F of $\varphi(D)F = f$ which are entire functions of c.r.g. with respect to the same proximate order as the proximate order of f?* If φ is a polynomial, the affirmative answer is an easy consequence of the results in [1] and of the integral representation of the solution, however the general case does not follow by passing to the limit.

REFERENCES

1. Goldberg A. A., Ostrovskii I. V., *On the derivatives and integrals of entire functions of a completely regular growth*, Teor. funkt., funkt. anal. i prilozh., Kharkov **18** (1973), 70–81. (Russian)
2. Ostrovskii I. V., *Operators preserving completely regular growth*, Lect. Notes Math. **1043** (1984), 600–603.
3. Gelfond A. O., *Calculation of Finite Differences*, Nauka, Moscow, 1967. (Russian)

DEPT. OF MATH., FTINT
PR. LENINA 47
KHARKOV, 310164
UKRAINE

CHAPTER EDITORS' NOTE

The conjecture has been proved by O. V. Epifanov and Yu. F. Korobeinik [4]. For the functions of exponential type, Epifanov [5] answered the last question in the affirmative.

REFERENCES

4. Epifanov O. V., Korobeinik Yu. F., *On preservation of completely regular growth by differential operators of infinite order*, Teor. funkt., funkt. anal. i prilozh., Kharkov, **47** (1987), 85–89. (Russian)
5. Epifanov O. V., *On preservation of non-completely regular growth by convolution operators*, Sib. Mat. Zh. **20** (1979), 420–422. (Russian)

PHRAGMEN–LINDELÖF INDICATORS
OF ENTIRE FUNCTIONS WITH REAL ZEROS

I. V. Ostrovskii,   M. L. Sodin

One of the commonly used classes of entire functions of exponential type is the class A which consists of functions whose zeros $\{z_k\}$ satisfy the Blaschke condition

$$\sum |\operatorname{Im} \frac{1}{z_k}| < \infty$$

(see [1], Ch. V). It is known that the indicator of growth

$$h_f(\theta) = \limsup_{r \to \infty} \frac{\log |f(re^{i\theta})|}{r}$$

of a function $f \in A$ satisfies the equation

(1) $$h_f(\theta) = h_f(-\theta) - \gamma \sin \theta,$$

or, in other words, the indicator diagram of $f(z)$ has the axis of symmetry $y = -\gamma/2$, which is parallel to the real axis (see [1], Ch. V, Sec. 5).

PROBLEM 1. *Describe the indicators or (what is the same) the indicator diagrams of entire functions of the class* A.

It is easy to see that this question is equivalent to the analogous question for functions with real zeros. It was quite surprising for the authors to discover that they do not know the answer to this question. Now consider the lower indicator

$$\underline{h}_f(\theta) = \liminf_{r \to \infty} \frac{\log |f(re^{i\theta})|}{r}, \qquad \theta \neq 0, \pi.$$

It also satisfies (1).

PROBLEM 2. *Describe the lower indicators of entire functions of exponential type with real zeros.*

Other questions are variations on the theme of the previous ones.

PROBLEM 3. *Describe the indicators of the following classes of entire functions of finite order:*
a) *functions of positive order* ρ *with negative zeros;*
b) *functions of positive order* ρ *with real zeros;*
c) *functions of positive order* $\rho < 2$ *without zeros in the open lower half-plane and such that* $|f(z)| < |f(\bar{z})|$, $\operatorname{Im} z > 0$. (*In the case* $\rho < 1$ *the first condition implies the second one.*)

It is possible that some information on this matter can be extracted from the known estimates of indicators in terms of upper and lower densities of zeros (see references in [1], Appendix VIII, Sec. 7).

REFERENCE

1. Levin B. Ya., *Distribution of Zeros of Entire Functions*, AMS, 1980.

DEPT. OF MATH., FTINT
PR. LENINA 47, KHARKOV, 310164
UKRAINE

DEPT. OF MATH., FTINT
PR. LENINA 47, KHARKOV, 310164
UKRAINE

ESTIMATES OF INDICATORS FROM BELOW

I. V. Ostrovskii,   M. L. Sodin

After publication of the large paper [1] by A. A. Goldberg and the paper [2] by E. D. Fainberg it might be regarded that, essentially, there remained no unsolved problems concerning estimates of the indicator of an entire function with prescribed zeros. To our opinion such problems exist. Here is one of them.

PROBLEM 1. *Denote by* $F(\Delta, \rho)$ *the class of all entire functions* f *of order* $0 < \rho < 1$ *whose zeros are positive and have the upper density*

$$\Delta = \limsup_{r \to \infty} r^{-\rho} n(r, 0, f).$$

Find the value

$$(1) \qquad H(\varphi, \Delta, \rho) = \inf\{ h(f, \varphi) : f \in F(\Delta, \rho) \},$$

where $h(f, \varphi) = \limsup_{r \to \infty} r^{-\rho} \ln |f(re^{i\varphi})|$ *is the indicator of the function* $f(z)$.

The following problem is more difficult.

PROBLEM 2. *Denote by* $F(\Delta, \delta, \rho)$ *the subclass of* $F(\Delta, \rho)$ *consisting of functions with lower density of zeros*

$$\delta = \liminf_{r \to \infty} r^{-\rho} n(r, 0, f).$$

Find the value

$$(2) \qquad H(\varphi, \Delta, \delta, \rho) = \inf\{ h(f, \varphi) : f \in F(\Delta, \delta, \rho) \}.$$

It is not difficult to find the values given by (1) and (2) for $\varphi \in [\frac{\pi}{2}, \frac{3\pi}{2}]$ (for example, $H(\varphi, \Delta, \rho) = \Delta \max\{ t^{-\rho} \ln |1 - te^{i\varphi}| : t > 0 \}$), so we are interested in finding these values only for $\varphi \in (-\frac{\pi}{2}, \frac{\pi}{2})$.

There are various generalizations of these problems. For example, it is possible to give up the condition $0 < \rho < 1$ and to replace the order by a proximate order. A more difficult generalization might be connected with omitting the condition for the zeros to be positive and with considerations, as in [1], of classes of functions with prescribed upper and lower angular density of zeros. Apparently, to study this question we need a notion of integral with respect to a semiadditive measure. It is possible that this notion will be different from that introduced in [1]. Finally, one more direction of generalization is connected with estimates of the lower indicator.

References

1. Goldberg A. A., *An integral with respect to a semiadditive measure and its application to the theory of entire functions*, I, Mat. Sbornik **58(100)** (1962), 289–334; part II, ibid. **61(103)** (1962); part III, ibid. **65(107)** (1964); part IV, ibid. **66(108)** (1965) (Russian); English transl. in Amer. Math. Soc. Transl. **88(2)** (1970), 105–289.
2. Fainberg E. D., *On the integral with respect to non-additive measure and on estimates of the indicator of entire function*, Sib. Mat. Zh. **24** (1984), 175–186. (Russian)

DEPT. MATH., FTINT,
PR. LENINA 47,
KHARKOV, 310164,
UKRAINE

DEPT. MATH., FTINT,
PR. LENINA 47,
KHARKOV, 310164,
UKRAINE

PHRAGMEN-LINDELÖF PRINCIPLE
IN CURVILINEAR STRIP

A. A. Goldberg, I. V. Ostrovskii

Despite the existence of a rather extensive literature devoted to the subject mentioned in the title, the following theorem seems to be new.

THEOREM 1. *Let us suppose that a function v is subharmonic in the curvilinear strip*

$$(1) \qquad S = \{ z = x + iy \,:\, y(x) < y < y(x) + 2\pi, \ -\infty < x < \infty \},$$
$$y(0) = 0, \ y(x) \in C(\mathbb{R}),$$

is bounded from above in $S \cap \{ z : \operatorname{Re} z < R \}$ for every $R > 0$ and satisfies

$$(2) \qquad \limsup_{z \in S, z \to \xi} v(z) \leqslant 0$$

for every $\xi \in \partial S$. If

$$(3) \quad \sup \{ v(R + iy) \,:\, y(R) < y < y(R) + 2\pi \}$$
$$= o \left\{ \exp \frac{R^2 + y^2(R) - 4\pi |y(R)|}{2R} \right\}, \quad R \to \infty,$$

then $v(z) \leqslant 0$ for every $z \in S$.

This result was obtained by the first author and was published in [1]. The second author obtained (see [1]) the following modification which was useful in the theory of the Riemann boundary problem with infinite index.

THEOREM 2. *Under the additional assumption of smoothness of the boundary of S ($y(x) \in C^1(\mathbb{R})$), Theorem 1 remains true if we replace condition (3) by the following one. There exists $p > 1$ such that*

$$(4) \quad \left\{ \int_{y(R)}^{y(R)+2\pi} \left(v^+(R+iy) \right)^p dy \right\}^{1/p}$$
$$= o \left\{ R^{1/p} \exp \left(\frac{p-1}{p} \frac{R^2 + y^2(R) - 4\pi |y(R)|}{2R} \right) \right\}, \quad R \to \infty.$$

We note that Theorems 1 and 2 are proved in [1] under the additional assumption $v(z) = \log |f(z)|$, the function f being analytic in S, but passing to the general case doesn't meet any difficulties.

Would it be possible to omit the additional assumption $y(x) \in C^1(\mathbb{R})$, Theorem 1 would follow from Theorem 2 by formal passing to the limit as $p \to +\infty$. Nevertheless formal passing to the limit as $p \to 1$ leads most likely to a too weak result.

PROBLEM. *Does the statement of Theorem 2 remain true if condition (4) is replaced by the following one?*

There exists $p \geqslant 1$ such that

$$(5) \quad \left\{ \int_{y(R)}^{y(R)+2\pi} \left(v^+(R+iy)\right)^p dy \right\}^{1/p}$$

$$= o\left\{ R^{1/p} \exp\left(\frac{R^2 + y^2(R) - 4\pi|y(R)|}{2R} \right) \right\}, \quad R \to \infty.$$

If we compare the last condition for $p = 1$ with the condition in the well-known extension of the Phragmen-Lindelöf principle due to R. Nevanlinna (an essential amplification of Nevanlinna's result was obtained recently in [2]), then we can conjecture that condition (5) can even be weakened in the following way. The quantity on the left-hand side of (5) can be replaced by

$$\left\{ \int_{y(R)}^{y(R)+2\pi} \left(v^+(R+iy)\right)^p d\mu_R(y) \right\}^{1/p},$$

where μ_R is a non-negative Borel measure on the segment $[y(R), y(R) + 2\pi]$ which can rather rapidly tend to zero near its ends (for example, $\mu_R([y(R), y(R)+\varepsilon] \cup [y(R)+2\pi-\varepsilon, y(R) + 2\pi]) = O(\exp(-1/\sqrt{\varepsilon}))$, $\varepsilon \to 0$, for every fixed $R > 0$).

REFERENCES

1. Ostrovskii I. V., *Conditions of solvability of homogeneous Riemann boundary problem with infinite index*, Teor. funkt., funkt. anal. i prilozh., Kharkov **55** (1991), 3–23. (Russian)
2. Rashkovskii A. Yu., *Majorants of harmonic measures and uniform boundedness of families of subharmonic functions*, "Analytic methods in probability theory and operator theory", Naukova dumka, Kiev, 1990, pp. 115–126. (Russian)

LVOV STATE UNIVERSITY
UNIVERSITETSKAYA 1
LVOV 290602
UKRAINE

DEPARTMENT OF MATHEMATICS
FTINT, PR. LENINA 47
KHARKOV 310164
UKRAINE

CONNECTION BETWEEN ARGUMENTS
OF ZEROS AND LACUNARITY

A. A. Goldberg, I. V. Ostrovskii

Let

$$f(z) = \sum_{k=0}^{\infty} a_k z^{n_k}, \qquad a_k > 0,$$

be an entire function with infinite set of zeros $r_k \exp(i\theta_k)$, $\theta_k \in [-\pi, \pi]$. Put

$$A(f) = \lim_{k \to \infty} \sup(n_{k+1} - n_k),$$

$$\delta(f) = \lim_{k \to \infty} \inf |\theta_k|,$$

$$b(f) = \lim_{r \to \infty} \sup(d/d\log r)^2 \log f(r).$$

The following inequalities can be found, respectively, in [1] and [2]:

(1)
$$b(f) \geqslant \frac{1}{4} A^2(f),$$

$$b(f) \geqslant \frac{1}{4} \csc^2\big(\delta(f)/2\big).$$

The desire to establish the connection between these results has led us to the following problem we are unable to solve.

Problem 1. *Put $G(n) = \{\, f : A(f) = n \,\}$, $n = 2, 3, \dots$. Compute the quantity*

$$g(n) = \sup\{\, \delta(f) : f \in G(n) \,\}, \qquad n \geqslant 2.$$

We do not even know whether the inequality $g(n) < \pi$ $(n > 2)$ is true. Since $\limsup \log f(r)/(\log r)^2 = \infty$ implies that $b(f) = \infty$, in the above question about connection between inequalities (1) we may restrict ourselves to the functions f with

(2)
$$\limsup_{r \to \infty} \log f(r)/(\log r)^2 < \infty.$$

This leads us to the following modification of Problem 1.

Problem 2. *Let $\widetilde{G}(n)$ be the subclass of $G(n)$ which consists of the functions f satisfying (2). Compute the quantity*

$$\tilde{g}(n) = \sup\{\, \delta(f) : f \in \widetilde{G}(n) \,\}, \qquad n \geqslant 2.$$

The second author has shown (unpublished) that $\tilde{g}(2) = \pi/2$. The values $\tilde{g}(n)$, $n \geqslant 3$, are unknown. The simple examples $f_n(z) = \prod_{k=1}^{\infty}(1 + z^n e^{-k})$ show that $\tilde{g}(n) \geqslant \pi/n$.

Problem 1 may be generalized in the following way.

Compute the quantity

$$g_\rho(n) = \sup\{\, \delta(f) : f \in G_\rho(n) \,\}, \qquad n \geqslant 2,$$

where $G_\rho(n)$ is the set of entire functions of order ρ.

PROBLEM 16.7

REFERENCES

1. Boichuk V. S., Goldberg A. A., *On the "three-lines" theorem*, Mat. Zametki **15** (1974), 45–53. (Russian)
2. Abi-Khuzam F. F., *Maximum modulus convexity and the location of zeros of an entire function*, Proc. Amer. Math. Soc. **106** (1989), 1063–1068.

LVOV STATE UNIVERSITY
UNIVERSITETSKAYA 1
LVOV, 290602
UKRAINE

DEPT. MATH., FTINT
PR. LENINA 47
KHARKOV, 310164
UKRAINE

CLUSTER SETS AND A PROBLEM OF LEONT'EV

V. S. Azarin,   A. E. Eremenko.   A. F. Grishin

1. Let f be an entire function of order ρ and normal type (this will be denoted $f \in A(\rho)$) and let $h_f(\varphi)$ be its indicator of growth. As it is known, $f' \in A(\rho)$ and $h_{f'}(\varphi) = h_f(\varphi)$ for $\varphi \in [0, 2\pi)$ such that $h_f(\varphi) > 0$. Let $A_{\text{reg}}(\rho) \subset A(\rho)$ be the subclass of functions of c.r.g. As it was proved in [2], if $f \in A_{\text{reg}}(\rho)$ and

$$(1) \qquad\qquad h_f(\varphi) > 0, \qquad \forall \varphi \in [0, 2\pi)$$

then $f' \in A_{\text{reg}}(\rho)$ too.

The following condition plays an important role in questions concerning interpolation by entire functions (see, for example, [1, p.202] and [3, p.156]):

$$(2) \qquad\qquad \liminf_{k \to \infty} \big[|\lambda_k|^{-\rho} \log |f'(\lambda_k)| - h_f(\arg \lambda_k) \big] \geqslant 0,$$

where λ_k are zeros of f and it is assumed that they are simple.

In the case $\rho = 1$ significant for applications A. F. Leont'ev proved that condition (2) with the additional condition

$$\limsup_{r \to \infty} r^{-1} \big(\min_{\varphi} \log |f(re^{i\varphi})| \big) > 0$$

implies $f \in A_{\text{reg}}(1)$. He also posed the question (see [5] and this collection) *whether the last assertion remains true if we assume only* (1) *and* (2) *without this additional condition.*

The following more general question appears to be natural: *whether* (1) *and* (2) *imply* $f \in A(\rho)$? In mathematical folklore this question was named Leont'ev's problem.

The above result by A. F. Leont'ev was generalized by Yu. I. Mel'nik [6] and A. V. Bratishchev [7]. Like Leont'ev, these authors used methods of interpolation theory. In particular, it was proved in the latter work that condition (2) implies $f \in A_{\text{reg}}(\rho)$ for $\rho < 1/2$ and examples of functions $f \in A(\rho)$, $1/2 \leqslant \rho \leqslant 1$ were constructed showing that condition (1) in Leont'ev's theorem is essential.

2. In [4] the notion of regular growth on a set was introduced and it was proved that (2) is equivalent to the fact that f' has regular growth on the set of zeros of f.

Later on we use some notions from [8]. Let $\operatorname{Fr} f$ be the limit set of a function $f \in A(\rho)$ and let

$$h(z) := \sup\{ v(z) : v \in \operatorname{Fr} f \}.$$

Then $h_f(\varphi) = h(e^{i\varphi})$. Denote by μ_v the Riesz measure of a subharmonic function v.

Theorem 1 ([4]). *If* $f \in A(\rho)$ *then condition* (2) *implies that for every function* $v \in \operatorname{Fr} f$ *one has* $v(z) = h(z)$ *everywhere on the set* $\operatorname{supp} \mu_v$, *and* $\mu_v \leqslant \mu_h$.

This theorem helped to prove some assertions connected with Leont'ev's problem.

3. Consider a subharmonic function $v(z)$ satisfying the identity

$$v(Tz) = T^\rho v(z), \qquad \forall z \in \mathbb{C}$$

for a certain number $T > 1$. We shall call it a self-similar function. The limit set generated by this function is of the form

$$\operatorname{Fr} f = \{ v(tz)t^{-\rho} : 1 \leqslant t \leqslant T \}.$$

It is called periodic limit set [8].

THEOREM 2 (V. S. Azarin, V. B. Giner, L. R. Podoshev). *Suppose $f \in A(\rho)$ satisfies conditions (1) and (2) and let $1/2 \leqslant \rho \leqslant 1$. Then every self-similar function $v \in \operatorname{Fr} f$ such that $v(z) = h(z)$ for some point $z \neq 0$ coincides identically with $h(z)$. If $\rho > 1$ then the same conclusion is true for every self-similar function satisfying the condition*

$$\min_\varphi v(re^{i\varphi}) \geqslant 0$$

for a certain r.

If $\operatorname{Fr} f$ consists of self-similar functions, in particular if it is periodic, then the assertion of Theorem 2 implies that $f \in A_{\mathrm{reg}}(\rho)$.

4. In A. V. Bratishchev's example mentioned above the limit set $\operatorname{Fr} f$ has the form $\operatorname{Fr} f = \{ v(re^{i\varphi}) = \tau r^\rho \cos^+ \rho\varphi : c_1 \leqslant \tau \leqslant c_2 \}$, $c_1 \geqslant 0$, $1/2 \leqslant \rho \leqslant 1$.

Let $\rho > 1$. Consider a 2π-periodic even ρ-trigonometric convex function defined as follows: $h(\varphi) = 1$ for $0 \leqslant \varphi < \Delta$ and $h(\varphi) = \cos \rho(\varphi - \Delta)$ for $\Delta \leqslant \varphi \leqslant \pi$, where $\Delta = \pi(1 - 1/\rho)$. Then, as was proved by V. S. Azarin, V. B. Giner and L. R. Podoshev, there exists a function $f \in A(\rho)$ with the limit set

$$\operatorname{Fr} f = \{ v(re^{i\varphi}) = h(\varphi - \tau)r^\rho : \tau \in [0, 2\pi] \}$$

satisfying conditions (1) and (2) which is not of completely regular growth.

5. DEFINITION. A limit set consisting of functions of the form $h(\varphi)r^\rho$ where $h(\varphi)$ is a 2π-periodic ρ-trigonometric convex function will be called an indicator limit set.

Entire functions with indicator limit set can be considered as a generalization of functions of completely regular growth.

CONJECTURE. *If $f \in A(\rho)$ satisfies (2) then $\operatorname{Fr} f$ is an indicator limit set.*

A periodic limit set $\operatorname{Fr} f$ can be an indicator only if $f \in A_{\mathrm{reg}}(\rho)$. Note that the indicator set consists of self-similar functions. Hence the affirmative answer to the question of A. F. Leont'ev (in the case $\rho = 1$) follows from the Conjecture and Theorem 2.

The examples of Section 4 do not contradict this Conjecture.

6. The following assertion also supports Conjecture

THEOREM 3 (A. F. Grishin). *Let a function $f \in A(1)$ satisfy condition (2) and $n_f(r) = o(1)$ as $r \to \infty$. Then*

$$\operatorname{Fr} f = \{ v(re^{i\varphi}) = [tB_1 \cos(\varphi - \varphi_1) + (1 - t)B_2 \cos(\varphi - \varphi_2)]r, \, 0 \leqslant t \leqslant 1 \}$$

for some φ_1, φ_2.

Chapter 16. ENTIRE AND SUBHARMONIC FUNCTIONS

REFERENCES

1. Levin B. Ya., *Distribution of Zeros of Entire Functions*, AMS, 1980.
2. Goldberg A. A., Ostrovskii I. V., *On derivatives and primitives of entire functions of completely regular growth*, Teor. funkt., funkt. anal. i prilozh. (Kharkov) **18** (1972), 185–201. (Russian)
3. Leont'ev A. F., *Entire Functions*, Exponential series, FML, Moscow, 1983. (Russian)
4. Grishin A. F., *On sets of regular growth of entire functions*, Teor. funkt., funkt. anal. i prilozh. (Kharkov) **40** (1983), 37–47. (Russian)
5. Leont'ev A. F., *On conditions of representability of analytic functions by Dirichlet series*, Izv. AN SSSR **36** (1972), 1282–1285. (Russian)
6. Mel'nik Yu. I., *On conditions of convergence of Dirichlet series representing regular functions*, Mat. anal. i teor. veroyatn., Nauk. Dumka, Kiev, 1978, pp. 120–123. (Russian)
7. Bratischev A. V., *On a problem of A. F. Leont'ev*, Dokl. Akad. Nauk SSSR **270** (1983), 265–267. (Russian)
8. Azarin V. S., *Limit sets of entire and subharmonic functions*, Itogi Nauki i Tehn. Sovr. Probl. Matem. Fundament. Napravl., vol. 85, VINITI, Moscow, 1991, pp. 52–68. (Russian)

KHARKOV INSTITUTE OF RAILROAD ENGINEERS,
KHARKOV 310050
UKRAINE

DEPARTMENT OF MATHEMATICS
PURDUE UNIVERSITY
WEST LAFAYETTE, IN 47907
USA

AND

DEPARTMENT OF MATHEMATICS
FTINT, PR. LENINA 47
KHARKOV 310164
UKRAINE

MATH. DEPT., KHARKOV STATE UNIVERSITY,
KHARKOV 310077
UKRAINE

MINIMAL SUBHARMONIC FUNCTION

V. S. AZARIN

1. Denote by $U[\rho, \sigma]$ the class of subharmonic functions $v(z)$, $z \in \mathbb{C}$, which satisfy the conditions: $v(0) = 0$; $v(z) \leqslant \sigma|z|^\rho$, $\forall z \in \mathbb{C}$. We shall denote $U[\rho] = \bigcup_{\sigma > 0} U[\rho, \sigma]$.

A function $v \in U[\rho]$ is called minimal if for an arbitrarily small $\varepsilon > 0$ the function $v(z) - \varepsilon|z|^\rho$ has no subharmonic minorants which belong to $U[\rho]$.

We formulate a conjecture concerning minimality and discuss it.

2. Let

$$L_\lambda = \frac{\partial^2}{\partial x^2} + \frac{\partial^2}{\partial y^2} + 2\lambda \frac{\partial}{\partial x} + \lambda^2$$

and let the notation $r(x, y)$, $(x, y) \in \mathbb{R} \times \mathbb{T}$, mean that $r(x, y)$ is 2π-periodic in y. Let us consider a homogeneous boundary value problem of the form

$$(1) \qquad L_\lambda r = 0; \quad (x, y) \in \mathcal{D} \subset \mathbb{R} \times \mathbb{T}; \quad r|_{\partial \mathcal{D}} = 0,$$

where $r(x, y)$ is bounded for $(x, y) \in \mathcal{D}$.

This problem has only trivial solution for $\lambda = 0$ because L_0 is the Laplace operator.

Let $\lambda(\mathcal{D})$ denote the infimum of those λ for which this problem has a nontrivial solution. If it is necessary (if $\partial \mathcal{D}$ is too bad), we understand the problem in some generalized sense.

3. Let

$$(2) \qquad \widehat{v}(x, y) = v(e^{x+iy})e^{-\rho x}, \qquad v \in U[\rho].$$

One can see that $\widehat{v}(x, y)$, $(x, y) \in \mathbb{R} \times \mathbb{T}$ is bounded from above. Moreover, it is subelliptic (see [1], p. 103) with respect to the operator L_λ for $\lambda = \rho$.

Every open set $H(v)$ of harmonicity of $v(z)$ is transformed to the set $\widehat{H} \subset \mathbb{R} \times \mathbb{T}$, where $L_\rho \widehat{v} = 0$. Then $\lambda(v) := \lambda(\widehat{H}(v))$ can be considered as a "measure" of the maximal connected component of the set $\widehat{H}(v)$.

CONJECTURE. *The following inequality is necessary and sufficient for minimality of* $v \in U[\rho]$:

$$(3) \qquad \lambda(v) \leqslant \rho.$$

Discussion. Let $h(\varphi)$ be a ρ-trigonometrical convex function (see [2, p. 53]) and let $v = h(\varphi)r^\rho$, $z = re^{i\varphi}$. It can be seen that $v \in U[\rho]$ and the set $H(v)$ consists of the angular domains $A_j = \{\arg z \in I_j \subset [0, 2\pi)\}$ such that

$$h(\varphi) = B_j \cos \rho(\varphi - \varphi_j), \qquad \varphi \in I_j.$$

THEOREM 1. *A criterion of minimality of $v = h(\varphi)r^\rho$ is the inequality*

$$\max_j \operatorname{mes} I_j \geqslant \frac{\pi}{\rho}.$$

This theorem may be proved directly or derived from Conjecture.

A function $\widehat{v}$ defined by (2) can be doubly periodic (not only in y but in x as well). This means that it is defined on the two-dimensional torus $\mathbb{T}^2$.

THEOREM 2. *Let $v \in U[\rho]$ and $\widehat{v}$ be doubly periodic. Then (3) is sufficient for minimality of v.*

Sketch of the proof. Suppose that $\exists w \in U[\rho]$, $\exists \varepsilon > 0$ such that $w(z) \leqslant v(z) - \varepsilon|z|^\rho$, $z \in \mathbb{C}$. It can be shown that there exists $w \in U[\rho]$ such that $\widehat{w}$ is doubly periodic and $\widehat{w}(x, y) \leqslant \widehat{v}(x, y) - \varepsilon$, $(x, y) \in \mathbb{T}^2$. Suppose $\lambda(v) \leqslant \rho$. We can find $\mathcal{D} \subset \widehat{H}(v)$ such that $\lambda(\mathcal{D}) = \rho$. Denote by R_0 any solution of problem (1). It preserves its sign in $\mathcal{D}$. The difference $\widehat{r} = \widehat{w} - \widehat{v}$ is negative in $\mathcal{D}$. One can find a constant K such that the function $\widehat{r} + KR_0$ is negative in $\mathcal{D}$ and it vanishes in some inner point of $\mathcal{D}$. But this contradicts to the maximum principle for such functions. $\quad\square$

It is important for this proof that the torus is compact.

It was shown in [3] that if $\rho = 1$ and v is nonnegative and doubly periodic, then condition (3) is necessary.

THEOREM 3 (M. L. Sodin, private communication). *Let $v \in U[\rho]$ have no masses in the angle $\{|\arg z| < \frac{\pi}{2\rho}\}$. Then v is a minimal function.*

This theorem can be derived from Conjecture too.

REFERENCES

1. Kondrat'ev V. A., Landis E. M., *Partial differential equations*, Modern Problems of Mathematics, vol. 32, VINITI, Moscow, 1988, pp. 99–216. (Russian)
2. Levin B. Ya., *Distribution of Zeros of Entire Functions*, AMS, 1980.
3. Azarin V. S., Giner V. B., *Limit sets of entire functions and completeness of exponentials*, Teor. funkt., funkt. anal.i prilozh. (to appear). (Russian)

KHARKOV INSTITUTE OF THE RAILROAD ENGINEERS,
KHARKOV, 310050
UKRAINE

ON APPROXIMATION OF A SUBHARMONIC FUNCTION
BY LOGARITHM OF THE MODULUS OF AN ANALYTIC ONE

M. L. Sodin

The approximation of subharmonic functions in $\mathbb{C}$ by logarithms of moduli of entire functions has lots of applications in Complex Analysis. For functions of finite order the first general result was obtained by V. S. Azarin in [1]. The best result belongs to R. S. Yulmukhametov [2]. In some special cases, when the subharmonic function in question is homogeneous, better results were obtained in [3] and [4].

QUESTION 1. *Is it true that for every subharmonic function $v(z)$, $z \in \mathbb{C}$ with an infinite Riesz measure one can find an entire function $f(z)$ such that*

$$\frac{1}{R^2} \int_0^{2\pi} \int_0^R \left| v(re^{it}) - \log|f(re^{it})| \right| r \, dr \, dt = O(1), \qquad R \to \infty.$$

The second question is connected with the previous one but also is of independent interest.

QUESTION 2. *Let μ be a probability measure supported by the square $Q = \{ z = x + iy : |x| \leqslant 1/2, |y| \leqslant 1/2 \}$. Is it possible to find a sequence of polynomials P_n, $\deg P_n = n$, such that*

$$\iint\limits_{|x| \leqslant 1, |y| \leqslant 1} \left| nU_\mu(z) - \log|P_n(z)| \right| dx \, dy = O(1),$$

as $n \to \infty$? Here $U_\mu(z) = \iint \log|z - \zeta| \, d\mu(\zeta)$ is the logarithmic potential of the measure μ.

REFERENCES

1. Azarin V. S., *On rays of completely regular growth of an entire function*, Math. USSR Sb. **8** (1969).
2. Yulmukhametov R. S., *Approximation of subharmonic functions*, Anal. Math. **11** (1985), 257–282. (Russian)
3. Lyubarskii Yu. I., Sodin M. L., *Analogues of functions of sine-type for convex domains*, Preprint of FTINT AN Ukr.SSR 17–86, Kharkov, 1986. (Russian)
4. Yulmukhametov R. S., *Approximation of homogeneous subharmonic functions*, Math. USSR Sb. **62** (1989), 507–529.

DEPT. OF MATH., FTINT
PR. LENINA 47
KHARKOV, 310164
UKRAINE

16.11

ON THE INVERSE PROBLEM OF BEST APPROXIMATION
OF ALMOST PERIODIC FUNCTIONS
BY ENTIRE FUNCTIONS OF EXPONENTIAL TYPE

M. I. KADEC

Let $B = B(\mathbb{R})$ be the Banach space of all bounded and uniformly continuous functions on $\mathbb{R}$. Let B_σ be its closed subspace which consists of entire functions of exponential type $\leqslant \sigma$. Define the distance between $f \in B$ and B_σ by

$$(2) \qquad A(f, \sigma) = \inf\{ \|f - g\| : g \in B_\sigma \}, \qquad 0 \leqslant \sigma < \infty.$$

For every fixed f the function $A(f, \sigma)$ in the variable σ satisfies the following properties:

1) $A(f, \rho) \geqslant A(f, \sigma)$ for $\rho < \sigma$;
2) $A(f, \sigma) = A(f, \sigma + 0)$;
3) $\lim\limits_{\sigma \to \infty} A(f, \sigma) = 0$.

In [1] the question was posed whether for every function $F(\sigma)$ satisfying 1)–3) there exists a function $f \in B$ such that $A(\sigma, f) \equiv F(\sigma)$, as well as some related questions.

This question, as it was proved by A. Ya. Gordon [1, Theorem 1], has an affirmative answer. Moreover, the quotient space $B_\sigma/B_{\sigma-0}$ ($B_{\sigma-0}$ is the closure of the union of all B_ρ for $\rho < \sigma$ in the space B) is infinite dimensional (even non-separable). This fact seems to give us very many degrees of freedom to choose the desired function $f \in B$. But the following problem remains unsolved.

Let $AP = AP(\mathbb{R})$ be the Banach space of all Bohr almost periodic functions with the usual sup-norm and let AP_σ be its subspace consisting of the entire functions of exponential type $\leqslant \sigma$ ($AP_\sigma = AP \cap B_\sigma$).

PROBLEM. *Suppose $F(\sigma)$ satisfies conditions 1)–3) and*

4) *F is a jump function.*

Is it true that there exists a function $f \in AP$ such that $A(f, \sigma) = F(\sigma)$?
Is it true that $A(f, \sigma)$ satisfies conditions 1)–4) for every function $f \in AP$?

REFERENCE

1. Kadec M. I., *The inverse problem of best approximation of bounded uniformly continuous functions by entire functions of exponential type*, Lect. Notes Math. **1043** (1984), 591–594.

INSTITUTE OF MUNICIPAL ECONOMY
KHARKOV, 310002
UKRAINE

SOME PROBLEMS ABOUT
UNBOUNDED ANALYTIC FUNCTIONS

JAMES LANGLEY,   LEE A. RUBEL

A famous theorem of Iversen [1, p.284a] says that if f is a transcendental entire function then there is a path γ along which f tends to ∞. Thinking about this theorem has led us to formulate the following four problems, which we would like to solve, but cannot yet solve.

PROBLEM 1. *If f is a transcendental entire function, must there exist the path γ along which* **every** *derivative of f tends to ∞? Short of that, how about just having both f and f' tend to infinity along γ?*

It follows, say, by Wiman–Valiron theory, that if f is a transcendental entire function, then there exists a **sequence** $\{z_k\}$ such that $f^{(k)}(z_n) \to \infty$ as $n \to \infty$ for each $k = 0, 1, 2, \ldots$, but to obtain a **path** on which this happens seems much more difficult.

PROBLEM 2. *If f is an unbounded analytic function in the open disk $\mathbb{D}$, must there exist a sequence $\{z_n\}$ of points of $\mathbb{D}$ such that for every $k = 0, 1, 2, \ldots$, $f^{(k)}(z_n) \to \infty$ as $n \to \infty$? Short of that, how about just having $f(z_n) \to \infty$ and $f''(z_n) \to \infty$?*

The authors have shown that one can always find $\{z_n\}$ so that both $f(z_n) \to \infty$ and $f'(z_n) \to \infty$. If f grows fast enough, i.e., if either

$$(1) \qquad \limsup_{r \to 1-} \frac{\log \log M(r, f)}{-\log(1-r)} = \rho > 0$$

or

$$(2) \qquad \limsup_{r \to 1-} \frac{\log T(r, f)}{\log \log \left(\frac{1}{1-r}\right)} = \infty$$

then we can show that there exists a $\{z_n\}$ with $f^{(k)}(z_n) \to \infty$ as $n \to \infty$ for $k = 0, 1, 2, \ldots$, but we do not know how necessary these growth assumptions are. Note that conditions (1) and (2) are not strictly comparable. The proof involving (2) uses some Nevanlinna theory.

PROBLEM 3. *Is there an easy elementary proof, using neither Wiman–Valiron theory nor advanced Nevanlinna theory, that if f is a transcendental entire function then there exists a sequence $\{z_n\}$ such that $f^{(k)}(z_n) \to \infty$ as $n \to \infty$ for $k = 0, 1, 2 \ldots$?*

The following possibility is suggested by many examples: the differential equation

$$P_0(y)P_1(y')\ldots P_n(y^{(n)}) = 1,$$

where the P_j are polynomials in one variable, not all constants, has no solutions $y(z)$ that are analytic and unbounded in the unit disk $\mathbb{D}$. For example, the equation $yy'' = 1$ can be solved explicitly by means of the standard substitution $v = y'$, $v^{\cdot} = v\frac{dv}{dy}$, and it is easily seen that it has no unbounded analytic solution in any disk, lending support to the above hypothetical statement. The question remains:

PROBLEM 4. *Whether that statement is true or not.*

REFERENCE

1. Titchmarsh E. C., *The Theory of Functions*, 2nd Edition, Oxford, 1932.

DEPARTMENT OF MATHEMATICS
UNIVERSITY OF ILLINOIS,
1409 WEST GREEN STR.
URBANA, IL 61801-2917
USA

CHAPTER EDITORS' NOTE

Recently A. Gordon has obtained the affirmative answer to Problems 2.3.4.

THEOREM 1 (A. Gordon). *Let f be an unbounded analytic function in the open disk D. Then there exists a sequence $\{z_n\}$ of points of D such that for every $k = 0, 1, \ldots$ we have $f^{(k)}(z_n) \to \infty$ as $n \to \infty$.*

Proof. It is sufficient to show that for every $k \in \mathbb{Z}_+$ and $M > 0$ there exists a point $\zeta \in D$ such that $|f^{(j)}(\zeta)| \geqslant M$ for every $j = 0, 1, \ldots, k$. Supposing the contrary, let us consider a smallest k for which it is false (evidently $k \geqslant 1$). Then for some $M > 0$ the inequalities

$$|f^{(j)}(\zeta)| \geqslant M, \quad j = 0, 1, \ldots, k-1, \tag{1}$$

yield

$$|f^{(k)}(\zeta)| < M. \tag{2}$$

Moreover, there exists a sequence $\{z_m\} \subset D$ such that

$$f^{(j)}(z_m) \to \infty, \quad j = 0, 1, \ldots, k-1, \quad m \to \infty. \tag{3}$$

Let us decompose f in the sum $f(z) = T_m(z) + R_m(z)$, where

$$T_m(z) = f(z_m) + \frac{f'(z_m)(z - z_m)}{1!} + \ldots + \frac{f^{(k-1)}(z_m)(z - z_m)^{k-1}}{(k-1)!},$$

$$R_m(z) = \frac{1}{(k-1)!} \int_{z_m}^{z} f^{(k)}(\zeta)(z - \zeta)^{k-1} d\zeta.$$

If on the segment $[z_m, z]$ inequalities (1) hold, then inequality (2) holds too, and hence we have

$$|f^{(j)}(z) - T_m^{(j)}(z)| < \frac{M|z - z_m|^{k-j}}{(k-j)!}, \quad j = 0, 1, \ldots, k-1.$$

Thus, if inequalities (1) hold on $[z_m, z]$, then

$$|f^{(j)}(z) - T_m^{(j)}(z)| < B, \quad j = 0, 1, \ldots, k-1, \tag{4}$$

$B = 2^k M.$

LEMMA. *Let Q be a polynomial of the degree s, $Q(\zeta_0) \neq 0$, and let b be a number satisfying the condition $0 < b < |Q(\zeta_0)|$. Then the set $E_b = \{\zeta : |Q(\zeta)| < b\}$ is contained in the union of certain angles with vertices at the point ζ_0 whose sum φ satisfies the inequality*

$$\varphi \leqslant Cs\left(\frac{b}{|Q(\zeta_0)|}\right)^{1/s},$$

C being an absolute constant.

Proof of Lemma. Without loss of generality we suppose $\zeta_0 = 0$. Let $z_0, z_1, \ldots, z_s$ be $s+1$ distinct points of E_b. Let us suppose $|\arg z_i - \arg z_j| > a$, $(0 < a \leqslant \pi/2)$, for $i \neq j$. Then we have $|(z_i/z_j) - 1| > q = \sin a \geqslant (2/\pi)a$ Let us consider the equalities

$$c_0 + c_1 z_j + \ldots + c_s z_j^s = Q(z_j), \quad j = 0, 1 \ldots s,$$

as a system of linear algebraic equations with respect to the coefficients $c_0, c_1, \ldots, c_s$ of the polynomial Q. We obtain

$$c_0 = \sum_{j=0}^{s} Q(z_j) \prod_{i \neq j} \left(1 - \frac{z_i}{z_j}\right)^{-1}.$$

Hence $|c_0| < (s+1)b/q^s$ and

$$a \leqslant \pi q/2 < (\pi/2)(s+1)^{1/s}(b/|c_0|)^{1/s} \leqslant \pi(b/|c_0|)^{1/s}.$$

Thus, for any $s+1$ points $z_0, \ldots, z_s$ of the set E_b the smallest angle between the segments $[0, z_j]$ does not exceed $\pi(b/|c_0|)^{1/s}$. Therefore the set E_b is contained in the union of at most s angles of size $2\pi(b/|c_0|)^{1/s}$.

Let us apply the lemma to the polynomials $Q_j = T_m^{(j)}$, $0 \leqslant j \leqslant k-1$. Taking into account $Q_j(z_m) = f^{(j)}(z_m)$ and condition (3), we conclude that the set

$$\{z : |T_m^{(j)}(z)| < B + M \text{ for some } j, \ 0 \leqslant j \leqslant k-1\}$$

is contained in the union of certain angles with vertices at the point z_m whose sum tends to zero as $m \to \infty$.

That is why for m sufficiently large there exists a point z', $|z'| = 1/2$ such that the following inequalities hold on the whole segment $[z_m, z']$:

$$|T_m^{(j)}(z)| \geqslant B + M, \quad j = 0, 1, \ldots, k-1. \tag{5}$$

Let $[z_m, z^*]$ be a largest segment contained in $[z_m, z']$ where inequalities (1) hold and, consequently (4) holds. By (4) and (5) we have on $[z_m, z^*]$ the following inequalities

$$|f^{(j)}(z)| > M, \quad j = 0, 1, \ldots, k-1.$$

Hence $z^* = z'$ (otherwise the segment $[z_m, z^*]$ can be enlarged). Thus (1) and (2) hold on the whole segment $[z_m, z']$, hence

$$|f^{(k-1)}(z_m) - f^{(k-1)}(z)| < B.$$

It follows that

$$\limsup_{m \to \infty} |f^{(k-1)}(z_m)| \leqslant \max_{|z|=1/2} |f^{(k-1)}(z)| + B.$$

This contradicts to (3).

This result can be strengthened.

THEOREM 2 (A. Gordon). *Let $g(r)$ be a nondecreasing positive continuous function on the half-segment $[0,1)$. Assume that for some sequence $\{z_n\} \subset D$ we have $f(z_n)/g(|z_n|) \to \infty$ as $n \to \infty$. Then there exists a sequence $\{z_n^*\} \subset D$ such that for every $j = 0, 1, \dots$ we have $f^{(j)}(z_n^*)/g(|z_n^*|) \to \infty$ as $n \to \infty$.*

The corresponding theorem for entire functions is as follows.

THEOREM 3 (A. Gordon). *Let f be an entire transcendental function and let $\{g_j(r)\}_{j=0}^{\infty}$ be a sequence of nondecreasing positive continuous functions on the positive ray satisfying the following conditions*

(i) $\lim_{r \to \infty} (g_j(2r)/g_j(r)) = \infty$, $\quad j = 0, 1, 2, \dots$,
(ii) $rg_{j+1}(r) \leqslant C_j g_j(r)$, $\quad r \geqslant r_j$, $\quad j = 0, 1, 2, \dots$,
(iii) *there exists a sequence $\{z_n\} \subset D$ such that*

$$f(z_n)/g_0(|z_n|) \to \infty, \quad n \to \infty.$$

Then there exists a sequence $\{z_n^*\} \subset D\}$ such that for every $j = 0, 1, 2, \dots$

$$f^{(j)}(z_n^*)/g_j(|z_n^*|) \to \infty, \quad n \to \infty.$$

Evidently, for every entire transcendental function there exists a sequence $\{g_j(r)\}_{j=0}^{\infty}$ satisfying (i)–(iii). That is why the last theorem (or, more precisely, its proof close to the proof of first one) gives the affirmative answer to Problem 3 concerning the existence of an elementary proof of the proposition by J. Langley and L.A. Rubel.

The affirmative answer to Problem 4 is contained in the following theorem, which is an immediate consequence of Theorem 1.

THEOREM 4. *The differential equation*

$$P_0(y)P_1(y') \cdots P_n(y^{(n)}) = 1, \tag{6}$$

where P_j are polynomials, not all being constant, has no solutions y that are analytic and unbounded in D.

Indeed, let $y(z)$ be an analytic unbounded function in D which satisfies (6). By Theorem 1 on the sequence $\{z_n\}$ we have $P_j(y^{(j)}(z_n)) \to \infty$ as $n \to \infty$, $j = 0, 1, \dots n$, which contradicts to (6).

ON GLOBAL SOLVABILITY
OF A LINEAR DIFFERENCE EQUATION
IN REAL-ANALYTIC FUNCTIONS

G. R. BELITSKII,   V. A. TKACHENKO

Consider a multidimensional vector equation

$$(1) \qquad \sum_{j=1}^{m} A_j(x)y(x + \alpha_j) = \gamma_{;,}$$

where $x \in \mathbb{R}^n$, $\alpha_j \in \mathbb{R}^n$, $j = 1, \ldots, m$, $y \colon \mathbb{R}^n \to \mathbb{R}^p$, $\gamma \colon \mathbb{R}^n \to \mathbb{R}^q$, $A_j \in \mathrm{Hom}(\mathbb{R}^p \to \mathbb{R}^q)$. Let us suppose that the functions $A_j(x)$, $j = 1, \ldots, m$, are real-analytic in x.

PROBLEM. *What are additional conditions for* (1) *to have a real-analytic solution* $y(x)$ *for any real-analytic right-hand side* $\gamma(x)$?

Let us suppose that α_1 and α_m are extremal shifts in the sense that there exists a linear functional l such that

$$l(\alpha_1) < l(\alpha_j) < l(\alpha_m), \qquad 1 < j < m.$$

It is easy to prove that if

$$(2) \qquad \mathrm{rang}\, A_1(x) = \mathrm{rang}\, A_m(x) = q,$$

then (1) has an infinitely-differentiable solution $y(x)$. The existence of a real-analytic solution under the same restrictions is known to us for some particular cases.

Namely, for $n = 1$ the existence of a real-analytic solution of the scalar equation ($p = q = 1, \ldots, m$ being any integer) follows from results of L. Gruman [1]. The same fact for the two-term vector equation with $n = 1$ ($m = 2$, p and q being any integers) follows from [2]. In [3] the existence of a real-analytic solution of (1) was proved for arbitrary m, p, q and n under some additional conditions related to analyticity and growth of the coefficients in some layer containing $\mathbb{R}^n$.

We do not know whether a real-analytic solution of (1) exists if one assumes only conditions (2).

REFERENCES

1. Gruman L., *Solutions of difference equations with nonconstant coefficients*, Lecture Notes in Math. **1094** (1984), 84–138.
2. Belitskii G. R., Tkachenko V. A., *On locally analytic solvability of multidimensional functional equations in the neighborhood of a nonsingular point*, Dokl. Akad. Nauk SSSR **314** (1990), no. 1, 15–18. (Russian)
3. Belitskii G. R., Tkachenko V. A. (to appear).

DEPT. MATH., BEN-GURION UNIV.
BEER-SHEVA, 84105
ISRAEL

DEPT. MATH., FTINT, PR. LENINA 47
KHARKOV, 310164
UKRAINE

A PROBLEM ON EXACT MAJORANTS

S. Ya. Havinson

Let G be a domain on the complex plane ($G = \mathbb{D}$ for example), and let h be a positive function on G. Consider the class B_h of all single-valued functions f analytic in G such that $|f(z)| \leqslant h(z)$ ($z \in G$), and define a function H by

$$H(z) = \sup\{|f(z)| : f \in B_h\}, \qquad z \in G.$$

OUR PROBLEM is to find conditions on h necessary and sufficient for the equality $h = H$.

If $h = H$ then the function h will be called an exact majorant (e.m.).

It is clear that for any e.m. h the function $\log h$ is subharmonic. But easy examples show that it is not a sufficient condition. On the other hand the equality $h = |F|$, F being an analytic function, implies that h is an e.m. But this condition is not necessary. When trying to solve the Problem one may impose some additional requirements on h, e.g., suppose that h is continuous in G or even (as the first step) in $\operatorname{clos}(G)$. Theoretically one may treat this problem using the concept of duality in the theory of extremal problems (cf., e.g., [1]). But I did not succeed to get any useful information concerning the description of e.m. under this approach. The fact that each e.m. h is also an e.m. in every subdomain of G is likely to be useful in this approach.

Let Q be the class of e.m.'s for G that are continuous in G (or even in $\operatorname{clos}(G)$). Here is ONE OF CONJECTURES concerning the description of Q:

$h \in Q \iff$ (h is in the closure of functions of the form $|f_1| + \cdots + |f_n|$, f_j being analytic in G), here the closure is either in $C(\operatorname{clos}(G))$ (if Q consists of functions continuous in $\operatorname{clos}(G)$) or in the projective limit of the spaces $C(\operatorname{clos}(G_n))$, where the domains G_n exhaust G (if Q consists of functions continuous only in G).

Using the approach of the convex analysis we can formulate the DUAL VERSION OF OUR CONJECTURE: let μ be a real Borel measure on G; does the condition $\int_G |f| \, d\mu \geqslant 0$ for all functions f analytic in G imply $\int_G h \, d\mu \geqslant 0$ for $h \in Q$? We may try to treat the question investigating the measure in the Riesz representation of the subharmonic function $|f|$ (not $\log|f|$!). The answers to the above questions may happen to yield an interesting contribution to the theory of extrema in spaces of analytic functions.

INSTITUTE OF BUILDING AND ENGINEERING
121352, MOSCOW, RUSSIA

CHAPTER EDITORS' NOTE

The Conjecture has been DISPROVED by A. Ya. Gordon (it was explained in the Commentary to [2]), but the Problem seems to remain open.

REFERENCES

1. Havinson S. Ya., *Theory of extremal problems for bounded analytic functions which satisfy some conditions inside the domain*, Uspekhi Mat. Nauk. **18** (1963), no. 2, 25–98. (Russian)
2. Havinson S. Ya., *A problem of exact majorants*, Lecture Notes Math. **1043** (1984), 611–613.

HARMONIC MEASURE FOR THREE DISJOINT DOMAINS IN $\mathbb{R}^n$

ALEXANDRE EREMENKO,   BENT FUGLEDE,   MIKHAIL SODIN

The following problem is known to have a positive answer in dimension $n = 2$, see [4], but it seems to be open for $n > 2$ (even in the case of bounded domains which are regular for the Dirichlet problem).

PROBLEM a. *Let D_1, D_2, D_3 be pairwise disjoint domains in $\mathbb{R}^n$. Let ω_j denote the harmonic measure for D_j at some point $x_j \in D_j$, $j = 1, 2, 3$. Is then $\omega_1 \wedge \omega_2 \wedge \omega_3 = 0$?*

Equivalently, the question is whether there exist Borel sets S_1, S_2, S_3 supporting ω_1, ω_2, ω_3, respectively, such that $S_1 \cap S_2 \cap S_3 = \emptyset$.

When confined to *Lipschitz domains* D_1, D_2, D_3, Problem a has an affirmative answer. In fact, let S_j denote the set of points of the boundary ∂D_j at which ∂D_j has a normal. Then $(\partial D_j) \setminus S_j$ has $(n-1)$-dimensional Hausdorff measure 0, hence also harmonic measure 0 with respect to D_j according to a theorem of Dahlberg [2]. It suffices therefore to prove that $S_1 \cap S_2 \cap S_3 = \emptyset$. Suppose that $x_0 \in S_1 \cap S_2 \cap S_3 = \emptyset$. Then D_j contains an open truncated cone Q_j with vertex at x_0 and opening angle close enough to $\pi/2$ so that more than one third (measured in surface area) of a small sphere $\{x \in \mathbb{R}^n : |x - x_0| = r\}$ belongs to Q_j; and this contradicts Q_1, Q_2, Q_3 being pairwise disjoint.

On the other hand we give in [4] an example of Brelot harmonic space in which the analogue to Problem a has a negative answer.

Using the result of Friedland and Hayman [7], Bishop [1] has shown that, for any dimension n, there exists a natural number p (depending only on n) such that, for any family of p pairwise disjoint domains D_j in $\mathbb{R}^n$, the harmonic measures ω_j satisfy $\omega_1 \wedge \cdots \wedge \omega_p = 0$. Problem a therefore asks whether one can take $p = 3$ also for $n > 2$. (Clearly $p \geqslant 3$ is necessary). More precisely Bishop finds that $p = 5$ will do in dimension 3, and $p = 11$ in any dimension n [1, p. 22].

We became interested in Problem a because a generalization of it (Problem **A** below) is equivalent to a certain problem about the lower envelope of δ-subharmonic functions (Problem **B** below), cf. [4] for the planar case. The study of this later problem leads to the use of the *fine topology* because, for general subharmonic functions u, v in a domain $\Omega \subset \mathbb{R}^n$, the set $\{x \in \Omega : u(x) < v(x)\}$ needs not be open in the standard topology on $\mathbb{R}^n$, but it is always open in the fine topology, introduced by H. Cartan as the weakest topology on $\mathbb{R}^n$ for which all subharmonic functions are continuous. The fine topology is stronger than the standard topology on $\mathbb{R}^n$. As to the fine topology, we refer to Doob [3] and to the survey [10].

The fine topology is locally connected. By a *fine domain* (in $\mathbb{R}^n$) it is understood a finely open, finely connected set. Any two points of a fine domain D can be joined by a finite polygonal path in D. See [10], where also references are given.

For a fine domain $D \subset \mathbb{R}^n$ we denote by $\omega(D, x)$ the (fine) *harmonic measure* at $x \in D$. It is a positive measure of total mass $\leqslant 1$. If $n \geqslant 3$, $\omega(D, x)$ is obtained by sweeping the Dirac measure ε_x on $\complement D$. If $n = 2$ and $\mathbb{R}^2 \setminus D$ is non-polar, $\omega(D, x)$ is the probability measure on $\mathbb{R}^2$ whose trace on any Green domain $\Omega \subset D$ in $\mathbb{R}^2$ is the swept-out on ε_x on $\Omega \setminus D$ relative to Ω. For any n, $\omega(D, x) = 0$ if and only if $\mathbb{R}^n \setminus D$ is polar.

The harmonic measure $\omega(D, x)$ is carried by the fine boundary $\partial_f D$ of D in $\mathbb{R}^n$ and does not charge the polar sets. The sets of measure 0 for $\omega(D, x)$ are the same for any choice of $x \in D$ [8, p. 150]. For a usual domain $D \subset \mathbb{R}^n$, $\omega(D, x)$ equals the usual harmonic measure.

The following problem generalizes Problem **a** to the case of fine domains.

PROBLEM A. *Let D_1, D_2, D_3 be pairwise disjoint, fine domains in $\mathbb{R}^n$. Let $\omega_j = \omega(D_j, x_j)$ denote the harmonic measure for D_j at some point $x_j \in D_j$, $j = 1, 2, 3$. Is then $\omega_1 \wedge \omega_2 \wedge \omega_3 = 0$?*

Equivalently, the question is again whether there exist Borel set S_1, S_2, S_3 supporting ω_1, ω_2, ω_3, respectively, such that $S_1 \cap S_2 \cap S_3 = \emptyset$. The answer to Problem **A** is therefore independent of the choice of the points of evaluation $x_j \in D_j$, $j = 1, 2, 3$, cf. above.

Without loss of generality we may assume, in Problem **A**, that D_1, D_2, D_3 are bounded (cf. Lemma 1 in [4]) and *regular* in the sense that their complements (within $\mathbb{R}^n$) have no finely isolated points. Let A_j denote the set of points of $\complement D_j$ for Brownian motion in $\mathbb{R}^n$ starting at x_j. The answer to Problem **A** will therefore be affirmative in case the set $X = A_1 \cap A_2 \cap A_3$ is polar, or just that X decomposes into 3 sets X_j such that $\omega_j(X_j) = 0$, $j = 1, 2, 3$; for then $S_j = A_j \setminus X_j$ supports ω_j, and $S_1 \cap S_2 \cap S_3 = \emptyset$.

In the planar case $n = 2$ the answer to Problem **A** and to Problem **B** is always "yes", see [4]. As to Problem **A**, this is because the above set X then has at most 2 points, as shown by an easy application of the Jordan curve theorem. An alternative, purely analytical proof of the affirmative answer to Problem **A** in the plane is included in [4]. The positive answer to Problem **B** in the plane is useful in the subharmonic approach to value distribution theory of meromorphic functions, cf. [5],[6].

PROBLEM B. *Let $w_1, \ldots, w_m$ be a δ-subharmonic functions in a domain $\Omega \subset \mathbb{R}^n$. Is then*

$$(1) \qquad \mu\left[\bigwedge_{j=1}^m w_j\right] \geqslant \bigwedge_{1 \leqslant j < k \leqslant m} \mu[w_j \wedge w_k] ?$$

Here $\mu[w]$ denotes the Riesz measure associated with a δ-subharmonic function w on Ω, that is, a function representable (off some polar set) as $w = u - v$ with u and v subharmonic in Ω. In particular, w is *subharmonic* if and only if $\mu[w] \geqslant 0$. Problem **B** reduces to the case of subharmonic functions w_j (just replace each $w_j = u_j - v_j$ by $W - j + \sum_i v_i$). One may further reduce Problem **B** to the case where $m = 3$ (cf. §5 in [4]), $\Omega = \mathbb{R}^n$, and where each w_j equals 0 off some bounded set.

PROPOSITION. *For any dimension n, Problems **A** and **B** are equivalent.*

Proof: If the answer to Problem **A** is "yes" then so is the answer to Problem **B**. This can be read off from the proof of Theorem 1 in [4] because Theorem 2 in [4] extends to arbitrary dimension with the same proof. Conversely, if the answer to **B** is "yes", then so is the answer to **A**. Adding to each D_j the finely isolated points of $\complement D_j$, we achieve that the D_j are regular. Let $v_j > 0$ denote the (fine) Green function for D_j with pole at x_j, see [9] (we may assume that $n \geqslant 3$). After extension by 0 in $\complement D_j$, v_j becomes subharmonic in $\Omega = \mathbb{R} \setminus \{x_1, x_2, x_3\}$, and the associated Riesz measure $\mu[v_j]$ equals $w_j = \omega(D_j, x_j)$. In the framework of Problem **B** we now take $m = 3$ and $w_j = -v_j$, $j = 1, 2, 3$. Since the 3 sets $\{x \in \Omega : w_j(x) < 0\} = D_j$ are pairwise disjoint, we have $w_j \wedge w_k = w_j + w_k$, $w_1 \wedge w_2 \wedge w_3 = w_1 + w_2 + w_3$. Adding $w_1 + w_2 + w_3$ to both members of (1) we therefore obtain $w_1 \wedge w_2 \wedge w_3 = 0$.

Remark. We do not know wether Problem **a** is equivalent to the apparently more general problem **A**.

REFERENCES

1. Bishop C., *A characterization of Poissonian domains*, Ark. Mat. **29** (1991), 1–24.
2. Dahlberg B. E. J., *On estimates of harmonic measure*, Arch. Rational Mech. Anal. **65** (1977), 275–288.
3. Doob J. L., *Classical Potential Theory and its Probabilistic Counterpart*, Springer-Verlag, Berlin–New York, 1984.
4. Eremenko A., Fuglede B., Sodin M., *On the Riesz charge of the lower envelope of δ-subharmonic functions*, preprint, Copenhagen, 1991, 16 pp. submitted to Potential Analysis.
5. Eremenko A., Sodin M., *Value-distribution of meromorphic functions and meromorphic curves from Potential Theory viewpoint*, Algebra & Analysis 3 (1991) (Russian); English translation: Leningrad Math. J. (to appear).
6. Eremenko A., Sodin M., *On value distribution of meromorphic functions of finite order*, Soviet Math. Doklady **316** (1991), 538–541.
7. Friedland S., Hayman W. K., *Eigenvalue inequalities for the Dirichlet problem on spheres and the growth of subharmonic functions*, Comment. Math. Helv. **51** (1976), 133–161.
8. Fuglede B., *Finely Harmonic Functions*, Lecture Notes Math., vol. 289, Springer-Verlag, Berlin–New York, 1972.
9. Fuglede B., *Sur la fonction de Green pour un domaine fin*, Ann. Inst. Fourier (Grenoble) **25** (1975), no. 3–4, 201–206.
10. Fuglede B., *Asymptotic paths for subharmonic functions and polygonal connectedness of fine domains*, Séminaire de Théorie du Potentiel, Paris, No. 5, Lecture Notes Math., vol. 814, Springer-Verlag, Berlin–New York, 1980, pp. 97–116.

DEPARTMENT OF MATHEMATICS
PURDUE UNIVERSITY
WEST LAFAYETTE, IN 47907
USA

AND

FTINT
PR. LENINA 47
KHARKOV, 310164
UKRAINE

MATHEMATICS INSTITUE
UNIVERSITETSPARKEN 5
2100 COPENHAGEN
DENMARK

FTINT
PR. LENINA 47
KHARKOV, 310164
UKRAINE

A UNIQUENESS PROBLEM
FOR POLYHARMONIC FUNCTIONS

W. K. HAYMAN*

We are concerned with the class $H^m(D)$ of functions, real analytic in a domain D in $\mathbb{R}^n$ and satisfying the equation

$$\Delta^m u = 0,$$

where

$$\Delta = \left(\frac{\partial}{\partial x_1}\right)^2 + \cdots + \left(\frac{\partial}{\partial x_n}\right)^2$$

is Laplace's operator. For the general properties of such functions see [1].

The following two results are proved in the paper [2].

We denote by ∂B the frontier and by $\overline{B} = B \cup \partial B$ the closure of a set B.

THEOREM 1. *Suppose that $B_1, \ldots, B_{m-1}$ are distinct balls and that B_m is a bounded open set such that $\partial B_m \setminus \bigcup_{\nu=1}^{m-1} \partial B_\nu$ is dense in ∂B_m. Suppose further that $\overline{B}_\nu \subset D$ for $1 \leqslant \nu \leqslant m$. Then if $u \in H^m(D)$ and*

$$u = 0 \quad \text{on} \quad \bigcup_{\nu=1}^m \partial B_\nu$$

we conclude $u \equiv 0$ in D.

THEOREM 2. *Suppose that $m \geqslant 2$ and that S_1 to S_{m-2} are preassigned circles (if $m > 2$). Then there exists u, $u \not\equiv 0$, in $H^m(\mathbb{R}^2)$ such that $u = 0$, on S_1 to S_{m-2} and on infinitely many analytic Jordan curves S_ν, for $\nu \geqslant m - 1$. Here S_ν surrounds the origin for $\nu \geqslant m - 1$ and $S_\nu \to \infty$ as $\nu \to \infty$.*

The PROBLEM is *whether in Theorem 1 we can replace hyperspheres by any more general class of surfaces such as ellipsoids?*

It follows from Theorem 2 that we cannot, in the plane, replace $m - 1$ circles and one curve by $m - 2$ circles and two curves in Theorem 1.

For instance can a function in $H^2(\mathbb{R}^2)$ vanish on two ellipses without being constant?

If one of the ellipses is a circle the answer is no by Theorem 1. On the other hand the function

$$u = (x_1/a)^2 + (x_2/b)^2 - 1$$

vanishes on an ellipse (but nowhere else).

---

*following on earlier question by Donald Kershaw

REFERENCES

1. Aronszajn N., Creese T., Lipkin L., *Polyharmonic Functions*. Clarendon Press, Oxford, 1983.
2. Hayman W. K., Korenblum B., *Representation and uniqueness theorem for polyharmonic functions*, (submitted to the Mandelbrojt centenary volume of Journal d'Analyse).

DEPT. MATH, UNIVERSITY OF YORK,
HESLINGTON, YORK, Y01 5DD
ENGLAND

SOME EXTENSION PROBLEMS

JOSEF KRÁL

Let $K \colon G \to K(G)$ and $\widetilde{K} \colon G \to \widetilde{K}(G)$ associate with each open set $G \subset \mathbb{C} \equiv \mathbb{R}^2$ a class of complex-valued functions on G. A set $E \subset \mathbb{C}$ will be termed negligible $(K, \widetilde{K})$ if, for each open set $G \subset \mathbb{C}$ and each $f \in K(G)$, the existence of an open set $\widetilde{G} \subset G$ such that $f \mid \widetilde{G} \in \widetilde{K}(\widetilde{G})$ and $f(G \setminus \widetilde{G}) \subset E$ implies that $f \in \widetilde{K}(G)$. For the case when $K = C$ (=sheaf of continuous functions) and $K = A$ (=sheaf of holomorphic functions), negligibility of finite sets was established by T. Radó in [4]. P. Lelong showed in [3] that also all polar sets are negligible (C, A).

PROBLEM 1. *What are necessary and sufficient conditions for $E \subset \mathbb{C}$ to be negligible (C, A)?*

For continuously differentiable functions some related results concerning harmonicity are known. If $\omega \geqslant 0$ is a continuous non-decreasing function on $\mathbb{R}_+$ with $\omega(t) > 0$ for $t > 0$, we denote by $C^\omega(G)$ the class of all functions f on G satisfying the condition

$$|f(u) - f(v)| = O\big(\omega(|u - v|)\big) \quad \text{as} \quad |u - v| \downarrow 0$$

locally in G; $C^\omega_*(G)$ will stand for the subclass of all $f \in C^\omega(G)$ enjoying the property

$$|f(u) - f(v)| = o\big(\omega(|u - v|)\big) \quad \text{as} \quad |u - v| \downarrow 0$$

locally in G. Further we denote by $C^{1,\omega}(G)$ and $C^{1,\omega}_*(G)$ the classes of all continuously differentiable real-valued functions whose first order partial derivatives belong to $C^\omega(G)$ and $C^\omega_*(G)$, respectively. If $H(G)$ denotes the class of all real-valued functions harmonic on G, then the following result holds (cf.[2]).

THEOREM. *A set $E \subset \mathbb{R}$ is negligible $(C^{1,\omega}, H)$ if (and also only if in case $\omega(0) = 0$) the Hausdorff measure corresponding to the measure function ω vanishes on all compact subsets of E. A necessary and sufficient condition for $E \subset \mathbb{R}$ to be negligible $(C^{1,\omega}_*, H)$ consists in σ-finiteness of the Hausdorff measure corresponding to ω on all compact subsets of E.*

For subharmonic functions similar question seems to be open. (Of course, necessity of the corresponding condition follows from the above theorem.) Let $S(G)$ denote the class of all subharmonic functions on G.

CONJECTURE. *Any set $E \subset \mathbb{R}$ with vanishing Hausdorff measure corresponding to the measure function ω is negligible $(C^{1,\omega}, S)$. If $E \subset \mathbb{R}$ has σ-finite Hausdorff measure corresponding to ω, then E is negligible $(C^{1,\omega}_*, S)$.*

PROBLEM 2. *What are necessary and sufficient conditions for $E \subset \mathbb{C}$ to be negligible (C^ω, A) or (C^ω_*, A)?*

Similar questions may be posed for various classes of functions in more spaces (compare [1]).

PROBLEM 16.17

References

1. Cegrell U., *Removable singularities for plurisubharmonic functions and related problem*, Proc. London Math. Soc. **XXXVI** (1978), 310–336.
2. Král J., *Some extension results concerning harmonic functions*, J. London Math Soc. **28** (1983), 62–70.
3. Lelong P., *Ensembles singuliers impropres des fonctions plurisousharmoniques*, J. Math. Pures Appl. **36** (1957), 263–303.
4. Radó T., *Über eine nicht fortsetzbare Riemannsche Mannigfaltigkeit*, Math. Z. **20** (1924), 1–6.

MATHEMATICAL INSTITUTE
CZECH ACADEMY OF SCIENCES
ŽITNÁ 25
115 67 PRAGUE 1
CZECH REPUBLIC

PARTITION OF SINGULARITIES
OF ANALYTIC FUNCTIONS

R. KAUFMAN

Let S be a closed set in $\mathbb{R}^2$, and let $A^n(S)$ be the class of functions f, holomorphic in $W = \mathbb{R}^2 \setminus S$, such that $f, \ldots, f^{(n)}$ can be extended continuously to $W^- = W \cup \partial S$ ($n = 0, 1, 2, \ldots$). Let $S = S_1 \cup S_2$, each S_i being closed, and moreover $S = S_1^0 \cup S_2^0$, where S_i^0 is the interior of S_i relative to S.

In this situation IT IS NATURAL TO GUESS

$$(1) \qquad A^n(S) = A^n(S_1) + A^n(S_2), \qquad (n = 0, 1, 2, 3, \ldots).$$

To explain the difficulties involved in (1), we suppose that f is continuous in $\mathbb{R}^2$ and use the operator $\bar{\partial}$ defined by $2\bar{\partial} \equiv \partial/\partial x + i\partial/\partial y$. Following the classical method, we choose a function $\varphi \in C^\infty(\mathbb{R}^2)$ such that $0 \leqslant \varphi \leqslant 1$ and $\varphi = 0$ on $S \setminus S_1^0$, $\varphi = 1$ on $S \setminus S_2^0$. Suppose that $\bar{\partial}h = f\bar{\partial}\varphi$ (in the sense of distributions, or the Cauchy–Green formula). Then $\bar{\partial}(\varphi f - h) = \varphi\bar{\partial}f$, and therefore $F = \varphi f - h$ belongs to $A^0(S_1)$; similarly $f - F = (1 - \varphi)f + h$ belongs to $A^0(S_2)$. Therefore (1) is true for $n = 0$ (a classical observation, to be sure) but the reasoning seems to fail when $n = 1$ since f is generally not C^1 (or even Hölder-continuous) on $\mathbb{R}^2$.

If (1) were true (for some $n \geqslant 1$) it would imply that the triviality of $A^n(S)$ is a local property of S. (Triviality of $A^n(S)$ means of course that all of its elements are restrictions to W of entire functions.) Even this much is unknown.

DEPARTMENT OF MATHEMATICS
UNIVERSITY OF ILLINOIS
URBANA, IL 61801
USA

COMPARISON OF SETS OF EXCEPTIONAL VALUES
IN THE SENSE OF NEVANLINNA AND
IN THE SENSE OF V. P. PETRENKO

A. A. GOLDBERG, A. E. EREMENKO

Let f be a meromorphic function in $\mathbb{C}$ and put

$$\beta(a, f) = \liminf_{r \to \infty} \max_{|z|=r} \log^+ [f(z), a]^{-1} / T(r, f), \quad a \in \widehat{\mathbb{C}},$$

where $[a, b]$ is the spherical distance between a and b. Denote by

$$E_P(f) \stackrel{\text{def}}{=} \{ a \in \widehat{\mathbb{C}} : \beta(a, f) > 0 \}$$

the set of exceptional values in the sense of V. P. Petrenko and by

$$E_N(f) \stackrel{\text{def}}{=} \{ a \in \widehat{\mathbb{C}} : \delta(a, f) > 0 \}$$

the set of deficient values of f. It is clear that $E_N(f) \subset E_P(f)$. The set $E_P(f)$ is at most countable if f is of finite order [1]. There are examples of f's of finite order with $E_N(f) \neq E_P(f)$ [1–4].

PROBLEM 1. *Let $E_1 \subset E_2 \subset \widehat{\mathbb{C}}$ be arbitrary at most countable sets. Does there exist a meromorphic function f of finite order with $E_N(f) = E_1$ and $E_P(f) = E_2$?*

It is known that $T(2r, f) = O(T(r, f))$, $r \to \infty$, implies $E_N(f) = E_P(f)$ [5].

PROBLEM 2. *Let f be an entire function of finite order, $\beta(0, f) > 0$. Is it true that $\delta(0, f) > 0$?*

REFERENCES

1. Petrenko V. P., *Growth of Meromorphic Functions*, "Vysha Shkola", Kharkov, 1978. (Russian)
2. Grishin A. F., *On comparison of deficiencies $\delta_p(a)$*, Teor. funkt., funkt. anal. i prilozh., Kharkov **25** (1976), 56–66. (Russian)
3. Goldberg A. A., *To the question on the connection between deficiencies and deviations*, Teor. funkt., funkt. anal. i prilozh., Kharkov **29** (1978), 31–35. (Russian)
4. Sodin M. L., *On the relation between the sets of deficiencies and deviations*, Sib. Mat. Zh. **22** (1981), no. 2, 198–206. (Russian)
5. Eremenko A. E., *On deficiencies and deviations of meromorphic functions*, Dokl. Akad. Nauk Ukr. SSR (1985), no. 1, 18–20. (Russian)

LVOV STATE UNIVERSITY
UNIVERSITETSKAYA 1
LVOV 290602
UKRAINE

DEPARTMENT OF MATHEMATICS
PURDUE UNIVERSITY
WEST LAFAYETTE, IN 47907
USA

AND

DEPT. OF MATH., FTINT
PR. LENINA 47
KHARKOV, 310164
UKRAINE

CHAPTER EDITORS' NOTE

Problems 1 and 2 have been solved completely in [6].

Given arbitrary at most countable sets $E_1 \subset E_2 \subset \overline{\mathbb{C}}$ and $0 < \rho < \infty$, there exists a meromorphic function f of order ρ such that $E_N(f) = E_1$ and $E_P(f) = E_2$.

There exists an entire function g of prescribed order ρ, $1/2 < \rho < \infty$, with the property $0 \in E_P(g) \setminus E_N(g)$.

REFERENCE

6. Goldberg A. A., Eremenko A. E., Sodin M. L., *Exceptional values in the sense of Nevanlinna and in the sense of V. P. Petrenko*, Teor. funkt., funkt. anal. i prilozh., Kharkov **47** (1987), 31–35; **48** (1987), 58–70. (Russian)

S.16.20
old

VALIRON EXCEPTIONAL VALUES OF ENTIRE FUNCTIONS OF COMPLETELY REGULAR GROWTH

A. A. GOLDBERG,   A. E. EREMENKO,   I. V. OSTROVSKII

Let E_ρ be the class of all entire functions of order ρ, $1/2 < \rho < \infty$, and let E_ρ^0 be its subclass of entire functions of c.r.g. [1]. Let $E_N(f)$ and $E_V(f)$ be the sets of exceptional values of a function f in the sense of R. Nevanlinna and of G. Valiron respectively. It is known that

$$\{ E_N(f) : f \in E_\rho^0 \} \neq \{ E_N(f) : f \in E_\rho \}.$$

Indeed, for every function $f \in E_\rho^0$ we have $\text{card}(E_N(f)) \leqslant [2\rho]^*$, where

$$[x]^* = \max\{ k \in \mathbb{Z} : k < x \}, \quad [2].$$

On the other hand there exists $f \in E_\rho$ with the property $\text{card}(E_N(f)) = \infty$, [3].

PROBLEM. *Is it true that*

$$\{ E_V(f) : f \in E_\rho^0 \} = \{ E_V(f) : f \in E_\rho \}?$$

There are examples of $f \in E_\rho^0$ for which the set $E_V(f)$ has the cardinality of the continuum [4].

REFERENCES

1. Levin B. Ya., *Distributions of Zeros of Entire Functions*, AMS, 1980.
2. Oum Ki-Choul, *Bounds for the number of deficient values of entire functions whose zeros have angular densities*, Pacif. J. Math. **29** (1969), no. 1, 187–202.
3. Arakelyan N. U., *Entire functions of a finite order with an infinite set of deficient values*, Dokl. Akad. Nauk SSSR **170** (1966), no. 2, 187–202. (Russian)
4. Goldberg A. A., Eremenko A. E., Ostrovskii I. V., *On a sum of entire functions of regular growth*, Izv. Armyan. SSR **18** (1983), no. 1, 3–14. (Russian)

LVOV STATE UNIVERSITY
UNIVERSITETSKAYA 1
LVOV 290602
UKRAINE

DEPARTMENT OF MATHEMATICS
PURDUE UNIVERSITY
WEST LAFAYETTE, IN 47907
USA

AND

DEPT. OF MATH., FTINT
PR. LENINA 47
KHARKOV 310164
UKRAINE

DEPARTMENT OF MATHEMATICS
FTINT, PR. LENINA 47
KHARKOV 310164
UKRAINE

CHAPTER EDITORS' NOTE

The answer is negative. This follows from the description of the set $E_V(f)$, $f \in E_\rho^0$, given in [5] which can be compared with the characterization of the set $E_V(f)$, $f \in E_\rho$, from [6].

REFERENCES

5. Eremenko A. E., *On Valiron's exceptional values of entire functions of completely regular growth*, Teor. funkt., funkt. anal. i prilozh., Kharkov **44** (1985), 48–52. (Russian)
6. Hyllengren A., *Valiron deficient values for meromorphic functions in the plane*, Acta Math. **124** (1970), 1–8.

S.16.21
v.old

ZERO-SETS OF SINE-TYPE FUNCTIONS

B. Ya. Levin,   I. V. Ostrovskii

An entire function S of exponential type π is called a *sine-type function* (s.t.f.) if there exist positive constants m, M, H such that

$$m < |S(z)|e^{-\pi|\operatorname{Im} z|} < M \quad \text{for } |\operatorname{Im} z| \geqslant H.$$

The class of s.t.f. was introduced in [1]. It has found applications in the theory of interpolation by entire functions and for bases of exponentials in $L^2(-\pi, \pi)$ [2–5].

In connection with the results of [2–6] the following problem seems to be interesting:

PROBLEM. *Describe the zero-sets* $\{\lambda_n\}_{n\in\mathbb{Z}}$ *of s.t.f.*

This problem is, of course, equivalent to the problem of identification of s.t.f. because

$$S(z) = \lim_{R\to\infty} \prod_{|\lambda_n|<R} \left(1 - \frac{z}{\lambda_n}\right)$$

for every s.t.f. S. Note that without loss of generality the sequence $\{\lambda_n\}_{n\in\mathbb{Z}}$ can be assumed to be real. A sequence $\{\lambda_n\}_{n\in\mathbb{Z}}$, $|\operatorname{Im} \lambda_k| < H$, is the sequence of zeros of s.t.f. iff $\{\operatorname{Re}\lambda_n\}_{n\in\mathbb{Z}}$ is (see [7]).

REFERENCES

1. Levin B. Ya., *On bases of exponentials in $L^2(-\pi, \pi)$*, Zap. Fiz.-mat. fak. KhGU i Khark. Mat. ob-va **27** (1961), 39–48. (Russian)
2. Levin B. Ya., *Interpolation by entire function of exponential type*, Trudy FTINT AN USSR, ser. "Matem. fizika i funkts. analiz" **1** (1969), 136–146. (Russian)
3. Golovin V. D., *On biorthogonal expansions in L^2 against linear combinations of exponentials*, Zap. Fiz.-mat. fak. KhGU i Khark. Mat. ob-va **30** (1964), 18–29. (Russian)
4. Katsnelson V. E., *On bases of exponentials in L^2*, Funkt. Anal. i Prilozh. **5** (1971), no. 1, 37–47. (Russian)
5. Levin B. Ya., Lubarskii Yu. I., *Interpolation by entire functions of special classes and exponential expansions*, Izv. Akad. Nauk SSSR, ser.matem. **39** (1975), no. 3, 657–702. (Russian)
6. Ostrovskii I. V., *On a class of entire functions*, Soviet Math. Dokl. **17** (1976), no. 4, 977–981.
7. Levin B. Ya., Ostrovskii I. V., *On small perturbations of the set of zeros of functions of sine type*, Math. USSR Izvest. **14** (1980), no. 1, 79–101.

DEPT.MATH., FTINT, PR.LENINA 47,
KHARKOV, 310164
UKRAINE

DEPT.MATH., FTINT, PR.LENINA 47,
KHARKOV, 310164
UKRAINE

CHAPTER EDITORS' NOTE

A complete solution of this problem was obtained by A. Eremenko and M. Sodin (oral communication, to be published). Their considerations were based on a parameterization of entire functions of the Laguerre–Polya class LP (i.e. limits of polynomials with real zeros) with the help of special conformal maps [8,9]. (These maps differ from those used in [6] to describe the zero-sets of functions of M. G. Krein class).

For any sequence $s = \{c_n\}$ alternating in sign (i.e. $(-1)^n c_n \geqslant 0$) consider the simply connected domain

$$\Omega(s) = \mathbb{C} \setminus \bigcup_n \{\zeta = \xi + in\pi : -\infty < \xi \leqslant \log|c_n|\}$$

(if $c_n = 0$ then the n-th slit is absent). Denote by $\psi(z)$ the conformal map of the lower half-plane $\mathbb{C}_-$ onto $\Omega(s)$ such that the real axis maps onto the slits. Then the function

$$(1) \qquad f(z) = e^{\psi(z)}$$

can be continued across the real axis, $f \in LP$ and $s = \operatorname{Cr} f$ is the sequence of critical values of f. (It means that $s = \{f(\mu_n)\}$, where $\ldots \leqslant \mu_{n-1} \leqslant \mu_n \leqslant \mu_{n+1} \leqslant \ldots$ are zeros of $f'(z)$).

THEOREM. *A real function $f(z)$ is a sine-type function with real zeros iff it can be represented in the form (1), where $s = \operatorname{Cr} f$ satisfies the following conditions:*

   a) *$\sup\limits_{n \in \mathbb{Z}} \log|c_n| < \infty$;*

   b) *for some $N \in \mathbb{N}$, $\inf\limits_{n \in \mathbb{N}} \max\limits_{k \in [n, n+N]} \log|c_k| > -\infty$.*

REFERENCES

8. MacLane G. R., *Concerning the uniformization of certain Riemann surfaces*, Trans. Amer. Math. Soc. **62** (1947), 99–111.
9. Vinberg E. B., *Real entire functions with prescribed critical values*, Voprosy teorii grup i gomologicheskoi algebry, Yaroslavl', 1989, pp. 127–138. (Russian)

S.16.22
old

AN EXTREMAL PROBLEM FROM THE THEORY
OF SUBHARMONIC FUNCTIONS

B. Ya. Levin

A closed subset E of $\mathbb{R}$ is said to be *relatively dense* (in measure) if there exist positive numbers N and δ such that every interval of length N contains a part of E of measure at least δ. In this case we write $E \in E(N, \delta)$. Suppose in addition that all points of E are regular boundary points of the domain $\mathbb{C} \setminus E$. It was proved in [1] that there exists a harmonic function v positive on $\mathbb{C} \setminus E$ and (continuously) vanishing on E. Such a function was constructed in [2] with the usage of some special conformal mappings. It can be shown that if we require in addition that $v(\bar{z}) = v(z)$ then E determines v uniquely up to a positive constant factor (cf. [3]); see also [4] for a more general result).

If E is relatively dense then the limit

$$A = \lim_{y \to +\infty} \frac{v(iy)}{y} < \infty$$

exists and is positive. Multiplying by a positive constant we may assume $A = 1$. This normalized v will be from now on denoted by v_E.

It was proved in [1] and [2] that v_E is bounded on $\mathbb{R}$ by a constant depending only on N and δ provided $E \in E(N, \delta)$, i.e.

$$E \in E(N, \delta) \implies \sup_{x \in \mathbb{R}} v_E(x) \leqslant C(N, \delta).$$

PROBLEM. *Find the best possible value of $C(N, \delta)$ for given N and δ.*

I conjecture that

$$C(1, \delta) = \frac{1}{\pi} \log \cot \frac{\pi \delta}{4}$$

and that $\max v_E(x)$ attains this value for

$$E = \bigcup_{n \in \mathbb{Z}} [n - \delta/2, n + \delta/2]$$

only.

REFERENCES

1. Schaeffer A. C., *Entire functions and trigonometrical polynomials*, Duke Math. J. **20** (1953), 77–88.
2. Ahiezer N. I., Levin B. Ya., *A generalization of S. N. Bernstein's inequality for derivatives of entire functions*, "Investigations in modern problems of theory of functions of complex variables", GIFML, Moscow, 1960, pp. 111–165. (Russian)
3. Benedicks M., *Positive harmonic functions vanishing on the boundary of certain domains in $\mathbb{R}^n$*, Ark. Mat. **18** (1980), no. 1, 53–72.

4. Levin B. Ya., *Majorants in the classes of subharmonic functions*, Teor. Funkt., Funkt. Anal. i Prilozh. **51** (1989), 3–17; **52** (1989), 3–33. (Russian)

DEPT. MATH., FTINT, PR. LENINA 47,
KHARKOV, 310164
UKRAINE

CHAPTER EDITORS' NOTE

The conjecture $C(1, \delta) = \pi^{-1} \log \cot(1/4\pi\delta)$ has been confirmed by A. E. Fryntov [5] and, independently and slightly later, by A. Baernstein [6]. The latter paper contains also some uniqueness results.

The proofs in the two papers are similar. They are based on subharmonicity properties of a "star-type" function

$$u_l^*(x + iy) = \sup \int_E u(t + iy) \, dt.$$

Here $l > 0$, u is defined in a strip $\mathbb{R} \times (B_1, B_2)$, u_l^* is defined in the rectangle $[0, l] \times (B_1, B_2)$, and the sup is taken over all sets $E \subset \mathbb{R}$ with Lebesgue measure $2x$ and diameter $\leqslant 2l$.

The validity of Levin's conjecture is obtained as the limit case of the solution of an extremal problem for harmonic measures of domains $S \setminus E$, where E is a closed subset of $\mathbb{R}$ and S is a strip of the form $\mathbb{R} \times (-B, B)$, $B > 0$. The method produces inequalities for all convex integral means as well as for maxima.

REFERENCES

5. Fryntov A. E., *An extremal problem of potential theory*, Soviet Math.– Doklady **37** (1988), 754–755.
6. Baernstein A., *An extremal problem for certain subharmonic functions in the plane*, Revista Matematika Iberoamericana **4** (1988), 199–218.

Chapter 17

$$\mathbb{C}^n$$

Edited by

L. A. Aizenberg
Institute of Physics
SO AN of Russia
Akademgorodok
660036, Krasnoyarsk
Russia

INTRODUCTION

Multidimensional Complex Analysis and even its part dealing with Functional Analysis is an immense field. This accounts for the fact that all problems of this chapter (except for those which are discussed by A. G. Sergeev) are very diverse. However, this does not make this chapter less interesting. Some problems which could be presented in this chapter are in the chapter called "Miscellanea" and other chapters.

I take an opportunity to inform the reader interested in $\mathbb{C}^n$ about the collection of works "Some Unsolved Problems of Multidimensional Complex Analysis", preprint No. 4IM of the Institute of Physics (USSR Academy of Sciences, Siberian Branch), Krasnoyarsk, 1987, which contains 52 unsolved problems (edited by E. M. Chirka). Many of them pertain to linear analysis.

PROPER MAPPINGS OF CLASSICAL DOMAIN

G. Henkin,   R. Novikov

A holomorphic mapping $\varphi \colon \Omega \to \Omega$ of a bounded domain $\Omega \subset \mathbb{C}^n$ is called *proper* if $\operatorname{dist}(\varphi(z_\nu), \partial\Omega) \to 0$ for every sequence $z_\nu \in \Omega$ with $\operatorname{dist}(z_\nu, \partial\Omega) \to 0$, $\nu \to \infty$.

A biholomorphism (automorphism) of Ω is called a *trivial proper mapping* of Ω. If Ω is the 1–dimensional disc $\mathbb{D}$ the non-trivial proper holomorphic mappings $\varphi \colon \mathbb{D} \to \mathbb{D}$ do exist. They are called finite Blaschke products.

The existence of nontrivial proper holomorphic mappings seems to be the characteristic property of the 1–dimensional disc in the class of all irreducible symmetric domains.

CONJECTURE 1. *For an irreducible bounded symmetric domain Ω in $\mathbb{C}^n$, $\Omega \neq \mathbb{D}$, every proper holomorphic mapping $\Omega \to \Omega$ is an automorphism.*

According to the E. Cartan's classification there are six types of irreducible bounded symmetric domains. The domain $\Omega_{p,q}$ of the first type is the set of complex $p \times q$ matrices Z, $p \geqslant q \geqslant 1$, such that the matrix $I - Z^*Z$ is positive. The following beautiful result of H. Alexander was the starting point for our conjecture.

THEOREM 1 (H. Alexander [1]). *Let Ω be the unit ball in $\mathbb{C}^p$, i.e. $\Omega = \Omega_{p,1}$ and let $p \geqslant 2$. Then every proper holomorphic mapping $\varphi \colon \Omega_{p,1} \to \Omega_{p,1}$ is an automorphism of the ball.*

Denote by S the distinguished boundary (Bergman's boundary) of the domain Ω. A proper holomorphic mapping $\varphi \colon \Omega \to \Omega$ is called *strictly proper* if $\operatorname{dist}(\varphi(z_\nu), S) \to 0$ for every sequence $z_\nu \in \Omega$ with the property $\operatorname{dist}(z_\nu, S) \to 0$. The next result generalizing Alexander's theorem follows from [2] and gives a convincing evidence in favour of Conjecture 1.

THEOREM 2 (G. M. Henkin, A. E. Tumanov [2]). *If Ω is an irreducible bounded symmetric domain in $\mathbb{C}^n$ and $\Omega \neq \mathbb{D}$, then any strictly proper holomorphic mapping $\varphi \colon \Omega \to \Omega$ is an automorphism.*

Only recently we managed to prove Conjecture 1 for some symmetric domains different from the ball, i.e. when $\partial\Omega \neq S$.

THEOREM 3 (G. M. Henkin, R. G. Novikov). *Let $\Omega \subset \mathbb{C}^n$, $n \geqslant 3$, be the classical domain of the 4-th type, i.e.*

$$\Omega = \big\{ z : z \cdot \bar{z}' + \big((z\bar{z}')^2 - |zz'|^2\big)^{1/2} < 1 \big\},$$

where $z = (z_1, \ldots, z_n)$ and z' stands for the transposed matrix. Then every proper holomorphic mapping $\Omega \to \Omega$ is an automorphism.

Note that the domain $\Omega_{2,2}$ of the first type is equivalent to a domain of the 4-th type. Hence Theorem 3 holds for $\Omega_{2,2}$.

We present now the scheme of the proof of Theorem 3 which gives rise to more general conjectures on the mappings of classical domains.

The classical domain of the 4-th type is known to have a realization as a tabular domain in $\mathbb{C}^n$, $n \geqslant 3$, over the round convex cone

$$\Omega = \left\{ z = x + iy \in \mathbb{C}^n : y_0^2 - y_1^2 - \cdots - y_{n-1}^2 > 0, \ y_0 > 0 \right\}.$$

The distinguished boundary of this domain coincides with the space $\mathbb{R}^n = \left\{ z \in \mathbb{C}^n : y = 0 \right\}$. The boundary $\partial\Omega$ contains together with each point $z \in \partial\Omega \setminus S$ the 1–dimensional analytic component $\mathcal{D}_z = \left\{ x + \lambda y : \lambda \in \mathbb{C}, \ \operatorname{Im}\lambda > 0 \right\}$. The boundary $\partial\mathcal{D}_z$ of this component is the nil-line in the pseudoeuclidean metric $ds^2 = dx_0^2 - dx_1^2 - \cdots - dx_{n-1}^2$ on the distinguished boundary S.

If φ is a map satisfying the hypotheses of the theorem, an appropriate generalization of H. Alexander's [1] arguments yields that outside of a set of zero measure on $\partial\Omega$ a boundary mapping (in the sense of nontangential limits) $\tilde{\varphi} \colon \partial\Omega \to \partial\Omega$ of finite multiplicity is well defined. This mapping possesses the following property: for almost every analytic component $\mathcal{D}_z$ the restriction $\tilde{\varphi}|\mathcal{D}_z$ is a holomorphic mapping of finite multiplicity of $\mathcal{D}_z$ into some component $\mathcal{D}_w$. Furthermore, almost all points of $\partial\mathcal{D}_z$ are mapped (in the sense of nontangential limits) into points of $\mathcal{D}_w$.

It follows then from the classical Frostman's theorem that $\tilde{\varphi}|\mathcal{D}_z$ is a proper mapping the half-plane $\mathcal{D}_z$ into the half-plane $\mathcal{D}_w$.

So it follows that the boundary map $\tilde{\varphi}$ defined a.e. on the distinguished boundary $S \subset \partial\Omega$ has the following properties:

   a) $\tilde{\varphi}$ maps S into S outside a set of zero measure;

   b) $\tilde{\varphi}$ restricted on almost any nil-line $\partial\mathcal{D}_z$ coincides (almost everywhere on $\partial\mathcal{D}_z$) with a piecewise continuous map of finite multiplicity of the nil-line $\partial\mathcal{D}_z$ into some nil-line $\partial\mathcal{D}_w$.

With the help of A. D. Alexandrov's paper [3] one can prove that the mapping $\tilde{\varphi} \colon S \to S$ satisfying a), b) is a conformal mapping with respect to the pseudoeuclidean metric on S. It follows that φ is an automorphism of the domain Ω.

To follow this sort of arguments, say, for the domains $\Omega_{p,2}$ where $p > 2$, one should prove a natural generalization of H. Alexander's and O. Frostman's theorems.

Let us call a holomorphic mapping φ of the ball $\Omega_{p,1}$ *almost proper* if φ is of finite multiplicity and for almost all $z \in \partial\Omega_{p,1}$ we have $\tilde{\varphi}(z) \in \partial\Omega_{p,1}$, where $\tilde{\varphi}(z)$ is the nontangential limit of the mapping φ defined almost everywhere on $\partial\Omega_{p,1}$.

CONJECTURE 2. *Let φ be an almost proper mapping of $\Omega_{p,1}$ and $p \geqslant 2$. Then φ is an automorphism.*

If we remove the words "φ is of finite multiplicity" from the above definition, the conclusion of Conjecture 2 may fail, in virtue of a result of A. B. Alexandrov [4].

Finally we propose a generalization of Conjecture 1.

CONJECTURE 3. *Let Ω be a symmetric domain in $\mathbb{C}^n$ different from any product domain $\Omega_1 \times G$, $G \subset \mathbb{C}^{n-1}$ and let S be its distinguished boundary. Let U' and U'' be two domains in $\mathbb{C}^n$ intersecting S and $\varphi \colon \Omega \cap U' \to \Omega \cap U''$ a proper mapping such that for some sequence $\{z_\nu\}$ with $z_\nu \in \Omega \cap U'$, $z_\nu \to z' \in S \cap U'$ we have: $\varphi(z_\nu) \to z'' \in S \cap U''$.*

Then there exists an automorphism Φ *of* Ω *such that* $\Phi|\Omega \cap U' = \varphi$.

The verification of Conjecture 3 would lead to a considerable strengthening of a result on local characterization of automorphisms of classical domains obtained in [2].

One can see from the proof of Theorem 3 that Conjecture 3 holds for classical domains of the fourth type. At the same time, it follows from results of [1] and [5] that Conjecture 3 holds also for the balls $\Omega_{p,1}$.

Remark. After the paper had been submitted the authors became aware of S. Bell's paper [6] that enables, in combination with [2], to prove Conjecture 1.

REFERENCES

1. Alexander H., *Proper holomorphic mappings in* $\mathbb{C}^n$, Indiana Univ. Math. J. **26** (1977), 137–146.
2. Tumanov A. E., Henkin G. M., *Local characterization of analytic automorphisms of classical domains*, Doklady Akad. Nauk SSSR **267** (1982), no. 4, 796–799 (Russian); English transl. in Soviet Math. Dokl. **26** (1982), no. 3, 702–705.
3. Aleksandrov A. D., *On the foundations of relativity theory*, Vestnik Leningr. Gos. Univ. **19** (1976), 5–28. (Russian)
4. Aleksandrov A. B., *The existence of inner functions in the ball*, Matem. Sb. **118** (1982), 147–163 (Russian); English transl. in Math. USSR Sbornik **46** (1983), no. 2 143–159.
5. Rudin W., *Holomorphic maps that extend to automorphism of a ball*, Proc. Amer. Soc. **81** (1981), 429–432.
6. Bell S. R., *Proper holomorphic mapping between circular domains*, Comm. Math. Helv. **57** (1982), 532–538.

ANALYSE COMPLEXE ET GÉOMÉTRIE
UNIVERSITÉ PARIS VI
4, PLACE JUSSIEU
75252 PARIS CEDEX 05
FRANCE

UNIVERSITÉ DE NANTES
DEPARTEMENT DE MATHÉMATIQUES
2 RUE DE LA HOUSSINIERE
44072 NANTES CEDEX 03
FRANCE

HOLOMORPHIC ENDOMORPHISMS OF REGIONS IN $\mathbb{C}^N$

LEE A. RUBEL

Let G be a region in complex N–space $\mathbb{C}^N$, and let $\operatorname{Aut} G$ denote the class of biholomorphic maps of G onto G, considered as an abstract group. Let $\operatorname{Aut}_0 G$ be the subgroup of $\operatorname{Aut} G$ consisting of those biholomorphisms whose Jacobian determinant is 1. Let $\mathbf{B}^N$ be the unit ball in $\mathbb{C}^N$,

$$\mathbf{B}^N = \left\{ (z_1, \ldots, z_n) \in \mathbb{C}^N : |z_1|^2 + \cdots + |z_N|^2 < 1 \right\}$$

Similarly, for a region G in $\mathbb{C}^N$, one could consider $\operatorname{End} G$ ('End' for 'endomorphism'), consisting of all holomorphic maps of G into G. $\operatorname{End} G$ is a semigroup under composition of maps. (Perhaps, in both these situations, G should be a domain of holomorphy).

To what extent does $\operatorname{End} G$ determine G? That is, if $\operatorname{End} G_1$ is isomorphic to $\operatorname{End} G_2$, must G_1 and G_2 be biholomorphically or antibiholomorphically equivalent? Aimo Hinkkanen has shown [4] that, in case $N = 1$, the answer is **no** if G_1 and G_2 are unrestricted, but Alexandre Eremenko [3] has shown that the answer is **yes** (for $N = 1$) if G_1 and G_2 admit bounded holomorphic functions. What is the situation when G_1 is, say, a bounded region in $\mathbb{C}^M$ and G_2 is a bounded region in $\mathbb{C}^N$, with M and N possibly different?

We also ask the corresponding question for $\operatorname{Aut} G$ and for $\operatorname{Aut}_0 G$, this time as **groups** under composition of maps. Even the question whether $\operatorname{Aut} \mathbb{C}^M$ is isomprphic to $\operatorname{Aut} \mathbb{C}^N$ seems challenging. Ideally, one would like to write down a group–theoretic characterization of $\operatorname{Aut} \mathbb{C}^N$, say. This would be one way of making quantitative the qualitative sentence in [6]: "But $\operatorname{Aut} \mathbb{C}^N$ is a huge and complicated group for $N > 1$."

Heuristically, say $\operatorname{Aut} \mathbb{C}^3$ seems to be a more complicated group than $\operatorname{Aut} \mathbb{C}^2$, so we would not expect them to be isomorphic. But this is very far from a proof of nonisomorphism.

Besides *isomorphism*, one could also consider *elementary equivalence*. For example, two groups Γ_1 and Γ_2 are said to be elementarily equivalent if every first order sentence in a language of group theory that is true in Γ_1 is true in Γ_2 and vice versa. A sentence is first–order means roughly that (see [8] for a precise definition) it is of finite length and quantifies over elements of the group, and not over subsets, functions, or relations. Thus, to say that a group is of order 5 **is** first–order, but to say that it is of finite order is apparently (and actually) **not** first–order, since it requires a sentence of infinite length. To say that a group is Abelian is indeed first–order, but to say that it is **simple** is apparently (and actually) **not** first–order, because it requires quantification over **subsets** (... there does not exist a normal semigroup ...). Isomorphic groups are elementary equivalent, but the converse is not true.

$\operatorname{Aut} \mathbf{B}^N$ has a concrete representation as a group of Möbius maps (see [7]). Could $\operatorname{Aut} \mathbb{C}^N$ be isomorphic to $\operatorname{Aut} \mathbf{B}^N$? Are $\operatorname{Aut} \mathbf{B}^M$ and $\operatorname{Aut} \mathbf{B}^N$ isomorphic if $M \neq N$? There remain many similar questions whose formulation is left for reader.

PROBLEM 17.2

References

1. Bedford E., *On the automorphism group of a Stein manifold*, Math. Ann. **266** (1983), 215–227.
2. Bedford E., Pinchuk S. I., *Domains in $\mathbb{C}^2$ with noncompact holomorphic automorphism group*, Mat. Sb. **135 (177)** (1988), 147–157. (Russian)
3. Eremenko A., *On the characterization of a Riemann surface by its semigroup of endomorphisms*, preprint, 1991.
4. Hinkkanen A. (1990), Private communications.
5. Peschl E., *Automorphismes holomorphes de l'éspace à n dimensions complexes*, C. R. Acad. Sci. Paris **242** (1956), 1836–1838.
6. Rosay J.-P., Rudin W., *Holomorphic maps from $\mathbb{C}^n$ to $\mathbb{C}^n$*, Trans. Amer. Math. Soc. **310** (1988), 47–86.
7. Rudin W., *Function Theory in the Unit Ball of $\mathbb{C}^n$*, Springer-Verlag, 1980.
8. Schoenfield J. R., *Mathematical Logic*, Addison-Wesley, Reading, Mass., 1967.

DEPARTMENT OF MATHEMATICS
UNIVERSITY OF ILLINOIS
1409 WEST GREEN STREET
URBANA, IL 61801
USA

THE CARATHÉODORY TOPOLOGY

M. Jarnicki, P. Pflug

Given a domain $G \subset \mathbb{C}^n$, we denote by c_G the Carathéodory pseudodistance for G. Assume that G is c-hyperbolic, i.e. c_G is a distance. It is known that:

(a) If $G \subset \mathbb{C}^1$ then the *Carathéodory topology* of G (i.e. the topology generated by c_G) coincides with the standard Euclidean topology of G.

(b) For any $n \geq 3$ there exists a domain of holomorphy $G \subset \mathbb{C}^n$ such that the Carathéodory topology is different from the standard one — cf. [4].

PROBLEM 1. *What about the topology for $n = 2$?*

Now, let G be a bounded balanced domain of holomorphy in $\mathbb{C}^n$. The following is known:

(a) If G is a Reinhardt domain then G is *finely compact* w.r.t. the Carathéodory distance, i.e. all c_G-balls are relatively compact in the Euclidean topology — cf. [5].

(b) If G is *complete* w.r.t. the Kobayashi distance then the Minkowski functional h_G of G is continuous — cf. [1].

(c) If h_G is continuous then G is finitely compact w.r.t. the Bergman distance — cf. [2].

(d) For any $n \geq 3$ there exists a bounded balanced domain of holomorphy $G \subset \mathbb{C}^n$ with continuous Minkowski functional such that G is not complete w.r.t. the Kobayashi distance, consequently, G is neither complete nor finitely compact w.r.t. the Carathéodory distance — cf. [3].

PROBLEM 2. *What about the completeness for $n = 2$?*

REFERENCES

1. Barth T., *The Kobayashi indicatrix at a center of a circular domain*, Proc. Amer. Math. Soc. **88(3)** (1983), 527–530.
2. Jarnicki M., Pflug P., *Bergman completeness of complete circular domains*, Ann. Pol. Math. **50** (1989), 219–222.
3. Jarnicki M., Pflug P., *A counterexample for Kobayashi completeness of balanced domains*, Proc. Amer. Math. Soc. (1991).
4. Jarnicki M., Pflug P., Vigué J.-P., *The Carathéodory distance does not define the topology – the case of domains*, C. R. Acad. Sci. Paris **312(I)** (1991), 77–79.
5. Pflug P., *About the Carathéodory completeness of all Reinhardt domains*, Functional Analysis, Holomorphy and Approximation Theory (G. I. Zapata, ed.), North Holland, Amsterdam, 1984, pp. 331–337.

JAGIELLONIAN UNIVERSITY
INSTITUTE OF MATHEMATICS
REYMONTA 4
30–059 KRAKÓW
POLAND

UNIVERSITÄT OSNABRÜCK
STANDORT VECHTA
FACHBEREICH NATURWISSENSCHAFTEN, MATHEMATIK
DRIVESTR. 22, W-2848 VECHTA
GERMANY

ON THE POSSIBILITY OF HOLOMORPHIC
EXTENSION INTO THE DOMAIN
OF FUNCTIONS DEFINED ON A PART OF ITS BOUNDARY

L. A. AIZENBERG

The question of the possibility of holomorphic extension into the domain of functions defined on the whole boundary is well studied [1,2]. Let us consider the problem of describing the functions defined on a part of the boundary which can be holomorphically extended into the domain. We do not assume that the given domain is the hull of holomorphy of this part of the boundary. For the review of the results on the solution of this problem both for complex one-dimensional and multidimensional cases, see Section 27 of the book [3]. We shall quote the simplest facts.

Suppose $n = 1$ and the domain Ω is bounded by a contour consisting of a segment of the real axis and a smooth open arc Γ lying in the upper half-plane. If $f \in L^1(\Gamma)$, then we can consider the moments

$$J_\sigma(f) = \int_\Gamma f(\zeta) \exp(-i\sigma\zeta)\, d\zeta.$$

If in [4] we could get rid of excessive restrictions, it will be possible to obtain the following result. Denote by $A(\Omega)$ the set of functions holomorphic in Ω.

THEOREM 1 (Fok–Kuni). *Let* $f \in L^1(\Gamma) \cap C(\Gamma)$. *Then a necessary and sufficient condition for the existence of a function* $F \in A(\Omega) \cap C(\Omega \cup \Gamma)$ *such that its restriction to* Γ *coincides with* f *is*

$$(1) \qquad \lim_{\sigma \to \infty} \frac{\ln^+ |J_\sigma(f)|}{\sigma} = 0.$$

We are going to give a result of close nature about the possibility of holomorphic extension from Γ to a given circle (for example, to the unit circle) which cannot be obtained from Theorem 1 by conformal mapping and is simpler both to formulate and to prove (see [5]).

Let Ω_1 be the domain bounded by the unit circle $\gamma_1 = \{ z : |z| = 1 \}$ and a smooth open arc Γ connecting two points of γ_1 and lying inside unit disc, with zero lying outside $\overline{\Omega_1}$. We denote

$$a_k = \int_\Gamma \frac{f(\zeta)\, d\zeta}{\zeta^{k+1}}, \qquad k = 0, 1, 2, \ldots.$$

THEOREM 2 (Aizenberg). *If* $f \in L^1(\Gamma) \cap C(\Gamma)$, *then there exists a function* $F \in A(\Omega_1) \cap C(\Omega_1 \cup \Gamma)$ *with* $F|_\Gamma = f$ *if and only if*

$$(2) \qquad\qquad \overline{\lim_{k \to \infty}} \sqrt[k]{|a_k|} \leqslant 1.$$

This result can be generalized to the case $n > 1$. Let us give one of possible facts in this direction [6]. Let Ω^n be a bounded convex n-circular domain, and let a hypersurface Γ divide Ω^n into two domains Ω^+ and Ω^-, with $0 \in \Omega^-$. Let $\psi(z) = \psi(|z|) = \psi(|z_1|, \ldots, |z_n|)$ be the Minkowski functional of the domain Ω^n and $d_\alpha(\Omega^n) = \max_{\overline{\Omega^n}} |z^\alpha|$, where $z^\alpha = z_1^{\alpha_1} \ldots z_n^{\alpha_n}$. Put $\langle u, z \rangle = u_1 z_1 + \cdots + u_n z_n$.

THEOREM 3 (Aizenberg–Kytmanov). *If* $\Gamma \in C^2$ *and* $f \in L^1(\Gamma) \cap C(\Gamma)$, *then there exists a function* $F \in A(\Omega^+) \cap C(\Omega^+ \cup \Gamma)$ *with* $F|_\Gamma = f$ *if and only if the following two conditions hold:*

(i) f *is a* CR *function on* Γ,

(ii)

$$(3) \qquad\qquad \overline{\lim_{|\alpha| \to \infty}} \sqrt[|\alpha|]{|c_\alpha| d_\alpha(\Omega^n)} \leqslant 1,$$

where

$$c_\alpha = \frac{(|\alpha| + n - 1)!}{\alpha!} \int_\Gamma f(\zeta) \left(\frac{\operatorname{grad}_\zeta \psi(\zeta)}{\langle \operatorname{grad}_\zeta \psi(\zeta), \zeta \rangle} \right)^\alpha \omega,$$

$$\omega = \frac{(n-1)!}{(2\pi i)^n} \frac{\sum_{k=1}^{n} (-1)^{k-1} \psi'_{\zeta_k} d \operatorname{grad} \psi[k] \wedge d\zeta}{\langle \operatorname{grad}_\zeta \psi(\zeta), \zeta \rangle^n},$$

$$d \operatorname{grad} \psi[k] = d\psi'_{\zeta_1} \wedge \cdots \wedge d\psi'_{\zeta_{k-1}} \wedge d\psi'_{\zeta_{k+1}} \wedge \cdots \wedge d\psi'_{\zeta_n},$$

$$d\zeta = d\zeta_1 \wedge \cdots \wedge d\zeta_n, \quad |\alpha| = \alpha_1 + \cdots + \alpha_n, \quad \alpha! = \alpha_1! \ldots \alpha_n!.$$

It should be noted that there does not exist a solution of the problem in question in the form of the requirement that a family of continuous linear functionals vanishes, therefore passing to the limit in conditions (1)–(3) is essential. These facts can be formulated without the condition $f \in L^1(\Gamma)$, but then requirements similar to (1)–(3) must be formulated for arcs (hypersurfaces) smaller than Γ. These facts can be extended to distributions (see [5,6]). The above results can also be interpreted from the viewpoint of harmonic analysis: the moments (generalized Fourier coefficients) $J_\sigma(f)$, a_k or c_α must not grow faster than it is prescribed.

All other results towards the solution of the problems posed are considerably more complicated (see [3]). In Theorems 1–3 the set Γ is a connected arc or a connected piece of a hypersurface. There are results for non-connected pieces of the boundary of a domain (sets of positive $(2n-1)$-dimensional measure), but they are complicated.

PROBLEM 1. *Find a simple description of the functions holomorphically extendable from a non-connected piece of the boundary of a domain into this domain.*

If the holomorphic extension in question exists, then it is possible to write down the Carleman formulas performing such an extension. For example, in the case of Theorem 2 such a formula has the form

$$(4) \qquad F(z) = \frac{1}{2\pi i} \lim_{m \to \infty} \int_\Gamma f(\zeta) \left(\frac{z}{\zeta}\right)^m \frac{d\zeta}{\zeta - z}$$

(see [3] for a survey on Carleman formulas for different cases). The problem of holomorphic extension is ill-posed (the extending functional is not continuous), therefore the limit in (4) (or something else that "spoils" the integral) is necessary. It should be noted that in the cases considered in Theorems 1–3 there are also simple Carleman formulas with holomorphic kernel, whereas in the general situation of a set of positive measure such formulas are rather complicated.

PROBLEM 2. *Do there exist simple Carleman formulas with kernel holomorphic in z for a non-connected piece of the boundary of the domain?*

CONJECTURE. *Problems 1 and 2 are closely connected. A positive solution of Problem 2 will yield a solution of Problem 1.*

REFERENCES

1. Privalov I. I., *Boundary properties of analytic functions*, Gostekhizdat, Moscow, 1950. (Russian)
2. Henkin G. M., Chirka E. M., *Boundary properties of holomorphic functions of several complex variables*, Modern problems of mathematics, Vol. 4, Itogi Nauki i Tekhniki, Vsesoyuz. Inst. Nauchn. i Tekhn. Inform., Moscow, 1975, pp. 13–142. (Russian)
3. Aizenberg L. A., *Carleman formulas in complex analysis. First applications*, Nauka, Novosibirsk, 1990. (Russian)
4. Fok V. A., Kuni F. M., *On the "cutting" function in dispersion relations*, Dokl. Akad. Nauk SSSR **127** (1959), 1195–1198. (Russian)
5. Aizenberg L. A., Kytmanov A. M., *On the possibility of holomorphic extension of functions defined on a connected piece of the boundary*, IF SOAN SSSR (Preprint no. 50M), Krasnoyarsk, 1990 (Russian); vol. 4, 1991.
6. Aizenberg L. A., Kytmanov A. M., *On the possibility of holomorphic extension of functions defined on a connected piece of the boundary. II*, Mat. Sbornik (to appear). (Russian)

INSTUTUTE OF PHYSICS,
AKADEMGORODOK,
KRASNOYARSK, 660036
RUSSIA

MAPPINGS BETWEEN CR MANIFOLDS

YU. V. KHURUMOV

Let $M_1, M_2 \subset \mathbb{C}^n$ be CR manifolds of class C^k ($2 \leqslant k \leqslant \infty$), and $f\colon M_1 \to M_2$ be a continuous CR mapping. We shall suppose further that at any point $z \in M_i$ ($i = 1, 2$) the values of Levi form of M_i fill a sharp cone on $N_z M_i = T_z M_i / T_z^c M_i$ with non-empty interior. If M_i is a hypersurface, this condition means that M_i is strongly pseudoconvex. The problems below imply that such manifolds may be considered as a natural generalization of strongly pseudoconvex hypersurfaces. The following conjecture is a generalization of Fefferman theorem on boundary regularity of biholomorphisms between strongly pseudoconvex domains.

CONJECTURE. *Suppose that* $\dim M_1 = \dim M_2$ *and* $CR \dim M_1 = CR \dim M_2$. *Then* $f \in C^{k-1/2-0}$.

If M_i are hypersurfaces, [1] and [2] give an affirmative answer to the Conjecture. Verification of this Conjecture may lead to the solution of three other problems.

PROBLEM 1. *Is it true that under the conditions of the Conjecture f is locally homeomorphic?*

PROBLEM 2 (Pinchuk [3]). *Is it true that if the mapping f of the Conjecture is homeomorphic, then f^{-1} is a CR mapping too?*

Problem 1 is a generalization of Pinchuk's theorem on local biholomorphy of proper holomorphic mappings between strongly pseudoconvex domains [4]. Problem 2 has been solved for hypersurfaces by Pinchuk in [3] under weaker Levi conditions. It should be noted that Pinchuk's formulation of Problem 2 ([3]) contains no assumptions on Levi forms of M_i. The third problem connected with the Conjecture is a generalization of the Alexander theorem about proper holomorphic mappings between balls.

PROBLEM 3. *Suppose M_i satisfy conditions of the Conjecture and, moreover, are quadratic manifolds of the form*

$$M_i = \big\{ (z_1, \ldots, z_l) = F_i(z_{l+1}, \ldots, z_n) \big\},$$

where F_i are Hermitian forms. Is then f rational with linear fractional components?

This problem has recently been settled by A. Abrosimov (oral communication) for some classes of quadratic manifolds of codimension 2. Moreover, for CR homeomorphisms rationality of f in Problem 3 was established by Tumanov and Khenkin [5].

The main Conjecture seems to be far from its complete verification now. Nevertheless, a plenty of methods developed for mappings between strongly pseudoconvex domains (see, for example, surveys [6–8]) provide efficient approaches to it, and Problems 1–3, as in the case of hypersurfaces, might be preliminary steps towards the Conjecture.

Anyway, verification of the Conjecture requires some new methods and so might be fruitful.

In case of CR manifolds of different dimensions the Conjecture fails: this follows from the existence of proper holomorphic mappings between balls of different dimensions, which are continuous up to the boundary and are smooth nowhere at the boundary (see [8]). Even if some initial smoothness of f is assumed, in case of different dimensions smoothness of the mapping may be two times smaller than the smoothness of manifolds [9]. It is interesting whether the Conjecture is valid for CR manifolds of equal CR dimensions, i.e. without the condition $\dim M_1 = \dim M_2$. However, there are also interesting questions concerning mappings between manifolds of different CR dimensions:

PROBLEM 4. *Suppose M_i are real analytic and f is sufficiently smooth (say, $f \in C^n$ or $f \in C^\infty$).*

    (1) *Does f extend holomorphically to a neighbourhood of M_1?*
    (2) *Does the family of all such mappings depend on finite number of parameters?*
    (3) *If M_i are algebraic, is then f algebraic?*
    (4) *If M_i are quadratic, is then f rational?*

As for part (1) of Problem 4, Pushnikov's theorem [10,11] guarantees the existence of a holomorphic extension of f into a neighbourhood in $\mathbb{C}^n$ of some open dense subset of M_1. The affirmative answer to (3) for hypersurfaces (algebraicity of f) was given by Rabotin [12,13]. Then this result was generalized to CR manifolds by Pushnikov with some restrictions on M_1 (implying, in particular, $CR \dim M_1 > n/2$). The affirmative answer to (4) is known in case of mappings between spheres (Rabotin [12,13], Forstnerič [14]); moreover, in concordance with (2) the family of such mappings depends on a finite number of parameters (Forstnerič [14]). However, the answer to question (2) is unknown even in case of CR isomorphisms between CR manifolds (see [7]).

REFERENCES

1. Pinchuk S. I., Tsyganov Sh. I., *Smoothness of CR-mappings of strongly pseudoconvex hypersurfaces*, Izv. Akad. Nauk SSSR Ser. Mat. **53** (1989), 1020–1029. (Russian)
2. Khurumov Yu. V., *Boundary smoothness of proper holomorphic mappings of strictly pseudoconvex domains*, Mat. Zametki **48** (1990), no. 6, 149–150. (Russian)
3. Pinchuk S., *CR transformations of real manifolds in $\mathbb{C}^n$*, preprint MPI/90-37, Max-Planck-Institut für Matematik, Bonn, 1990.
4. Pinchuk S. I., *Holomorphic non-equivalence of some classes of domains in $\mathbb{C}^n$*, Mat. Sb. **111** (1980), 67–94. (Russian)
5. Tumanov A. E., Henkin G. M., *Local characterization of holomorphic automorphisms of Siegel domains*, Funktsional. Anal. i Prilozhen. **17** (1983), no. 4, 49–61. (Russian)
6. Pinchuk S. I., *Holomorphic mappings in $\mathbb{C}^n$ and the holomorphic equivalence problem*, Current Problems in Mathematics. Fundamental directions, Vol. 9, Itogi Nauki i Tekhniki, Akad. Nauk SSSR, Vsesoyuz. Inst. Nauchn. i Tekhn. Inform., Moscow, 1986, pp. 195–223 (Russian); English transl. in Encyclopaedia of Mathematical Sciences, Vol. 9 Springer-Verlag, 1989.
7. Tumanov A. E., *Geometry of CR-manifolds*, Current Problems in Mathematics. Fundamental directions, Vol. 9, Itogi Nauki i Tekhniki, Akad. Nauk SSSR, Vsesoyuz. Inst. Nauchn. i Tekhn. Inform., Moscow, 1986, pp. 225–246 (Russian); English transl. in Encyclopaedia of Mathematical Sciences, Vol. 9 Springer-Verlag, 1989.
8. Forstnerič, *Proper holomorphic mappings: a survey*, Preprint Series Dept. Math. University E.K. Ljubljana **27** (1989), no. 268, 1–48.

9. Khurumov Yu. V., *Smoothness of proper holomorphic mappings and invariant metrics of strictly pseudoconvex domains*, In: Actual questions of complex analysis, Tashkent, 1989, p. 136. (Russian)

10. Pushnikov A. Yu., *Holomorph CR-mappings into a space of greater dimension*, Mat. Zametki **48** (1990), no. 3, 147–149. (Russian)

11. _____, *Symmetry principle and regularity of CR- mappings of real-analytic manifolds*, Thesis, Ufa, 1990. (Russian)

12. Rabotin V. V., *Rationality of proper holomorphic mappings of balls of different dimensions*, In: Complex analysis and mathematical physics, Krasnoyarsk, 1987, p. 91. (Russian)

13. _____, *Holomorphic mappings of complex manifolds and related extremal problems of CR-mappings of real-analytic manifolds*, Thesis, Novosibirsk, 1988. (Russian)

14. Forstnerič, *Extending proper holomorphic mappings of positive codimension*, Invent. Math. **95** (1989), 31–62.

CHAIR OF FUNCTION THEORY
MATHEMATICAL DEPARTMENT
KRASNOYARSK STATE UNIVERSITY
PR. SVOBODNYI 79
KRASNOYARSK, 660062
RUSSIA

POLYNOMIAL CONVEXITY OF TOTALLY REAL DISCS

B. JÖRICKE

Up to now is not understood even in quite simple cases which compact sets $K \subset \mathbb{C}^n$ ($n > 1$) are polynomially convex. (For the definition of polynomially convex sets and the discussion of interesting related problems see also [1]). So it is natural to consider the problem for special classes of compact sets. For example, the fact that K is contained in a submanifold of $\mathbb{C}^n$ makes the problem easier. Indeed, Stolzenberg [13] showed that a compact contractable subset of a smooth submanifold of dimension one is polynomially convex. For interesting generalizations and related problems see [1]. The next step would be deal with *manifolds of dimension* 2.

A submanifold M of $\mathbb{C}^n$ is called totally real if M has no holomorphic tangent vectors. Smooth curves are of course totally real. Totally real manifolds are of special interest in this connection by two reasons. First such manifolds are locally polynomially convex in the following sense: for an arbitrary point $p \in M$ and for any sufficiently small ball B in $\mathbb{C}^n$ about p the set $M \cap \overline{B}$ is polynomially convex [11]. The second reason comes from approximation theory: continuous functions on compact subsets K of M can be uniformly approximated by functions holomorphic in some neighborhood of K ([9]) and if K is polynomially convex then uniform approximation can be realized by polynomials (for example [11], [16]; in any case we will not give here a complete list of references).

So we pose the following

PROBLEM 1. *Which closed totally real discs in $\mathbb{C}^2$ are polynomially convex?*

By a closed totally real disc we mean a compact subset of a C^1 totally real submanifold, diffeomorphic to the closed planar disc. Wermer gave an example [11] of a closed totally real disc which is not polynomially convex. Recently Duval [4] obtained an interesting sufficient condition for a totally real disc Δ to be polynomially convex. Put $\mathbb{D} = \{z \in \mathbb{C} : |z| < 1\}$, $\overline{\mathbb{D}}$ its closure and $\mathbb{T}$ its boundary.

THEOREM 1. *Put $\Delta = \{(z, f(z)) \in \mathbb{C}^2 : z \in \overline{\mathbb{D}}\}$, where $f \in C^1$ in a neighborhood of $\overline{\mathbb{D}}$ and $\left|\left(\frac{\partial}{\partial \overline{z}} f\right)(a)\right| > \left|\left(\frac{\partial}{\partial z} f\right)(a)\right|$ for each $a \in \overline{\mathbb{D}}$. Then Δ is polynomially convex.*

New examples of totally real polynomially convex discs not defined as graphs over $\mathbb{D}$ were obtained from another point of view, namely from the description of removable singularities of analytic functions in multi-dimensional domains ([14], [10], [8]).

THEOREM 2. *Closed totally real discs of class C^1 in the unit sphere in $\mathbb{C}^2$ are polynomially convex.*

This was first obtained for totally real discs of class C^2 as a combination of the results of Stout and Lupacciolu [14] and Jöricke [10] and then sharpened in [8]. At the beginning it was surprising that the theorem comes from results on removable singularities and one asked for a direct proof. But analyzing, on the one hand, Oka's characterization principle for the polynomially convex hull $\hat{K}$ of a compact set K ([12], p. 264), and on the other hand, the Kontinuitätssatz [15] as a tool of analytic continuation one can easily understand that the two problems are closely related at least in $\mathbb{C}^2$.

So it is not hard to see that the proof in [10] is indeed a direct proof of polynomial convexity of closed totally real discs of class C^2 in the unit sphere in $\mathbb{C}^2$. Even a bit more is proved in [10].

THEOREM 3. *Let K be a closed totally real disc on a totally real C^2 manifold M in $\mathbb{C}^2$ and let K' be a closed totally real disc in M with K contained in the (relative) interior of K'. If there exists a neighborhood V of K' (in $\mathbb{C}^2$) with $\hat{K} \cap V$ contained in a wedge with edge K' then K is polynomially convex.*

Lemma 1 of [10] means that the condition of Theorem 3 is satisfied for closed totally real C^2 discs on the unit sphere in $\mathbb{C}^2$ and Theorem 2 follows from Theorem 3. Note that it is useful to analyze the proof in [8] from the point of view of using Oka's characterization principle for proving polynomial convexity of given sets.

Recently Alexander proved that smooth closed totally real discs in the real hypersurface $M = \{(z,w) : |z| = 1\} \subset \mathbb{C}^2$ are polynomially convex [2].

The second problem concerns the description of the polynomially convex hull of a totally real disc. In Wermer's example the failure of polynomial convexity of the disc Δ has a simple reason: Δ bounds an analytic disc (that is, there is a non-constant holomorphic map $f : \mathbb{D} \to \mathbb{C}^2$ with f continuous on $\overline{\mathbb{D}}$ and $f(\mathbb{T}) \subset \Delta$). Recently Duval [5] proved that this is not the only obstruction for polynomial convexity of discs. He constructed a disc which is rationally convex (defined in the same way as polynomially convex with polynomials replaced by rational functions), does not bound analytic discs but is not polynomially convex. This makes the following problem of special interest.

PROBLEM 2. *Describe polynomial hulls of totally real discs.*

PROBLEM 3. *Which totally real discs do not bound analytic discs?*

For Problem 3 see also [5], [7], [3].

PROBLEM 4. *Suppose a totally real disc Δ is not polynomially convex but does not bound analytic discs. Is $\hat{\Delta} \setminus \Delta$ a limit of analytic varieties (in an appropriate sense)?*

(Compare also with Problem 6 in [1]).

Note that the problems can be posed in the more general situation of discs which are totally real outside a finite number of points ([8], [7], [6]).

REFERENCES

1. Alexander H., *This collection*, 17.12.
2. Alexander H., *Totally real sets in $\mathbb{C}^2$*, Proc. Amer. Math. Soc. **111** (1991), 131–133.
3. Duchamp T., *Intersections of totally real and holomorphic discs*, preprint, 1990.
4. Duval J., *Un example de disque polynômialement convexe*, Ann. Math. **281** (1988), 583–588.
5. Duval J., *Convexité rationelle des surfaces Lagrangiennes*, Invent. Math. **104** (1991), 581–599.
6. Forstnerič F., *Complex tangent of real surfaces in complex surfaces*, preprint, 1991.
7. Forstnerič F., *A smooth holomorphically convex disc in $\mathbb{C}^2$ that is not locally polynomially convex*, preprint, 1991.
8. Forstnerič F., Stout E. L., *A new class of polynomially convex sets*, Ark. Mat. **29** (1991), 51–62.

9. Hörmander L., Wermer J., *Uniform approximation on compact subsets in* $\mathbb{C}^n$, Math. Scand. **22** (1968), 5–21.
10. Jöricke B., *Removable singularities of CR-functions*, Ark. Mat. **26** (1988), 117–143.
11. Nirenberg R., Wells R. O., *Approximation theorems on differentiable submanifolds of a complex manifold*, Trans. Amer. Math. Soc. **142** (1969), 15–35.
12. Stolzenberg G., *Polynomially and rationally convex sets*, Acta Math **109** (1963), 259–289.
13. Stolzenberg G., *Uniform approximation on smooth curves*, Acta Math. **115** (1966), 185–198.
14. Stout E. L., *Removable singularities for the boundary values of holomorphic functions* (to appear).
15. Vladimirov V. S., *Fonctions de plusieurs variables complexes*, Durord, Paris, 1982.
16. Wermer J., *Polynomially convex discs*, Math. Ann. **158** (1965), 6–10.

MAX-PLANCK-INSTITUT FÜR MATHEMATIK
GOTTFRIED-CLAREN-STR. 26
W-5300 BONN 3
GERMANY

$\bar{\partial}$ IN THE FUTURE TUBE

A. G. Sergeev

Future tube in $\mathbb{C}^{n+1}$ is a tube domain

$$\tau^+ = \{\, z = x + iy \in \mathbb{C}^{n+1} : y_0^2 > y_1^2 + \cdots + y_n^2,\ y_0 > 0 \,\} = \mathbb{R}^{n+1} + iV^+$$

over the future light cone V^+ in $\mathbb{R}^{n+1}$. For $n = 0$ it is just the upper halfplane in $\mathbb{C}^1$, for $n = 1$ it coincides with the domain $\{\, z \in \mathbb{C}^2 : y_0 > |y_1| \,\}$ linearly isomorphic to the product of two upper halfplanes. So the non-trivial case corresponds to $n \geqslant 2$ (what is always assumed here). The importance of the future tube (besides its relation to the quantum field theory) is due, on the one hand, to the fact that it is one of the main examples of tubes over convex homogeneous cones or Siegel domains of the 1st kind. On the other hand, it is the simplest example of domains which do not belong to the class of (strictly) pseudoconvex polyhedra introduced in [1]. This follows from an interesting complex-geometric property of τ^+ formulated in the next section.

Structure of the boundary. The Shilov (or distinguished) boundary of τ^+ coincides with $\mathbb{R}^{n+1} = \{\, z = x + iy \in \mathbb{C}^{n+1} : y = 0 \,\}$. The non-real part of the boundary $S = \partial\tau^+ \setminus \mathbb{R}^{n+1}$ is smooth and through any point $\zeta = \xi + i\eta \in S$ it goes a complex halfline $\lambda_\zeta = \{\, z = \xi + \alpha\eta : \alpha \in \mathbb{C},\ \operatorname{Im}\alpha \geqslant 0 \,\}$, called complex light ray, lying entirely on the boundary. In the other complex tangential directions through ζ the Levi form of τ^+ is positive, so $\partial\tau^+$ looks locally as a direct product of the complex line $\mathbb{C}^1$ with a strictly pseudoconvex boundary in $\mathbb{C}^n$. Moreover, the map φ given by the formula $\varphi(z_0, z_1, \ldots, z_n) = (z_0, z_1/y_0, \ldots, z_n/y_0)$ transforms τ^+ into the direct product $\mathbb{C}^1 \times D$, where $D = \{\, w \in \mathbb{C}^n : (\operatorname{Im} w_1)^2 + \cdots + (\operatorname{Im} w_n)^2 < 1 \,\}$ is a convex, strictly pseudoconvex domain in $\mathbb{C}^n$. So for any $\zeta \in S$ the map φ gives a local diffeomorphism of a neighbourhood of ζ in $\partial\tau^+$ onto an open subset of $\mathbb{C}^1 \times \partial D$. But there exist no local biholomorphisms with that property. More precisely, let ζ be a point in S and U an open neighbourhood of ζ in $\mathbb{C}^{n+1}$.

Theorem [2]. *There exist no biholomorphisms ψ of U onto an open set $U \subset \mathbb{C}^{n+1}$ mapping biholomorphically $\tau^+ \cap U$ onto $(\mathbb{C}^1 \times D') \cap V$, where D' is a strictly pseudoconvex domain in $\mathbb{C}^n$.*

In other words, $\partial\tau^+$ cannot be (locally) biholomorphically straightened along complex light rays. It follows from the theorem that τ^+ is not (locally) a strictly pseudoconvex polyhedron. As was proved by Tsyganov [3], the above theorem is true also for biholomorphic maps of $\tau^+ \cap U$ continuous up to $\partial\tau^+ \cap U$ and for CR-homeomorphisms of $\partial\tau^+ \cap U$.

$\bar{\partial}$-equation. Consider the $\bar{\partial}$-equation in τ^+

$$(1) \qquad\qquad \bar{\partial}v = u, \qquad \bar{\partial}u = 0,$$

where u is a smooth $(0,1)$-form in τ^+ which is bounded in τ^+ and vanishes identically outside some ball $\{|z| < R\}$. We are looking for smooth functions v in τ^+ satisfying (1).

CONJECTURE 1. *The $\bar{\partial}$-equation (1) in τ^+ cannot be solved with uniform estimates. In other words, it is impossible to find for any smooth bounded form u of the above type a solution v of (1) satisfying the estimate $\|v\|_\infty < \text{const} \|u\|_\infty$ with a constant in dependent of u.*

This problem was first formulated in [1]. It is important here that we are considering the local solvability of the $\bar{\partial}$-equation in τ^+ (because $u \equiv 0$ outside some ball), otherwise the problem is solved by a counterexample of N. Sibony.

Explicit integral representations. The conjecture can be checked for some particular solutions of the $\bar{\partial}$-equation. Denote by K the Cauchy–Bochner integral operator (an analogue of the Cauchy integral operator for the polydisc) given by

$$(2) \qquad (Kf)(z) = \frac{1}{(2\pi i)^{n+1}} \int_{\mathbb{R}^{n+1}} \mathcal{K}(z - \xi) f(\xi) \, d^{n+1}\xi,$$

where

$$\mathcal{K}(z) = \frac{2^n \Gamma\left(\frac{n+1}{2}\right)}{\pi^{(n-1)/2}} \frac{1}{(z_0^2 - z_1^2 - \cdots - z_n^2)^{(n+1)/2}}$$

is the Cauchy–Bochner kernel. The operator K is well-defined for Sobolev functions $f \in \mathcal{H}_S(\mathbb{R}^{n+1})$. It projects $\mathcal{H}_S$ onto the space of boundary values of functions f holomorphic in τ^+ and having uniformly bounded $\mathcal{H}_S$-norms $\|f(\cdot + iy)\|_{\mathcal{H}_S}$ (cf. Vladimirov [4]). We call by the Cauchy–Bochner solution a solution v of the $\bar{\partial}$-equation (1) orthogonal to holomorphic functions w.r. to K, i.e. such that $Kv(z) \equiv 0$.

THEOREM 2 [5]. *For $n > 3$ there exists a right hand side u of the $\bar{\partial}$-equation (1) in τ^+ bounded in τ^+ such that the Cauchy–Bochner solution $v(z)$ of (1) grows as a power $\left[\text{dist}(z, \mathbb{R}^{n+1})\right]^{-(n-3)/2}$ when $z \to \mathbb{R}^{n+1}$.*

The same phenomenon occurs for the other known particular solutions of the $\bar{\partial}$-equation. The question for the general solution of the $\bar{\partial}$-equation remains open, we can construct only the following counterexample of Sibony's type.

Sibony-type counterexample. Consider the future tube τ^+ in $\mathbb{C}^3$ (i.e. $n = 2$). We construct a polynomially convex compact X^+ on the boundary of τ^+ such that $X^+ = X \cap \tau^+$, where $X = \bigcup_{k=1}^{\infty} X_k$ is a union of complex discs X_k lying on complex light lines going through the origin. The compact X^+ has the following property. There exists a sequence of smooth strictly pseudoconvex neighbourhoods U_k of X^+

$$U_1 \supset U_2 \supset \cdots \supset U_k \supset \cdots \supset X^- = \bigcap_{k=1}^{\infty} U_k$$

and a uniformly bounded sequence of r.h.s. u_k of the $\bar{\partial}$-equations

$$\bar{\partial} v_k = u_k \qquad \text{in } U_k$$

such that any sequence of solutions v_k of these equations if unbounded, i.e. $\|v_k\|_{X^+} \to \infty$ for $k \to \infty$. Shortly, $\bar{\partial}$-equation cannot be solved uniformly in a neighbourhood of X^+. The construction of X^+ given in [6] is close to the construction of Sibony [7].

One can think that the above counterexample is very near to the counterexample to the conjecture. Unfortunately this is not so because we have an example of a domain containing the same Sibony-type compact on the boundary, where the $\bar{\partial}$ can be solved with uniform estimates. This domain is a multidimensional analogue of the Hartogs triangle so we call it the Hartogs cone:

$$H_n = \{ z \in \mathbb{C}^{n+1} : |z_0|^2 > |z_1|^2 + \cdots + |z_n|^2 \}.$$

Note that H_n is a strictly pseudoconvex polyhedron (it coincides with τ^+ up to the 2nd order in the origin).

Some general remarks. Assuming that the conjecture is true, i.e. that the $\bar{\partial}$-equation in τ^+ cannot be solved uniformly, we can ask which properties of τ^+ can be responsible for this phenomena. As we have seen, our domain τ^+ has the following specific properties:

(1) the boundary $\partial\tau^+$ is not smooth along all Shilov boundary;

(2) the smooth part of the boundary is foliated by complex submanifolds of the boundary and the boundary cannot be (locally) holomorphically straightened along these submanifolds;

(3) in any point of the boundary there exists a compact at least countable set (parameterized by the points of the sphere S^{n-1} in our domain) of complex submanifolds of the boundary intersecting only in this point.

We are interested in the following question. Suppose a pseudoconvex D has one (or several) of the listed properties. Can we expect that the $\bar{\partial}$-equation in this domain cannot be solved uniformly (in the sense of the conjecture)? Property (1) cannot imply the uniform non-solvability because the direct product of upper halfplanes (biholomorphic to a polydisc) has the same property. Also, Property (2) cannot imply the uniform non-solvability. To see this it is sufficient to consider as D the intersection of τ^+ with the ball $B(z^0, 1/2)$ with the center $z^0 = (i, i, 0, \ldots, 0)$ and radius $1/2$. The part $\partial\tau^+ \cap B(z^0, 1/2)$ of the boundary of this domain is foliated by the holomorphically non-straightenable system of complex light rays but the $\bar{\partial}$ in this domain can be solved uniformly by the result of Sh. A. Dautov and G. M. Henkin (cf. [8], Ch. 4, Ex. 7c). If Property (3) is satisfied in some point of the Shilov boundary we can construct a Sibony-type compact at this point lying on the boundary of the domain. But, as the example of the Hartogs cone shows, this fact cannot imply the uniform non-solvability. Still, we think that the existence of Sibony-type compacts in all points of an open subset of the Shilov boundary of D can be the reason for uniform non-solvability of $\bar{\partial}$ in D (at least, we do not know any counterexamples to this assertion).

Lie ball. To avoid technical problems connected with the unboundedness of the domain τ^+, we can consider instead the analogue of the above conjecture for the bounded domain, Lie ball. Recall (cf., e.g. [9]) that τ^+ is biholomorphically equivalent to the bounded Cartan domain of the IVth type, Lie ball defined as follows

$$L = \{ w \in \mathbb{C}^{n+1} : |w_0^2 + \cdots + w_n^2|^2 + 1 > 2|w_0|^2 + \cdots + 2|w_n|^2, \ |w_0^2 + \cdots + w_n^2| < 1 \}.$$

Hence we can formulate a bounded analogue of the Conjecture 1 as follows.

CONJECTURE 2. *The $\bar{\partial}$-equation in L cannot be solved uniformly, i.e. it is impossible to find for any $\bar{\partial}$-closed smooth bounded $(0,1)$-form u in L a smooth solution v of $\bar{\partial}v = u$ in L satisfying the estimate $\|v\|_\infty < \text{const} \|u\|_\infty$ with a constant not depending on u.*

References

1. Sergeev A. G., Henkin G. M., *Uniform estimates of solutions of $\bar\partial$-equation in pseudoconvex polyhedra*, Mat. Sb. **112** (1980), no. 4, 522–567. (Russian)
2. Sergeev A. G., *Complex geometry and integral representations in the future tube*, Izv. Akad. Nauk SSSR Ser. Mat. **50** (1986), no. 6, 1241–1275. (Russian)
3. Tsyganov Sh. I., *Holomorphic non-rectifiability of the boundary of the future tube*, In: Actual questions of complex analysis, Tashkent, 1989, p. 136. (Russian)
4. Vladimirov V. S., *Generalized functions in mathematical physics*, Nauka, Moscow, 1979. (Russian)
5. Sergeev A. G., *On complex analysis in tube cones*, In: Proc. Summer Inst. on Several Compl. Variables, AMS, Providence, 1990.
6. Sergeev A. G., *On behaviour of solutions of the $\bar\partial$-equation on the boundary of the future tube*, Dokl. Akad. Nauk SSSR **298** (1988), no. 2, 294–298. (Russian)
7. Sibony N., *Un exemple de domaine pseudoconvexe régulier ou l'équation $\bar\partial u = f$ n'admet pas de solution bornée pour f bornée*, Invent. Math. **62** (1980), no. 2, 235–242.
8. Henkin G. M., Leiterer J., *Theory of Functions on Complex Manifolds*, Akademie-Verlag, Berlin, 1984.
9. Vladimirov V. S., Sergeev A. G., *Complex analysis in the future tube*, Current Problems in Mathematics. Fundamental directions, Vol. 8, Itogi Nauki i Tekhniki, Akad. Nauk SSSR, Vsesoyuz. Inst. Nauchn. i Techn. Inform., Moscow, 1985, pp. 191–266 (Russian); English transl. in Encyclopedia of Mathematical Sciences, Vol. 8 Springer-Verlag, 1989.

V. A. Steklov Mathematical Institute
Russian Academy of Sciences
Vavilova 42, Moscow, 117966
Russia

AROUND THE EXTENDED FUTURE TUBE CONJECTURE

A. G. SERGEEV

Formulation. Let τ^+ (resp. τ^-) be the future (resp. past) tube in $\mathbb{C}^4$, i.e. a tube domain $\tau^\pm = \mathbb{R}^4 + iV^\pm$ over the future (resp. past) light cone

$$V^\pm = \{ z = x + iy \in \mathbb{C}^4 : y_0^2 > y_1^2 + y_2^2 + y_3^2, \pm y_0 > 0 \}.$$

Denote by L the Lorentz group of linear transformations of $\mathbb{R}^4$ fixing the origin and preserving the Lorentz metric

$$x \cdot x' = x_0 x_0' - x_1 x_1' - x_2 x_2' - x_3 x_3', \qquad x, x' \in \mathbb{R}^4.$$

This group has four components, we denote by $L_+^\uparrow$ the identity component of L. Note that both τ^+ and τ^- are invariant under the natural extension of the action of $L_+^\uparrow$ to $\mathbb{C}^4$. Consider now the complex Lorentz group $L(\mathbb{C})$ which is the complexification of L isomorphic to the complex orthogonal group $O(4, \mathbb{C})$; $L(\mathbb{C})$ consists of complex linear transformations of $\mathbb{C}^4$ fixing the origin and preserving the complex Lorentz metric: $z \cdot z' = z_0 z_0' - z_1 z_1' - z_2 z_2' - z_3 z_3'$, $z, z' \in \mathbb{C}^4$. This group has two connected components, we denote by $L_+(\mathbb{C})$ the identity component of $L(\mathbb{C})$ consisting of transformations $A \in L(\mathbb{C})$ with $\det A = 1$.

Define the future (resp. past) N-tube as

$$\tau^\pm = \tau^\pm \times \cdots \times \tau^\pm \ (N \text{ times}) \subset \mathbb{C}^{4N}$$

and consider the diagonal action of the complex Lorentz group $L(\mathbb{C})$ on $\mathbb{C}^{4N}$ given by

$$\Lambda : (z^1, \ldots, z^N) \to (\Lambda z^1, \ldots, \Lambda z^N),$$

where $\Lambda \in L(\mathbb{C})$, $z^1, \ldots, z^N \in \mathbb{C}^4$. The extended future tube τ_N' is the image of τ_N^+ under the action of $L(\mathbb{C})$, i.e.

$$\tau_N' = \{ (\Lambda z^1, \ldots, \Lambda z^N) \in \mathbb{C}^{4N} : (z^1, \ldots, z^N) \in \tau_N^+, \Lambda \in L_+(\mathbb{C}) \}.$$

EXTENDED FUTURE TUBE (EFT-) CONJECTURE. *The extended future tube τ_N' is a domain of holomorphy in $\mathbb{C}^{4N}$.*

This conjecture is known in the quantum field theory (QFT) for more than 30 years (cf. survey articles Vladimirov [1], Vladimirov, Zharinov [2], Wightman [3] and references therein).

Why the EFT-conjecture is interesting? According to the axiomatic QFT (cf. Jost [4], Streater, Wightman [5]) we can describe physical properties of a quantum system using the Wightman functions W_N which are holomorphic functions in τ_N^+ invariant under the diagonal action of the Lorentz group $L_+^\uparrow$. Functions of this sort have the following extension property.

BARGMANN–HALL–WIGHTMAN (BHW-) THEOREM (cf.[4]). *Any holomorphic $L_+^\uparrow$-invariant function on τ_N^+ can be extended to a holomorphic $L_+(\mathbb{C})$-invariant function on τ_N'.*

This theorem explains how the extended future tube enters the QFT. Note that τ_N^+, being a convex domain, is a domain of holomorphy but it has a non-trivial hull with respect to L_N^+-invariant holomorphic functions. So the natural question connected with the EFT-conjecture is the following.

QUESTION. *Is the extended future tube τ_N' holomorphically convex with respect to the class of $L_+(\mathbb{C})$-invariant holomorphic functions on τ_N'?*

The importance of the EFT-conjecture is also due to the fact that there are some assertions in QFT, such as the finite covariance theorem of Bogolubov–Vladimirov [6], proved only assuming that this conjecture is true.

Jost points. In contrast with τ_N^+ the extended future tube τ_N' contains many real points. They are called the Jost points and can be described as follows. The point $(x_1, \ldots, x^N) \in \mathbb{R}^{4N}$ is a Jost point if and only if the N-tuple $\{x^1, \ldots, x^N\}$ is spacelike, i.e., the convex cone

$$k(x^1, \ldots, x^N) = \left\{ x = \sum_{i=1}^N \lambda_i x^i : \lambda \geqslant 0, \sum \lambda_i > 0 \right\}$$

in $\mathbb{R}^{4N}$ generated by $x^1, \ldots, x^N$ contains only spacelike points, i.e. the points

$$x = (x_0, x_1, x_2, x_3) \text{ with } x \cdot x = x_0^2 - x_1^2 - x_3^2 < 0.$$

GLASER–STREATER THEOREM (cf.[4]). *The holomorphically convex hull of the set $\tau_N^+ \cup \tau_N^- \cup J_N$, where J_N is the set of Jost points in τ_N', contains τ_N'.*

By this theorem, to prove the EFT-conjecture it is sufficient to prove that the holomorphic hull of $\tau_N^+ \cup \tau_N^- \cup J_N$ is contained in τ_N'. Note that the holomorphic hull of $\tau_N^+ \cup \tau_N^- \cup J_N$ is necessarily a domain of Dyson type, i.e. is of the form $\tau_N^+ \cup \tau_N^- \cup \tilde{J}_N$, where $\tilde{J}_N$ is a domain in $\mathbb{C}^{4N}$ such that $\tilde{J}_N \cap \mathbb{R}^{4N} = J_N$.

Particular cases. For $N = 1$ it is easy to show that

$$\tau_1' = \left\{ z \in \mathbb{C}^4 : z \cdot z = z_0^2 - z_1^2 - z_2^2 - z_3^2 \text{ does not belong to } \mathbb{R}^+ = [0, +\infty) \right\}.$$

In other words, τ_1' is the inverse image of $\mathbb{C} \setminus \mathbb{R}^+$ under the holomorphic mapping $z \cdot z : \mathbb{C}^4 \to \mathbb{C}^1$, so it is a domain of holomorphy. For $N = 2$ the EFT-conjecture can be proved by using the explicit description of the boundary of τ_2'. There are also results describing the boundary of τ_N' in the general case (cf.[1-3], [7] and references therein).

It is interesting to consider also the 2-dimensional analogue of the EFT-conjecture. For the formulation of this analogue replace the future tube τ^+ in $\mathbb{C}^4$ in the definition of τ'_N by the future tube $\tilde{\tau}^+$ in $\mathbb{C}^2$ given by

$$\tilde{\tau}^+ = \{ z = x + iy \in \mathbb{C}^2 \colon y_0 > |y_1| \},$$

and the complex Lorentz group $L_+(\mathbb{C})$ in $\mathbb{C}^4$ by its 2-dimensional analogue $\tilde{L}_+(\mathbb{C})$. Defining the 2-dimensional extended future tube $\tilde{\tau}'_N$ as before, it is easy to prove (cf.[4]) that $\tilde{\tau}'_N$ is a domain of holomorphy in $\mathbb{C}^{2N}$ for any N.

Compact version. In [8] a conjecture was proposed which can be considered as a compact version of the EFT-conjecture. To formulate this compact version we need to do the following changes in the definition of τ'_N. The future tube τ^+ has to be replaced by its bounded counterpart, the matrix disc

$$\Delta = \{ Z \in \mathbb{C}[2 \times 2] \colon I - Z^* Z \text{ is positive definite} \},$$

which is a domain in the space $\mathbb{C}[2 \times 2]$ of complex 2×2 matrices. The Lorentz group $L_+^\uparrow$ has to be replaced by its compact analogue $K = \mathrm{SU}(2) \times \mathrm{SU}(2)$ and the complex Lorentz group $L_+^\uparrow(\mathbb{C})$ by the group $K^{\mathbb{C}} = \mathrm{SL}(2, \mathbb{C}) \times \mathrm{SL}(2, \mathbb{C})$ acting on $\mathbb{C}[2 \times 2]$ by the formula

$$K^{\mathbb{C}} \colon Z \to AZB^{-1}, \qquad A, B \in \mathrm{SL}(2, \mathbb{C}).$$

The resulting domain (the extended matrix disc Δ'_N) is defined as follows:

$$\Delta'_N = K^{\mathbb{C}} \cdot \Delta^N = \{ (AZ_1 B^{-1}, \ldots, AZ_N B^{-1}) \colon (Z_1, \ldots, Z_N) \in \Delta^N, \ A, B \in \mathrm{SL}(2, \mathbb{C}) \}.$$

The extended matrix disc conjecture, which can be considered as a compact version of the EFT-conjecture, asserts that Δ'_N is a domain of holomorphy. This conjecture was proved in [9] and [10].

REFERENCES

1. Vladimirov V. S., *Analytic functions of several complex variables and axiomatic quantum field theory*, Actes Congres Int. Math. (Nice, 1970), vol. 3, Gauthier–Villars, Paris, 1971, pp. 21–26.
2. Vladimirov V. S., Zharinov V. V., *Analytic methods in mathematical physics*, Trudy Math. Inst. Steklov **175** (1988), 117–138. (Russian)
3. Wightman A. S., *Quantum field theory and analytic functions of several complex variables*, J. Indian Math. Soc. **24** (1961), 625–677.
4. Jost R., *General theory of quantum fields*, Springer–Verlag, Berlin, Heidelberg, New York, 1960.
5. Stritter R., Wightman A., *Spin, statistics and all that*, Springer–Verlag, Berlin, Heidelberg, New York, 1960.
6. Bogolyubov N. N., Vladimirov V. S., *Representation of n-points functions*, Trudy Math. Inst. Steklov **112** (1971), 5–21. (Russian)
7. Zavyalov B. I., Trushin V. B., *On extended n-poins tube*, Teoret. Mat. Fiz. **27** (1976), no. 1, 3–15. (Russian)
8. Sergeev A. G., *On matrix Reinhardt domains*, preprint, Mittag–Leffler Inst., Stockholm, 1989.
9. Heinzner P., *Invariantentheorie in der komplexen Analysis*, preprint, Ruhr Univ., Bochum, 1990.
10. Heinzner P., Sergeev A., *The extended future tube conjecture for the compact model*, preprint, Ruhr Univ., Bochum, 1990.

V. A. STEKLOV MATHEMATICAL INSTITUTE
RUSSIAN ACADEMY OF SCIENCES
VAVILOVA 42, MOSCOW, 117966
RUSSIA

JESSEN FUNCTION OF A HOLOMORPHIC
ALMOST PERIODIC FUNCTION OF SEVERAL VARIABLES

L. I. RONKIN

For a set $E \subset \mathbb{R}^n$ let $T_E = \{ z \in \mathbb{C}^n : \operatorname{Re} z \in \mathbb{R}^n, \operatorname{Im} z \in E \}$, $\|f\|_E = \sup\{ |f(z)| : z \in T_E \}$. A function $f(z)$ holomorphic in a domain T_G is said to be almost periodic (a.p.) in T_G if for every $\varepsilon > 0$ and every set $E \subset\subset G$ a number $\ell > 0$ exists such that the set $\{ h \in \mathbb{R}^n : \|f(z + h) - f(z)\|_E < \varepsilon \}$ (the set of (ε, E)-almost periods) has a non-empty intersection with any cube $Q_\ell(a) = \{ x \in \mathbb{R}^n : |x_j - a_j| < \ell; j = 1, \ldots, n \}$, $a = (a_1, \ldots, a_n) \in \mathbb{R}^n$.

The fundamentals of the theory of holomorphic a.p. functions of one variable, i.e. a.p. functions in a strip, can be found, for example, in [1]. One of the main results of the theory is the following

THEOREM (Jessen). *Let $f(z)$, $z \in \mathbb{C}$ be a holomorphic almost periodic function in the strip $a < \operatorname{Im} z < b$. Then*

(1) *for every $y \in (a, b)$ the limit*

$$\lim_{t \to \infty} \frac{1}{2t} \int_{-t}^{t} \ln |f(x + iy)| \, dx \overset{\text{def}}{=} \varphi_f(y)$$

exists;

(2) *the function $\varphi_f(y)$ is convex;*

(3) *if the function $\varphi_f(y)$ is differentiable at points α and β then*

$$\exists \lim_{t \to \infty} \frac{1}{2t} n_f(t, \alpha, \beta) = \varphi'_f(\beta) - \varphi'_f(\alpha),$$

where $n_f(t, \alpha, \beta)$ is the number of zeros of $f(z)$ in the rectangle $\{ z \in \mathbb{C} : |\operatorname{Re} z| < t, \alpha < \operatorname{Im} z < \beta \}$ counted with their multiplicities.

A multidimensional analog of this theorem is given in [2]. To formulate it we introduce the necessary notation.

For a convex function Ψ let μ_Ψ denote its Riesz measure. As an element of the space of distributions it is defined by $\mu_\Psi = \frac{1}{\Theta_n} \Delta \Psi$, where $\Theta_n = (n - 2) \int_{|x|=1} dS$, $n > 2$, and $\Theta_2 = 2\pi$. Let Z_f be the divisor of a holomorphic function $f(z)$ in T_G and $n_f(t, G^\circ)$ be its volume in the domain $\{ z \in \mathbb{C}^n : \operatorname{Re} z \in Q_t(0), \operatorname{Im} z \in G^\circ \}$, where $G^\circ \subset\subset G$.

THEOREM 1 [2]. *Let $f(z)$ be a holomorphic a.p. function in a domain $T_G \subset \mathbb{C}^n$, $n \geq 1$. Then*

(1) $\exists \varphi_f(y) = \lim\limits_{t \to \infty} \frac{1}{(2\pi)^n} \int_{Q_t(0)} \ln |f(x + iy)| \, dx$, $\forall y \in G$;

(2) *the function $\varphi_f(y)$ is convex;*

(3) $\exists \lim\limits_{t \to \infty} \frac{1}{(2t)^n} n_f(t, G^\circ) = \mu_{\varphi_f}(G^\circ)$ *for any domain $G^\circ \subset\subset G$ with $\mu_{\varphi_f}(\partial G^\circ) = 0$.*

For $n = 1$ Theorem 1 coincides with Jessen's theorem, up to some negligible equivalent changes in its statement. The function $\varphi_f(y)$ appearing in these theorems is said to be the Jessen function of the a.p. function $f(z)$. A convex function $\varphi(y)$ is called a Jessen function (simply) if it is the Jessen function of some holomorphic a.p. function. Due to the connection between the function φ_f and the distribution of zeros (points of divisor) of the function $f(z)$, it is of great interest to study properties of the function φ_f. The main problem here is as follows.

PROBLEM. *Describe the class of Jessen functions directly, i.e. in terms which are not connected with a.p. functions generating Jessen functions.*

In the case of one variable the problem was solved by Jessen and Tornehave [3]. It follows from their results that for $n = 1$:

   a) not every convex function $\varphi(y)$ is a Jessen function;
   b) any piecewise linear convex function $\varphi(y)$ is a Jessen function;
   c) any convex function $\varphi(y)$ which is not linear on any interval is a Jessen function.

For functions of several variables only one particular case of the general theory has been considered. It is the case of a convex function $\varphi(y)$ which is assumed a priori to be piecewise linear. The following result is true.

THEOREM 2 [4]. *A convex piecewise linear function $\varphi(y)$ in a convex domain $G \subset \mathbb{R}^n$ is a Jessen function iff* *

$$\varphi(y) = \sum_j \varkappa_j \langle y, a_j \rangle^+ + \langle y, a_0 \rangle,$$

where $a_j \in \mathbb{R}^n$ and $\varkappa_j \geqslant 0$. Moreover, if a holomorphic a.p. function $f(z)$ satisfies $\varphi_f = \varphi$ then the divisor of $f(z)$ is a union of complex hyperplanes.

It follows in particular that (in contrast with the case $n = 1$, see b)) for $n > 1$ not every convex piecewise linear function is a Jessen function.

PROBLEM. *Let a convex function $\varphi(y)$ be non-linear in any interval (or any open set) in its domain of definition. Is $\varphi(y)$ a Jessen function?*

Let us note that a.p. function $f(z)$ has no zeros in T_G if and only if $\varphi_f(y)$ is linear in G. That is why the following question is closely related with the problem of the description of Jessen functions.

PROBLEM. *Let G be a convex domain in $\mathbb{R}^n$. Is it possible to construct an entire a.p. function $f(z)$ such that $Z_f \cap T_G = \emptyset$ and $Z_f \cap T_{G'} \neq \emptyset$ $\forall G' \supsetneqq G$?*

REFERENCES

1. Levitan B. M., *Almost Periodic Functions*, GITTL, Moscow, 1953. (Russian)
2. Ronkin L. I., *Jessen's theorems for holomorphic almost periodic functions in tube domains*, Siberian Math. J. **28** (1987), no. 3, 199–204. (Russian)
3. Jessen B., Tornehave H., *Means motions and zeros of almost periodic functions*, Acta Math. **77** (1945), 137–279.

---

*where $A^+ = \max(A, 0)$.

4. Ronkin L. I., *On a class of holomorphic almost periodic functions of several variables* (in preparation).

BARABASHOVA 38, 255
KHARKOV, 310168
UKRAINE

C-CONVEX SETS

S. V. Znamenskij

Let $0 \in \mathcal{D} \subset \mathbb{C}^n$ be a bounded domain of holomorphy in $\mathbb{C}^n$, $\mathcal{D}' = \{ w : w_1 z_1 + \ldots + w_n z_n \neq 1 \ \forall z \in \mathcal{D} \}$ and k is natural number, $1 \leqslant k < n$. *The following conditions are equivalent* [1–5]:

1. Any intersection of K with a k-dimensional complex line is connected and simply connected in any dimension.
2. The correspondence between analytic functionals and their indicatrices realizes an isomorphism of spaces $H'(\mathcal{D})$ and $H(\mathcal{D}')$.
3. Kergin interpolation polynomials are well defined for any holomorphic function and for any finite set in $\mathcal{D}$.
4. Any linear differential equation with constant coefficients and right side holomorphic in $\mathcal{D}$ has a solution holomorphic in $\mathcal{D}$.

If one of these conditions holds, domain $\mathcal{D}$ is called *strongly linearly convex* [4] or $\mathbb{C}$-*convex* [2].

The following conditions are equivalent [5–7] *for any holomorphically convex compact set* $0 \in K \subset \mathbb{C}^n$ *with non–empty interior*:

1. Any intersection of K with a k-dimensional complex line is acyclic.
2. The correspondence between analytic functionals and their indicatrices realize an isomorphism of spaces $H'(K)$ and $H(K')$.
3. Any $f \in H(K)$ can be represented by the series

$$\sum_m (a_0 + a_{1m} z_1 + \cdots + a_{nm} z_n)^{-n}$$

absolutely convergent in some neighbourhood of K.
4. The domain $\mathcal{D} = K'$ is $\mathbb{C}$-convex.

That is why $\mathbb{C}$-convexity can be regarded as an important complex analogue of convexity. $\mathbb{C}$-convexity is preserved under any projective map. Any $\mathbb{C}$-convex domain is homeomorphic to a ball. However, unlike common convexity, the intersection of two $\mathbb{C}$-convex domains is not necessarily $\mathbb{C}$-convex. This is the cause of difficulties in understanding geometry of such domains.

Problem 1 (L. A. Aizenberg, [8]). *Can every $\mathbb{C}$-convex bounded domain be approximated by an increasing sequence of $\mathbb{C}$-convex domains with a smooth boundary?*

The answer to a more special question is also unknown:

Problem 1'. *Can every $\mathbb{C}$-convex bounded domain be approximated by an increasing sequence of $\mathbb{C}$-convex compacts?*

The following two problems are closely connected with a certain result of L. Lempert [9]:

PROBLEM 2 (D. Pflug). *Is every $\mathbb{C}$-convex domain biholomorphically equivalent to a convex domain?*

PROBLEM 3. *Are the balls on $\mathbb{C}$-convex domains (with respect to Kobajashi metric) also $\mathbb{C}$-convex?*

PROBLEM 4. *Describe possible intersection of any complex line with arbitrary bounded $\mathbb{C}$-convex domain.*

PROBLEM 5. *Find (if possible) any local condition on a domain, necessary and sufficient for its $\mathbb{C}$-convexity.*

We can conjecture this condition to be similar to a non-local one, contained in [9]. More precisely, let $\Gamma_z = \Gamma_z(\mathcal{D})$ denote the set of all hyperplanes H, for which

(1) $z \in H$,
(2) there exists a neighbourhood U_z of z, such that $\Gamma_z \cap (U_z \cap \mathcal{D}) = \emptyset$.

PROBLEM 6. *Prove that an arbitrary domain $\mathcal{D} \subset \mathbb{C}^n$ is $\mathbb{C}$-convex if and only if for any $z \in \partial\mathcal{D}$ the set $\Gamma_z(\mathcal{D})$ is connected.*

PROBLEM 7 (M. Passare). *Does there exist a class of functions which is connected with a class of $\mathbb{C}$-convex sets as closely as the class of $\mathbb{C}$-convex is connected with the class of convex sets?*

REFERENCES

1. Znamenskij S. V., *A geometric criterion for strong linear convexity*, Functional Anal. Appl. **13** (1989), 83–84.
2. Anderson M., Passare M., *Complex kergin interpolation*, Reports Department of Mathematics University of Stockholm (1989), no. 14, 1–11.
3. Martineau A., *Sur la notion d'ensemble fortement linéelment convexe*, Anaise. Acad. Brasil. Ciênc. **40** (1968), no. 4, 427–435.
4. Znamenskij S. V., *On the existence of holomorphic primitives in all directions*, Mat. Zametki **45** (1989), 16–19. (Russian)
5. Zelinskii Yu. B., *On relations between properties of sets and properties of their sections and projections*, Russian Math. Surveys **34** (1979), 261–266.
6. Znamenskij S. V., *Strong linear convexity. 1. The duality in spaces of holomorphic functions*, Sibirsk. Mat. Zh. **26** (1985), no. 3, 415–422. (Russian)
7. Trutnev V. M., *On an analog of the Lauorent series for functions of several complex variables holomorphic on strongly linearly convex sets*, In: Holomorphic functions of several complex variables, Krasnoyarsk, 1972, pp. 139–152. (Russian)
8. Aizenberg L. A., This collection, Problem 1.18.
9. Zelinskii Yu. B., *On the geometrical criteria of strong linear convexity*, Complex analysis and applications. Proceedings of the International conference on complex analysis and applications. Varna, September 20–27, 1981. Sofia, 1984, p. 333.
10. Lempert L., *Intrinsic distances and holomorphic retracts*, Complex analysis and applications. Proceedings of the International conference on complex analysis and applications. Varna, September 20–27, 1981. Sofia, 1984, pp. 341–364.

V. A. STEKLOV MATHEMATICAL INSTITUTE
RUSSIAN ACADEMY OF SCIENCES
VAVILOVA 42, MOSCOW, 117966
RUSSIA

OUTER FUNCTIONS IN THE BALL ALGEBRA

Håkan Hedenmalm

Let $\mathbb{B}$ be the open unit ball in $\mathbb{C}^2$, and let $A(\mathbb{B})$ denote the space of functions continuous on $\overline{\mathbb{B}}$ and holomorphic on $\mathbb{B}$. This space $A(\mathbb{B})$ is known as the *ball algebra*, and supplied with the supremum norm on $\mathbb{B}$ and pointwise multiplication, it is a commutative Banach algebra. Suppose that we have a function $f \in A(\mathbb{B})$ whose zero set

$$Z(f) = \left\{ z \in \overline{\mathbb{B}} : f(z) = 0 \right\}$$

is contained in the sphere $\mathbb{S} = \partial\mathbb{B}$, and let $I(f)$ denote the closure of the ideal generated by f in $A(\mathbb{B})$. One question which comes to mind is for which f the ideal $I(f)$ coincides with the collection of all functions in $A(\mathbb{B})$ that vanish on $Z(f)$. This seems to be a difficult problem; the answer in $\mathbb{C}^1$ is that f should be an outer function in the sense of Beurling. If $Z(f)$ consists of a single point only, say $(1,0)$, the question should be easier to answer.

PROBLEM. *Suppose that $f \in A(\mathbb{B})$ has $Z(f) = \{(1,0)\}$. Give a necessary and sufficient condition on f for $I(f)$ to coincide with the maximal ideal of functions vanishing at $(1,0)$.*

In [1] (see p.14), it was shown that, if

$$\log|f(z)| = o\left(\frac{1}{1 - |z_1|}\right) \text{ as } z \to (1,0),$$

then $I(f)$ equals the maximal ideal of functions vanishing at $(1,0)$. On the other hand, a necessary condition for this to happen is that

$$\log|f(z)| = o\left(\frac{1}{1 - |z|}\right) \text{ as } z \to (1,0).$$

Maybe this condition is sufficient as well?

REFERENCE

1. Hedenmalm H., *Outer functions of Several Complex Variables*, J. Funct. Anal. **80** (1988), 9–15.

UPPSALA UNIVERSITET
MATEMATISKA INSTITUTIONEN
BOX 480
S–751 06 UPPSALA
SWEDEN

POLYNOMIALLY CONVEX HULLS

H. ALEXANDER

We shall denote Hausdorff one-dimensional measure ("linear measure") by $\mathcal{H}_1$. For X a compact subset of $\mathbb{C}^n$, $\widehat{X}$ will be its polynomially convex hull: $\{ z \in \mathbb{C}^n : |p(z)| \leqslant \sup\{|p(\zeta)|, \zeta \in X\} \}$ for all polynomials p in $\mathbb{C}^n$. The unit ball in $\mathbb{C}^n$ will be denoted by B. As usual, $P(X)$ will be the uniform closure in $C(X)$ of the polynomials.

In [1] it was shown that if X is (or lies in) a connected set of finite linear measure, then $\widehat{X} \setminus X$ is a one-dimensional analytic variety. Recently V. M. Golovin [4] claimed that the connectedness assumption could be dropped. We find his argument unconvincing and shall list a special case as a first question.

PROBLEM 1[1]. *Does $\mathcal{H}_1(X) < \infty$ imply that $\widehat{X} \setminus X$ is a one-dimensional subvariety of $\mathbb{C}^n \setminus X$?*

Known methods to solve this kind of problem involve the classical F. and M. Riesz theorem for Jordan domains with rectifiable boundaries. One may to treat Problem 1 would be to generalize this. Namely, let Ω be a bounded domain in $\mathbb{C}$ with $\mathcal{H}_1(\partial\Omega) < \infty$ (do *not* assume that Ω is even finitely connected). Suppose that the outer boundary Γ of Ω is a Jordan curve. Let $z_0 \in \Omega$ and let μ be a Jensen measure for the algebra $P(\overline{\Omega})$ supported on $\partial\Omega$ with respect to z_0.

PROBLEM 2. *Is $\mu|\Gamma$ absolutely continuous with respect to $\mathcal{H}_1|\Gamma$?*

The F. and M. Riesz theorem is the case $\partial\Omega = \Gamma$. A variant is:

PROBLEM 2′. *Let u be subharmonic on Ω and u.s.c. on $\overline{\Omega}$. Let $E \subset \Gamma$ with $\mathcal{H}_1(E) > 0$. If $u(z) \to -\infty$ as $z \in \Omega \to E$, does it follow that $u \equiv -\infty$?*

Examples of non–polynomially convex sets X which are totally disconnected have long been known; a recent example was given by Vitushkin [6]. It is known that such a set cannot lie in a torus $\mathbb{T}^n$.

PROBLEM 3. *Find a set $X \subset \partial B$ which is totally disconnected such that $0 \in \widehat{X}$.*

One possible approach is to approximate such a set $X \subset \mathbb{C}^2$ by sets $V \cap \partial B$ where V is an analytic (or algebraic) curve in $\mathbb{C}^2$ passing through the origin. Then $V \cap \partial B$ would be required to have arbitrary small components. On the other hand, this will not be possible if there is a lower bound on the size of this components — it is known that the sum of their lengths, $\mathcal{H}_1(\partial V \cap B)$, is at least 2π.

---

[1] This problem was misstated in the previous edition of the Problem Book

PROBLEM 4. *Is there a lower bound for* $\{\mathcal{H}_1(\gamma) : \gamma$ *a connected component of* $V \cap \partial B\}$?

In [3], V. K. Belošapka conjectured, for "$\mathcal{H}_1$" replaced by "diameter", that one is a lower bound. He showed that if B is replaced by the bidisc $\mathbb{D}^2$ there is some component of diameter at least one.

There exist sets $X \subset \mathbb{C}^2$ such that $\widehat{X} \setminus X$ is non-empty but contains no analytic structure. This phenomenon was discovered by Stolzenberg [5]. A recent example of such a set X has been given by Wermer [7] with the additional property that $X \subset \mathbb{T} \times \mathbb{D}$.

PROBLEM 5. *Find a set* $X \subset \mathbb{T}^2$ *such that* $0 \in \widehat{X}$ *and* $\widehat{X} \setminus X$ *contains no analytic structure.*

One interesting property of such a set would be that it could be "reflected" in $\mathbb{T}^2$ which would then become a "removable singularity".

The Stolzenberg and Wermer sets both arise from limits of analytic varieties. A well-known question asks if this must necessarily hold. Our last problem is a special case of this.

PROBLEM 6. *Let* $X \subset \mathbb{T} \times \mathbb{D}$ *with* $(\widehat{X} \setminus X) \cap \mathbb{D}^2$ *non-empty. Is* $\widehat{X} \setminus X$ *a limit of analytic subvarieties of* $\mathbb{D}^2$?

A rather particular case of this was considered by Alexander–Wermer [2].

REFERENCES

1. Alexander H., *Polynomial approximation and hulls in sets of finite linear measure in* $\mathbb{C}^n$, Amer. J. Math. **93** (1971), 65–74.
2. Alexander H., Wermer J., *On the approximation of singularity sets by analytic varieties*, Pacific J. Math. **104** (1983), 263–268.
3. Belošapka V. K., *On a metric property of analytic sets*, Izv. Akad. Nauk SSSR **40** (1976), no. 6, 1409–1415 (Russian); English transl. in Math. USSR Izvestiya **10** (1976), 1333–1338.
4. Golovin V. M., *Polynomial convexity and sets of finite linear measure in* C^Λ, Sibirsk. Matem. Zhurn. **20** (1979), no. 5, 990–996 (Russian); English transl. in Siberian Math. J. **20** (1979), no. 5, 700–704.
5. Stolzenberg G., *A hull with no analytic structure*, J. of Math. and Mech. **12** (1963), 103–112.
6. Vituškin A. G., *On a problem of W. Rudin*, Dokl. Akad. Nauk SSSR **213** (1973), 14–15 (Russian); English transl. in Soviet Math. Doklady **14** (1973), no. 6, 1618–1619.
7. Wermer J., *Polynomially convex hulls and analyticity*, Arkiv för mat. **20** (1982), 129–135.

DEPARTMENT OF MATHEMATICS
UNIVERSITY OF ILLINOIS AT CHICAGO
P. O. BOX 4348
CHICAGO, ILLINOIS 60680
U.S.A.

COMMENTARY BY THE AUTHOR

Problem 1 (corrected) is answered negatively in [8]. The construction of Beurling employed there also answers Problems 2 and 2′ negatively. It seems likely that Problem 3 can be answered positively by the methods of [9].

REFERENCES

8. Alexander H., *The polynomial hull of a set of finite linear measure in* $\mathbb{C}^n$, J. d'Analyse Mathématique **47** (1986), 238–242.
9. Globevnik J., *A disk in the ball whose end is an arc*, preprint, 1990.

THE EXTREME RAYS OF THE
POSITIVE PLURIHARMONIC FUNCTIONS

Frank Forelli

1. Let $n \geqslant 2$ and consider the class $N(\mathbb{B})$ of all holomorphic functions g on $\mathbb{B}$ such that $\mathrm{Re} g > 0$ and $g(0) = 1$, where $\mathbb{B}$ is the open unit ball in $\mathbb{C}^n$. Thus $N(\mathbb{B})$ is convex (and compact in the compact open topology). We think that the structure of $N(\mathbb{B})$ is of interest and importance. Thus we ask:

What are the extreme points of $N(\mathbb{B})$?

Very little is known. Of course if $n = 1$, and if

$$(1) \qquad\qquad g = (1+f)/(1-f),$$

then g is extreme if and only if $f(z) = cz$, where $c \in \mathbb{T}$. It is proved in [1] that if $f(z) = \sum_1^n z_j^2$ and if g is the Cayley transform (1) of f, then $g \in E(\mathbb{B})$, where $E(\mathbb{B})$ is the class of all extreme points of $N(\mathbb{B})$.

Let $k = (k_1, \ldots, k_n)$ be a multi-index and consider monomials $f(z) = cz^k$ in $\mathbb{C}^n$ such that $|f(z)| \leqslant 1$ if $z \in \mathbb{B}$. Thus $|c| \leqslant \left(\frac{k}{|k|}\right)^{-k/2}$ where by $|k|$ we mean $k_1 + \cdots + k_n$. Let $c_k = \left(\frac{k}{|k|}\right)^{-k/2}$ and let $g(z) = (1 + c_k z^k)/(1 - c_k z^k)$. It is proved in [2] that $g \in E(\mathbb{B})$ if and only if the components of k are relatively prime and positive.

2. We have $1 \notin E(\mathbb{B})$, however it is a corollary of the just mentioned theorem of [2] that $1 \in \mathrm{clos}\, E(\mathbb{B})$, where the closure is in the compact open topology. Thus $E(\mathbb{B}) \neq \mathrm{clos}\, E(\mathbb{B})$. (If $n = 1$, then $E(\mathbb{B}) = \mathrm{clos}\, E(\mathbb{B})$).

It is also known that if g is an extreme point of $N(\mathbb{B})$ and if (1) holds (that is to say if $f = (g-1)/(g+1)$), then f is irreducible. This is a special case of Theorem 1.2 of [3]. The term "irreducible" is defined in [4]. If $n = 1$, then g is extreme if and only if f is irreducible. However for $n \geqslant 2$, the fact that f is irreducible does not imply that g is extreme.

3. The extreme points g in section 1 and the extreme points that can be obtained from them by letting $\mathrm{Aut}(\mathbb{B})$ act on $N(\mathbb{B})$ have the property that the Cayley transform $f = (g-1)/(g+1)$ is holomorphic on $\mathbb{B} \cup \partial\mathbb{B}$.

Is this the case for every g in $E(\mathbb{B})$?

If the answer is yes, then it would follow (since $n \geqslant 2$) that the F. and M. Riesz theorem holds for those Radon measure on $\partial\mathbb{B}$ whose Poisson integrals are pluriharmonic. In particular there would be no singular Radon measures $\neq 0$ with this property, which in turn would imply that there are no nonconstant inner functions on $\mathbb{B}$.

Chapter 17. $\mathbb{C}^n$

REFERENCES

1. Forelli F., *Measures whose Poisson integrals are pluriharmonic* II, Illinois J. Math. **19** (1975), 584–592.
2. Forelli F., *Some extreme rays of the positive pluriharmonic functions*, Canad. J. Math. **31** (1979), 9–16.
3. Forelli F., *A necessary condition on the extreme points of a class of holomorphic functions*, Pacific J. Math. **73** (1977), 81–86.
4. Ahern P., Rudin W., *Factorizations of bounded holomorphic functions*, Duke Math. J. **39** (1972), 767–777.

UNIVERSITY OF WISCONSIN
DEPARTMENT OF MATHEMATICS
MADISON, WISCONSIN 53106
U.S.A.

COMMENTARY

The second question has been answered in the negative. See Commentary in Problem S.10 of the previous edition of the Problem Book.

ON BIHOLOMORPHY OF HOLOMORPHIC MAPPINGS
OF COMPLEX BANACH SPACES

S. L. KRUSHKAL

Let f be a function holomorphic in a domain D, $D \subset \mathbb{C}$. It is well known that if f is univalent, then $f'(z) \neq 0$ in D. Holomorphic mappings f of domain $D \subset \mathbb{C}^n$ for $n > 1$ also possess a similar property: if $f: D \to \mathbb{C}^n$ is holomorphic and one-to-one then $\det f'(z) \neq 0$ at every point z of D $\left(f'(z) = \left(\frac{\partial f_j}{\partial z_k} \right)^n_{j,k=1} \right.$ is the Jacobi matrix$)$, or equivalently the differential $df(z)h = f'(z)h$ is an automorphism of $\mathbb{C}^n$, and then, by the implicit function theorem, f itself is biholomorphic.

Note that the continuity and the injective character of f immediately imply that f is a homeomorphism because $\dim \mathbb{C}^n = n < \infty$. The proof of the diffeomorphic property depends essentially on the holomorphic properties of f. Such a result, as is known, for real spaces and mappings is wrong, which is clear from the example $(x_1, \ldots, x_n) \to (x_1^3, \ldots, x_n^3)$: $\mathbb{R}^n \to \mathbb{R}^n$.

Let now X and Y be the complex Banach spaces and f be a holomorphic mapping of the domain $D \subset X$ into Y. Remind that the mapping f is called holomorphic in D if it is continuous and weakly Gâteaux differentiable, i.e. for any $x \in D$ and $h \in X$ there exists

$$(*) \qquad \lim_{t \to 0} \frac{f(x + th) - f(x)}{t} = f'(x)h \qquad (t \in \mathbb{C}).$$

Then $df(x)h = f'(x)h$ is a linear operator $X \to Y$. It is proved that in the complex case $(*)$ implies the strong Fréchet differentiability of f:

$$\left\| f(x + h) - f(x) - df(x)h \right\|_Y \leqslant c(x) \left\| h \right\|_X^2$$

(for small $\|h\|_X$).

PROBLEM. *Let f be a holomorphic one-to-one mapping of the domain $D \subset X$ onto the domain $D' \subset Y$. Is the differential $df(x)$ an isomorphism (i.e. an injective and surjective mapping) of the space X onto Y at every point $x \in D$?*

The positive answer and the implicit function theorem would imply that all one-to-one holomorphic mappings of domains of complex Banach spaces are diffeomorphisms.

Positive solution of the problem would allow to obtain, for instance, some important corollaries in the geometric theory of functions of the complex variable (in problems concerning univalency and quasi-conformal extendability of holomorphic functions, characterization of boundary properties of functions starting from the interior properties; all that can be reduced to the consideration of some Banach spaces of holomorphic functions).

The author does not know any general result in this direction. It seems likely that the problem in general statement must have a negative solution. The following conjecture can be formulated (at least as a stimulus to refute it).

CONJECTURE. *Suppose conditions of the Problem are fulfilled. Then the mapping $df(x)$ $(x \in D)$ is injective but there exists spaces X, Y for which it is not surjective.*

Then a QUESTION arises under what additional conditions of, maybe, geometric character, concerning the structure of the spaces X and Y, the mapping $df(x)$ is an isomorphism (for f satisfying assumptions of the Problem); will this be so at least for Hilbert spaces or spaces possessing some special convexity properties, etc?

INSTITUTE OF MATHEMATICS SO RAN
NOVOSIBIRSK 630072
RUSSIA

SOME QUESTIONS ABOUT RESIDUE CURRENTS
AND MELLIN TRANSFORM

C. A. BERENSTEIN,   R. GAY,   A. YGER

Given m holomorphic functions of n variables $(m \leqslant n)$, $f_1, \ldots, f_m$, in some open set $\Omega \subset \mathbb{C}^n$, and some $(n, n - m)$ smooth test form with compact support in Ω, let $\lambda \mapsto I(\lambda; \varphi)$ be the application

$$\lambda \in \mathbb{C}^m \mapsto I(\lambda; \varphi) = (-1)^{m(m-1)/2} \int_\Omega \bar\partial\left(\frac{|f_1|^{2\lambda_1}}{f_1}\right) \wedge \cdots \wedge \bar\partial\left(\frac{|f_m|^{2\lambda_m}}{f_m}\right) \wedge \varphi .$$

As a consequence of Atiyah's theorem [1], this function, which is a priori defined in $\mathrm{Re}(\lambda_j) >> 0$, $j = 1, \ldots, m$, can be extended to a meromorphic function in $\mathbb{C}^n$, with poles on a countable collection of hyperplanes

$$< \lambda, \alpha_j > + \gamma = 0, \quad \alpha_1, \ldots, \alpha_N \in \mathbb{N}^n, \gamma \in \mathbb{N}$$

(see for example [3], [4]). Moreover, it can be shown that, in some open domain of the form $\Lambda_\delta = \{\lambda, \ \mathrm{Re}(\lambda_j) > -\delta, \ j = 1, \ldots, m\}$ ($\delta > 0$, depending on $\mathrm{supp}(\varphi)$), this meromorphic function can be written as

$$I(\lambda; \varphi) = \sum_s \frac{\lambda_1 \ldots \lambda_m}{\pi_s(\lambda)} h_s(\lambda)$$

where the h_s are holomorphic functions in Λ_δ and the π_s are products of at most m linear forms with coefficients in $\mathbb{N}$. This implies that, for $t \in (\mathbb{R}^+)^m$, the value at $\mu = 0$ of $\mu \in \mathbb{C} \mapsto I(\mu t; \varphi)$ is well defined; it will be denoted $I_t(\varphi)$. In the particular case where $f_1, \ldots, f_m$ define a complete intersection in Ω, $I_t(\varphi)$ is independent of t and equals

$$(2i\pi)^m < \bar\partial(1/f_1) \wedge \cdots \wedge \bar\partial(1/f_m), \ \varphi >$$

that is $(2i\pi)^m$ times the action on φ of the multidimensional residue introduced by Coleff–Herrera [6], Dolbeault [7], or Passare ([9], [10]). The fact that I_t remains constant obviously fails in the non-complete intersection case. Since all proofs we know involve a resolution of singularities for the hypersurface $f_1 \ldots f_m = 0$ (instead of one for the analytic set $\{f_1 = \ldots f_m = 0\}$), the following questions remain unsolved:

QUESTION 1. *In the complete intersection case, adding some auxiliary conditions (as for example in the case of nondegenerated systems in the sense of Khovanskii [8] or Varchenko [14]), give some geometric interpretation (in terms of Newton polyedra) for the slopes α_j of the affine hyperplanes along which $\lambda \mapsto I(\lambda; \varphi)$ is singular. The methods initiated by Sabbah in [12] should give some insight into this question; the main* difference is that one has to deal here with analytic continuation of currents instead of distributions. A lot of polar terms are cancelled by the differential form $\overline{\partial f}_1 \wedge \cdots \wedge \overline{\partial f}_m$; as an example, when $m = 2$ and f_1, f_2 define a complete intersection, the function $\lambda \mapsto I(\lambda; \varphi)$ is holomorphic in some Λ_δ. Does this result remain true in the case of higher codimension?

QUESTION 2. In the complete intersection case, if $f_1, \ldots, f_m$ vanish at the origin, and if $\text{supp}(\varphi) \subset B(0, \epsilon)$, with ϵ small enough, does the function $\xi \in \mathbb{R}^m \mapsto I(c + i\xi; \varphi)$ satisfy rapid decrease properties at infinity for any c in $(\mathbb{R}^+)^m$? When $m = 1$, this is a consequence of a result by Kashiwara (see [2]). The answer is also obviously yes in the normal crossing case. Nothing seems to be known in general.

QUESTION 3. When the complete intersection hypothesis is not fulfilled, what is the interpretation of the value $I_t(\varphi)$ for t in $(\mathbb{R}^+)^m$?

Furthermore, in the complete intersection case, one may show (see [4]) that the function

$$\lambda \in \mathbb{C}^m \mapsto J(\lambda; \varphi) = (-1)^{m(m-1)/2}(m-1)! \left(\sum_{j=1}^{j=m} \lambda_j \right) \int_\Omega \prod_{j=1}^{j=m} |f_j|^{2\lambda_j} \frac{\overline{\partial f}_1 \wedge \cdots \wedge \overline{\partial f}_m \wedge \varphi}{\left(\sum_{j=1}^m |f_j|^2 \right)^m}$$

is holomorphic in some Λ_δ and takes the same value than $\lambda \mapsto I(\lambda; \varphi)$ at the origin. One can ask Question 2 for J instead of I. The reason for this is that J can be written for $\text{Re}(\lambda_j) >> 0$ as

$$J(\lambda; \varphi) = (\lambda_1 + \cdots + \lambda_m) \int_{(\mathbb{R}^+)^m} \frac{(m-1)! s_1^{\lambda_1} \ldots s_m^{\lambda_m}}{(s_1 + \cdots + s_m)^m} \theta(s) ds_1 \ldots ds_m$$

where θ is the almost everywhere defined function on $(\mathbb{R}^+)^m$

$$\theta(s) = \int_{\{|f_j|^2 = s_j, \, j=1,\ldots,m\}} \frac{\varphi}{f_1 \cdots f_m} .$$

The problem of the existence of a limit for θ at the origin remains the most untractable question in the multidimensional residue theory. The behaviour of θ along particular curves has been studied by Coleff–Herrera [6], M. Passare [10] and A. Tsikh [13]. An inversion formula for the transform $\theta \mapsto J$ would be useful to study the behavior of θ at the origin.

REFERENCES

1. Atiyah M.F., *Resolution of singularities and division of distributions*, Commun. pure Appl. Math. **23** (1970), 145–150.
2. Barlet D., Maire H.M., *Développements asymptotiques, Transformation de Mellin Complexe et intégration sur les fibres*, Lect. Notes in Math. **1295** (1987).
3. Berenstein C.A., Gay R., Yger A., *Analytic continuation of currents and division problems*, Forum Math. **1** (1989), 15–51.
4. Berenstein C.A., Gay R., Vidras A., Yger A., *Residue currents and Bezout identities*, 1993, expository manuscript.

5. Berenstein C.A., Yger A., *Une formule de Jacobi et ses conséquences*, Ann. Sci. Ec. Norm. Sup. Paris **24** (1991), 363–377.
6. Coleff N., Herrera M., *Les courants résiduels associés à une forme méromorphe*, Lect. Notes in Math. **633** (1978).
7. Dolbeault P., *Theory of residues and homology*, Lect. Notes in Math. **116** (1970).
8. Khovanskii A.G., *Newton polyedra and toroidal varieties*, Funct. Anal. Appl. **11** (1978), 289–295.
9. Passare M., *Residues, currents, and their relation to ideals of holomorphic functions*, Math. Scand. **62** (1988), 75–152.
10. Passare M., *Courants méromorphes et égalité de la valeur principale et de la partie finie*, Lect. Notes in Math. **1295** (1987), 157–166.
11. Passare M., *A calculus for meromorphic currents*, J. reine angew. Math. **392** (1988), 37–56.
12. Sabbah C., *Proximité évanescente I, II*, Compositio Math. **62** (1987), 283–328; **64** (1988), 213–241.
13. Tsikh A., *Multidimensional Residues and Their Applications*, Transl. Amer. Math. Soc. **103** (1992).
14. Varchenko A.N., *Newton polyedra and estimation of oscillating integrals*, Funct. Anal. Appl. **10** (1976), 175–196.

DEPARTMENT OF MATHEMATICS
AND SYSTEM RESEARCH CENTER
UNIVERSITY OF MARYLAND
MD 20742, USA

CeReMaB, UA ASSOCIÉE 226
UNIVERSITÉ BORDEAUX 1
33405 TALENCE, FRANCE

CeReMaB, UA ASSOCIÉE 226
UNIVERSITÉ BORDEAUX 1
33405 TALENCE, FRANCE

17.16

POMPEIU'S PROBLEM AND SCHIFFER'S CONJECTURE

C. A. Berenstein, R. Gay, A. Yger

Let $X = G/K$ be a symmetric space of non compact type. If Ω is a compact subset of X, let P be the transformation which associates to any ψ in $L^1_{loc}(X)$ the function on G defined by

$$P(\psi)(g) = \int_{g\Omega} \psi \, dx \qquad g \in G$$

where dx is the G-invariant measure on X. One says that Ω satisfies Pompeïu's property if P is injective.

It is known ([1],[2],[5],[8]) that when $X \backslash \overline{\Omega}$ is connected and $\partial\Omega$ Lipschitz, Ω fails Pompeïu's property if and only if there exists a strictly positive eigenvalue α for the mixed overdetermined problem

$$(1) \qquad\qquad \Delta u + \alpha u = 0 \ \text{in} \ \Omega \qquad \frac{\partial u}{\partial n} = 0, \ u = \text{const on} \ \partial\Omega$$

(when X has rank 1).

In ([3],[4]) C.A. Berenstein and P. Yang prove that, when $X = \mathbb{R}^n$ or X hyperbolic, then the existence of an infinite number of strictly positive eigenvalues for which (1) admits a non trivial solution implies that Ω is a ball. The conjecture which asserts that the existence of just one $\alpha > 0$ (such that (1) admits a non trivial solution) implies Ω is a ball is known as Schiffer's conjecture ([6],[7]). The question is to know, even in $\mathbb{R}^n$, if the only Ω (with $X \backslash \overline{\Omega}$ connected) which fails Pompeïu's property is a ball.

Very recently, P. Ebenfelt [6] showed that, in $\mathbb{R}^2$, the only bounded domains which are quadrature domains and fail Pompeïu's property are the planar discs. By definition, a domain is a quadrature domain if there exists a point distribution ν with finite support in Ω such that the distribution $\chi_\Omega dx - \nu$ is orthogonal to the space of holomorphic integrable functions in Ω.

References

1. Berenstein C. A., *An inverse spectral theorem and its relation to the Pompeïu problem*, J. Analyse Math. **37** (1980), 128–144.
2. Berenstein C. A. and Shahshahani M., *Harmonic analysis and the Pompeïu problem*, Amer. J. Math. **105** (1983), 1217–1229.
3. Berenstein C. A. and Yang P., *An overdetermined Neumann problem in the unit disk*, Advances in Mathematics **44** (1982), 1–17.
4. Berenstein C. A. and Yang P., *An inverse Neumann problem*, J. Reine. Angew. Math. **382** (1987), 1–21.
5. Berenstein C. A. and Zalcman L., *Pompeïu's problem on symmetric spaces*, Comment. Math. Helvetici **55** (1980), 593–621.
6. Ebenfelt P., *Some results on the Pompeïu problem*, Ann. Acad. Sc. Fenn. (to appear).

7. Garofalo N. and Segala F., *New results on the Pompeïu problem*, Trans. Amer. Math. Soc **325** (1991), 273–286.
8. Williams S. A., *A partial solution for the Pompeïu problem*, Math. Ann. **223** (1976), 183–190.

DEPARTMENT OF MATHEMATICS AND SYSTEM RESEARCH CENTER
UNIVERSITY OF MARYLAND
MD 20742, USA

CEREMAB, UA ASSOCIÉE CNRS 226.
UNIVERSITÉ BORDEAUX 1
33405 TALENCE, FRANCE

CEREMAB, UA ASSOCIÉE CNRS 226.
UNIVERSITÉ BORDEAUX 1
33405 TALENCE, FRANCE

ABOUT E. FISCHER'S PROBLEM

R. GAY, A. MERIL, AND A. YGER

Let $P \in \mathbb{C}[z_1, \ldots, z_n]$ and P^* be the polynomial which is obtained from P when replacing each coefficient by its conjugate. In [1], E. Fischer proved the following

THEOREM. *Let P be some homogeneous polynomial in n variables with complex coefficients. Any element Q in $\mathbb{C}[z_1, \ldots, z_n]$ can be written in an unique way $Q = PQ_1 + R_1$, where Q_1, R_1 are polynomials such that $P^*(D)(R_1) = 0$.*

In [2] A. Méril and D. C. Struppa, in the spirit of the work of D. J. Newman and H. S. Shapiro [4], show that the existence of a splitting

$$(1) \qquad H(\mathbb{C}^n) = P.H(\mathbb{C}^n) \oplus \operatorname{Ker}(P^*(D))$$

(where $H(\mathbb{C}^n)$ denotes the space of entire functions) is equivalent to the fact that there exists a splitting

$$(2) \qquad \operatorname{Exp}(\mathbb{C}^n) = P^*. \operatorname{Exp}(\mathbb{C}^n) \oplus \left(\operatorname{Ker}(P(D)) \cap \operatorname{Exp}(\mathbb{C}^n) \right)$$

(where $\operatorname{Exp}(\mathbb{C}^n)$ denotes the space of entire functions with exponential growth).

In [3] A. Méril and A. Yger show that the operator $f \mapsto P^*(D)(P(f))$ is injective from $H(\mathbb{C}^n)$ into itself when $\deg(P) \leqslant 2$; furthermore, they show that this operator is surjective (which is equivalent to the existence of a splitting as in (1)) if $\deg(P) \leqslant 2$ and $n = 2$.

In general, the question concerning the existence of a splitting of type (1) seems to remain open. For instance, this is surely the case for the following PROBLEM: *is the "Fourier transform"*

$$\left(\frac{H(\mathbb{C}^n)}{P.H(\mathbb{C}^n)} \right)' \longrightarrow \frac{\operatorname{Exp}(\mathbb{C}^n)}{P^*. \operatorname{Exp}(\mathbb{C}^n)},$$

associated to

$$T \in \left(\frac{H(\mathbb{C}^n)}{P.H(\mathbb{C}^n)} \right)' = (P.H(\mathbb{C}^n))^{\perp}$$

the class its Fourier–Borel transform in $\operatorname{Exp}(\mathbb{C}^n)/P^*. \operatorname{Exp}(\mathbb{C}^n)$, a bijective map? Recall that the Fourier–Borel transform of T is the entire function of n variables defined by

$$\mathcal{F}(T)(z) = <T(\zeta), \exp(<\zeta, z>)>, \quad <\zeta, z> = \sum_{1 \leqslant j \leqslant n} \zeta_j z_j \, .$$

REFERENCES

1. Fischer E., *Über die Differentiations prozesses der Algebra*, J. Math. **148** (1917), 1–78.
2. Méril A., Struppa D. C., *Equivalence of Cauchy problems for entire and exponential type functions*, Bull. London Math. Soc. **17** (1985), 469–473.
3. Méril A., Yger A., *Problèmes de Cauchy globaux*, Bull. Soc. Math. France **120** (1992), 87–111.
4. Newman D. J., Shapiro H., *Fischer spaces of entire functions*, Proc. Sympos. Pure Math. **11** (1968), 360–369.

CeReMaB, UA associée CNRS 226
Université Bordeaux 1
33405 Talence France

Departement de Mathématiques
Université Antilles–Guyane
97159 Pointe a Pitre, Guadeloupe
French West Indies

CeReMaB, UA associée CNRS 226
Université Bordeaux 1
33405 Talence France

ITERATION OF HOLOMORPHIC MAPPINGS ON $\mathbb{C}^p$

JEAN ESTERLE

The problem can be stated as follows

PROBLEM 1. *Let $p \geqslant 2$ and let $(F_n)_{n \geqslant 1}$ be a sequence of holomorphic mappings $\mathbb{C}^p \to \mathbb{C}^p$. Is it always true that $\underset{n \geqslant 1}{\cap} (F_1 \circ \cdots \circ F_n)(\mathbb{C}^p) \neq \emptyset$?*

This question is related to the following famous open problem, raised by Michael [7]:

PROBLEM 2. *Are characters on Fréchet algebras necessarily continuous?*

In fact, a negative answer to Problem 1 would imply a positive answer to Problem 2 [3]. Problem 1 is open, even in the case where the sequence $(F_n)_{n \geqslant 1}$ is constant, i.e. $F_n = F$ for every n.

It is an obvious consequence of Picard's theorem that the answer to Problem 1 is yes when $p = 1$. For $p \geqslant 2$ it follows from classical constructions of Poincaré [9], Fatou [4] and Bieberbach [1] that there exist one-to-one holomorphic mappings $F : \mathbb{C}^p \to \mathbb{C}^p$, of Jacobian 1, such that $F(\mathbb{C}^p)$ is not dense in $\mathbb{C}^p$.

Recently, some literature has been devoted to the study of such mappings. See for example [2], [3], [5], [6], [8], [10], [11]. Much more information seems to be needed in order to solve Problem 1.

REFERENCES

1. Bieberbach L., *Beispiel zweier ganzer Funktionen zweier komplexer Variablen, welche eine schlichte volumtreue Abbildung des $\mathbb{R}^4$ auf einen Teil seiner selbst vermitteln*, S.B. Preuss Akad. Wiss. **14/15** (1933), 476-479.
2. Couty D., *Formes réduites des automorphismes de $\mathbb{C}^n$ à variété linéaire fixe et répulsive*, Séminaire d'Algèbre de Paris 6, Lect. Notes in Math., vol. 1404, Springer-Verlag, Berlin, 1989, pp. 346-410.
3. Dixon P. G., Esterle J., *Michael's problem and the Poincaré – Fatou – Bieberbach phenomenon*, Bull. Amer. Math. Soc. **15** (1986), 127-187.
4. Fatou P., *Sur certaines fonctions uniformes de deux variables*, C.R. Acad. Sci. Paris **175** (1922), 1030-1033.
5. Fornaess J. E., Sibony N., *Complex Henon mappings in $\mathbb{C}^2$ and Fatou – Bieberbach domains*, Duke Math. J. **65** (1992), no. 2, 345-380.
6. Gruman L., *L'image d'une application holomorphe*, Ann. Fac. Sc. Toulouse **12** (1991), no. 1, 75-101.
7. Michael E. A., *Locally multiplicatively-convex topological algebras*, Mem. AMS **11** (1952).
8. Nishimura Y., *Applications holomorphes injectives de $\mathbb{C}^2$ dans lui-même qui exceptent une droite complexe*, J. Math. Kyoto-Univ. **24** (1984), no. 4, 755-761.
9. Poincaré H., *Sur une classe nouvelle de transcendantes uniformes*, J. de Math. **6** (1890), 315-365.
10. Rosay J. P., Rudin W., *Holomorphic maps from $\mathbb{C}^n$ to $\mathbb{C}^n$*, Trans. Amer. Math. Soc. **310** (1988), no. 1, 47-86.
11. Sibony N., Pit-Mann-Wong, *Some remarks on the Casorati – Weierstrass theorem*, Ann. Polon. Math. **39** (1981), 165-174.

MATHÉMATIQUES PURES
UNIVERSITÉ BORDEAUX 1
33405 TALENCE, FRANCE

Chapter 18

GEOMETRIC FUNCTION THEORY

Edited by

P. Duren
Department of Mathematics
University of Michigan
Ann Arbor, MI 48109
USA

INTRODUCTION

Geometric function theory was represented in the last edition of the Problem Book only by a few miscellaneous contributions. Now it has its own chapter. The editor made an attempt to shape a more unified collection of problems, but those actually received are unified only by relating in some way to analytic functions of a complex variable. The only remarkable theme is their diversity. To some extent this may reflect the breadth of the subject and its tendency to permeate other areas of analysis. In any event, the problems will speak for themselves.

HARMONIC MEASURE AND HAUSDORFF DIMENSION

C. J. Bishop

Let Ω be a plane domain and let ω denote harmonic measure on Ω. Results of Makarov (for simply connected Ω) and Jones and Wolff (for general Ω) show that ω always gives full measure to a subset $E \subset \partial\Omega$ of Hausdorff dimension at most 1. Here we will consider several questions which are all related to the connections between harmonic measure and α–dimensional measure, Λ_α.

1. The lower density conjecture. For $x \in \partial\Omega$, define a continuous branch of $\arg(z - x)$ on Ω. We say $x \in \partial\Omega$ is a twist point of Ω if both

$$\liminf_{z \to x, z \in \Omega} \arg(z - x) = -\infty, \qquad \limsup_{z \to x, z \in \Omega} \arg(z - x) = +\infty.$$

On the other hand, we say Ω has an inner tangent at z if there is a unique $\theta_0 \in [0, 2\pi)$ such that for every $0 < \varepsilon < \pi/2$ there is a $\delta > 0$ such that:

$$\left\{ x + re^{i\theta} : 0 < r < \delta, \ |\theta - \theta_0| < \pi/2 - \varepsilon \right\} \subset \Omega.$$

Almost every (with respect to ω) boundary point of Ω is one of these two types. Combining results of Makarov, McMillan and Pommerenke one sees that (see [7], [8]),

$$\lim_{r \to 0} \frac{\omega\big(D(x, r)\big)}{r} \text{ exists and } \neq 0, \infty \text{ for a.e. inner tangent } x,$$

$$\limsup_{r \to 0} \frac{\omega\big(D(x, r)\big)}{r} = \infty \text{ for a.e. twist point } x.$$

This implies $\omega \ll \Lambda_1 \ll \omega$ on the inner tangents and $\omega \perp \Lambda_1$ on the twist points.

CONJECTURE 1. *At almost every (ω) twist point*

$$\liminf_{r \to 0} \frac{\omega\big(D(x, r)\big)}{r} = 0$$

2. A Lavrentiev type estimate. Conjecture 1 and the Besicovitch covering lemma give an easy proof of the following result from [3].

THEOREM 1. *If Ω is simply connected and $E \subset \partial\Omega$ lies on a rectifiable curve then $\Lambda_1(E) = 0$ implies $\omega(E) = 0$.*

This is analogous to the F. and M. Riesz theorem which states that on a domain with rectifiable boundary $\omega \ll \Lambda_1 \ll \omega$. A more quantitative version was given by Lavrentiev who showed that

$$\omega(E) \leqslant C \frac{\log \Lambda_1(\partial\Omega)}{|\log \Lambda_1(E)| + 1}.$$

The next conjecture is that this holds in general.

CONJECTURE 2. *If Ω is simply connected and $E \subset \partial\Omega$ lies on a rectifiable curve Γ then*

$$\omega(E) \leqslant C \frac{\log \Lambda_1(\Gamma)}{|\log \Lambda_1(E)| + 1}.$$

In [4] this is proven with the left hand side replaced by $\omega(E)/(|\log \omega(E)| + 1)$.

3. Characterizing $\omega \ll \Lambda_1 \ll \omega$ for general domains. Jones and Wolff [6] have shown that for a general planar domain Ω, ω gives full mass to some set of dimension $\leqslant 1$ and Wolff (unpublished) has shown this set can be taken to have sigma finite Λ_1 measure. We would like to geometrically characterize the sets where $\omega \ll \Lambda_1 \ll \omega$ and $\omega \perp \Lambda_1$ in a way analogous to the inner tangent/twist point characterization for simply connected domains. For $x \in \partial\Omega$ we say x is a weak cone point of Ω if there are $\delta > 0$, $\varepsilon > 0$ and $\theta \in [0, 2\pi)$ so that if

$$W(x, \delta, \varepsilon, \theta) = \left\{ x + re^{i\psi} : 0 < r < \delta, |\psi - \theta| < \varepsilon \right\} \setminus D(x, \delta/2),$$

then

$$\sum_{k=1}^{\infty} \text{cap}\left(2^{k-2}\left(W(x, 2^{-k}, \varepsilon, \theta) \setminus \Omega\right)\right) < \infty.$$

Here "cap" means logarithmic capacity and the condition says x is the vertex of a cone which contains very little of Ω^c in terms of capacity.

CONJECTURE 3. *For a general planar domain Ω, $\omega \ll \Lambda_1 \ll \omega$ on the set of weak cone points and $\omega \perp \Lambda_1$ elsewhere.*

One consequence of Conjecture 3 may be of interest in its own right.

CONJECTURE 4. *If $\Omega \subset \mathbb{R}^2$ is a domain and $E \subset \partial\Omega$ is Besicovitch irregular then there exists $F \subset E$ with $\omega(F) = \omega(E)$ and $\Lambda_1(F) = 0$.*

Peter Jones has pointed out that this is true in the case when $E = \partial\Omega$ satisfies a capacitary "thickness" condition: there exists $\varepsilon > 0$ such that for every $x \in \partial\Omega$ and $0 < r < r_0$, $\text{cap}\left(r^{-1}(D(x, r/4) \cap \partial\Omega)\right) \geqslant \varepsilon$.

4. Harmonic measures on disjoint domains. The firs claim of Conjecture 3 is proven in [2] as part of the proof of

THEOREM 2. *Suppose Ω_1 and Ω_2 are disjoint subdomains in $\mathbb{R}^2$ and let ω_1 and ω_2 be their harmonic measures. Then $\omega_1 \perp \omega_2$ iff the set of points in $\partial\Omega_1 \cap \partial\Omega_2$ satisfying a weak cone condition with respect to both Ω_1, Ω_2 has zero Λ_1 measure. If ω_1 and ω_2 are mutually absolutely continuous on a set E then there is Besicovitch regular $F \subset E$ with $\omega_i(F) = \omega_i(E)$ and ω_i mutually absolutely continuous with Λ_1 on F for $i = 1, 2$.*

The higher dimensional version is unknown.

CONJECTURE 5. *If $\Omega_1, \Omega_2 \subset \mathbb{R}^n$ are disjoint domains with harmonic measures ω_1, ω_2 that are mutually absolutely continuous on a set $E \subset \partial\Omega_1 \cap \partial\Omega_2$, of positive ω_1 measure, then there exists $F \subset E$ of positive Λ_{n-1} measure such that ω_1 and ω_2 are mutually absolutely continuous with Λ_{n-1} on F.*

The related question of whether harmonic measure for $\Omega \subset \mathbb{R}^n$ always gives full measure to a set of dimension $n-1$ has been disproved by Wolff in [9], though Bourgain has shown it always gives full mass to some set of dimension $\leqslant n - \varepsilon(n)$ [5].

QUESTION. *What is $d_n = \sup\limits_{\Omega \subset \mathbb{R}^n} \left(\inf \{ \dim(E) : \omega(E, \Omega) = 1 \} \right)$?*

In $\mathbb{R}^2$ one of the key facts is that $\log |\nabla u|$ is subharmonic for any harmonic function u. In $\mathbb{R}^n$ this is false, but $|\nabla u|^p$ is subharmonic if $p \geqslant (n-2)/(n-1)$. Perhaps $d_n = n - 1 + \frac{n-2}{n-1}$. Suggestions for a better guess are welcome.

5. Looping of Brownian motion. Harmonic measure can be defined as the first hitting distribution of Brownian motion on $\partial\Omega$ so perhaps it is not out of place to end with a question on Brownian motion. We say that a Brownian path γ in $\mathbb{R}^2$ separates a set E if there are times $0 \leqslant s < t$ before γ hits E so that $\gamma(s) = \gamma(t)$ and loop formed separates E. If E is a continuum this happens with probability less than one, but for the middle third Cantor set it happens almost surely.

CONJECTURE 6. *If $E \subset \mathbb{R}^2$ is compact and $\dim(E) < 1$ then Brownian paths separate it almost surely.*

This has also been conjectured by Terr Lyons. The conjecture is true if $E \subset \mathbb{R}$ or if we assume $\dim(E) < 1/2$ [1]. Makarov [7] has proved that if Ω is simply connected and

$$\Lambda_\varphi(E) = 0, \qquad \varphi(f) = t \exp \left(C \sqrt{\log \frac{1}{t} \log\log\log \frac{1}{t}} \right),$$

then $\omega(E) = 0$ (and this is sharp expect for C). In particular this shows that $\dim(E) < 1$ implies $\omega(E) = 0$. This is also implied by Conjecture 6 because if $E \subset \delta\Omega$ is separated a.s. by Brownian paths then a.e. such path hits $\partial\Omega \setminus E$ before it hits E. The two problems are not quite the same however. There is a compact $E \subset \mathbb{R}$ of length zero (and thus has zero harmonic measure in any simply connected domain) but which is not a.s. separated by Brownian paths [1]. Perhaps there is also a sharper version of Conjecture 6 in terms of Λ_φ, as in Makarov's theorem.

REFERENCES

1. Bishop C. J., *Brownian motion in Denjoy domains*, Ann. Prob. (to appear).
2. Bishop C. J., *A characterization of Poissonian domains*, Arkiv Mat. (to appear).
3. Bishop C. J., Jones P. W., *Harmonic measure and arclength*, Ann. Math. **132** (1990), 511–547.
4. Bishop C. J., Jones P. W., *Harmonic measure, L^2 estimates and the Swarzian derivative*, preprint, 1990.
5. Bourgain J., *On the Hausdorff dimension of harmonic measure in higher dimensions*, Inv. Math **87** (1987), 477–483.
6. Jones P. W. and Wolff T. H., *Hausdorff dimension of harmonic measures in the plane*, Acta. Math. **161** (1988), 131–144.
7. Makarov N. G., *On distortion of boundary set under conformal mappings*, Proc. London Math. Soc. **51** (1985), 369–384.
8. Pommerenke Ch., *On conformal mapping and linear measure*, J. Analyse Math. **46** (1986), 231–238.
9. Wolff T. H., *Counterexamples with harmonic gradients*, preprint, 1987.

DEPARTMENT OF MATHEMATICS
STATE UNIVERSITY OF NEW YORK AT STONY BROOK
STONY BROOK, NY 11794
U.S.A.

MAJORIZATION AND DOMINATION
IN THE BERGMAN SPACE

B. KORENBLUM

Let A^2 denote the set of all functions f analytic in the unit disk $\mathbb{D} = \{z \in \mathbb{C} : |z| < 1\}$ such that

$$\|f\|^2 = \pi^{-1} \int_{\mathbb{D}} |f|^2 \, dm < \infty,$$

where dm is the Lebesgue area measure. A^2 is the "analytic" subspace of the Hilbert space $L^2(\mathbb{D})$; it is known as the Bergman space.

We say that f *majorizes* g in a region G if $|g(z)| \leq |f(z)|$ for all $z \in G$. Clearly, if f majorizes g in $\mathbb{D}$ then $\|g\| \leq \|f\|$.

PROBLEM. *What other cases of majorization will imply* $\|g\| \leq \|f\|$? *In particular, when does majorization in an annulus* $c < |z| < 1$ *imply* $|g| \leq \|f\|$?

Let Z_f and Z_g denote the zeros, repeated according to multiplicity, of f and g, respectively. If $Z_f \subseteq Z_g$ then g/f is analytic in $\mathbb{D}$ and majorization $|g| \leq |f|$ in any annulus $c < |z| < 1$ implies majorization in $\mathbb{D}$, and $\|g\| \leq \|f\|$. The following results holds for the opposite case $Z_f \supseteq Z_g$:

THEOREM 1 [1, 2]. *If* $Z_f \supseteq Z_g$ *and* $|g(z)| \leq |f(z)|$ *for all* $e^{-2}/2 < |z| < 1$, *then* $\|g\| \leq \|f\|$.

CONJECTURE. *There is an absolute constant* c, $0 < c < 1$, *such that whenever* f *majorizes* g *in the annulus* $c < |z| < 1$ $(f, g \in A^2)$, f *also dominates* g *in norm, i.e.* $\|f\| \geq \|g\|$.

That c must not exceed $2^{-1/2}$ follows from the example of $f(z) = z$, $g(z) \equiv c$ in which f majorizes g in $c < |z| < 1$ but $\|f\| = 2^{-1/2}$ and $\|g\| = c$. It was originally conjectured [1] that the sharp (i.e., largest possible) constant c is $2^{-1/2}$; however, R. Martin (see [2]) later disproved this by the following counter-example.

EXAMPLE. Let $\varepsilon = 2^{1/2} - 1$ and define

$$f(z) = 2^{1/2} z, \qquad g(z) = (1 + \varepsilon z^{20})/(1 + \varepsilon 2^{-10}).$$

Then $|f(z)| > |g(z)|$ for $2^{-1/2} < |z| < 1$ but $\|f\| = 1$ and $\|g\| = \frac{(1 + \varepsilon^2/21)^{1/2}}{1 + \varepsilon 2^{-10}} > 1$.

The proof of Theorem 1 is based upon the following inequality for functions $f(z)$ holomorphic in $\mathbb{D}$ $(\gamma > 0)$:

$$\int_{\mathbb{D}} \left| \exp\{-2\gamma(1+z)/(1-z)\} f(z) \right|^2 dm \leq \int_{\mathbb{D}} \left| z^\gamma f(z) \right|^2 dm.$$

This inequality, as well as some others of a similar nature [3], suggests the following

DEFINITION. Let $h_1, h_2 \in L^2(\mathbb{D})$. We write $h_1 \prec h_2$ and say that h_1 is dominated by h_2 if for all $f \in H^\infty$ $\|h_1 f\| \leqslant \|h_2 f\|$.

In [3] necessary and sufficient conditions were found for $|B_1|^{\gamma_1} \prec |B_2|^{\gamma_2}$ where B_1 and B_2 are Blaschke factors and $\gamma_1, \gamma_2 > 0$. Using these domination relations it is possible to generalize Theorem 1:

THEOREM 2 [4]. Let $f, g \in A^2$. There is an absolute constant c_0, $0 < c_0 < 1$, such that if $c < c_0$ and (i) $|g(z)| < |f(z)|$ $(c < |z| < 1)$, (ii) $Z_g \setminus Z_f \subset \{z \in \mathbb{C} : c^{1/3} < |z| < 1\}$, then $g \prec f$ (in particular, $\|g\| \leqslant \|f\|$).

Another case when the Conjecture is proved to be true is

THEOREM 3 [5]. If one of the two functions $f, g \in A^2$ is a monomial (i.e., $f = az^n$ or $g = az^n$) and f majorizes g in $1/\sqrt{3} < |z| < 1$, then $\|f\| \geqslant \|g\|$.

It is not known whether the conclusion $\|f\| \geqslant \|g\|$ can be replaced by $f \succ g$. However, the assumption in Theorem 3 can be replaced by a weaker assumption $\|f_r\|_2 \geqslant \|g_r\|_2$ $(1/\sqrt{3} < r < 1)$, where $f_r(t) = f(re^{it})$ and $\| \cdot \|_2$ is the L^2 norm on $(0, 2\pi)$. A simple example shows that such a replacement would invalidate the Conjecture.

REFERENCES

1. Korenblum B., AMS Abstracts, 855-30-04 (1990).
2. Korenblum B., *A maximum principle for the Bergman space*, Publicacions Matemátiques (Barcelona) (to appear).
3. Korenblum B., *Transformations of zero sets by contractive operators in the Bergman space*, Bull. Sc. Math. 2e série **114** (1990), 385–394.
4. Korenblum B., Richards K., *Majorization and domination in the Bergman space*, preprint.
5. Korenblum B., O'Neil R., Richards K., Zhu K., *Totally monotone functions with applications to the Bergman space*, preprint.

DEPARTMENT OF MATHEMATICS AND STATISTICS
STATE UNIVERSITY OF NEW YORK AT ALBANY
ALBANY, NY 12222
U.S.A.

SCHATTEN CLASS COMPOSITION OPERATOR
ON THE BERGMAN SPACE

KEHE ZHU

Let $\mathbb{D}$ be the open unit disk in the complex plane $\mathbb{C}$. The Bergman space $L^2_a(\mathbb{D})$ is the closed subspace of $L^2(\mathbb{D}, dA)$ where dA is the normalized area measure. For an analytic function φ mapping the unit disc $\mathbb{D}$ into itself the composition operator C_φ is defined by $C_\varphi = f \circ \varphi$. It follows easily from Littlewood's subordination theorem that each C_φ is a bounded linear operator on $L^2_a(\mathbb{D})$. It is proved in [2] that the operator C_φ is compact on $L^2_a(\mathbb{D})$ if and only if

$$\lim_{|z| \to 1^-} \frac{1 - |z|^2}{1 - |\varphi(z)|^2} = 0.$$

By a well-known theorem of Carathéodory [3], the above condition holds if and only if the function φ does not have a finite angular derivative at any point of the unit circle. (This is how the result is actually stated in [2]).

Our problem here is to determine when the composition operator C_φ (acting on $L^2_a(\mathbb{D})$) is in the Schatten class S_p of $L^2_a(\mathbb{D})$. Recall that for any $p > 0$ a bounded linear operator T on a Hilbert space H belongs to the Schatten class S_p of H if and only if the operator $(T^*T)^{p/2}$ is in the trace class of H.

CONJECTURE. *For any $p > 1$ the composition operator C_φ on $L^2_a(\mathbb{D})$ belongs to the Schatten class S_p of $L^2_a(\mathbb{D})$ if and only if*

$$\int_{\mathbb{D}} \left[\frac{1 - |z|^2}{1 - |\varphi(z)|^2} \right]^p \frac{dA(z)}{\left(1 - |z|^2\right)^2} < +\infty.$$

We know that when $p \geqslant 2$ the above integral condition is necessary for the composition operator C_φ to be in S_p of $L^2_a(\mathbb{D})$; and the condition is sufficient when $0 < p \leqslant 2$. It is easy to see that the above integral condition is not necessary when $0 < p \leqslant 1$ as shown by the function $\varphi(z) = rz$, where r is any positive number less than 1, because in this case the integral clearly diverges but the composition operator is in all Schatten classes.

A necessary and sufficient condition for C_φ to be in S_p of $L^2_a(\mathbb{D})$ is given in [1]. We consider the result there unsatisfactory because it involves certain generalized Nevanlinna counting functions which are not yet well understood. Schatten class composition operators are studied in [1] on the Hardy space H^2 in terms of the (usual) Nevanlinna counting function, extending results of J. Shapiro [4].

PROBLEM 18.3

References

1. Luecking D., Zhu K., *Composition operators belonging to Schatten classes*, Amer. J. Math (to appear).
2. MacCluer B., Shapiro J., *Angular derivatives and compact composition operators on the Hardy and Bergman spaces*, Canadian J. Math. **38** (1986), 878–906.
3. Sarason D., *Angular derivatives via Hilbert space*, Complex Variables **10** (1988), 1–10.
4. Shapiro J., *The essential norm of a composition operator*, Ann. Math. **12** (1987), 375–404.

DEPARTMENT OF MATHEMATICS
STATE UNIVERSITY OF NEW YORK
ALBANY, NY 12222
U.S.A.

ON AN INTERPOLATION PROBLEM FOR $|dz|$ PERIODS

SABUROU SAITOH

Let G denote a bounded regular region whose boundary components $\{C_j\}_{j=1}^n$ are disjoint analytic Jordan curves. Let $H_p(G)$ $(1 \leqslant p \leqslant \infty)$ denote the analytic Hardy class on G. We first note the following

LEMMA. *Let* $\{I_1, I_2, \ldots, I_N\}$ *denote a family of measurable subsets of* ∂G *such that each* I_j *is of positive Lebesgue measure. Let* I_0 *denote* $\partial G \setminus \cup_{j=1}^N I_j$ *and* $\rho \in L_q(\partial G)$ *$(\frac{1}{p} + \frac{1}{q} = 1; p = 1, q = \infty)$. Then, for arbitrary given complex numbers* $\alpha_1, \alpha_2, \ldots, \alpha_N$, *there exists an* $H_p(G)$ *function* $f(z)$ *such that*

$$\int_{I_j} f(z)\rho(z)\,dz = \alpha_j, \qquad j = 1, 2, \ldots, N,$$

if and only if the following condition is satisfied: (C) *If for some* $F \in H_q(G)$ *and for some complex numbers* $\{\lambda_j\}_{j=0}^N$ *($\lambda_0 = 0$),*

$$\lambda_j \rho(z) = F(z) \qquad \text{a.e. on } I_j; \ j = 0, 1, 2, \ldots, N,$$

then all the λ_j *are zero.*

Proof. We consider the linear functionals T_j on $H_p(G)$ such that

$$T_j(f) = \int_{I_j} f(z)\rho(z)\,dz; \qquad j = 1, 2, \ldots, N.$$

Then, the functionals $\{T_j\}_{j=1}^N$ are linearly independent on $H_p(G)$ if and only if for given complex numbers $\{\alpha_j\}_{j=1}^N$, there exists an $f \in H_p(G)$ such that $T_j(f) = \alpha_j$; $j = 1, 2, \ldots, N$. Let ω_j denote the harmonic measure of I_j on G. Suppose that for some complex numbers $\{\lambda_j\}_{j=1}^N$,

$$\sum_{j=1}^N \lambda_j T_j(f) = \int_{\partial G} f(z) \left(\sum_{j=1}^N \lambda_j \omega_j(z) \right) \rho(z)\,dz = 0 \qquad \forall f \in H_p(G).$$

Then, by Cauchy–Read theorem on measures, we have, for some $F \in H_p(G)$

$$\left(\sum_{j=1}^N \lambda_j \omega_j(z) \right) \rho(z) = F(z) \qquad \text{a.e. on } \partial G$$

and vice versa. We thus have the desired result.

In our Lemma, if $\{C_1, C_2, \ldots, C_N\}$ is taken instead of $\{I_1, I_2, \ldots, I_N\}$, our interest will be in the maximal number N for which there exists an $H_p(G)$ function with arbitrary given weighted ρ-periods:

$$\int_{C_j} f(z)\rho(z)\,dz = \alpha_j, \qquad j = 1, 2, \ldots, N.$$

In particular, our interest will be in whether $N = n$ of $N = n - 1$. If $\rho(z) = 0$ on a subset of ∂G of positive measure, then we are able to find the maximum number N, immediately. Therefore, we assume that $|\rho(z)| > 0$ a.e. on ∂G. We see immediately that for $\rho(z) = A(z)$ ($\neq 0$, $\in H_q(G)$), $N = n - 1$, and for $\rho(z) = M(z)$; M is a meromorphic function with poles on G such that $M \in L_q(\partial G)$, $N = n$. At this point, we have the following

CONJECTURE. For $\rho(z) = |dz|/dz$, we have $N = n$ expect for the case of annulus regions whose boundary components are two concentric circles.

DEPARTMENT OF MATHEMATICS
FACULTY OF ENGINEERING
GUNMA UNIVERSITY
KIRYU 376
JAPAN

COEFFICIENT PROBLEMS IN GEOMETRIC FUNCTION THEORY

A. W. GOODMAN

Recently de Branges proved that if

$$(1) \qquad f(z) = \sum_{n=1}^{\infty} a_n z^n$$

is univalent in $E : |z| < 1$, then $|a_n| \leqslant n|a_1|$ for $n = 2, 3, 4, \ldots$. This theorem settled the most famous conjecture in geometric function theory, but there are many other coefficient problems that are still open and may be more difficult. Here we mention four such problems.

Suppose that for fixed integer $p > 1$, the function defined by (1) is at most p–valent in E. The author conjectured that then

$$(2) \qquad |a_n| \leqslant \sum_{k=1}^{p} \frac{2k(n+p)!}{(p+k)!(p-k)!(n-p-1)!(n^2-k^2)} |a_k|$$

for $n = p+1, p+2, p+3, \ldots$. If true, the inequality (2) would be sharp for every $n > p$ and every nonzero set $(a_1, a_2, \ldots, a_p)$. The proof when $p = 1$ (univalent functions) rested heavily on Loewner's theory. There is no corresponding Loewner Theory (at present) for p–valent functions, $p > 1$. For more details and references, see [2]

Let $a_1 = 1$ in (1) and suppose that $f(E) = D$ a convex domain. We say that $f(z)$ is a convex function of bounded type and we write $f(z) \in \mathrm{CV}(R_1, R_2)$ if at each point of ∂D the radius of curvature ρ satisfies $R_1 \leqslant \rho \leqslant R_2$. The terminology was introduced by the author in [3], but, in fact, the class $\mathrm{CV}(R_1, R_2)$ had been studied earlier without the name (see [12]). In [2] and [3] the author obtained a few results about the new class, but as Wirths [12] pointed out, the sharp inequality

$$(3) \qquad |a_2| \leqslant \left(1 - 1/R_2\right)^{1/2}$$

was already known earlier. Further Wirths [12] proved that $|a_3| \leqslant (1 - 1/R_2)$, but the sharp upper bound for $|a_n|$ when $n > 3$ poses an open problem. Goodman [3, 4] showed that the function $F(z) = z/(1 - Az)$ which is an extremal function for $|a_2|$ and $|a_3|$ cannot be the extremal function for $|a_n|$ for all R_2 if $n \geqslant 4$.

Ma, Mejia, and Minda [7] studied a closely related set of functions in which the upper bound R_2 on the radius of curvature on ∂D is replaced by an equivalent lower bound $1/R_2$ on κ the curvature on ∂D. They also obtained the bound (3) an $|a_2|$. In four other papers [8, 9, 10, 11] these results were generalized to include the curvature when the euclidean metric was replaced by the hyperbolic or spherical metric.

Next, we consider the class UST of uniformly starlike functions. We say that $f(z)$ given by (1) with $a_1 = 1$ is in UST if for each point ζ in E and each circular arc γ in E with center at ζ, the image arc $f(\gamma)$ is starlike with respect to $f(\zeta)$. Investigation of this class was suggested by B. Pinchak who asked if $f(\gamma)$ is always starlike with respect to $f(\zeta)$ when γ is a complete circle in E. In [5] the author proved that $f(z) \in$ UST iff

$$(4) \qquad \operatorname{Re} \frac{f(z) - f(\zeta)}{(z - \zeta)f'(z)} \geqslant 0$$

for all $(z, \zeta) \in E \times E$. The basis for (4) is covered in [1]. C. Horowitz proved that if $f(z) \in$ UST, then $|a_n| \leqslant 2/n$ but we do not know if this bound is sharp. Still worse, we do not have a reasonable conjecture for the sharp upper bound for $|a_n|$ for any $n > 1$.

The class UCV of uniformly convex functions is defined in a similar way. We say that $f(z)$ given by (1), with $a_1 = 1$, is in UCV if for each ζ in E and each circular arc γ in E with center at ζ, the arc $f(\gamma)$ is a convex arc. It was proved in [6] that $f(z) \in$ UCV iff

$$(5) \qquad 1 + \operatorname{Re}\left[\frac{f''(z)}{f'(z)}(z - \zeta)\right] \geqslant 0$$

for all $(z, \zeta) \in E \times E$. Further we have the bound $|a_n| \leqslant 1/n$ for every $n > 1$. However, this bound is not sharp and we do not have a reasonable conjecture for the sharp upper bound for $|a_n|$ for any $n > 1$.

REFERENCES

1. Goodman A. W., *Univalent functions*, Vols. I and II, Polygonal Publishing House, Washington, New Jersey U.S.A., 1983.
2. Goodman A. W., *Topics in mathematical analysis*, World Scientific Publishing Co., Singapore, 1989.
3. Goodman A. W., *Convex functions of bounded type*, Proc. Amer. Math Soc. **92** (1984), 541–546.
4. Goodman A. W., *More on convex functions of bounded type*, Proc. Amer. Math Soc. **97** (1986), 303–306.
5. Goodman A. W., *On uniformly starlike functions*, Jour. of Math. Analysis and Appl. **155** (1991), 364–370.
6. Goodman A. W., *On uniformly convex functions*, Ann. Polo. Math. (to appear).
7. Ma Wancang, Mejia Diego, Minda David, *Distortion theorems for euclidean k-convex functions*, Complex Var. Theor. and Appl. (to appear).
8. Ma Wancang, Mejia Diego, Minda David, *Distortion theorems for hyperbolically and spherically k-convex functions*, Proc. Inter. Conf. New Trends in Geom. Func. Th. and App. (to appear).
9. Mejia Diego, Minda David, *Hyperbolic geometry in k-convex regions*, Pac. Jour. of Math. **141** (1990), 333–354.
10. Mejia Diego, Minda David, *Hyperbolic geometry in spherically k-convex regions*, Comp. Methods and Func. Th. Proc., Lecture notes in Math, vol. 1435, Springer, 1990.
11. Mejia Diego, Minda David, *Hyperbolic geometry in hyperbolically k-convex regions*, submitted.
12. Wirths K. J., *Coefficient bounds for convex functions of bounded type*, Proc. Amer. Math. Soc. **103** (1988), 525–530.

DEPARTMENT OF MATHEMATICS
UNIVERSITY OF SOUTH FLORIDA
TAMPA, FLORIDA 33620
U.S.A.

QUASISYMMETRIC STARLIKE FUNCTIONS

J. G. Krzyż

A quasisymmetric automorphism h of the unit disk $\mathbb{D}$ with $h(0) = 0$ generates an automorphism of $\mathbb{T} = \partial\mathbb{D}$ characterized by the condition: *There exists $M \geqslant 1$ such that*

$$(*) \qquad\qquad |h(\alpha_1)|/|h(\alpha_2)| \leqslant M$$

holds for any pair of adjacent subarcs α_1, α_2 *of* $\mathbb{T}$ *with* $|\alpha_1| = |\alpha_2| \leqslant \frac{1}{2}$, cf. [1]. Here $|\alpha|$ denotes the harmonic measure $\omega(0, \alpha; \mathbb{D})$.

Let v be harmonic in $\mathbb{D}$ with boundary values $v(e^{i\theta}) - \theta$. Then, due to (*), $\theta + v(e^{i\theta})$ is a strictly increasing M-quasisymmetric function on $\mathbb{R}$. Consequently

$$F(z) = z \exp q(z) = z + A_2(F)z^2 + \dots, \qquad \operatorname{Im} q(z) = v(z),$$

is starlike univalent in $\mathbb{D}$ and (*) has an obvious geometrical interpretation so that F may be called an M-quasisymmetric function.

PROBLEM. *Find the sharp upper bound $c(M)$ of $|A_2(F)|$ for M-quasisymmetric starlike functions.*

Remark. M-quasisymmetric functions $\theta \mapsto \theta + v(e^{i\theta})$ characterize the boundary correspondence under quasiconformal automorphisms of $\mathbb{D}$. The Fourier series of $v(e^{i\theta})$ may be written in the form $\sum_{n=1}^{\infty} \rho_n \sin(n\theta + \theta_n)$, $\rho_n \geqslant 0$ for all $n \in \mathbb{N}$. We have $\rho_n \leqslant c(M)/n$ and the bound is sharp for any n, cf. [2].

References

1. Krzyż J. G., *Quasicircles and harmonic measure*, Ann. Acad. Sci. Fenn. Ser. A I Math. **12** (1987), 19–24.
2. Krzyż J. G., *Harmonic analysis and boundary correspondence under quasiconformal mappings*, Ann. Acad. Sci. Fenn. Ser. A I Math. **14** (1989), 225–242.

Department of Mathematics
Maria Curie-Skłodowska University
Plac M. Curie-Skłodowskiej 1
PL 20031 Lublin
Poland

ADVANCED COEFFICIENT PROBLEMS
FOR UNIVALENT FUNCTIONS

A. Z. Grinshpan

The class S of functions $f(z) = z + c_2 z^2 + \cdots$ regular and univalent in the unit disk E is of primary concern for the theory of univalent functions. For many years the coefficient problems for the class S and its subclasses have been an active field of research (see e.g. [4, 11, 3]). The following problems are connected with the author's papers in recent years.

1. Coefficient estimates depending on Grunsky operator norm. Each function $f \in S$ generates its Grunsky operator

$$G_f = \left\{ \sqrt{nm} \left\{ \log \frac{f(z) - f(\zeta)}{z - \zeta} \right\}_{n,m} \right\}_{n,m=1}^{\infty} : \ell^2 \to \ell^2, \qquad \text{where } \|G_f\| \leqslant 1$$

(see e.g. [3]). The stratification of the class S by Grunsky operator norm is connected with the natural stratification of analytical and geometric properties of univalent functions [9]. Let for $k \in [0, 1]$

$$S(k) = \{ f \in S : \|G_f\| \leqslant k \},$$

then $S(1) = S$ and $S(0) = \{ z/(1 - \zeta z) : \zeta \in \overline{E} \}$. The function $f_k(z) = z/(1-z)^{1+k} \in S(k)$, $\|G_{f_k}\| = k$, plays the role in $S(k)$ $(k < 1)$ which is largely similar to the known extremal role of the Koebe function f_1 in the whole class S. Namely, though f_k is not extremal for principal functionals on $S(k)$ for all $k \in (0,1)$, it realizes such values of many functionals which don't differ essentially from their supremums on $S(k)$. It is important since the exact solutions of extremal problems with regard to Grunsky operator norm are usually difficult to obtain and they are unlikely to be applied. As $S(k)$ $(k < 1)$ contains all the functions in S which have k-quasiconformal extension in $\widehat{C} \setminus E$, f_k included, where $\widehat{C}$ is the extended complex plane, all the above-said concerns these functions too [9]. Let

$$V_n(k) = \sup_{S(k)} \frac{|\{f\}_n|}{\{f_k\}_n} \quad (n = 2, 3, \ldots), \qquad V(k) = \sup_n V_n(k)$$

(here $\{f\}_n$ is the n-th Taylor coefficient of f). All functions $V_n(k)$ are unlikely to be found in the nearest future. However it is known that $V(0) = 1$; $V(k) \in (1, \sqrt{2})$ (and what is more, $V(k) < \left(\frac{1.87}{k}\right)^{k/2}$) for all $k \in (0, 1)$ [9] and $V(1) = 1$ (the famous Bieberbach conjecture) [1].

PROBLEM 1. *Find the function $V(k)$ for $k \in (0,1)$.*

CONJECTURE 1. *$V(k)$ is a continuous one-modal function in $[0,1]$ with a maximum (within $1.04 \div 1.08$) at $k \approx 0.5$ [9].*

PROBLEM 2. *Find for all $k \in (0,1)$ an exact (in the sense of the growth order by n) estimate of values*
$$d_n(k) = \sup_{S(k)} ||\{f\}_{n+1}| - |\{f\}_n||.$$

W. K. Hayman first solved such problem in the case $k = 1$ (see details e.g. in [3]). It is known that for any $k \in [0,1]$

$$d_n(k) < 4n^{(k^2-1)/2} \qquad (n = 1, 2, \ldots) \; [9].$$

CONJECTURE 2. *For any $k \in (0.5, 1)$ $d_n(k) = O(n^{k-1})$ as $n \to \infty$.*

2. Logarithmic coefficients and admissible vectors. For each function $f \in S$ its logarithmic coefficients γ_n are defined by

$$\log \frac{f(z)}{z} = 2 \sum_{n=1}^{\infty} \gamma_n z^n.$$

For each natural n, a real vector $X_n = (x_1, \ldots, x_n) \neq 0$ will be called admissible if $\sup_S M_{X_n}(f) = 0$, where $M_{X_n}(f) = \sum_{m=1}^{n} x_m (m|\gamma_m|^2 - \frac{1}{m})$. That is for admissible vector X_n the Koebe function maximizes the functional M_{X_n} on the class S. The fact that vectors $(n, n-1, \ldots, 1)$ are admissible for all n corresponds to the assertion of the former Milin conjecture which results from the de Branges theorem about univalent functions [1]. Basing on it, the Bieberbach conjecture was proved in [1]. The approach used there showed the deep connection between univalent and some special functions. Therefore the following problem can be considered as stimulating one not only for the theory of univalent functions.

PROBLEM 3. *Describe admissible vectors for all n.*

Almost elementary variation of the Koebe function [7, 10] shows that each admissible vector X_n necessarily satisfies the condition

(1)
$$\min_{\theta \in [0,\pi]} \sum_{m=1}^{n} x_m \sin(m\theta) = 0,$$

which for $n \leqslant 2$ gives a complete description of the admissible vectors [8]. For each star function $f_* \in S$ and any real vector $X_n \neq 0$ satisfying (1) the strict inequality $M_{X_n}(f_*) < 0$ holds unless f_* is $f_1(\zeta z)\bar{\zeta}$, $|\zeta| = 1$. Nevertheless, two-parameter family of functions

$$F_{\varepsilon,t}(z) = W(z)[1 - W(z)]^{-2}/t \in S,$$

where $t \cdot g(z) = g \circ W(z)$, $g(z) = z/[(1 - z)^{1+\varepsilon} \cdot (1 + z)^{1-\varepsilon}]$, for some $t \in (0,1)$ and $\varepsilon \in (-1,1)$ shows that the fascinating condition (1) is not sufficient for describing admissible vectors in general case [8]. As for other individual cases, it is unknown e.g.

PROBLEM 18.7

whether vectors $(1, \frac{1}{2}, \ldots, \frac{1}{n})$ are admissible for $n \geqslant 3$. The Duren–Loeng conjecture assumes it and some author's numerical research shows arguments in favor of it. It is known that unit vectors $X_n^1 = (1, \ldots, 1)$ are not admissble for $n \geqslant 2$ but for any $f \in S$

$$M_{X_n^1}(f) \leqslant \delta, \tag{2}$$

where δ is Milin's constant ($\delta < 0.312$) [11].

CONJECTURE 3. *One can replace δ with $\log(4|c|)$ in the inequality (2) if $f \neq c$* [5].

3. Coefficients of powers of univalent functions. It is known that for any $\lambda \geqslant 1$ and $n = 1, 2, \ldots$

$$\sup_S \left| \left\{ \left[\frac{f(z)}{z} \right]^\lambda \right\}_n \right| = \left\{ \left[\frac{f_1(z)}{z} \right]^\lambda \right\}_n. \tag{3}$$

In the case $\lambda = 1$, (3) corresponds to the former Bieberbach conjecture [1]. In the case $\lambda > 1$, (3) (conjectured in [6]) has been proved independently by several authors (see e.g. [10] or the paper of W. K. Hayman and J. A. Hummel in the same issue) due to the mentioned theorem from [1]. Let

$$A_n = \sup\{ \lambda : \lambda \geqslant 0, (3) \text{ doesn't hold} \}.$$

For all even $n = 2, 4, \ldots$ $A_n = 1$ [7].

CONJECTURE 4. *The strict inequality $A_n < 1$ holds for all odd n.*

For $n \leqslant 3$ it is true: $A_1 = 0$, $A_3 \in (0.618, 0.84)$ [7]. It is known that all $A_{2n-1} \geqslant 0.5$ ($n \geqslant 3$) (see Schaeffer–Spencer example e.g. in [3] and [7]). Some stronger numerical lower bounds for A_{2n-1} (up to $n = 51$) [2] allow us to formulate a deeper conjecture.

CONJECTURE 5. *The sequence $\{A_{2n-1}\}_1^\infty$ strictly increases and*

$$\lim_{n \to \infty} A_{2n-1} = 1.$$

The odd functions $f \in S$ are worth mentioning (the case $\lambda = 0.5$). It is known [11] that they satisfy

$$|\{f\}_{2n-1}| \leqslant \exp\{\delta/2\} \qquad (n = 2, 3, \ldots). \tag{4}$$

CONJECTURE 6. *One can replace $\exp\{\delta/2\}$ with $2|c|$ in the inequality (4) if $f \neq c$* [5].

This conjecture is weaker than that of [3].

REFERENCES

1. de Branges L., *A proof of the Bieberbach conjecture*, Acta Math. **154** (1985), 137–152.
2. Bshouty D., Hengartner W., *Criteria for the extremality of the Koebe mapping*, Proc. Amer. Math. Soc. **111** (1991), no. 2, 403–411.
3. Duren P. L., *Univalent Functions*, Springer Verlag, Heidelberg–New York, 1983.

4. Golusin G. M., *Geometric Theory of Functions of a Complex Variable*, "Nauka", Moscow, 1966 (Russian); English transl. Amer. Math. Soc., 1969.
5. Grinshpan A. Z., *On the Taylor coefficients of certain classes of univalent functions*, Metric Questions in the Theory of Functions, "Naukova Dumka", Kiev, 1980, pp. 28–32. (Russian)
6. Grinshpan A. Z., *On coefficients of powers of univalent functions*, Sibirsk. Mat. Zh. **22** (1981), no. 4, 88–93 (Russian); English transl. in Siberian Math. J. **22** (1981), 551–555.
7. Grinshpan A. Z., *On the power stability for the Bieberbach inequality*, Zap. Nauchn. Sem. LOMI **125** (1983), 58–64. (Russian)
8. Grinshpan A. Z., *Method of exponentiation for univalent functions*, Theory of Functions and Applications, Proc. Conf. Saratov, 1988, Part 2, Izdat. Sar. Univ., Saratov, 1990, pp. 72–74. (Russian)
9. Grinshpan A. Z., *Univalent functions with prescribed logarithmic restrictions*, Annals Polonici Mathematici (to appear).
10. Milin I. M., Grinshpan A. Z., *Logarithmic coefficients means of univalent functions*, Complex Variables: Theory and Appl. **7** (1986), no. 1–3, 139–147.
11. Milin I. M., *Univalent Functions and Orthonormal Systems*, "Nauka", Moscow, 1971 (Russian); English transl. Amer. Math. Soc., Providence, R.I., 1977.

INENCO
2, TRUDA SQUARE
ST. PETERSBURG, 190000
RUSSIA

AND

DEPARTMENT OF MATHEMATICS
UNIVERSITY OF SOUTH FLORIDA
4202 EAST FOWLER AVENUE, PHY 114
TAMPA, FLORIDA 33620-5700
USA

AN EXTENSION PROBLEM
FOR THE COEFFICIENTS OF RIEMANN MAPPINGS

JAMES ROVNYAK

By a *normalized Riemann mapping* we mean an analytic function which is defined and univalent on the unit disk and has value zero and positive derivative at the origin. We are concerned with normalized Riemann mappings $B(z)$ on the unit disk into itself.

For such $B(z)$, the operator $f(z)$ into $f(B(z))$ maps the Dirichlet space contractively into itself. Elementary examples show, however, that other functions also have this property [4]. More generally, if $B(z)$ is a normalized Riemann mapping of the unit disk into itself, the operator $f(z)$ into $f(B(z))$ acts as a contraction on a family of spaces which generalize the Dirichlet space. Given any real number ν, let $\mathcal{D}^\nu$ be the Kreĭn space of series $f(z) = \sum_{n=1}^\infty a_n z^{\nu+n}$ with finite self-product

$$\langle f(z), f(z) \rangle_{\mathcal{D}^\nu} = \sum_{n=1}^\infty (\nu + n) |a_n|^2.$$

Constant terms, if present, make no contribution to the sum and are identified to zero. The operator $f(z)$ into $f(B(z))$ is an everywhere defined contraction of $\mathcal{D}^\nu$ into itself for every real number ν. This property is characteristic of the class of normalized Riemann mappings of the unit disk into itself.

THEOREM (de Branges [3]). *Let $B(z) = B_1 z + B_2 z^2 + \ldots$ $(B_1 > 0)$ be a formal power series such that for every nonpositive integer ν, the operator $f(z)$ into $f(B(z))$ is an everywhere defined and contractive mapping of $\mathcal{D}^\nu$ into itself. Then $B(z)$ represents a normalized Riemann mapping of the unit disk into itself.*

The proof given in [3] makes an interesting link between geometric function theory and operator theory. We outline the main steps. Since $f(z)$ into $f(B(z))$ is contractive on the Dirichlet space, $B(z)$ represents an analytic function which is bounded by 1 in the unit disk. The problem is to show that $B(z)$ takes distinct values at distinct points.

Consider the operator $T: f(z) \to f(B(z))$ on the Dirichlet space $\mathcal{D} = \mathcal{D}^0$. A unique Hilbert space $\mathcal{G}(B)$ exists which is contained contractively in $\mathcal{D}$ such that the adjoint of the inclusion of $\mathcal{G}(B)$ in $\mathcal{D}$ coincides with the selfadjoint operator $I - TT^*$. A contraction operator G on $\mathcal{G}(B)$ into the Dirichlet space $\mathcal{D}$ is constructed using the hypothesis that for every nonpositive integer ν, the operator $f(z)$ into $f(B(z))$ is an everywhere defined contraction of $\mathcal{D}^\nu$ into itself. The operator has the property that

(1) $$G: f(z) - g(B(z)) \to h(z)$$

whenever $f(z)$ and $g(z)$ are polynomials such that $T^*: f(z) \to g(z)$ and $h(z)$ is a power series with constant term zero such that

$$f(1/z) - g(1/B^*(z)) = \text{const} + h(z).$$

The construction of the operator G is related to the Lagrange–Bürmann identity [5]. A calculation of the action of G on reproducing kernel functions for $\mathcal{G}(B)$ yields

$$G: \log \frac{1 - \bar{B}(w)B(z)}{1 - \bar{w}z} \rightarrow \log \frac{B'(0)/B^*(z) - B'(0)/\bar{B}(w)}{1/z - 1/\bar{w}}$$

for any w in the unit disk. The formula is verified first for restricted values of w and z and then extended to arbitrary values in the unit disk. The result follows because the expression on the right side is meaningful only when $B(z)$ is a normalized Riemann mapping. Details of this argument, which is due to de Branges, will be supplied upon the request by the author. A proof has also been constructed by N. K. Nikol'skiĭ and V. I. Vasyunin (private communication).*

The contractive substitution property has a local version which depends only on initial segments of the coefficients of a normalized Riemann mapping $B(z) = B_1 z + B_2 z^2 + \ldots$ of the unit disk into itself. For any real number ν and positive integer r, let $\mathcal{D}_r^\nu$ be the finite-dimensional Kreĭn space of generalized power series such that

$$\langle f(z), f(z) \rangle_{\mathcal{D}_r^\nu} = \sum_{n=1}^{r} (\nu + n)|a_n|^2$$

and with terms with $n > r$ or $\nu + n = 0$ are identified to zero. The inequality

$$(2) \qquad \langle f(B(z)), f(B(z)) \rangle_{\mathcal{D}_r^\nu} \leqslant \langle f(z), f(z) \rangle_{\mathcal{D}_r^\nu}$$

holds for every $f(z)$ in $\mathcal{D}_r^\nu$ [1, 5]. See [4] for a variant on the proof of the inequality and a symmetry involving ν. The inequality depends only on the coefficients $B_1, \ldots, B_r$, and the question arises if this property is characteristic of initial segments of coefficients [1]?

An analogy with the Carathéodory–Fejér interpolation problem [2] is suggestive. However, it is difficult to see how to proceed in parallel fashion because of the dependence of the condition (2) on ν. It may be possible instead to construct the operator G defined by (1) by an extension process that uses initial segments of coefficients to define finite-dimensional approximations of the operator.

PROBLEM: *Study approximations of the operator G defined by* (1) *that depend only on initial segments of the coefficients of* $B(z)$.

The point is to try to isolate conditions needed to make a step-by-step construction of the operator from local data.

REFERENCES

1. De Branges L., *Unitary linear systems whose transfer functions are Riemann mapping functions*, Integral Equations and Operator Theory **19** (1986), 105–124.
2. De Branges L., *Underlying concepts in the proof of the Bieberbach conjecture*, Proceedings of the International Congress of Mathematicians 1986, Berkeley, California, 1986, pp. 25–42.
3. De Branges L., *Square Summable Power Series*, Springer–Verlag, in preparation.
4. Li Kin Y., Rovnyak James, *On the coefficients of Riemann mappings of the unit disk into itself*, 1991, preprint.
5. Rovnyak J., *Coefficient estimates for Riemann mapping functions*, J. Anal. Math. **52** (1989), 53–93.

DEPARTMENT OF MATHEMATICS
MATHEMATICS-ASTRONOMY BUILDING
UNIVERSITY OF VIRGINIA
CHARLOTTESVILLE, VIRGINIA 22903–3199
U.S.A.

---

*See *Operator-valued measures and coefficients of univalent functions*, Algebra i Analiz **3** (1991), no.6, 1–75 (Russian); English transl. in St. Petersburg Math. J. **3**, (1992), no.6, 1199–1270. – Ed.

MODULI OF SMOOTHNESS
OF RIEMANN MAPPING FUNCTIONS

V. I. Belyĭ

Let G be a simply connected region, L its boundary, $a \in G$ a fixed point of G, $\overline{G} = G \cup L$. Consider the Riemann mapping function $v = \varphi(z)$ that gives a conformal mapping of G onto the unit disk $D = \{|w| < 1\}$, $\varphi(a) = 0$, $\varphi'(a) > 0$. The inverse function is denoted by $z = \psi(w)$.

There are some known reasonable definitions of moduli of smoothness for continuous functions in the complex domain (see [1]).

Let $H^{\omega_2}(\mathcal{M})$ be the class of functions f continuous on a compact set $\mathcal{M}$ and satisfying $\omega_2(f, t) \leqslant M_f \omega_2(t)$, $t \geqslant 0$, where $\omega_2(f, t)$ is the second order modulus of smoothness of f and ω_2 is a prescribed function.

PROBLEM. *What are the necessary and sufficient conditions ensuring $\varphi(z) \in H^{\omega_2}(\overline{G})$? What are the conditions ensuring $\psi(w) \in H^{\omega_2}(\overline{D})$?*

The answer can be given in terms of conformal invariants or some other terms.

It should be noted that for Hölder continuity and even for an arbitrary modulus of continuity of the first order the problem is solved (see [2], [3]).

REFERENCES

1. Tamrazov P. M., *Smoothness and Polynomial Approximations*, Naukova Dumka, Kiev, 1975. (Russian)
2. Näkki R., Palka B., *Extremal length and Hölder continuity of conformal mappings*, Comment. Math. Helvetici **61** (1986), 389–414.
3. Belyĭ V. I., *On moduli of continuity of exterior and interior conformal mappings of the unit disk*, Ukranian Math. J. **41** (1989), no. 4, 469–475. (Russian)

INST. OF APPLIED MATH. AND MECH.
UKRANIAN ACADEMY OF SCIENCES
UL. ROZA LUXEMBURG 74
340114 DONETSK 114
UKRAINE

LANDAU'S CONSTANT, AND EXTREMAL PROBLEMS
INVOLVING DISCRETE SUBSETS OF $\mathbb{C}$

ALBERT BAERNSTEIN II

Let $A = \{a_1, a_2, \dots\}$ denote a discrete subset of $\mathbb{C}$. Define

$$\sigma(A) = \sup_{z \in \mathbb{C}} \inf_{a \in A} |z - a|.$$

Thus, $\sigma(A)$ is the radius of the largest open disk contained in the domain $\mathbb{C} \setminus A$. We will be interested in sets A for which $\sigma(A) < \infty$.

Next, assume that $\varphi \colon \mathbb{R} \to \mathbb{R}$ is non-negative, nondecreasing and convex, and that $u_A = u$ is a solution, in the distribution space $\mathcal{D}'(\mathbb{C})$, of the p.d.e.

$$\Delta u = \varphi(u) - \sum_{a \in A} \delta_a$$

where δ_a denotes a unit mass of a.

PROBLEM. *Suppose that φ and $\sigma(A)$ are given. Find A such that $\inf_{\mathbb{C}} u_A$ is minimal.*

Results in the theories of packing, covering, and geometry of numbers, see, e.g., [10], [2], suggest that the extremal set should be an "equilateral lattice". Let $\mathcal{L}(e) = \{m + ne^{\pi i/3} : (m, n) \in \mathbb{Z}^2\}$. Then the points of $\mathcal{L}(e)$ are the vertices of a tiling of $\mathbb{C}$ by equilateral triangles. Let $b = 3^{-1/2}e^{i\pi/6}$. Then b is the barycenter of the triangle with vertices $0, 1, e^{\pi i/3}$. We have $\sigma(\mathcal{L}(e)) = |b| = 3^{-1/2}$, and thus $\sigma(3^{-1/2}\sigma(A)\mathcal{L}(e)) = \sigma(A)$. Set $U = u_{\sqrt{3}\sigma(A)\mathcal{L}(e)}$.

CONJECTURE 1. $\inf_{z \in \mathbb{C}} u_A(z) \geqslant U(b)$.

In the case when $\varphi(x) = e^{2x}$ we have $u_A = \log \rho$, where ρ is the density of the Poincaré metric for the domain $\mathbb{C} \setminus A$. Using the principle of subordination, see, e.g. [4, §9.4], and an argument which shows that arbitrary domains in $\mathbb{C}$ can be appropriately approximated by domains of the form $\mathbb{C} \setminus A$, one can show that in the special case $\varphi(x) = e^{2x}$ Conjecture 1 is equivalent to the following conjecture about covering properties of holomorphic functions. Suppose that f is holomorphic in the unit disk $\mathbb{D}$. Let $\sigma(f)$ denote the radius of the largest disk contained in $f(\mathbb{D})$, and set $L = \Gamma(\frac{1}{3})\Gamma(\frac{5}{6})/\Gamma(\frac{1}{6}) = 0.54\dots$.

CONJECTURE 2. $\sigma(f) \geqslant L|f'(0)|$.

Equality holds when f is a universal covering map f_1 of $\mathbb{D}$ onto $\mathbb{C} \setminus \mathcal{L}(e)$ with $f_1(0) = b$. Thus, Conjecture 1 asserts that the exact value of "Landau's constant" is L. This conjecture appears in a paper by H. Rademacher (1943), where the calculation of $f_1'(0)$ is carried out. The calculation, and the observation that f_1 provides an upper bound for Landau's constant, is also in unpublished work by R. M. Robinson (1937). Discussions of the constant of Landau and of Bloch can be found in [5] and [6, see esp. pp. 80, 83]. The analogue of Rademacher's conjecture for Bloch's constant, due to Ahlfors and Grunsky (1937), is also unresolved. I think that a positive solution to Conjecture 1 would also lead to one for the Ahlfors–Grunsky conjecture.

THEOREM 1. *Let $A \subset \mathbb{C}$ be discrete. Let $B = \{ |a| : a \in A \}$. Then*

$$u_A(z) \geqslant u_B(-|z|), \qquad \forall z \in \mathbb{C}.$$

THEOREM 2. *Let $A \subset \mathbb{R}$ be discrete. Suppose that $\sup_{x \in \mathbb{R}} \inf_{a \in A} |x - a| = \frac{1}{2}$. Then*

$$u_A(z) \geqslant u_{\mathbb{Z}} \left(\frac{1}{2} + iy \right), \qquad \forall z = x + iy \in \mathbb{C}.$$

Theorem 1 is contained in a theorem of Weitsman [11]. See also [4]. Its proof is based on subharmonicity considerations involving the function

$$u^*(re^{i\theta}) = \sup_{|E|=2\theta} \int_E u(re^{it}) \, dt = \int_{-\theta}^{\theta} u_0(re^{it}) \, dt$$

where u_0, for fixed r, denotes the symmetric decreasing rearrangement on $[-\pi, \pi]$ of $u(re^{it})$. Theorem 2 can be proved by use of a variant of u^* which was used to solve Problem 11.7 in the 1984 edition of this collection.* See [1] and [3]. One avenue for attacking Conjecture 1 is to seek rearrangement type processes which transform functions $f : \mathbb{C} \to \mathbb{R}$ with singularities on A top functions $f_0 : \mathbb{C} \to \mathbb{R}$ with singularities on $\sqrt{3}\sigma(A)\mathcal{L}(e)$.

I cannot settle Conjecture 1 even in the linear case $\varphi(x) = x$. The key to understanding this complex of problems might come from a study of the heat equation. Define

$$v_A(z,t) = \frac{1}{2\pi t} \sum_{a \in A} \exp \left(-\frac{|z-c|^2}{2t} \right).$$

The $(v_A)_t = \frac{1}{2} \Delta v_A$ in $\mathbb{C} \times (0, \infty)$, and $v_A(z,0) = \sum_{a \in A} \delta_a$. Write $V = v_{\sqrt{3}\sigma(A)\mathcal{L}(e)}$.

CONJECTURE 3. $v_A(z,t) \geqslant V(b,t), \qquad \forall (z,t) \in \mathbb{C} \times (0, \infty).$

This conjecture is open even in the case when A is itself a lattice $\mathcal{L}$, so that $u_{\mathcal{L}}$ is a theta function and may be viewed as the heat kernel for the torus $\mathbb{C}/\mathcal{L}$. I have found no comparison results of any kind involving analysis on tori when $\sigma(\mathcal{L})$, the diameter of $\mathbb{C}/\mathcal{L}$, is fixed. If instead one fixes the area $|\mathcal{L}|$ on $\mathbb{C}/\mathcal{L}$ then results do exist. Suppose that

$$\mathcal{L} = \{ m\omega_1 + n\omega_2 : (m,n) \in \mathbb{Z}^2 \}$$

with $\omega_1, \omega_2 \in \mathbb{C} \setminus \{0\}$, and that $\tau := \frac{\omega_1}{\omega_2}$ belongs to the open upper half plane $\mathbb{H}$. Then $|\mathcal{L}| = \operatorname{Im} \bar{\omega}_1 \omega_2$. Define $W = v_{c\mathcal{L}(e)}$, where $c^2 = 2 \cdot 3^{-1/2} \cdot |\mathcal{L}|$. Then $|c\mathcal{L}(e)| = |\mathcal{L}|$.

---

*See Problem S.16.22 in this edition. – Ed.

THEOREM 3. (H. Montgomery [7]). $\forall t > 0$ we have

$$(1) \qquad v_{\mathcal{L}}(0, t) \geqslant W(0, t).$$

Now set

$$Z_{\mathcal{L}}(s) = \sum a \in \mathcal{L}(2\pi |a|)^{-2s}.$$

Except for the factor 2π, $Z_{\mathcal{L}}$ is the Epstein zeta function for the quadratic form $Q(m, n) = |m\omega_1 + n\omega_2|^2$. The series converges for $s \in \mathbb{C}$ with $\operatorname{Re} s > 1$. Moreover, $Z_{\mathcal{L}}$ has a meromorphic continuation to $\mathbb{C}$, which is finite on the interval $[0, 1)$. Montgomery shows that his theorem implies that (1) holds when $v_{\mathcal{L}}$ and W are replaced by $Z_{\mathcal{L}}(s)$ and $Z_{c\mathcal{L}(e)}(s)$, for $s \in [0, 1) \cup (1, \infty)$. This result had been proved in papers between 1953 and 1964 by Rankin, Ennola, Diananda, and Cassels.

The quantity $\exp(-Z_{\mathcal{L}}'(0))$ is the "determinant of the Laplacian" for the torus $\mathbb{C}/\mathcal{L}^*$, with the flat metric, where $\mathcal{L}^*$ is the dual lattice of $\mathcal{L}$. From the Kronecker limit formula, see [8] or [9], it follows that

$$(2) \qquad \exp(-Z_{\mathcal{L}}'(0)) = |\mathcal{L}|^{-1}(\operatorname{Im} \tau)|\eta^4(\tau)|,$$

where $\tau = \omega_2/\omega_1$ and

$$\eta(\tau) = e^{\pi i \tau/12} \prod_{k \in \mathbb{Z}} (1 - e^{2\pi i k \tau})$$

is the Dedekind eta function.

THEOREM 4. $\displaystyle\sup_{\tau \in \mathbb{H}} (\operatorname{Im} \tau)|\eta^4(\tau)| = (\operatorname{Im} e^{\pi i/3})|\eta^4(e^{\frac{\pi i}{3}})|.$

There appear to be at least three independent proofs of Theorem 4. A numerical one is described in [8]. It completes the proof there that among all Riemannian metrics of fixed area on a torus the determinant of the associated Laplacian is maximal for the flat metric on $\mathbb{C}/\mathcal{L}(e)$. Moreover, since $Z_{\mathcal{L}}(0) = -1 \,\forall \mathcal{L}$, see [8], Theorem 4 follows also from (2) and the zeta function analogue of Montgomery's Theorem.

The third proof hints at a mysterious link between Conjectures 1, 2, and of Montgomery's theorem. Let u_A be as in Conjecture 1, and take $A = \kappa\mathcal{L}(e)$, where κ is a positive constant. I can prove that

$$\inf_{z \in \mathbb{C}} u_{\kappa\mathcal{L}(e)}(z) = u_{\kappa\mathcal{L}(b)}.$$

This shows that Conjecture 1 is true at least for $A = \kappa\mathcal{L}(e)$. When $\varphi(x) = e^{2x}$, (3) is equivalent to an inequality for the hyperbolic derivative of the universal covering map $F: \mathbb{H} \to \mathbb{C}/\mathcal{L}(e)$ with $F(e^{\pi i/3}) = b$:

$$\sup_{z \in \mathbb{H}} (\operatorname{Im} z)|F'(z)| = (\operatorname{Im} e^{\pi i/3})|F'(e^{i\pi/3})|.$$

But considerations with automorphic forms shows that F' is a constant multiple of η^4! Thus, Theorem 4 is a consequence of (3).

PROBLEM 18.10

REFERENCES

1. Baerstein A., *An extremal problem for certain subharmoric functions in the plane*, Rev. Mat. Iberoamericana **4** (1988), 199–219.
2. Cassels J. W. S., *An introduction to the Geometry of numbers* (2nd ed.), Springer, Berlin, 1972.
3. Fryntov A. E., *An extremal problem of potential theory*, Dokl. Akad. Nauk USSR **300** (1988), no. 4 (Russian); English transl. in Soviet Math. – Doklady **37** (1988), 754–755.
4. Hayman W. K., *Subharmonic functions*, vol. 2, Academic Press, London, 1989.
5. Hille E., *Analytic Function Theory*, vol. 2, Ginn, Boston, 1962.
6. Minda C. D., *Bloch constants*, J. Analyse Math. **41** (1982), 54–84.
7. Montgomery H., *Minimal theta functions*, Glasdow Math. J. **30** (1988), 75–83.
8. Osgood B., Phillips R., Sarnak P., *Extremals of determinants of Laplacians*, J. Funct. Anal. **80** (1988), 148–211.
9. Quine J. R., Heydari S. H., Song R. Y., *Zeta regularized products*. Trans. Amer. Math. Soc. (to appear).
10. Rogers C. A., *Packing and Covering*, Cambridge U. P., Cambridge, 1964.
11. Weitsman A., *Symmetrization and the Poincaré metric*, Annals of Math. **124** (1986), 159–169.

MATHEMATICS DEPARTMENT
WASHINGTON UNIVERSITY
ST.LOUIS MO 63130
USA

HARMONIC MAPPINGS IN THE PLANE

PETER DUREN

A harmonic mapping of a domain $\Omega \subset \mathbb{C}$ is a complex-valued harmonic function $f = u + iv$ which is univalent in Ω. The real and imaginary parts of f are not required to satisfy the Cauchy-Riemann equations. It can be shown (see [2], [7]) that much of the classical theory of analytic univalent functions extends to harmonic mappings, but there are many open problems.

By a theorem of Lewy [5], a harmonic mapping has a nonvanishing Jacobian: $J_f(z) = |f_z(z)|^2 - |f_{\bar{z}}(z)|^2 \neq 0 \in \Omega$. The mapping is said to be sense-preserving if $J_f(z) > 0$. Then the second complex dilatation $\omega = \overline{f_{\bar{z}}}/f_z$ is an analytic function with $|\omega(z)| < 1$. If Ω is simply connected, there is no loss of generality (in view of the Riemann mapping theorem) in assuming that $\Omega = \mathbb{D}$, the unit disk. Each harmonic function in $\mathbb{D}$ has a unique representation $f = h + \bar{g}$, where h and g are analytic in $\mathbb{D}$ and $g(0) = 0$. After suitable normalizations (see [2]) it may be further assumed that $h(0) = 0$, $h'(0) = 1$, and $g'(0) = 0$. Then it is said that $f \in S_H^0$. It can be shown (see [2]) that S_H^0 is a compact normal family. It appears to be an appropriate generalization of the classical family S of analytic univalent functions in the disk. For instance, the class S_H^0 contains the "harmonic Koebe function" $K = H + \overline{G}$, where

$$H(z) = \frac{z - \frac{1}{2}z^2 + \frac{1}{6}z^3}{(1-z)^3}, \qquad G(z) = \frac{\frac{1}{2}z^2 + \frac{1}{6}z^3}{(1-z)^3}.$$

It maps $\mathbb{D}$ harmonically onto the full plane slit along the negative real axis from $-\infty$ to $-1/6$. Clunie and Sheil-Small [2] showed that the range of each $f \in S_H^0$ covers the disk $|w| < 1/16$, and it is conjectured that the radius can be improved to $1/6$.

Let $f = h + \bar{g} \in S_H^0$ and write

$$h(z) = z + \sum_{n=2}^{\infty} a_n z^n, \qquad g(z) = \sum_{n=2}^{\infty} b_n z^n.$$

The coefficients of the harmonic Koebe function then suggest the conjecture that the inequalities

$$|a_n| \leqslant \frac{1}{6}(2n+1)(n+1), \qquad |b_n| \leqslant \frac{1}{6}(2n-1)(n-1)$$

hold, but at present only $|b_2| \leqslant 1/2$ is a theorem. (It follows from the Schwarz inequality, since $\omega = g'/h'$ vanishes at the origin). The conjecture that $|a_2| \leqslant 5/2$ is of special importance because of its geometric consequences (see [8]), but so far the best known estimate is $|a_2| < 49$. Another attractive conjecture is that $||a_n| - |b_n|| \leqslant n$ for $n = 2, 3, ...$, a generalization of the Bieberbach conjecture. Clunie and Sheil-Small [2] have verified these conjectures for typically real functions, but for the full class S_H^0 they remain unsettled. In fact, they are not even known to be correct in order of magnitude as $n \to \infty$.

Another class of problems comes from a proposed generalization of the Riemann mapping theorem to harmonic mappings. The problem is to map the disk harmonically onto a given simply connected region by a function f with prescribed dilatation $\omega = \overline{f_{\bar{z}}}/f_z$. (Riemann's theorem handles the case $\omega(z) \equiv 0$). Hengartner and Schober [3] have proved the existence of such a mapping in the quasiconformal case where $|\omega(z)| \leqslant k < 1$, but they have shown by examples that a solution does not always exist For instance, if f is a harmonic mapping of $\mathbb{D}$ into itself with dilatation $\omega(z) = z$, then (see [7]) the area of $f(\mathbb{D})$ must be less than $\pi/2$. In particular, there is no harmonic mapping of $\mathbb{D}$ *onto* itself with dilatation $\omega(z) = z$. On the other hand, there is such a mapping of $\mathbb{D}$ onto an equilateral triangle inscribed in the unit circle, and its boundary values lie on the unit circle almost everywhere. This phenomenon of "collapsing" needs to be better understood. Is the area of the image the only obstruction to the existence of a mapping in the usual sense? And what about *uniqueness* of the mapping with prescribed dilatation? That question is wide open.

In a recent paper, Abu-Muhanna and Lyzzaik [1] have shown that in the canonical representation $f = h + \bar{g}$ of a harmonic mapping, both h and g belong to the Hardy space H^p for some $p > 0$. It seems likely that h and g actually belong to H^p for all $p < 1/3$. The harmonic Koebe function shows that nothing better is possible.

There are many other open questions concerning harmonic mappings in the plane. Some relate to estimates of (Gaussian) curvature of minimal surfaces, which are represented parametrically by harmonic functions. (See [4], [7]). Another interesting problem concerns the distortion of the modulus of an annulus under harmonic mapping (see [6] or [7]). Limitations of space prevent a fuller discussion.

REFERENCES

1. Abu-Muhanna Y., Lyzzaik A., *The boundary behaviour of harmonic univalent maps*, Pacific J. Math. **141** (1990), 1–20.
2. Clunie J., Sheil-Small T., *Harmonic univalent functions*, Ann. Acad. Sci. Fenn. Ser. **A.I 9** (1984), 3–25.
3. Hengartner W., Schober G., *Harmonic mappings with given dilatation*, J. London Math. Soc. **33** (1986), 473–483.
4. Hengartner W., Schober G., *Curvature estimates for some minimal surfaces*, Complex Analysis: Articles dedicated to Albert Pfluger on the occasion of his 80th birthday (Hersch J., Huber A., eds.), Birkhäuser Verlag, Basel, 1988, pp. 87–100.
5. Lewy H., *On the non-vanishing of the Jacobian in certain one-to-one mappings*, Bull. Amer. Math. Soc. **42** (1936), 689–692.
6. Nitsche J. C. C., *On the module of double-connected regions under harmonic mappings*, Amer. Math. Monthly **69** (1962), 781–782.
7. Schober G., *Planar harmonic mappings*, Computational Methods and Function Theory, Lecture Notes in Math., Springer–Verlag, Berlin–Heidelberg, 1990, pp. 171–176.
8. Sheil-Small T., *Constants for planar harmonic mappings*, J. London Math. Soc. **42** (1990), 237–248.

DEPARTMENT OF MATHEMATICS
UNIVERSITY OF MICHIGAN
ANN ARBOR, MICHIGAN 48109
USA

CIRCLE PACKINGS
AND DISCRETE ANALYTIC FUNCTIONS

KENNETH STEPHENSON

Circle packings were introduced Thurston in 1985 in a conjecture regarding the approximation of conformal maps, a conjecture proven by Rodin/Sullivan [RS]. (See also [St1], [St2]). However, circle packings may have more to offer: they appear to provide a comprehensive *discrete analytic function theory* paralleling the classical continuous case. In particular, this discrete theory seems to reflect very faithfully the fundamental geometric aspects of complex function theory. We introduce this developing analogy by posing two questions: one concerns a discrete version of Koebe's 1/4 theorem, the other, a parallel to the "type" problem. See especially [BSt1] and [BSt2] for relevant details.

Definitions/Notation. A *circle packing* is a collection P (finite or infinite) of circles with specified tangencies. Its combinatorics are best described using an (abstract, oriented, simplicial) complex: specifically, the *complex K* for P consists of vertices for the circles of P, edges for pairs of (externally) tangent circles, and faces for mutually tangent triples of circles. Assume in the sequel that K forms a closed topological disc when P is finite and an open topological disc when P is infinite. In the former case, circles on the outer edge of P (associated with vertices in ∂K) are *boundary* circles. Fig. 1 is a finite packing, Fig. 2 suggests an infinite packing; in each of these, the circles have mutually disjoint interiors, but that is not a requirement.

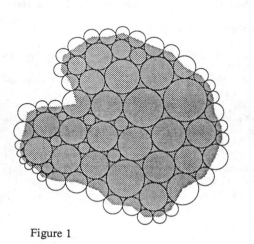

Figure 1

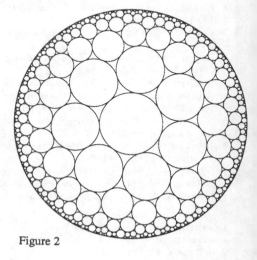

Figure 2

The *carrier* of P, carr(P), is the natural geometric realization of K: that is, identify vertices of K with the (euclidean) centers of corresponding circles of P, edges with segments between centers of tangent circles, and so forth. Our assumptions on K imply that carr(P) is a triangulated topological disc. (E.g., shaded in Fig. 1.)

DEFINITION. *Assume circle packings P_1 and P_2 have the same complex K. The map $f: P_1 \longrightarrow P_2$ which identifies each circle of P_1 with the corresponding circle of P_2 will be called a discrete analytic function.*

This notion reflects the fact that the tangency relationships coded in K provide a certain geometric rigidity which all associated circle packings must share. The "rigidity" preserved in going from one packing to another is reminiscent of that preserved by an analytic function mapping one domain onto another (after all, analytic functions preserve infinitesimal circles). This formal definition of f is rather abstract; when a point mapping is preferable, we abuse notation and write $f : \text{carr}(P_1) \longrightarrow \text{carr}(P_2)$ for the piecewise-affine map which identifies the centers of circles corresponding under f and then extends to carr(P_1) via barycentric coordinates.

Basic Theory. The remarkable result which initiated this topic was Thurston's interpretation of work of E. M. Andreev (see [T, §13]), though it also follows in part from a 1936 paper of Koebe [K] which has been overlooked until recently:

THEOREM 1. *Let K be a finite simplicial complex which is a closed topological disc. Then there exists an (essentially) unique circle packing P_K in the unit disc $\mathbb{D}$ whose boundary circles are horocycles and whose complex is K.*

We assume henceforth that a designated circle, C_0, is centered at the origin, another on the positive real axis (apply a Möbius transformation of $\mathbb{D}$, if needed). So normalized, P_K is unique and we will call it the *complete* packing for K. (See Fig. 3 below).

One of the fundamental parallels between circle packings and classical function theory is manifest in the Discrete Schwarz–Pick Lemma ([BSt1]):

THEOREM 2. *Let P be a packing lying in $\mathbb{D}$ and having complex K. The discrete analytic function $f : P_K \longrightarrow P$ is a hyperbolic contraction, with each circle $C'_j = f(C_j)$ having hyperbolic radius no greater than that of C_j. In particular, if $C'_0 = f(C_0)$ is centered at the origin, then the (euclidean) radii satisfy $\text{rad}(C'_0)/\text{rad}(C_0) \leqslant 1$, with equality iff f is a rotation (i.e. P is a rotation of P_K).*

The first part of this is Pick's Lemma. For the second part, note that the ratio of radii reflects the "stretching" factor in going from P to P_K and should be interpreted as $|f'(0)|$; so we get the precise analogue of the classical Schwarz Lemma.

If K is infinite, one can apply Theorem 1 to an exhausting sequence of finite subcomplexes. Using Theorem 2 in arguments precisely parallel to the classical case, one can prove:

THEOREM 3. *Let K be an infinite complex which is an open topological disc. Then there exists an (essentially) unique extremal circle packing P_K having complex K. carr(P_K) is either the unit disc $\mathbb{D}$ or the plane $\mathbb{C}$; we say K is hyperbolic or parabolic, respectively.*

Many other parallels with classical results flow from this theorem — the discrete uniformization theorem, discrete covering maps and groups, a discrete Liouville theorem, and so forth (see [BSt2]).

411

Koebe's Theorem. The Schwarz–Pick Lemma is a geometric result regarding classical analytic functions which is already present in the discrete setting. Is the Koebe 1/4-theorem another?

Assume P is a circle packing with a finite complex K. The map $f : P_K \longrightarrow P$ will be of *class S* if P is normalized so that $C_0' = f(C_0)$ is centered at the origin and $\mathrm{rad}(C_0')/\mathrm{rad}(C_0) = 1$ (i.e., $f(0) = 0$ and $|f'(0)| = 1$) and if f, considered as a pointwise mapping $f : \mathrm{carr}(P_K) \longrightarrow \mathrm{carr}(P)$, is univalent.

QUESTION 1. *Let $f : P_K \longrightarrow P$ be of class S and let r_f be the radius of the largest open disc centered at the origin and lying in $\mathrm{carr}(P)$. Is $r_f \geqslant 1/4$?*

The existence of some universal lower bound seems likely, since there are only finitely many "small" complexes K, and the functions f associated with "large" ones begin to approximate analytic functions (see [RS], [St1]). This approach is wrong in spirit, however: one would prefer to establish the discrete result and obtain the classical one as a consequence. Preliminary computer experiments suggest a bound close, if not equal, to 1/4. The mapping f from Fig. 3 to Fig. 4 is of class S and has $r_f \approx 0.386$; note that f begins to mimic the Koebe function, which is extremal for the classical case.

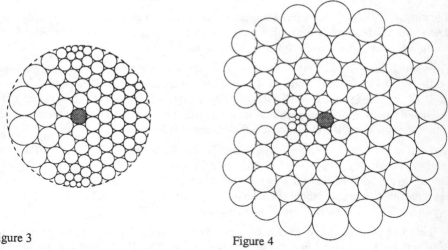

Figure 3   Figure 4

The Type Problem. The classical type problem asks whether a given (i.e., concretely defined) simply connected (open) Riemann surface is conformally equivalent to $\mathbb{D}$ or to $\mathbb{C}$. The discrete version: *Given complex K, an open topological disc, is K hyperbolic or parabolic?* This is intriguing because it raises apparently deep connections between the combinatorics and the geometry of circle packings. For instance, the constant 6-degree complex (6 edges from every vertex) is parabolic (the regular hexagonal circle packing of $\mathbb{C}$ is extremal), while the constant 7-deg complex is hyperbolic (see Fig. 2).

Of the many classical type criteria, perhaps most pleasing is Kakutani's: a Riemann surface R is parabolic iff Brownian motion on R is recurrent. Examples hint at a discrete version. In particular, arguments in [St2] establish a close connection between circle packings $P \subset \mathbb{D}$ and certain random walks on the 1-skeleton, $K^{(1)}$, of K. The random walks mimic Brownian motion, suggesting:

QUESTION 2. *Given an infinite complex K, let Z_K denote the simple random walk on $K^{(1)}$. Is it the case that K is parabolic iff Z_K is recurrent?*

REFERENCES

[BSt1] Alan F. Beardon and Kenneth Stephenson, *The Schwarz–Pick lemma for circle packings*, Ill. J. Math. **141** (1991), 577–606.

[BSt2] Alan F. Beardon and Kenneth Stephenson, *The uniformization theorem for circle packings*, Indiana Univ. Math. J. **39** (1990), 1383–1425.

[K] Koebe P., *Kontaktprobleme der Konformen Abbildung*, Ber. Sächs. Akad. Wiss. Leipzig, Math.--Phys. Kl. **88** (1936), 141–164.

[RS] Burt Rodin and Dennis Sullivan, *The convergence of circle packings to the Riemann mapping*, J. Differential Geometry **26** (1987), 349–360.

[St1] Kenneth Stephenson, *Circle packings in the approximation of conformal mappings*, Bulletin, Amer. Math. Soc. (Research Announcements) **23** (1990), no. 2, 407–415.

[St2] Kenneth Stephenson, *Thurston's conjecture on circle packings in the nonhexagonal case*, preprint.

[T] William Thurston, *The Geometry and Topology of 3-Manifolds*, preprint, Princeton University Notes.

DEPARTMENT OF MATHEMATICS
UNIVERSITY OF TENNESSEE, KNOXVILLE
KNOXVILLE, TN 37996-1300
U.S.A.

A CONJECTURE ON LOGARITHMIC COEFFICIENTS
OF UNIVALENT FUNCTION

I. M. MILIN

Let S be the class of functions $f(z) = z + c_2 z^2 + \ldots$, regular and univalent in the disc $|z| < 1$. Coefficients γ_k of the expansion

$$\log \frac{f(z)}{z} = \sum_{k=1}^{\infty} 2\gamma_k z^k, \quad |z| < 1,$$

are called *logarithmic coefficients* of $f(z) \in S$.

For a function $\psi(z) = \sum_{k=0}^{\infty} c_k z^k$, regular in $|z| < 1$ and for a number $r \in (0,1)$ we put

$$M(r, \psi) = \max_{|z|=r} |\psi(z)|,$$

$$\sigma(r, \psi) = \frac{1}{\pi} \iint_{|z|<r} |\psi'(z)|^2 \, dm_2(z) = \sum_{k=1}^{\infty} k|c_k|^2 r^{2k}.$$

It was conjectured in [1] that for any $f(z) \in S$

$$(1) \qquad \sigma\left(r, \log \frac{f}{z}\right) = \sum_{k=1}^{\infty} 4k|\gamma_k|^2 r^{2k} \leqslant 2\log \frac{M(r^2, f)}{r^2}, \quad r < 1$$

This estimate becomes an equality iff $f(z) = z$ or $f(z) = z(1 - e^{-i\beta}z)^{-2/m}$ $(m = 1, 2, \ldots;$ $0 \leqslant \beta \leqslant 2\pi)$. The equality (1) is equivalent to

$$(1') \qquad \sum_{k=1}^{\infty} k|\gamma_k| r^{2k} \leqslant \max_{|z|=r^2} \operatorname{Re} \sum_{k=1}^{\infty} \gamma_k z^k,$$

since

$$\log \frac{M(r, f)}{r} = \max_{|z|=r} \log \frac{|f(z)|}{|z|} = \max_{|z|=r} \operatorname{Re} \sum_{k=1}^{\infty} 2\gamma_k z^k.$$

Suppose $f(z)$ belongs to S and is bounded:

$$|f(z)| < M \quad \text{for } |z| < 1.$$

Then the following sharp estimate holds:

$$(2) \qquad \sigma\left(1, \log \frac{f}{z}\right) = \sum_{k=1}^{\infty} 4k \left|\gamma_k\right|^2 \leqslant 2 \log M$$

This inequality corroborates our conjecture (1). It is implied by a generalized area theorem proved for the first time by N. A. Lebedev in [2]. In [1] the estimate (1) is proved for starlike functions $f(z) \in S$ using the standard integral representation. Moreover, it is shown in [1] that (1) is true for any $f(z) \in S$, but for sufficiently small r's, $0 < r < r_f$. Some results concerning (1) can be found in [3].

REFERENCES

1. Milin I. M., *On a property of logarithmic coefficients on univalent functions*. Metric questions of the function theory, Naukova Dumka, Kiev, 1980, pp. 86–90. (Russian)
2. Lebedev N. A., *An application of the area principle to problems on non overlapping domains*, Trudy Mat. Inst. Akad. Nauk SSSR **60** (1961), 211–231. (Russian)
3. Milin I. M., *On a conjecture on a logarithmic coefficients of univalent functions*, Zapiski nauchn. semin. LOMI **125** (1983), 135–143 (Russian); English transl. in J. Soviet Math. **26** (1984), no. 6.

18.14

QUASICIRCLES, A_∞, AND HARMONIC MEASURE

J. Heinonen,   J.-M. Wu

Let D be a simply connected domain in the extended complex plane $\widehat{\mathbb{C}}$ and let f be a conformal mapping from the unit disk Δ onto D. Assume further that Γ is a K–quasicircle in $\widehat{\mathbb{C}}$.

The following conjecture was advanced in [FHM]:

DIAMETER CONJECTURE.

$$(DC) \qquad\qquad \sum_i \operatorname{diam} f^{-1}(\Gamma_i) \leqslant A(\Gamma) < \infty,$$

where $\{\Gamma_i\}$ are the connected components of $\Gamma \cap D$.

This conjecture was motivated by the theorem which states that if Γ is a circle, the sum of the lengths of $f^{-1}(\Gamma_i)$ is bounded by an absolute constant; see [HW], [GGJ], [FHM], [Ø].

Recently (DC) was disproved by Astala, Fernández, and Rohde [AFR] with a counterexample based on a careful analysis of harmonic measure. Let Ω^+ and Ω^- denote the complementary components of Γ with ω^+ and ω^- the corresponding harmonic measures; the choice of the base points is irrelevant here.

THEOREM 1 [AFR]. *If ω^+ and ω^- are A_∞ equivalent, then (DC) is true. If ω^+ and ω^- are mutually singular, then (DC) is false.*

Väisälä [V] has showed that $\sum_i \big(\operatorname{diam} f^{-1}(\Gamma_i)\big)^{1+\varepsilon} \leqslant A(K,\varepsilon) < \infty$ for all $\varepsilon > 0$. Moreover, we can show that $\sum_i \operatorname{diam} f^{-1}(\Gamma_i) \leqslant A(K)N < \infty$ when summed over those Γ_i which can be reached inside D from $f(0)$ by crossing Γ at most N times. In particular, (DC) is always true if one of the complementary components of Γ lies in D. This counterexamples such as in [AFR] necessarily require complicated constructions with domain D turning around Γ infinitely many times.

The interesting feature of Theorem 1 is that the validity of (DC) can be described in fairly simple terms, depending only on the geometry of Γ. There is, however, a big gap between the necessary and the sufficient condition in Theorem 1 so that the following question is still open.

PROBLEM 1. *Characterize quasicircles for which (DC) is true.*

Our attempts to solve Problem 1 remained unsuccessful but some interesting questions emerged along the way.

By a *partition* $\mathcal{P}$ of Γ we mean a disjoint collection of open subarcs whose union is dense in Γ. Every arc I in a given partition has two distinguished sides: the one that

---

Both authors supported by an NSF Grant

faces Ω^+ and the other that faces Ω^-. We say that a partition $\mathcal{P}$ is *marked* if every arc $I \in \mathcal{P}$ is marked by attaching the value 1 on one of the distinguished sides and the value 0 on the other. If we remove an arc I from $\mathcal{P}$, it makes sense to speak about the harmonic measure of the "1-part" of the boundary of the domain $\widehat{\mathbb{C}} \setminus (\Gamma \setminus I)$. Then

$$\omega^z = \omega^z\left(\text{1-part}, \widehat{\mathbb{C}} \setminus (\Gamma \setminus I)\right)$$

for $z \in \widehat{\mathbb{C}} \setminus (\Gamma \setminus I)$ describes the probability that a Brownian traveler starting at z exits $\widehat{\mathbb{C}} \setminus (\Gamma \setminus I)$ on those sides of the arcs in $\mathcal{P}$ that are marked as 1.

This construction is of course valid on an arbitrary Jordan curve Γ, and next we describe three properties that a curve may have based on this construction.

PROPERTY P_0. *There is $\beta < 1$ such that for any partition $\mathcal{P}$ consisting of exactly two arc it holds that*

$$\omega^z\left(\text{1-part}, \widehat{\mathbb{C}} \setminus (\Gamma \setminus I)\right) < \beta$$

for all $z \in I$, where I is either of the two arcs.

Condition P_0 says that a Brownian traveler starting from a point on I has roughly the same probability to fall either side of $\Gamma \setminus I$. It is unpublished observation made some years ago by Fernández, Hamilton and the first author that a Jordan curve has Property P_0 if and only if it is a quasicircle.

PROPERTY P_1. *There is $\beta < 1$ such that if $\mathcal{P}$ is any marked partition, if I is any arc from $\mathcal{P}$, and if z is any point on I, then*

$$\omega^z\left(\text{1-part}, \widehat{\mathbb{C}} \setminus (\Gamma \setminus I)\right) < \beta$$

for all $z \in I$.

Obviously P_1 is a stronger condition than P_0. In fact, a quasicircle has Property P_1 if and only if the complementary harmonic measures ω^+ and ω^- are A_∞ equivalent [AFR]. In particular, P_1 implies (DC). Quasicircles with Property P_1 have also arisen in recent works [AZ], [BJ].

The next condition is harder to describe. Let $\mathcal{P}$ be a partition of Γ and suppose that in $\mathcal{P}$ there is given a partial ordering in the following manner. One distinguished arc from $\mathcal{P}$ is called I. It has immediate successors $I_1, \ldots, I_{k_1}$, where $0 \leqslant k_1 \leqslant \infty$ with the understanding that $k_1 = 0$ implies I_0 has no successors. The arc I has immediate predecessors $I_{-1}, \ldots, I_{-k_{-1}}$, where $0 \leqslant k_{-1} \leqslant \infty$ and $k_{-1} = 0$ implies no predecessors. In general, for $j \geqslant 1$ an arc $I_{i_1,\ldots,i_j}$ with positive indices has only one immediate predecessor $I_{i_1,\ldots,i_{j-1}}$, and it has immediate successors $I_{i_1,\ldots,i_j,1}, \ldots, I_{i_1,\ldots,i_j,k(i_1,\ldots,i_j)}$ with $0 \leqslant k(i_1,\ldots,i_j) \leqslant \infty$ and no successors if $k(i_1,\ldots,i_j) = 0$. Similarly, an arc $I_{-i_1,\ldots,-i_j}$ with negative indices has only one immediate successor $I_{-i_1,\ldots,-i_{j-1}}$, and its immediate predecessors are denoted by $I_{-i_1,\ldots,-i_j,-1}, \ldots, I_{-i_1,\ldots,-i_j,-k(-i_1,\ldots,-i_j)}$ with $0 \leqslant k(-i_1,\ldots,-i_j) \leqslant \infty$.

We denote any such partial ordering in $\mathcal{P}$ by $\mathcal{O}$.

For a given partition $\mathcal{P}$ and its associated ordering $\mathcal{O}$ we attach a simply connected Riemann surface $\mathcal{R} = \mathcal{R}(\mathcal{P}, \mathcal{O})$ as follows. Start with $\mathcal{R}_1 = \widehat{\mathbb{C}} \setminus (\Gamma \setminus I)$. We regard the arcs $I_1, \ldots, I_{k_1}$ as *gates* for $\mathcal{R}_1$ opened from Ω^+, and $I_{-1}, \ldots, I_{-k_{-1}}$ as gates for $\mathcal{R}_1$

opened from Ω^-. The surface $\mathcal{R}_2$ is obtained by opening all this gates and attaching a copy of Ω^- to each gate with positive index, and a copy of Ω^+ to each gate of negative index. After we cross a gate $J \in \{I_1, \dots, I_{k_1}, I_{-1}, \dots, I_{-k_{-1}}\}$ from $\mathcal{R}_1$ and enter a new sheet Ω^- (or Ω^+), we find new gates on the boundary of this sheet which are exactly the immediate successors (or predecessors) of J if J has a positive (or a negative) index.

We continue by opening new gates. The surface $\mathcal{R}_2$ is a subsurface in $\mathcal{R}_3$ which is a subsurface in $\mathcal{R}_4$, etc. In general we call the arcs $\{I_{i_1,\dots,i_j}\}$ and $\{I_{-i_1,\dots,-i_j}\}$ gates for $\mathcal{R}_j$. At each stage we cross a gate to a new sheet which lies above a different component of $\widehat{\mathbb{C}} \setminus \Gamma$ from the one we just came from. We also regard gates $\{I_{i_1,\dots,i_j}\}$ and $\{I_{-i_1,\dots,-i_j}\}$ as cross-cuts of the surface $\mathcal{R}_{j+1}$, and as such they are a part of the boundary of $\mathcal{R}_j$. Finally, set

$$\mathcal{R} = \bigcup_{j=1}^{\infty} \mathcal{R}_j,$$

where the convention is made that $\mathcal{R}_k = \mathcal{R}_{k+1} = \dots$ if after some k there are neither successors nor predecessors.

Graphically $\mathcal{R}$ is an infinite tree starting from a fixed point I going forwards and backwards and having an arbitrary number of branches at each vertex. The surface $\mathcal{R}$ is (usually) an infinitely sheeted Riemann surface over the extended plane such that under the natural projection $\pi\colon \mathcal{R} \to \widehat{\mathbb{C}}$ each point on Γ is covered at most once. Clearly $\mathcal{R}$ is simply connected and to reach the part of the boundary of $\mathcal{R}$ which does not lie on the boundary of any $\mathcal{R}_j$ one has to wind around Γ infinitely many times.

PROPERTY P_N. *There is $N \geqslant 1$ and $\beta < 1$ such that for any partition $\mathcal{P}$ and for any ordering $\mathcal{O}$ we have*

$$\omega^z\big(S_N \cup S_{-N}, \mathcal{R}_N\big) < \beta$$

for all $z \in I$, where $S_N \subset \partial \mathcal{R}_N$ consists of the gates $\{I_{i_1,\dots,i_N}\}$ and $S_{-N} \subset \partial \mathcal{R}_N$ consists of the gates $\{I_{-i_1,\dots,-i_N}\}$.

We recognize that for $N = 1$ this is exactly Property P_1 which was introduced earlier. Property P_N means that a Brownian traveler on $\mathcal{R}$, starting at I, has at most β chance to survive to a gate for $\mathcal{R}_N$ before perishing somewhere on the way.

We can prove

THEOREM 2. *If a quasicircle has a property P_N, then (DC) is true with constant $A = A(K, \beta, N)$.*

If we change Conjecture (DC) a bit, we obtain a characterization. Let $\mathfrak{K}_\Gamma$ denote the class of all simply connected Riemann surfaces $\mathcal{R}$ that lie over the extended plane such that each point on Γ has at most one preimage on $\mathcal{R}$ under the canonical projection $\pi\colon \mathcal{R} \to \widehat{\mathbb{C}}$.

THEOREM 3. *A quasicircle Γ has Property P_N for some $N \geqslant 1$ if and only if there is a constant A such that*

(DCR)
$$\sum_i \operatorname{diam} f^{-1}(\Gamma_i) \leqslant A < \infty$$

for all $\mathcal{R} \in \mathfrak{K}_\Gamma$ and conformal $f \colon \Delta \to \mathcal{R}$, where $\{\Gamma_i\}$ are the connected components of $\pi^{-1}(\Gamma) \cap \mathcal{R}$.

To prove Theorems 2 and 3 we use standard estimates for harmonic measure, results from [JM], and the fact that quasicircles have Property P_0. However, our results are unsatisfying because it seems difficult to verify whether or not a given quasicircle has Property P_N. It is also not clear how (DC) and (DCR) in Theorem 3 are related. So we wish to pose the following problems.

PROBLEM 2. *Does P_N for some $N \geqslant 2$ imply P_1?*

PROBLEM 3. *Does* (DC) *hold for a quasicircle Γ if and only if* (DCR) *hold for Γ?*

If the answer to both problems is "yes", then (DC) holds if and only if Γ is an "A_∞ quasicircle", i.e. if the complementary harmonic measures are A_∞ equivalent.

REFERENCES

[AFR] Astala K., Fernández J. L., Rohde S. (1991) (to appear).

[AZ] Astala K., Zinsmeister M., *Teichmüller spaces and BMOA*, Mittag–Leffler Report 20 (1989–90).

[BJ] Bishop C. J., Jones P. W., *Harmonic measure, L^2-estimates and the Schwarzian derivative*, preprint (1990).

[FHM] Fernández J. L., Heinonen J., Martio O., *Quasilines and conformal mappings*, J. Analyse Math. 52 (1989), 117–132.

[GGJ] Garnett J. B., Gehring F. W., Jones P. W., *Conformally invariant length sums*, Indiana Univ. Math. J. 32 (1983), 809–829.

[HW] Haiman W. K., Wu J.-M., *Level sets of univalent functions*, Comment. Math. Helv. 56 (1981), 366–403.

[JM] Jones P. W., Marshall D. E., *Critical points of Green's function, harmonic measure and the corona problem*, Ark. Math. 23 (1985), 281–314.

[Ø] Øyma K., *Harmonic measure and conformal length*, Proc. Amer. Math. Soc. (to appear).

[V] Väisälä J, *Bounded turning and quasiconformal maps*, Monatsch. Math. (to appear).

UNIVERSITY OF MICHIGAN
DEPARTMENT OF MATHEMATICS
ANN ARBOR, MI 48109
U.S.A.

UNIVERSITY OF ILLINOIS
DEPARTMENT OF MATHEMATICS
URBANA, IL 61801
U.S.A.

SUPPORT POINTS OF UNIVALENT FUNCTIONS

P. L. DUREN

Let $H(\mathbb{D})$ be the linear space of all functions analytic in the unit disk $\mathbb{D}$, endowed with the usual topology of uniform convergence on compact subsets. Set S be the class of functions $f \in H(\mathbb{D})$ which are univalent and normalized by the conditions $f(0) = 0$ and $f'(0) = 1$. Thus each $f \in S$ has an expansion of the form

$$f(z) = z + a_2 z^2 + a_3 z^3 + \cdots , \qquad |z| < 1.$$

Let L be a complex-valued continuous linear functional on $H(\mathbb{D})$ not constant on S. Because S is a compact subset of $H(\mathbb{D})$, the functional $\mathrm{Re}\{L\}$ attains a maximum value on S. The extremal functions are called *support points* of S. In view of the Krein–Milman theorem, the set of support points associated with each linear functional L must contain an extreme point of S. It is NOT KNOWN *whether every support point is an extreme point, or whether every extreme point is a support point.*

The support points of S have a number of interesting properties. It is known that each support point f maps $\mathbb{D}$ onto the complement of an analytic arc Γ which extends with increasing modulus from a point w_0 to ∞, satisfying

(1) $$L\left(\frac{f^2}{f - w}\right) \frac{dw^2}{w^2} > 0.$$

The *radial angle* $\alpha(w) = \arg\left\{\dfrac{dw}{w}\right\}$ of Γ has the property $|\alpha(w)| \leqslant \dfrac{\pi}{4}$, $w \neq w_0$. The bound $\dfrac{\pi}{4}$ is best possible and in fact there are support points for which $|\alpha(w_0)| = \dfrac{\pi}{4}$. It is also known that $L(f^2) \neq 0$, from which it follows that Γ is asymptotic at infinity to the half-line

(2) $$W = \frac{L(f^3)}{3L(f^2)} - L(f^2)t, \qquad t \geqslant 0.$$

An exposition of these properties, with further references to the literature, may be found in [4].

Evidence obtained from the study of special functionals [1,2,6,3] suggests the CONJECTURE that *the omitted arc Γ always has monotonic argument.* This is true for point-evaluation functionals $L(f) = f(\zeta)$, where $\zeta \in \mathbb{D}$; for derivative functionals $L(f) = e^{-i\sigma} f'(\zeta)$; for coefficient functionals $L(f) = a_3 + \lambda a_2$, where $\lambda \in \mathbb{C}$, and of course for coefficient functionals $L(f) = a_n$ with $2 \leqslant n \leqslant 6$, where the Bieberbach conjecture has been proved. A STRONGER CONJECTURE, supported by somewhat less evidence, is that *the radial angle $\alpha(w)$ tends monotonically to zero as $w \to \infty$ along Γ.*

The Bieberbach conjecture asserts that $|a_n| \leqslant n$, with strict inequality for all n unless f is a rotation of the Koebe function $k(z) = z(1-z)^{-2}$. A geometric reformulation is that the arc Γ corresponding to each coefficient functional $L(f) = a_n$ is a radial half-line. It is essentially equivalent to say that the asymptotic half-line (2) is a trajectory of the quadratic differential (1). A weak form of the Bieberbach conjecture if that for each coefficient functional $L(f) = a_n$ the asymptotic half-line is radial. It is interesting to ask what relation this conjecture may bear to other weak forms of the Bieberbach conjecture, such as the asymptotic Bieberbach conjecture and Littlewood's conjecture on omitted values, now known [5] to be equivalent.

References

1. Brown J. E., *Geometric properties of a class of support points of univalent functions*, Trans. Amer. Math. Soc. **256** (1979), 371–382.
2. Brown J. E., *Univalent functions maximizing* Re$\{a_3 + \lambda a_2\}$, Illinois J.Math. **25** (1981), 446–454.
3. Duren P. L., *Arcs omitted by support points of univalent functions*, Comment. Math. Helv. **56** (1981), 352–365.
4. Duren P. L., *Univalent Functions*, Springer–Verlag, New York, 1983
5. Hamilton D. H., *On Littlewood's conjecture for univalent functions*, Proc. Amer. Math. Soc. **86** (1982), 32–36.
6. Pearce K., *New support points of S and extreme points of HS*, Proc. Amer. Math. Soc. **81** (1981), 425–428.

DEPARTMENT OF MATHEMATICS
UNIVERSITY OF MICHIGAN
ANN ARBOR, MI 48109
USA

MORE PROBLEMS BY ALBERT BAERNSTEIN

A. BAERNSTEIN

Let Ω be a simply connected domain in $\mathbb{C}$ and F a conformal mapping from Ω onto $\mathbb{D}$, normalized by $|F'(a)| = 1$ when $F(a) = 0$. Hayman and Wu [1] proved that

$$\int_{\mathbb{R} \cap \Omega} |F'(x)| \, dx \leqslant A < \infty$$

for some constant A. A simpler proof has been given by Garnett, Gehring and Jones [2]. *Is it true that*

$$(1) \qquad \int_{\mathbb{R} \cap \Omega} |F'(x)|^p \, dx \leqslant A_p < \infty$$

for some constant A_p when $1 < p < 2$? The example $f(z) = (1-z)^{1/2}$, $F(z)$ the inverse of f, shows that $\displaystyle\int_{\mathbb{R} \cap \Omega} |F'(x)|^p \, dx = \infty$ is possible when $2 \leqslant p < \infty$.

Using the technique of [1] or [2] together with classical harmonic measure estimates, it can be shown that (1) is true for $\frac{2}{3} < p \leqslant 1$. The inverse function of $f(z) = (1-z)^{-2}$ shows that $\displaystyle\int_{\mathbb{R} \cap \Omega} |F'(x)|^p \, dx = \infty$ is possible when $0 < p \leqslant \frac{2}{3}$.

Inequality (1) would follow if a symmetrization type inequality, (2) below, is true. Let $\{I_j\}$ be a Whitney type decomposition of $\mathbb{R} \cap \Omega$ as described in [2, §3]. Denote by x_j the center of I_j, l_j the length of I_j, L_j the vertical half line starting from $x_j + il_j$, and let Ω' be the domain obtained by deleting from $\mathbb{C}$ all the half lines L_j. *Is it true that*

$$(2) \qquad g(x, a) \leqslant CG(x, a)$$

for every $x \in \mathbb{R}$ and $a \in \mathbb{C}$ with $\operatorname{Im} a \leqslant 0$, where g and G denote the Green's functions of Ω and Ω' respectively?

REFERENCES

1. Hayman W. K., Wu J.-M. G., *Level sets of univalent functions*, Comm. Math. Helv. **56** (1981), 366–403.
2. Garnett J. B., Gehring F. W., Jones P. W., *Conformally invariant length sums*, Indiana Univ. Math. J. **32** (1983), 809–829.

DEPARTMENT OF MATHEMATICS
WASHINGTON UNIVERSITY
ST.LOUIS, MO 63130
USA

Chapter 19

HOLOMORPHIC DYNAMICS

Edited by

B. Bielefeld
Institute for Math. Sciences
SUNY at Stony Brook
Stony Brook, NY 11794
USA

M.Lyubich
Institute for Math. Sciences
SUNY at Stony Brook
Stony Brook, NY 11794
USA

INTRODUCTION

Holomorphic dynamics posed several closely related key problems going back to Fatou and Julia: density of axiom A maps, local connectivity of the Julia and Mandelbrot sets, measure and dimension of the above sets. A great deal of progress has been achieved in these problems during the last decade, but they are still in the focus of modern research. Most of the discussion in the chapter is concentrated on these problems. We will not quote any particular results: the reader will find references and a variety of viewpoints on the subject inside the articles.

For a general introduction to holomorphic dynamics we recommend one of the following surveys: [Be], [Bl], [C], [EL], [L] and [M].

The chapter can be divided into five sections which cover a good part of the field:

(1) Quasiconformal Surgery and Deformations (Problems 19.1–19.4)
(2) Geometry of Julia Sets (Problems 19.5, 19.6)
(3) Measurable Dynamics (Problems 19.7, 19.8)
(4) Iterates of Entire Functions (Problems 19.9, 19.10)
(5) Newton's Method (Problem 19.11)

The chapter partially arose from an earlier problem list [Bi]. We hope it will help to fix up the present state of affairs of the field and to stimulate further development. We thank everybody who has contributed to the chapter. We would also like to thank M. Herman and J. Milnor for many helpful comments.

REFERENCES

[Be] Beardon A., *Iteration of Rational Functions*, Springer-Verlag, 1990.

[Bi] Bielefeld B. (editor), *Conformal Dynamics Problem List*, SUNY Stony Brook Institute for Mathematical Sciences, preprint #1990/1.

[Bl] Blanchard P., *Complex analytic dynamics on the Riemann sphere*, Bull. Amer. Math. Soc. **11** (1984), 85–141.

[C] Carleson L., *Complex dynamics*, UCLA course notes, 1990.

[EL] Eremenko A., Lyubich M., *The dynamics of analytic transformations*, Leningrad Math J. **1:3** (1990).

[L] Lyubich M., *The dynamics of rational transforms: the topological picture*, Russian Math. Surveys **41:4** (1986), 43–117.

[M] Milnor J., *Dynamics in one complex variable: Introductory Lectures*, SUNY Stony Brook Institute for Mathematical Sciences, preprint #1990/5.

QUESTIONS IN QUASICONFORMAL SURGERY

BEN BIELEFELD

It is possible to investigate rational functions using the technique of quasiconformal surgery as developed in [DH2], [BD] and [S]. There are various methods of gluing together polynomials via quasiconformal surgery to make new polynomials or rational functions. The idea of quasiconformal surgery is to cut and paste the dynamical spaces for two polynomials so as to end up with a branched map whose dynamics combines the dynamics of the two polynomials. One then tries to find a conformal structure that is preserved under this branched map of the sphere to itself, so that using the Ahlfors–Bers theorem the map is conjugate to a rational function. There are several topological surgeries which experimentally seem to exist, but for which no one has yet been able to find a preserved complex structure.

The first such kind of topological surgery is *mating* of two monic polynomials with the same degree. (Compare [TL]). The first step is to think of each polynomial as a map on a closed disk by thinking of infinity as a circle worth of points, one point for each angular direction. The obvious extension of the polynomial at the circle at infinity is $\theta \mapsto d\theta$ where d is the degree of the polynomial. Now glue two such polynomials together at the circles at infinity by mapping the θ of the first polynomial to $-\theta$ in the second. Finally, we must shrink each of the external rays for the two polynomials to a single point. The result should be conjugate to a rational map of degree d. (Surprisingly this construction sometimes seems to make sense even when the filled Julia sets for both polynomials have vacuous interior).

For instance we can take the rabbit to be the first polynomial, that is $z^2 + c$ where the critical point is periodic of period 3 ($c \sim -.122561 + .744862i$). The Julia set appears in the following picture.

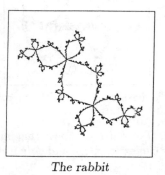

The rabbit

Then for the second polynomial we could take the basilica, that is $z^2 - 1$ (it is named after the Basilica San Marco in Venice. One can see the basilica on top and its reflection in the water below). The Julia set for the basilica appears in the following figure.

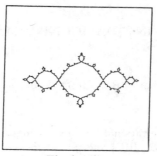

The basilica

Next we show the basilica inside-out ($\frac{z^2}{z^2 - 1}$) which is what we will glue to the rabbit.

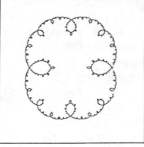

The inside-out basilica

And finally we have the Julia set for the mating ($\frac{z^2 + c}{z^2 - 1}$ where $c = \frac{1 + \sqrt{-3}}{2}$).

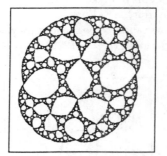

The basilica mated with the rabbit

QUESTION 1. *Which matings correspond to rational functions? There are some known obstructions. For example, Tan Lei has shown that matings between postcritically finite quadratic polynomials can exist only if and only if they do not belong to complex conjugate limbs of the Mandelbrot set.*

QUESTION 2. *Can matings be constructed with quasiconformal surgery? Tan Lei uses Thurston's topological characterization of rational maps to do this. It would be nice to have a cut and paste type of construction, giving results for the case when the orbit of critical points is not finite.*

QUESTION 3. *If one polynomial is held fixed and the other is varied continuously, does the resulting rational function vary continuously? Is mating a continuous function of two variables?*

The second type of topological surgery is *tuning*. First take a polynomial P_1 with a periodic critical point ω of period k, and assume that no other critical points are in the entire basin of this superattractive cycle. Let P_2 be a polynomial with one critical point whose degree is the same as the degree of ω. We also assume that the Julia sets of P_1 and P_2 are connected. We assume the closure $\bar{B}$ of the immediate basin of ω is homeomorphic to the closed unit disk $\bar{D}$, and that the Julia set for P_2 is locally connected. Now, P_1^k maps $\bar{B}$ to itself by a map which is conjugate to the map $z \mapsto z^d$ of $\bar{D}$, where d is the degree of the critical point. (In fact, if $d > 2$, then there are $d - 1$ possible choices for the conjugating homeomorphism, and we must choose one of them). Intuitively the idea is now the following. Replace the basin B by a copy of the dynamical plane for P_2, gluing the "circle at infinity" for this plane onto the boundary of B so that external angles for P_2 correspond to internal angles in $\bar{B}$. Now shrink each external ray for P_2 to a point. Also, make an analogous modification at each pre-image of B. The map from the modified B to its image will be given by P_2, and the map on all other inverse images of the modified B will be the identity. The result, P_3, called P_1 tuned with P_2 at ω, should be conjugate to a polynomial having the same degree as P_1. Conversely P_2 is said to be obtained from P_3 by renormalization.

In the case of quadratic polynomials, the tunings can be made also in the case when P_2 is not locally connected.

As an example we can take P_1 to be the rabbit polynomial. Then we can take $P_2(z) = z^2 - 2$ which has the closed segment from -2 to 2 as its Julia set. The following figure shows the resulting quadratic Julia set tuning the rabbit with the segment ($z^2 + c$ where $c \sim -.101096 + .956287i$).

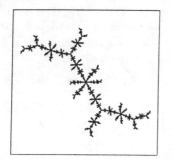

The rabbit tuned with the segment

In the picture we see each ear of the rabbit replaced with a segment.

QUESTION 4. *Does the tuning construction always give a result which is conjugate to a polynomial? This is true when P_1 and P_2 are quadratic.*

QUESTION 5. *Can tunings be constructed with quasiconformal surgery?*

QUESTION 6. *Does the resulting polynomial vary continuously with P_2? This is true when P_1 and P_2 are quadratic* [DH2].

QUESTION 7. *Does the resulting tuning vary continuously with P_1?* (here we consider only polynomials P_1 of degree greater than 2 with a superstable orbit of fixed period).

QUESTION 8. *Let $P_{1,k}$ be a sequence of polynomials with a superstable orbit whose period tends to infinity. If $P_{1,k}$ tends to a limit $P_{1,\infty}$, do the tunings of P_2 with $P_{1,k}$ also tend to $P_{1,\infty}$?*

The third kind of surgery is *intertwining surgery*.

Let P_1 be a monic polynomial with connected Julia set having a repelling fixed point x_0 which has a ray landing on it with combinatorial rotation number p/q. Look at the cycle of q rays which are the forward images of the first. Cut along these rays and we get q disjoint wedges. Now let P_2 be a monic polynomial with a ray of the same combinatorial rotation number landing on a repelling periodic point of some period dividing q (such as 1 or q). Slit this dynamical plane along the same rays making holes for the wedges. Fill the holes in by the corresponding wedges above making a new sphere. The new map will be given by P_1 and P_2 except on a neighborhood of the inverse images of the cut rays where it will have to be adjusted to make it continuous. This construction should be possible to do quasiconformally using the methods in [BD] together with Shishikura's new (unpublished) method of presurgery in the case where the rays in the P_2 space land at a repelling orbit. This construction doesn't seem to work when the rays land at a parabolic orbit.

For instance we can take $P_1(z) = z^2$ and $P_2(z) = z^2 - 2$. The Julia set for P_1 is the unit circle with repelling fixed point at 1 and the ray at angle 0 lands on it with combinatorial rotation number 0. The Julia set for P_2 is the closed segment from -2 to 2 with repelling fixed point 2 and the ray at angle 0 lands on it with combinatorial rotation number 0. We cut along the 0 ray in both cases. Opening the cut in the first dynamical space gives us one wedge. The space created by opening the cut in the second space is the hole into which we put the wedge. The resulting cubic Julia set is shown in the following picture (the polynomial is $z^3 + az$ where $a \sim 2.55799i$).

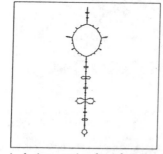

A circle intertwined with a segment

We see in the picture the circle and the segment, and at the inverse image of the fixed point on the segment we see another circle. At the other inverse of the fixed point on the circle we see a segment attached. All the other decorations come from taking various inverses of the main circle and segment.

As a second example we can intertwine the basilica with itself. The ray 1/3 lands at a fixed point and has combinatorial rotation number 1/2. The following is the Julia set for the basilica intertwined with itself (the polynomial here is $z^3 - \frac{3}{4}z + \frac{\sqrt{-7}}{4}$).

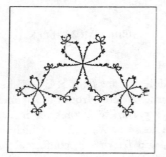

A basilica intertwined with itself

QUESTION 9. *When does an intertwining construction give something which is conjugate to a polynomial?*

QUESTION 10. *Can intertwinings be constructed with quasiconformal surgery?*

QUESTION 11. *Does the resulting polynomial vary continuously in P_2?*

REFERENCES

[BD] Branner B., Douady A., *Surgery on Complex Polynomials*, Proc. Symp. of Dynamical Systems Mexico (1986).

[DH2] Douady A., Hubbard J. H., *On the Dynamics of Polynomial Like Mappings*, Ann. Sc. E.N.S., 4ème Séries **18** (1966).

[S] Shishikura M., *On the Quasiconformal Surgery of Rational Functions*, Ann. Sc. E.N.S., 4ème Séries **20** (1987).

[STL] Shishikura M., Tan Lei, *A Family of Cubic Rational Maps and Matings of Cubic Polynomials*, preprint of Max–Plank–Institute, Bonn **50** (1988).

[TL1] Tan Lei, *Accouplements des polynômes quadratiques complexes*, CRAS Paris (1986), 635–638.

[TL2] Tan Lei, *Accouplements des polynômes complexes*, Thèse, Orsay (1987).

[W] B. Wittner, *On the Bifurcation Loci of Rational Maps of Degree Two*, Ph.D thesis, Cornell Univ., Ithaca N.Y. (1986).

INSTITUTE FOR MATH. SCIENCES
SUNY AT STONY BROOK
STONY BROOK, NY 11794
USA

RATIONAL MAPS AND TEICHMÜLLER SPACE

Curt McMullen

Let X be a complex manifold and let $f : X \times \widehat{\mathbb{C}} \to \widehat{\mathbb{C}}$ be a holomorphic map. Then f describes a family $f_\lambda(z)$ of rational maps from the Riemann sphere to itself, depending holomorphically on a complex parameter λ ranging in X.

By [MSS], there is an open dense set $X_0 \subset X$ on which the family is structurally stable near the Julia set: in fact f_a and f_b are quasiconformally conjugate on their respective Julia sets whenever a and b lie in the same component U of X_0. The mappings in X_0 are said to be J-stable.

In this note we will record some problems concerning the boundaries of components U, and consequently concerning limits of quasiconformal deformations of a given rational map.

Example I. Quadratic polynomials. The most famous such problem is the following. Let $X = \mathbb{C}$, and let $f_\lambda(z) = z^2 + \lambda$. Then X_0 contains a unique unbounded component U.

PROBLEM. *Is the boundary of U locally connected?*

This is equivalent to the question:

Is the Mandelbrot set M locally connected?

Indeed, X_0 is just the complement of the boundary of the Mandelbrot set.

The importance of this question is twofold. First, if M is locally connected, then existing work provides detailed information about its combinatorial structure, and one has a good understanding of the "bifurcations" of a quadratic polynomial and many related maps. Secondly, the local connectivity of M implies the density of hyperbolic dynamics ("Axiom A") for degree two polynomials, another well-known conjecture which has eluded proof for many years. For more details see [Dou1], [Dou2], [DH1], [DH2], [Lav], [Th].

Compactifying the space of proper maps. We now turn to a second example motivated by an analogy with Bers' embedding of Teichmüller space. Let A and B be two proper holomorphic maps of the unit disk Δ to itself, both of degree $n > 1$ and fixing zero. (A and B are finite Blaschke products). Then it is well known that A and B are topologically conjugate on the unit circle S^1, and the conjugacy h is unique once we have chosen a pair of fixed points (a, b) for A and B such that $h(a) = b$. Moreover h is quasisymmetric; this is a general property of conjugacies between expanding conformal dynamical systems [Sul].

Now glue two copies of the disk together by h and transport the dynamics of A and B to the resulting Riemann surface, which is a sphere. We obtain in this way an expanding (i.e. hyperbolic) rational map $f(A, B)$. The Julia set J of $f(A, B)$ is a quasicircle, and f

is holomorphically conjugate to A and B on the components of the complement of J. The mapping $f(A, B)$ is determined by h up to conformal conjugacy.

We will loosely speak of spaces of mappings as being "the same" if they represent the same conformal conjugacy classes. It is often useful to require that the conjugacy preserves some finite amount of combinatorial data, such as a distinguished fixed point. For simplicity we will gloss over such considerations below.

Example II. Let X be the space of degree n polynomials, X_0 the open dense subset of J-stable polynomials and U the component of X_0 containing z^n. Then U is the same as the set of maps of the form $f(z^n, B)$. Equivalently, U consists of those polynomials with an attracting fixed point with all critical points in its immediate basin.

Let us denote this set of polynomials by $\mathcal{B}(z^n)$. It is easy to see that $\mathcal{B}(z^n)$ is an open set of polynomials with compact closure. Thus this construction supplies both a complex structure for the space of Blaschke products, and a geometric compactification of that space.

PROBLEM. *Describe the boundary of $\mathcal{B}(z^n)$ in the space of polynomials of degree n.*

For degree $n = 2$ this is easy (the boundary is a circle) but for $n = 3$ it is already subtle.

To explain the kind of answer one might expect, we consider not one boundary but many. More precisely, let $\mathcal{B}(A)$ denote the space of rational maps $f(A, B)$ for some other *fixed* A and varying B. This space also inherits a complex structure and the map $f(z^n, B) \mapsto f(A, B)$ gives an biholomorphic map

$$F : \mathcal{B}(z^n) \to \mathcal{B}(A).$$

The closure of $\mathcal{B}(A)$ in the space of rational maps provides another geometric compactification of this complex manifold.

PROBLEM. *Show that for $n > 2$ and $A \neq z^n$, F does not extend to a homeomorphism between the boundaries of $\mathcal{B}(z^n)$ and $\mathcal{B}(A)$.*

Thus we expect that the complex space $\mathcal{B}$ (whose complex structure is independent of A) has many natural geometric boundaries. But perhaps the lack of uniqueness can be accounted for by the presence of *complex submanifolds* of the boundary, i.e. by the presence of rational maps in the compactification which admit quasiconformal deformations.

To make this precise, let $\partial(A)$ denote the quotient of the boundary of $\mathcal{B}(A)$ by the equivalence relation $f \sim g$ if f and g are quasiconformally conjugate (equivalently, if f and g lie in a connected complex submanifold of the boundary). The resulting space (in the quotient topology) still forms a boundary for $\mathcal{B}(A)$, but it is non-Hausdorff when $n > 2$.

CONJECTURE. *The holomorphic isomorphism $F : \mathcal{B}(z^n) \to \mathcal{B}(A)$ extends to a homeomorphism from $\partial(z^n)$ to $\partial(A)$.*

PROBLEM. *Give a combinatorial description of the topological space $\partial(z^n)$.*

Such a description may involve laminations, as discussed in [Th].

An analogy with Teichmüller theory. The "mating" of A and B has many similarities with the mating of Fuchsian groups uniformizing a pair of compact genus g Riemann surfaces X and Y. Such a mating is provided by Bers' simultaneous uniformization theorem [Bers]. The result is a Kleinian group $\Gamma(X, Y)$ whose limit set is a quasicircle. Moreover, fixing X, the map $Y \mapsto \Gamma(X, Y)$ provides a holomorphic embedding of the Teichmüller space of genus g into the space of Kleinian groups. One can then form a boundary for Teichmüller space by taking the closure.

It has recently been shown that *this* boundary does indeed *depend* on the base point X [KT]. However Thurston has conjectured that the space $\partial(X)$, obtained by identifying quasiconformally conjugate groups on the boundary, is a (non-Hausdorff) boundary which is independent of X.

Moreover a combinatorial model for $\partial(X)$ is conjecturally constructed as follows. Let $\mathbb{PML}$ denote the space of projective measured laminations on a surface of genus g; then $\partial(X)$ is homeomorphic to the quotient of $\mathbb{PML}$ by the equivalence relation which forgets the measure. (See [FLP] for a discussion of $\mathbb{PML}$ as a boundary for Teichmüller space).

Remarks.

1. We do not expect that one can give a combinatorial description of the "actual" boundary of $\mathcal{B}(z^n)$ (in the space of polynomials). For similar reasons, we believe it unlikely that one can describe the uniform structure induced on the space of critically finite rational maps by inclusion into the space of all rational maps.

2. It is known that Teichmüller space is a domain of holomorphy. So it is natural to ask the following intrinsic:

QUESTION. *Is $\mathcal{B}(z^n)$ a domain of holomorphy? More generally, is every component of the space of expanding rational maps (or polynomials) a domain of holomorphy?*

Density of cusps. The preceding discussion becomes interesting only when the space of rational maps under consideration has two or more (complex) dimensions. We conclude with two concrete questions about boundaries in a one-parameter family of rational maps.

Example III. Let

$$f_\lambda(z) = \lambda z^2 + z^3$$

where λ ranges in $X = \mathbb{C}$, and let U denote the component of X_0 containing the origin. That is, U is the set of λ for which both finite critical points are in the immediate basin of zero.

A *cusp* on ∂U is an f_λ with a parabolic periodic cycle.

CONJECTURE. *Cusps are dense in ∂U.*

This conjecture is motivated by the density of cusps on the boundary of Teichmüller space [Mc]. It is not hard to show that it is implied by the following:

CONJECTURE. *The boundary of U is a Jordan curve.*

REFERENCES

[Bers] Bers L., *Simultaneous uniformization*, Bull. AMS **66** (1960), 94–97.

[Dou1] Douady A., *Systèmes dynamiques holomorphes*, Astérisque **105–106** (1983), 39–64.

[Dou2] Douady A., *Algorithms for computing angles in the Mandelbrot set*, Chaotic Dynamics and Fractals (Barnsley M. F., Demko S. G., eds.), Academic Press, 1986, pp. 155–168.

[DH1] Douady A., Hubbard J., *Étude dynamique des polynômes complexes*, Pub. Math. d'Orsay, 1984.

[DH2] Douady A., Hubbard J., *On the dynamics of polynomial-like mappings*, Ann. Sci. Éc. Norm. Sup. **18** (1985), 287–344.

[FLP] Fathi A., Laudenbach F., Poénaru V., *Travaux de Thurston sur les surfaces*, volume 66-67, Astérisque, 1979.

[KT] Kerckhoff S., Thurston W., *Non-continuity of the action of the modular group at Bers' boundary of Teichmüller space*, Invent. math. **100** (1990), 25–48.

[Lav] Lavaurs P., *Une description combinatoire de l'involution définie par M sur les rationnels à dénominateur impair*, CRAS Paris **303** (1986), 143–146.

[MSS] Mañé R., Sad P., Sullivan D., *On the dynamics of rational maps*, Ann. Sci. Éc. Norm. Sup. **16** (1983), 193–217.

[Mc] McMullen C., *Cusps are dense*, Annals of Math. **133** (1991), 217–247.

[Sul] Sullivan D., *Quasiconformal homeomorphisms and dynamics III: Topological conjugacy classes of analytic endomorphisms*, Preprint.

[Th] Thurston W. P., *On the combinatorics and dynamics of iterated rational maps*, Preprint.

DEPARTMENT OF MATHEMATICS
UNIVERSITY OF CALIFORNIA, BERKELEY
EVANS HALL
BERKELEY, CA 94720
USA

THURSTON'S ALGORITHM WITHOUT CRITICAL FINITENESS

JOHN MILNOR

Thurston's algorithm is a powerful method for passing from a topological branched covering $S^2 \to S^2$ to a rational map having closely related dynamical properties. (See [DH]). The same method can be used to pass from a piecewise monotone map of the interval to a closely related polynomial map of the interval.

Suppose that we start with an orientation preserving branched covering map $f_0 : S^2 \to S^2$. We identify S^2 with the Riemann sphere $\overline{\mathbb{C}} = \mathbb{C} \cup \infty$. In order to anchor this sphere, choose three base points. (For best results, choose dynamically significant base points, for example periodic points of f_0, or critical points, or critical values).

LEMMA. *There is one and only one homeomorphism* $h_0 : S^2 \to S^2$ *which fixes the three base points, and which has the property that the composition* $r_0 = f_0 \circ h_0$ *is holomorphic, or in other words is a rational map.*

Proof. Let σ_0 be the standard conformal structure on the 2–sphere, and let $\sigma = f_0^*(\sigma_0)$ be the pulled back conformal structure, so that f_0 maps (S^2, σ) holomorphically onto (S^2, σ_0). Then h_0 must be the unique conformal isomorphism from (S^2, σ_0) onto (S^2, σ) which fixes the three base points.

Now consider the map $f_1 = h_0^{-1} \circ f_0 \circ h_0$, which is topologically conjugate to f_0. In this way, we obtain a commutative diagram

$$
\begin{array}{ccc}
S^2 & \xrightarrow{f_1} & S^2 \\
h_0 \downarrow & r_0 \searrow \quad h_0 \downarrow & \\
S^2 & \xrightarrow{f_0} & S^2 .
\end{array}
$$

Continuing inductively, we produce a sequence of branched coverings f_n, and a sequence of homeomorphisms h_n fixing the base points, so that $f_{n+1} = h_n^{-1} \circ f_n \circ h_n$, and so that each composition $r_n = f_n \circ h_n$ is a rational map. The marvelous property of this construction is that in many cases the homeomorphisms h_n seem to tend uniformly to the identity, so that the successive maps f_n, which are all topologically conjugate to f_0, come closer and closer to the rational maps r_n. In fact the sequence of compositions $\phi_n = (h_0 \circ \cdots \circ h_n)^{-1}$ may converge uniformly to a limit map ϕ, at least on the non-wandering set. In this case, it follows that the rational limit map is topologically semi-conjugate (or perhaps even conjugate) to f_0,

$$
r_\infty \circ \phi = \phi \circ f_0
$$

on the non-wandering set.

PROBLEM: *Under what conditions will this sequence of rational maps r_n converge uniformly to a limit map r_∞? Under what conditions, and on what subset of S^2, will the maps ϕ_n converge uniformly to a limit?*

In the post-critically finite case, Thurston defines an obstruction, which vanishes if and only if the restriction of the ϕ_n to the post-critical set converges uniformly to a one-to-one limit function. If this obstruction vanishes, then it follows that the r_n converge.

However, there would be interesting applications where f_0 is not post-critically finite, so that no such criterion is known. A typical example is provided by the problem of "mating". (Compare Bielefeld's discussion, as well as [Ta], [Sh]). Let p and q be monic polynomial maps having the same degree $d \geqslant 2$. Conjugating p by the diffeomorphism $z \mapsto z/\sqrt{1+|z|^2}$ from $\mathbb{C}$ onto the unit disk D, we obtain a map p^* which extends smoothly over the closed disk $\bar{D}$. Similarly, conjugating q by $z \mapsto \sqrt{1+|z|^2}/z$ we obtain a map q^* which extends smoothly over the complementary disk $\overline{\mathbb{C}} \setminus D$. Now p^* and q^* together yield a C^1–smooth map $f_0 : \overline{\mathbb{C}} \to \overline{\mathbb{C}}$ and we can apply Thurston's method as described above. If this procedure converges to a well behaved limit, then the resulting rational map r_∞ of degree d may be called the "mating" of p and q.

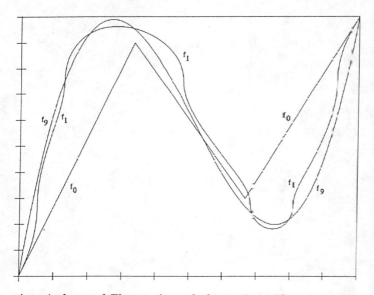

A typical run of Thurston's method, starting with a piece-wise linear map f_0 of the interval. (Horizontal scale exaggerated). The graphs of f_0, f_1 and f_9 are shown. The latter seems indistinguishable from $f_\infty = p_\infty$.

Maps of the interval. The situation here is quite similar. Let f_0 be a piecewise–monotone map of the interval $I = [0,1]$ with d alternately ascending and descending laps, and suppose that f_0 carries the boundary points 0 and 1 to boundary points. Then there is one and only one orientation preserving homeomorphism h_0 of the interval such that the composition $p_0 = f_0 \circ h_0$ is a polynomial map of degree d. Setting

$f_1 = h_0^{-1} \circ f_0 \circ h_0$, we can proceed inductively, constructing homeomorphisms h_n, polynomials $p_n = f_n \circ h_n$, and topologically conjugate maps $f_{n+1} = h_n^{-1} \circ f_n \circ h_n$. Again the problem is to decide when and where this procedure converges.

REFERENCES

[DH] Douady A., Hubbard J. H., *A proof of Thurston's topological characterization of rational functions*, Mittag–Leffler, 1984, preprint.

[Sh] Shishikura M., *On a theorem of M. Rees for the matings of polynomials*, IHES, 1990, preprint.

[Ta] Tan Lei, *Accouplements des polynômes complexes*, Thèse, Orsay, 1987; *Mating of quadratic polynomials* (to appear).

INSTITUTE FOR MATH. SCIENCES
SUNY AT STONY BROOK
STONY BROOK, NY 11794
USA

A POSSIBLE APPROACH TO A COMPLEX
RENORMALISATION PROBLEM

MARY REES

Preliminary definitions. For a branched covering f $\overline{\mathbb{C}} \to \overline{\mathbb{C}}$, we define

$$X(f) = \{ f^n(c) : c \text{ critical}, n > 0 \}.$$

Then f is *critically finite* if $\sharp(X(f))$ is finite. Two critically finite branched coverings f_0, f_1 are (*Thurston*) *equivalent* if there is a path f_t through critically finite branched coverings connecting them with $X(f_t)$ constant in t.

We are only concerned, here, with orientation-preserving degree two branched coverings for which one critical point is fixed and the other is periodic. By a theorem of Thurston's ([T], [D-H]), any such branched covering f_0 is equivalent to a unique degree two polynomial f_1 of the form $z \mapsto z^2 + c$ (some $c \in \mathbb{C}$).

Now let f_1, f_2 be two degree two polynomials of the form $z \mapsto z^2 + c_i$ ($i = 1, 2$), with 0 periodic of periods m, n respectively. Then we define *the tuning of f_1 about* 0 *by* f_2, written $f_1 \vdash f_2$, as follows. This is simply a branched covering defined up to equivalence. Let D be an open topological disc about 0 such that the discs $f_1^i(D)$ ($0 \le i < m$) are all disjoint, $f_1^m(D) \subset D$ and $f_1 : f_1^i(D) \to f_1^{i+1}(D)$ is a homeomorphism for $1 \le i < m$. Let g be a rescaling of f_2, and V a closed bounded topological disc with $V \subset gV \subset f_1^m(D)$ whose complement is in the attracting basin of ∞ for g. Then we define

$$f_1 \vdash f_2 \quad \begin{aligned} &= f_1 \quad \text{outside } D, \\ &= f_1^{-(m-1)} \circ g \quad \text{in } V, \end{aligned}$$

and extend to map the annulus $D \setminus V$ by a two-fold covering to $f_1^m(D) \setminus g(V)$. Then

$$(f_1 \vdash f_2)^m = g \text{ in } V.$$

Thus $f_1 \vdash f_2$ is critically finite with 0 of period $n \cdot m$, and is equivalent to a unique polynomial $z \mapsto z^2 + c$.

For any sequence $\{f_i\}$ of polynomials, we can also define $f_1 \vdash \cdots \vdash f_n$ for all n.

For concreteness, we consider the following renormalization problem, but different versions are possible.

Let $\{f_i\}$ be any sequence of polynomials of the form $z \to z^2 + c_i$, where the f_i (and c_i) take only finitely many different values, and 0 is of period m_i under f_i. Write g_n for the polynomial $z \mapsto z^2 + c$ equivalent to $f_1 \vdash \cdots \vdash f_n$. and

$$n_k = \prod_{i \le k} m_i.$$

PROBLEM. *Prove geometric properties of $X(g_n)$. Specifically, show that the set*

$$\left\{ g_n^{n_k \ell + i}(0) : 0 \le \ell < m_{k+1} \right\} \tag{1}$$

has uniformly bounded geometry for all $i \le n_k$, $k < n$ and all n.

Of course, this problem (and stronger versions) is not new, has been the focus of much effort, and, in the real case, has been resolved by Sullivan [S]. The most obvious method of approach (which was not, in the end, efficacious in the real case) is through analysis of the main technique used to prove Thurston's theorem mentioned above. We now recall this.

Thurston's Pullback Map on Teichmüller space.
To simplify, we stick to orientation-preserving degree two critically finite branched coverings with fixed critical value v_2 and periodic critical value v_1. Let g be one such. Let $X = X(g)$. We let $s : \overline{\mathbb{C}} \to \overline{\mathbb{C}}$ be given by $s(z) = z^2$. Let $\mathcal{T} = \mathcal{T}(X)$ be the Teichmüller space of the sphere with set of marked points X, so that

$$\mathcal{T} = \left\{ [\varphi] : \varphi \text{ is a homeomorphism of } \overline{\mathbb{C}} \right\}$$

and $[\varphi]$ denotes the quotient of the isotopy class under isotopies constant on X by left Möbius composition, that is, $[\varphi] = [\sigma \circ \varphi \circ \psi]$ for any Möbius transformation σ and ψ isotopic to the identity rel X. It is convenient to choose representatives φ so that $\varphi(v_1) = 0$, $\varphi(v_2) = \infty$. Then

$$\tau : \mathcal{T} \to \mathcal{T}$$

is defined by

$$\tau([\varphi]) = [s^{-1} \circ \varphi \circ g].$$

(The righthand side makes perfectly good sense as a homeomorphism).
By Thurston's theorem (in this setting), τ is a contraction with respect to the Teichmüller metric d on $\mathcal{T}$, and has a unique fixed point $[\varphi]$. Then there are ψ isotopic to the identity via an isotopy fixing $X(g)$ (unique given φ) and a Möbius transformation σ such that

$$g = \varphi^{-1} \circ s \circ \sigma \circ \varphi \circ \psi.$$

In particular, g and $s \circ \sigma$ are equivalent, and $X(s \circ \sigma) = \varphi(X(g))$.

The "Obvious" Method of Approach.
We can choose h_n equivalent to g_n so that the sets of (1), with h_n replacing g_n, have uniformly bounded geometry for $i \le n_k$, $k < n$, and all n. Then let $\mathcal{T}_n = \mathcal{T}(X(h_n))$, and $\tau_n : \mathcal{T}_n \to \mathcal{T}_n$ be the associated pullback. It suffices (!) to prove convergence, as $m \to \infty$, and uniform in n, of the sequences $\{ \tau_n^m (\text{identity}) \}$. Of course, for fixed n, the convergence would be with respect to the Teichmüller metric d_n on $\mathcal{T}_n$. This seems to be impossible to implement. An alternative is suggested below. One virtue — and probably the only one — of this alternative is that it has not yet been tried (so far as I know). Before making this precise, we need to clarify some properties of the Teichmüller metric.

The Teichmüller metric and its Derivative.

Let $\mathcal{T} = \mathcal{T}(X)$ (for any finite set $X \subset \overline{\mathbb{C}}$) and let d denote the Teichmüller metric. Let $[\varphi]$, $[\psi] \in \mathcal{T}$. Assume without loss of generality that $\infty \in \varphi(X)$, $\psi(X)$. Then there is a unique quasiconformal homeomorphism $\chi : \overline{\mathbb{C}} \to \overline{\mathbb{C}}$ with the following properties.

1. $\chi(\varphi(X)) = \psi(X)$ and $[\chi \circ \varphi] = [\psi]$.
2. There is a rational function q with at most simple poles in $\mathbb{C}$, all occurring at points of $\varphi(X)$, and at least three more poles than zeros in $\mathbb{C}$, such that the directions of maximal stretch and contraction of χ are tangent to the vector fields $i\sqrt{q}$, $\sqrt{q}$ respectively, and the dilatation (ratio of infinitesimal stretch to contraction) is constant.
3. The images under χ_* of these vector fields are of the form $i\sqrt{p}$, $\sqrt{p}$, for a rational function p with at most simple poles in $\mathbb{C}$, all occurring at points of $\psi(X)$.

The function q is then also unique, up to a positive scalar multiple, and becomes unique if we normalize so that

$$\int |q| \frac{d\overline{z} \wedge dz}{2i} = 1.$$

Similarly, we normalize p. (Of course, q represents a quadratic differential $q(z)dz^2$, but it is convenient to keep the representing rational function in the foreground).

Let $h = (h(x)) \in \mathbb{C}^X$ be small, taking $h(x) = 0$ if $\varphi(x) = \infty$. Then by abuse of notation, we write $\varphi + h$ for a homeomorphism near φ with $(\varphi + h)(x) = \varphi(x) + h(x)$. Then the following holds, where q, p are determined by $[\varphi]$, $[\psi]$ as above [R].

$$d([\varphi + h],[\psi + k]) = d([\varphi],[\psi])$$
$$+ 2\pi \operatorname{Re}\left(\sum_{x \in X} (\operatorname{Res}(q, \varphi(x))h(x) - \operatorname{Res}(p, \psi(x))k(x)) \right)$$
$$+ o(h) + o(k).$$

Now we consider the case $X = X(g)$ and $y = \tau x$. As before we consider only specific g and take $s(z) = z^2$ (as before). The *pushforward* $s_* q$ of a rational function (or quadratic differential) q is defined by

$$s_* q(z) = \sum_{s(w) = z} \frac{q(w)}{s'(w)^2}$$

if $s'(w) \neq 0$ for $s(w) = z$. If q has only simple poles in $\mathbb{C}$, and at least 3 more poles than zeros in $\mathbb{C}$, then $s_* q$ extends to a rational function on $\mathbb{C}$ with the same properties. Then if q, p are determined by $[\varphi]$, $\tau([\varphi])$, taking $\varphi(v_1) = 0$, $\varphi(v_2) = \infty$ as above,

$$d([\varphi + h],\tau([\varphi + h])) = d([\varphi], \tau([\varphi]))$$
$$+ 2\pi \operatorname{Re}\left(\sum_{x \in X} (\operatorname{Res}(q, \varphi(x)) - \operatorname{Res}(s_* p, \varphi(x)))h(x) \right)$$
$$+ o(h).$$

The Suggested Alternative Approach to the Problem. Take $\mathcal{T} = \mathcal{T}(X(g))$, $\tau : \mathcal{T} \to \mathcal{T}$. Let

$$F([\varphi]) = d([\varphi], \tau([\varphi])).$$

Then the derivative formula for F above theoretically enables us to construct flows for which F decreases along orbits. It can be shown that the only critical point of F occurs where $F = 0$. So if we can find a compact subset B of $\mathcal{T}$ with smooth boundary and a vector field v pointing inward on ∂B with $DF(v) < 0$, then the (unique) zero of F must be inside B. Now put a subscript n on everything. Conceivably we can find $B_n \subset \mathcal{T}_n$ and vector field v_n pointing inward on ∂B_n such that if $A \subset X(g_n)$ is any of the sets in (1) and $[\varphi] \in \partial B_n$ then $\varphi(A)$ has bounded geometry (uniformly in A, n) and $DF_n(v_n) < 0$?

References

[D-H] Douady A., Hubbard J. H., *A proof of Thurston's topological characterization of rational functions*, preprint, Mittag–Leffler, 1985.

[R] Rees M., *Critically-defined spaces of branched coverings*, In preparation.

[S] Sullivan D., *Bounds, Quadratic differentials and renormalization conjectures*, Preprint, 1990.

[T] Thurston W. P., *On the combinatorics of iterated rational maps*, Preprint, Princeton University and I.A.S., 1985.

DEPARTMENT OF PURE MATHEMATICS
THE UNIVERSITY OF LIVERPOOL
LIVERPOOL L69 3BX
UNITED KINGDOM

GEOMETRY OF JULIA SETS

LENNART CARLESON

The geometry of connected Julia sets for hyperbolic quadratic polynomials is now well understood. Bounded components of the Fatou set are quasi-circles while the unbounded component is a John domain.

The geometry of a flower (for a rational fixed point) is also known. If the flower has more than one petal, each component is a quasi-disk. The 1-petal flower is a John domain (see a forthcoming paper by P. Jones and L. Carleson in Boletin de Brasil).

For Siegel disks S, a basic result by M. Herman is that the critical point belongs to the boundary of S if the rotation number $\lambda = e^{2\pi i \theta}$ is a Siegel number, i.e.

$$\left| \theta - \frac{p}{q} \right| > \frac{c}{q^n} \text{ for some } c > 0, \, n < \infty,$$

or more generally when the arithmetic condition of J.-C. Yoccoz's global theorem on conjugacy of analytic diffeomorphisms of the circle is satisfied.

Another remarkable result of M. Herman is that when the critical point belongs to ∂S, then S is a quasi-disk if and only if λ is of bounded type, i.e.

$$\left| \theta - \frac{p}{q} \right| > \frac{c}{q^2}.$$

With J.-C. Yoccoz he has also proved that ∂S is a Jordan curve for almost all θ. It is not known which arithmetic condition implies this. E.g., is there $\gamma_0 > 2$ so that $|\theta - p/q| > C/q^\gamma$ implies that S is a Jordan domain for $\gamma < \gamma_0$ but not for $\gamma > \gamma_0$?

A particularly interesting question concerns the geometry of ∂S at the critical point. Computer experiments show that in many cases ∂S has an angle of about $120°$ opening at the critical point. Prove this at least for $\theta = \theta_0 = (\sqrt{5} - 1)/2$. For this value there should also exist a renormalization at the critical point.

There is also a very interesting regularity of the Taylor coefficients of the conjugating map. Consider more generally the family $\mathcal{P}_\rho(z)$, $\text{Re}(\rho) > 0$, with

$$\mathcal{P}'_\rho = \lambda(1 - z)^\rho, \quad \mathcal{P}_\rho(0) = 0$$

so that $\rho = 1$ corresponds to $\lambda(z - z^2/2)$. Let $h(\zeta)$ be the conjugating map in $|\zeta| < 1$ with $h(1) = 1$. (For general ρ the proof that $1 \in \partial S$ is not known, but should be rather similar to the case $\rho = 1$). Form

$$f(\zeta) = \frac{h'(\zeta)}{1 - h(\zeta)} = \sum_0^\infty a_\nu \zeta^\nu.$$

Then

$$f' - f^2 = f\rho \sum_0^\infty \left(\frac{1}{2} + \frac{i}{2} \cot(\nu + 1)\pi\theta \right) a_\nu \zeta^\nu.$$

If the imaginary part in the parenthesis is dropped we obtain

$$f_0' = \left(1 + \frac{\rho}{2}\right) f_0^2 \,, \quad f_0 = \frac{1}{(1 + \rho/2)(1 - z)}, \quad h_0 = (1 - z)^{2/(\rho+2)}.$$

Computer experiments indicate for $\theta = \theta_0$, $\rho = 1$

$$\left| a_\nu - \frac{2}{3} \right| < 0.1 \quad \text{(say) for all } \nu,$$

where $2/3$ corresponds to f_0. It would be interesting to make the approximation rigorous at least for small ρ.

In the non-hyperbolic case very little is known (and very little can be probably said in general). The simplest case of a strictly preperiodic critical point leads to John domains (the Julia set is called a dendrite). It should be possible to analyse the general Misiurewicz case when the critical point never returns close to itself. In the case of $1 - az^2$, a is real, this condition is equivalent to the Fatou set being a John domain. To which extent does this hold for general Misiurewicz points?

INSTITUTE MITTAG-LEFFLER
AURAVAGEN 17
S-18262, DJURSHOLM 1
SWEDEN

PROBLEMS ON LOCAL CONNECTIVITY[1]

JOHN MILNOR

If the Julia set $J(f)$ of a quadratic polynomial is connected, then Yoccoz has proved[2] that $J(f)$ is locally connected, unless either:

(1) f has an irrationally indifferent periodic point, or
(2) f is infinitely renormalizable.

Cremer Points. To illustrate case (1), consider the polynomial

$$P_\alpha(z) = z^2 + e^{2\pi i \alpha} z$$

with a fixed point of multiplier $\lambda = e^{2\pi i \alpha}$ at the origin. Take α to be real and irrational. For generic choice of α (in the sense of Baire category), Cremer showed that there is no local linearizing coordinate near the origin. We will say briefly that the origin is a *Cremer point*, or that P_α is a *Cremer polynomial*. According to Sullivan and Douady, the existence of such a Cremer point implies that the Julia set is not locally connected. More explicitly, let $t(\alpha)$ be the angle of the unique external ray which lands at the corresponding point of the Mandelbrot set. For generic choice of α, Douady has shown that the corresponding ray in the dynamic plane does not land, but rather has an entire continuum of limit points in the Julia set. (Compare [Sø]). Furthermore, the $t(\alpha)/2$ ray in the dynamic plane accumulates both at the fixed point 0 and its pre-image $-\lambda$.

PROBLEM 1. *Is there an arc joining 0 to $-\lambda$, in the Julia set of such a Cremer polynomial?*

PROBLEM 2. *Give a plausible topological model for the Julia set of a Cremer polynomial.*

PROBLEM 3. *Make a good computer picture of the Julia set of some Cremer polynomial.*

PROBLEM 4. *Can there be any external rays landing at a Cremer point?*

PROBLEM 5. *Can the critical point of a Cremer polynomial be accessible from $\mathbb{C} \setminus J$?*

PROBLEM 6. *If we remove the fixed point from the Julia set of a Cremer polynomial, how many connected components are there in the resulting set $J(P_\alpha) \setminus \{0\}$, i.e., is the number of components countably infinite?*

PROBLEM 7. *The Julia set for a generic Cremer polynomial has Hausdorff dimension two. Is it true for an arbitrary Cremer polynomial? Do Cremer Julia sets have measure zero?* (*Compare* [Sh], [L1], [L2].)

In the quadratic polynomial case, Yoccoz has shown that every neighborhood of a Cremer point contains infinitely many periodic orbits. On the other hand, Perez-Marco [P-M1] has described non-linearizable local holomorphic maps for which this is not true.

---

[1] Based on questions by a number of participants in the 1989 Stony Brook Conference
[2] Compare [Hu].

PROBLEM 8. *For a Cremer point of an arbitrary rational map, does every neighborhood contain infinitely many periodic orbits?*

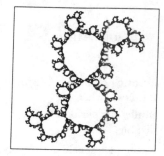

Figure 1. Julia set of P_α where $\alpha = .78705954039469$ has been randomly chosen.

Siegel Disks. (Compare Carleson's discussion.) If α satisfies a Diophantine condition (in particular, for Lebesgue almost every α), Siegel showed that there is a local linearizing coordinate for the polynomial $P_\alpha(z) = z^2 + e^{2\pi i \alpha} z$ in some neighborhood of the origin. Briefly we say that the origin is the center of a *Siegel disk* Δ, or that P_α is a *Siegel polynomial*. Yoccoz has given a precise characterization of which irrational angles yield Siegel polynomials and which yield Cremer polynomials. ([Y], [P-M2]).

Herman, making use of ideas of Ghys, showed that there exists a value α_0 so that P_{α_0} has a Siegel disk whose boundary $\partial \Delta$ does not contain the critical point. It follows that the Julia set $J(P_{\alpha_0})$ is not locally connected. On the other hand if α satisfies a Diophantine condition, then Herman showed that $\partial \Delta$ does contain the critical point.

PROBLEM 9. *Give any example of a Siegel polynomial whose Julia set is provably locally connected. Is $J(P_\alpha)$ locally connected for Lebesgue almost every choice of α? (Compare Figure 1). What can be said about the Hausdorff dimension of $J(P_\alpha)$?*

PROBLEM 10. *Can a Siegel disk have a boundary which is not a Jordan curve?*

PROBLEM 11. *Does any rational function have a Siegel disk with a periodic point in its boundary? Such an example would be extremely pathological. (In the polynomial case, Poirier has pointed out that at least there cannot be a Cremer point in the boundary of a Siegel disk. See* [GM]).

Infinitely Renormalizable Polynomials. A quadratic polynomial $f_c(z) = z^2 + c$ is *renormalizable* if there exists an integer $p \geqslant 2$ and a neighborhood U of the critical point zero so that the orbit of zero under $f^{\circ p}$ remains in this neighborhood forever, and so that the map $f^{\circ p}$ restricted to U is polynomial-like of degree 2. (Thus the closure $\bar{U}$ must contain no other critical points of $f^{\circ p}$, and must be contained in the interior of $f^{\circ p}(U)$). Let M be the Mandelbrot set, and let $H \subset M$ be any hyperbolic component of period $p \geqslant 2$. Douady and Hubbard [DH2] show that H is contained in a small copy of M. This small copy is the image of a homeomorphic embedding of M into itself, which I will denote by $c \mapsto H * c$. The elements of these various small copies $H * M \subset M$ (possibly with the root point $H * \frac{1}{4}$ removed) are precisely the renormalizable elements of M.

Now consider an infinite sequence of hyperbolic components $H_1, H_2, \ldots \subset M$. If the H_i converge to the root point $1/4$ sufficiently rapidly, then Douady and Hubbard (unpublished) show that the intersection $\bigcap_k H_1 * \cdots * H_k * M$ consists of a single point c_∞ such that the corresponding Julia set $J(f_{c_\infty})$ is not locally connected.

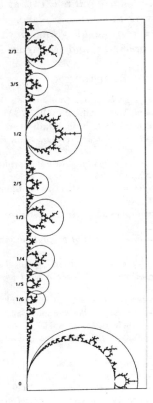

Figure 2. Picture of the $\log \lambda$-plane, showing the Yoccoz disks of radius $\log(2)/q$. (Heights in units of 2π).

PROBLEM 12. Suppose that f_{c_∞} is infinitely renormalizable of bounded type. For example, suppose that $c_\infty \in \bigcap_k H_1 * \cdots * H_k * M$, where the H_i are all equal. Does it then follow that $J(f_{c_\infty})$ is locally connected? As the simplest special case, if we take $H_1 = H_2 = \cdots$ to be the period two component centered at -1, then f_{c_∞} will be the quadratic Feigenbaum map. Is the Julia set for the Feigenbaum map locally connected?

PROBLEM 13. More generally, if c is real (belonging to the intersection $M \cap \mathbb{R} = [-2, 1/4]$), does it follow that the Julia set $J(f_c)$ is locally connected?

The Mandelbrot Set. Here the most basic remaining question is the following.

PROBLEM 14. Does every infinite intersection of the form $\bigcap_k H_1 * \cdots H_k * M$ reduce to a single point? Equivalently, is the set of infinitely renormalizable points totally

445

disconnected? Does this set have measure zero? Does it in fact have small Hausdorff dimension?

PROBLEM 15. *For each rational number $0 < p/q < 1$ let $M(p/q)$ be the limb of the Mandelbrot set with interior angle p/q. Is the diameter of $M(p/q)$ less than k/q^2 for some constant k independent of p and q? If not, is it at least less than $k\log(q)/q^2$? (It is actually more natural to work in the $\log \lambda$ plane, where $f(z) = z^2 + \lambda z$. The Yoccoz inequality asserts that the corresponding limb in this $\log \lambda$ plane is contained in a disk of radius $\log(2)/q$. Compare [P], and see Figure 2).*

REFERENCES

[B]   Bielefeld B., *Conformal dynamics problem list*, Stony Brook IMS, Preprint #1990/1.

[C1]   Cremer H., *Zum Zentrumproblem*, Math. Ann. **98** (1927), 151–163.

[C2]   Cremer H., *Über die Häufigkeit der Nichtzentren*, Math. Ann. **115** (1938), 573–580.

[D]   Douady A., *Disques de Siegel et anneaux de Herman*, Sém. Bourbaki, 1986-87; vol. 152–153, 1987-88.

[DH1]   Douady A., Hubbard J. H., *Systèmes dynamiques holomorphes I,II: itération des polynômes complexes*, Math. Orsay, **84.02** and **85.04**.

[DH2]   Douady A., Hubbard J. H., *On the dynamics of polynomial-like mappings*, Ann. Sci. Ec. Norm. Sup. **18** (1985), Paris, 287–343.

[G]   Ghys E., *Transformations holomorphes au voisinage d'une courbe de Jordan,*, CRAS Paris **298** (1984), 385-388.

[GM]   Goldberg L., Milnor J., *Fixed point portraits of polynomial maps,*, Stony Brook IMS, preprint 1990/14.

[He]   Herman M., *Recent results and some open questions on Siegel's linearization theorem of germs of complex analytic diffeomorphisms of C^n near a fixed point*, Proc 8th Int. Cong. Math. Phys., World Sci., 1986, pp. 138–198.

[Hu]   Hubbard J. H., *Puzzles and quadratic tableaux*, preprint 1990 (according to Yoccoz).

[L1]   Lyubich M., *An analysis of the stability of the dynamics of rational functions*, Funk. Anal. i. Pril. **42** (1984), 72–91 (Russian); Selecta Math. Sovietica **9** (1990), 69–90.

[L2]   Lyubich M., *On the Lebesgue measure of the Julia set of a quadratic polynomial*, Stony Brook IMS, preprint 1991/10.

[P]   Petersen C., *On the Pommerenke–Levin–Yoccoz inequality*, IHES, 1991, preprint.

[P-M1]   Perez-Marco R., *Sur la dynamique des germes de difféomorphismes holomorphes de* $(\mathbf{C}, 0)$ *et des difféomorphismes analytiques du cercle*, Paris-Sud, 1990, Thèse.

[P-M2]   Perez-Marco R., *Solution complète au Problème de Siegel de linéarisation d'une application holomorphe au voisinage d'un point fixé (d'apres J.-C. Yoccoz)*, Sém. Bourbaki, Feb. 1992.

[R]   Rogers J. T., *Singularities in the boundaries of local Siegel disks* (to appear).

[Sh]   Shishikura M., *The Hausdorff dimension of the boundary of the Mandelbrot set and Julia sets*, Stony Brook IMS, preprint 1991/7.

[Si]   Siegel C. L., *Iteration of analytic functions,*, Ann. of Math. **43** (1942), 607–612.

[Sø]   Sørensen D. E. K., *Local connectivity of quadratic Julia sets*, Tech. Univ. Denmark, Lyngby, 1992, preprint.

[Su]   Sullivan D., *Conformal dynamical systems*, Geometric Dynamics (Palis, ed.), Lecture Notes Math. **1007**, Springer, 1983, pp. 725–752.

[Y]   Yoccoz J.-C., *Linéarisation des germes de difféomorphismes holomorphes de* $(\mathbb{C}, 0)$, CRAS Paris **306** (1988), 55–58.

INSTITUTE FOR MATH. SCIENCES
SUNY AT STONY BROOK
STONY BROOK, NY 11794
USA

MEASURE AND DIMENSION OF JULIA SETS

MIKHAIL LYUBICH

PROBLEM 1. *Can it happen that a nowhere dense Julia set has positive Lebesgue measure?*

The corresponding Ahlfors problem in Kleinian groups is also still unsolved.

So far it is known that the Julia set has zero measure in the following cases:

(i) hyperbolic, subhyperbolic and parabolic cases [DH], [L1].
(ii) a cubic polynomial with one simple non-escaping critical point and with a "non-periodic tableau" (McMullen, see [BH]);
(iii) a quadratic polynomial which is only finitely renormalizable and has no neutral irrational cycles (Lyubich [L2] and Shishikura (unpublished)).

Let us say that a polynomial with one non-escaping critical point c is *renormalizable* if there is a quadratic-like map $f^n : U \to V$, $c \in U \subset V$, $n > 1$ with connected Julia set. It corresponds to the case of periodic tableau. Cases (i) and (ii) can be generalized in the following way:

(iv) a polynomial of any degree but with only one non-escaping critical point which does not have irrational neutral points and which is only finitely renormalizable.

In higher degrees one can describe a wide class of combinatorics for which the Julia set has zero measure (non-recurrent and "reluctantly recurrent" cases). The basic examples for which the answer is still unclear are

1. The Feigenbaum quadratic polynomial.
2. The Fibonacci polynomial $z \mapsto z^d + c$ with $d > 2$ (see [BH] or [LM] for the definition of the Fibonacci polynomial).
3. A polynomial with a Cremer point or Siegel disk (see the discussions in Milnor's notes).

In the case when the Julia set coincides with the whole sphere the corresponding question is the following.

PROBLEM 2. *Is it true for all f with $J(f) = \overline{\mathbb{C}}$ that the following hold?*

(i) $\omega(z) = \overline{\mathbb{C}}$ *for almost all $z \in \overline{\mathbb{C}}$?*
(ii) f *is conservative with respect to the Lebesgue measure? (Conservativity means that the Poincaré Return Theorem holds).*

Note that for the interval maps (replacing $\overline{\mathbb{C}}$ by an interval on which f is topologically mixing) (i) and (ii) are equivalent [BL2]. Moreover, both of them hold for the quadratic-like maps of the interval [L3].

PROBLEM 3. *Let again* $J(f) = \overline{\mathbb{C}}$. *Is it true that* f *is ergodic with respect to the Lebesgue measure? Is it at least true that it has at most* $2 \deg f - 2$ *ergodic components?*

The answer to the first question is yes for a large set of rational maps [R]. The answer to the second one is yes for interval maps [BL1].

The discussed problems are closely related to the deformation theory of rational maps. The link between them is given by the notion of *measurable invariant line field on the Julia set* (see [MSS]). Each such field generates a quasi-conformal deformation of f supported on the Julia set. There is a series of Lattes examples having an invariant line field on the Julia set, and in these examples $J(f) = \overline{\mathbb{C}}$. Such a phenomenon is impossible at all for finitely generated Kleinian groups [S].

PROBLEM 4. (Sullivan) *Are the Lattes examples the only ones having measurable invariant line fields on the Julia sets?*

Let us consider now an analytic family A of rational maps, and denote by $Q \subset A$ the set of J-unstable maps.

A recent remarkable result by Shishikura [Sh] says that in the quadratic family $z \mapsto z^2 + c$ there are a lot of Julia sets with Hausdorff dimension 2.

PROBLEM 5. *Find an explicit example of a Julia set of Hausdorff dimension 2. What is a natural geometric measure in the case when* $J(f)$ *has Hausdorff dimension 2 but zero Lebesgue measure?*

A more general program is to develop an appropriate Thermodynamical Formalism in non-hyperbolic situations.

PROBLEM 6.

  (i) *What is the Lebesgue measure of* Q*?*
  (ii) *Is the Hausdorff dimension of* Q *equal to* $\dim A$*? The answer is yes in the quadratic case* [Sh]

Mary Rees proved that the Lebesgue measure of Q is positive [R] in the case when A is the whole space of rational maps of degree d. On the other hand, Shishikura claims that in the quadratic family $z \mapsto z^2 + c$ the measure of the set of only finitely renormalizable points in Q is equal to zero (here Q is just the boundary of the Mandelbrot set). How do these results fit?

REFERENCES

[BH]   Branner B., Hubbard J. H., *The iteration of cubic polynomials, Part II : patterns and parapatterns*, Acta Math. (to appear).

[BL1] Blokh A., Lyubich M., *The decomposition of one-dimensional dynamical systems into ergodic components*, Leningrad Math. J. **1** (1990), 137–155.

[BL2] Blokh A., Lyubich M., *Measurable dynamics of S-unimodal maps*, Ann. Sci. École Norm. Sup. **24** (1991), no. 4, 545–573.

[DH]   Douady A., Hubbard J. H., *Études dynamique des polynômes complexes*, I., Math Orsay, 84-02.

[L1]   Lyubich M., *On the typical behavior of trajectories of a rational mapping of the sphere*, Soviet Math. Dokl. **27** (1983), 22–25.

[L2]   Lyubich M., *On the Lebesgue measure of the Julia set of a quadratic polynomial*, 1991/10, Preprint IMS.

[L3]   Lyubich M., *Combinatorics, geometry and attractors of quadratic polynomials*, 1992, Preprint.

PROBLEM 19.7

[LM]   Lyubich M., Milnor J., *The Fibonacci unimodal map*, Stony Brook, 1991/15, Preprint IMS.

[MSS] Mané R., Sad P., Sullivan D., *On the dynamics of rational maps*, Ann. Sci. École Norm. Sup. **16** (1983), no. 4, 193–217.

[R]   Rees M., *Positive measure sets of ergodic rational maps* Ann. Sci. École Norm. Sup. **19** (1986), no. 4, 383–407.

[Sh]   Shishikura M., *On the quasiconformal surgery of rational functions*, Ann. Sci. École Norm. Sup. **20** (1987), no. 4, 61–77.

[S]   Sullivan D., *The ergodic theory at infinity of a discrete group of hyperbolic isometries*, Ann. of Math. Studies **97** (1981), Princeton Univ. Press, 465–497.

INST. FOR MATH. SCI.
STATE UNIV. AT STONY BROOK
STONY BROOK, NY 11794-3660
USA

ON INVARIANT MEASURES FOR
ITERATIONS OF HOLOMORPHIC MAPS

Feliks Przytycki

Let U be an open subset of the Riemann sphere $\hat{C}$. Consider any holomorphic mapping $f: U \to \hat{C}$ such that $f(U) \supset U$ and $f: U \to f(U)$ is a proper map, (for a more general situation see [PS]). Consider any $z \in f(U)$. Let $z^1, z^2, ..., z^d$ be some of the f-preimages of z in U where $d \geqslant 2$. Consider curves $\gamma^i: [0,1] \to \hat{C}$, $i = 1, ..., d$, also in $f(U)$, joining z with z^i respectively (i.e. $\gamma^i(0) = z$, $\gamma^i(1) = z^i$).

Let $\Sigma^d := \{1, ..., d\}^{Z^+}$ denote the one-sided shift space and σ the shift to the left, i.e. $\sigma((\alpha_n)) = (\alpha_{n+1})$. For every sequence $\alpha = (\alpha_n)_{n=0}^{\infty} \in \Sigma^d$ we define $\gamma_0(\alpha) := \gamma^{\alpha_0}$. Suppose that for some $n \geqslant 0$, for every $0 \leqslant m \leqslant n$, and all $\alpha \in \Sigma^d$, the curves $\gamma_m(\alpha)$ are already defined. Suppose that for $1 \leqslant m \leqslant n$ we have $f \circ \gamma_m(\alpha) = \gamma_{m-1}(\sigma(\alpha))$, and $\gamma_m(\alpha)(0) = \gamma_{m-1}(\alpha)(1)$.

Define the curves $\gamma_{n+1}(\alpha)$ so that the previous equalities hold (by taking f-preimages of curves already existing; if there are no critical values for iterations of f in $\bigcup_{i=1}^d \gamma^i$ one has a unique choice). For every $\alpha \in \Sigma^n$ and $n \geqslant 0$ denote $z_n(\alpha) := \gamma_n(\alpha)(1)$.

The graph with the vertices z and $z_n(\alpha)$ and edges $\gamma_n(\alpha)$ is called a *geometric coding tree* with the root at z. For every $\alpha \in \Sigma^d$ the subgraph composed of z, $z_n(\alpha)$ and $\gamma_n(\alpha)$ for all $n \geqslant 0$ is called a *geometric branch* and denoted by $b(\alpha)$. The branch $b(\alpha)$ is called *convergent* if the sequence $z_n(\alpha)$ is convergent in clU. We define the *coding map* $z_\infty: \mathcal{D}(z_\infty) \to \text{cl}U$ by $z_\infty(\alpha) := \lim_{n \to \infty} z_n(\alpha)$ on the domain $\mathcal{D}(z_\infty)$ of all such α's for which $b(\alpha)$ is convergent.

There are two basic examples:

1. $f: U \to U$ where U is a simply-connected domain in $\hat{C}$, $\deg f \geqslant 2$, and the iterates f^n converge to a constant in U, in particular U is an immediate basin of attraction of a sink for f a rational map on $\hat{C}$.

2. $U = \hat{C}$, f is a rational mapping.

It is known that except for a "thin" set in Σ^d all branches are convergent (i.e. $\Sigma^d \setminus \mathcal{D}(z_\infty)$ is "thin" and for every $x \in \text{cl}U$, the set $z_\infty^{-1}(x)$ is "thin"). These hold under very mild assumptions about the tree even allowing the existence of critical values in it. Proofs and a discussion of various possibilities of "thinness" can be found in [PS]. In particular one obtains the classical Beurling's Theorem that a holomorphic univalent function R on the unit disc $\mathbb{D}$ has radial limits everywhere except on a set of logarithmic capacity zero, and for every limit point, the set in $\partial\mathbb{D}$ to which radii converge is also of logarithmic capacity 0. One just transports the map $z \mapsto z^2$ to $U := R(\mathbb{D})$, and gets a type 1 situation. There is a 1-to-1 correspondence between the radii and geometric branches.

General Problem. How large is the image: $z_\infty(\mathcal{D}(z_\infty))$?

We shall specify this Problem separately in the basin of attraction case (the situation 1 above) and in the general situation.

To simplify the notation we have restricted ourselves to trees and codings from the full shift space. In the general situation it might be useful to consider also a topological Markov chain, see [PS].

THE CASE OF THE BASIN OF ATTRACTION

PROBLEM 1.1. *If f extends holomorphically to a neighbourhood of clU, is every periodic point in ∂U accessible from U ?*

Comment. Accessible means being $\varphi(1)$ for a continuous curve $\varphi : [0,1] \to$ clU where $\varphi([0,1)) \subset U$ what is equivalent to being in the radial limit (i.e. $\lim_{r \nearrow 1} R(r\zeta)$ for $\zeta \in \partial\mathbb{D}$, R denoting a univalent map from $\mathbb{D}$ onto U). For g denoting the holomorphic extension of $R^{-1} \circ f \circ R$ to a neighbourhood of cl$\mathbb{D}$ and $\bar{R}$ the radial limit of R wherever it exists, it is known that at every g-periodic $\zeta \in \partial\mathbb{D}$, $\bar{R}$ exists and f at $\bar{R}(\zeta)$ is f-periodic (equivalently we could speak about σ-periodic points in Σ^d and the mapping z_∞, for a tree in U). Are there other periodic points in ∂U? It seems it does not matter if one assumes here that f is defined only on a neighbourhood of ∂U. This is the case of an RB-domain U (the boundary is repelling on the U side) considered in [PUZ]. Problem 1.1 has a positive answer in the case where f is a polynomial on $\mathbb{C}$ and U is the basin of attraction to ∞, (Douady, Yoccoz, Eremenko, Levin), even if U is not simply-connected, see [EL]. Here the fact $f^{-1}(U) \subset U$ helps.

PROBLEM 1.2. *In the situation of Problem 1.1 is every point $x \in \partial U$ of positive Lyapunov exponent (i.e. such that $\liminf_{n\to\infty} \frac{1}{n} \log |(f^n)'(x)| > 0$) accessible from U?*

PROBLEM 1.3. *In the situation of Problem 1.1 is it true that the topological entropy $h_{top}(f|_{\partial U}) = \log \deg(f|_U)$?*

Comment. The $\geqslant$ inequality is known and easy. The problem is with the opposite one. It would be true if every point $x \in \partial U$ had at most $\deg(f|_U)$ pre-images in ∂U.

A positive answer to Problem 1.2 would give a positive answer to 1.3. The reason is that topological entropy is approximated by measure-theoretic entropies for f-invariant measures which having positive entropies would have positive Lyapunov exponents (Ruelle's inequality). Then they would be images under $\bar{R}$ of g-invariant measures on $\partial\mathbb{D}$ which all have entropies upper bounded by $\log d$ (as g is a degree d expanding map on $\partial\mathbb{D}$).

PROBLEM 1.4. *Can there be periodic points or points with positive Lyapunov exponents in the boundary of a Siegel disc S? Is it always true that $h_{top}(f|_{\partial S}) = 0$?*

THE GENERAL CASE

We suppose here only that f extends holomorphically to a neighbourhood of the closure of the limit set Λ of a tree, $\Lambda = z_\infty \mathcal{D}(z_\infty)$. Then Λ is called a quasi-repeller, see [PUZ]. Denote the space of all probability f-invariant ergodic measures on the closure of a quasi-repeller Λ by $M(\Lambda)$. The space of measures in $M(\Lambda)$ which have positive entropy will be denoted by $M^+(\Lambda)$.

PROBLEM 2.1. *Is it true that every $m \in M(\Lambda)$ is the image of a measure on the shift space Σ^d through a geometric coding tree with z in a neighbourhood of clΛ. What* *about measures in $M^+(\Lambda)$? The same questions for f a rational mapping of degree d on $U = \hat{\mathbb{C}}$ and measures on the Julia set $J(f)$.*

Comment. It is easy to see at least, due to the topological exactness of f on the Julia set $J(f)$ (for every open V in $J(f)$ there exists $n > 0$ so that $f^n(V) = J(f)$), that for every z except at most two, $z_\infty(\mathcal{D}(z_\infty))$ is dense in $J(f)$. The answer is of course positive in the case f is expanding on Λ because then z_∞ is well defined and continuous on Σ^d, hence Λ is closed.

PROBLEM 2.2. *For which $m \in M^+(\Lambda)$ for every "reasonable" function $\varphi \colon \Lambda \to \mathbb{R} \cup \pm\infty$ (for example Hölder, into $\mathbb{R}$ or allowing isolated values $-\infty$ with $\exp \varphi$ nonflat there, as $\log |g|$, g holomorphic) do the probability laws like Almost Sure Invariance Principle, Law of Iterated Logarithm, or Central Limit Theorem hold for the sequence of sums $S_n(\varphi) = \sum_{j=0}^{n-1} t_j$ of the random variables $t_j := \varphi \circ f^j - \int \varphi dm$ provided $\sigma^2(\varphi) = \lim \frac{1}{n} \int S_n(f)^2 dm > 0$?*

Comment. If the measure is a z_∞-image of a measure on Σ^d with a Hölder continuous Jacobian (a Gibbs measure for a Hölder continuous function) then the probability laws hold, see [PUZ]. The positive answer in Problem 2.1 would be very helpful in solving Problem 2.2.

The class of measures for which Problem 2.2 has not been solved, but does not seem out of reach, are equilibrium states for Hölder continuous functions, say on the Julia set in the case f is rational. In this case the transfer (Ruelle–Perron–Frobenius) operator is already understood to some extent [DU], [P]. A proof seems to depend on finding an appropriate space of functions on which the maximal eigenvalue has modulus strictly larger than supremum over the rest of the spectrum (by the analogy to the expanding case, [B]).

Actually these equilibrium states are z_∞-images of measures on Σ^d. The Jacobians of these equilibrium states have modulus of continuity bounded by $\mathrm{Const}(m)(\log(1/t))^{-m}$ for any $m > 0$ (I don't know if it is Hölder). The Jacobian of the pull-back of the equilibrium measure to Σ^d is not wild. This gives a chance to prove that mixing in Σ^d is polynomially fast.

PROBLEM 2.3. *Is it true for every $m \in M^+(\Lambda)$ that m is absolutely continuous with respect to H_κ (where H_κ is the Hausdorff measure in dimension $\kappa = \mathrm{HD}(m)$) iff $HD(m) = HD(\mathrm{cl}\Lambda)$?*

Comment. In such a generality I would expect a negative answer. One should probably restrict the family of measures under consideration and/or impose additional assumptions on the mapping f.

If f is expanding on Λ then the answer is positive for all measures in $M^+(\Lambda)$ with Hölder continuous Jacobian. This is basically Bowen's theorem.

In the discussion here we assume that on every set E on which f is 1-to-1 the measure $(f|_E)^{-1}(m)$ is equivalent to m, and we write $\mathrm{Jac}_m f(z) = \frac{d(f|_E)^{-1}(m)}{dm}(z)$.

When the Jacobian exists in this sense we can replace the absolute continuity hypothesis $m \ll H_\kappa$ or the alternative singularity hypothesis $m \perp H_\kappa$ with another pair of alternative hypotheses.

PROBLEM 2.4. *In what class of measures in $M^+(\Lambda)$ does the property: the family $S_n(\log \mathrm{Jac}_m(f) - \kappa \log |f'|)$ is not uniformly bounded in $L^2(m)$, imply $m \perp H_\kappa$ and $HD(m) < HD(\mathrm{cl}\Lambda)$.*

Comment. The answer is positive for f expanding and Jacobian Hölder continuous.

It is positive also if $m = z_\infty(\mu)$ for any Gibbs measure μ for a Hölder continuous function on Σ^d. The singularity $\perp$ follows then from the positive answer to Problem 2.2 in this special case, see [PUZ]. From the probability laws one can deduce a stronger singularity, for example with respect to the measure $H_\Phi(\kappa, c)$ which is the Hausdorff measure for the function

$$\Phi(\kappa, c)(t) = t^\kappa \exp c \sqrt{\log \frac{1}{t} \log\log\log \frac{1}{t}}$$

for all

$$c < \sqrt{2\sigma^2 (\log \mathrm{Jac}_\mu(s) - \kappa \log |f'| \circ z_\infty) / \int \log |f'| dm}.$$

The inequality $HD(m) < HD(\mathrm{cl}\Lambda)$ follows from [Z1].

PROBLEM 2.5. *In what class of measures in $M^+(\Lambda)$ s does the property: the family $S_n(\log \mathrm{Jac}_m(f) - \kappa \log |f'|)$ is uniformly bounded in $L^2(m)$, imply $m \ll H_\kappa$?*

Comment. Again the answer is positive for f expanding and Jacobian Hölder continuous.

If $m = z_\infty(\mu)$ then the boundedness of the family $S_n(\varphi)$ where $\varphi := \log \mathrm{Jac}_\mu(f) - \kappa \log |f'| \circ z_\infty$ occurs precisely when $\sigma^2(\log \mathrm{Jac}_\mu(f) - \kappa \log |f'| \circ z_\infty) = 0$ assuming the series $\sum_{n=1}^\infty n \int |\varphi \cdot (\varphi \circ s^n)| d\mu$ is convergent. This is equivalent to the existence of a function u in $L^2(\mu)$ so that $\varphi = u \circ s - u$. Then we say that we can solve the cohomology equation for φ. Then we can also solve the cohomology equation for $\log \mathrm{Jac}_m(f) - \kappa \log |f'|$ on Λ. The naive way to compare m with H_κ is to prove that the sequence $S_n(\log \mathrm{Jac}_m(f) - \kappa \log |f'|)(z)$ is bounded at almost every $z \in \Lambda$. In the expanding case this allows comparison of the m-measure and the radius to the κ power of little discs, so the naive method happens to be successful. In the general case we do not have even pointwise boundedness, because the function u is only in $L^2(\mu)$.

The problem has the positive answer in the following special cases:

1. In the RB-domain case, where m is equivalent to a harmonic measure on the boundary of a simply-connected domain U, see PUZ] and [Z2]. Then $m = \bar{R}(\mu)$ where μ is equivalent to the Lebesgue measure on $\partial\mathbb{D}$. $\log |R'|$ happens to be within a bounded distance from any harmonic extension of u to a neighbourhood of $\partial\mathbb{D}$, in particular radial limits for $\log |R'|$ exist a.e. In [Z2] it is proved in fact that all this implies that ∂U is analytic, giving the answer to Problem 2.3 in this case.

2. In the case where f is a rational map on $\hat{\mathbb{C}}$ and m is a measure with maximal entropy (in which case Jacobian $\equiv \deg f$). Then again a careful look at u proves that f is either $z \mapsto z^n$ or is a Tchebysheff polynomial (in respective holomorphic coordinates on $\hat{\mathbb{C}}$) or else $J(f) = \hat{\mathbb{C}}$ and f has a parabolic orbifold, see [Z1].

In the general case it seems hopeful to treat any harmonic extension of u as a logarithm of a derivative of a "Riemann mapping". In the case $m = z_\infty(\mu)$ one can average u over cylinders in Σ^d extending u to the vertices $z_n(\alpha)$ of the tree.

The mapping z_∞ can be viewed as a dynamical version of a Riemann mapping. We can formulate the following problem:

PROBLEM 2.6. *Which theorems about the boundary behaviour of Riemann maps hold for geometric coding trees?*

Comment. Beurling Theorems hold, see the discussion in Section 1.

One has a natural dictionary:

For R:	For z_∞:
prime end	a geometric branch
impression	$I(\alpha) = \cap_{n=0}^{\infty} z_\infty \{\beta : \beta_i = \alpha_i, i = 0, ..., n\}$
the set of principal points	the limit set for the vertices $z_n(\alpha)$ of $b(\alpha)$.

PROBLEM 2.7. *Is it true that* $\sup\limits_{m \in M^+(\Lambda)} \mathrm{HD}(m) = \mathrm{HD}(\mathrm{cl}\Lambda)$? *Does* $\sup\limits_{m \in M(\Lambda)}$ *help?*

Comment. Of course a negative answer to this Problem for some Λ and positive to Problem 2.3 would mean that $m \perp H_{\mathrm{HD}(m)}$ for all m.

Problem 2.7 has positive answer in the expanding and subexpanding cases where sup is attained, it is so even for a positive measure set of rational mappings on $\hat{\mathbb{C}}$ for which absolutely continuous invariant measures exist (with respect to the Lebesgue), see [R]. The problem has also a positive answer for rational mappings with neutral points but without critical points in the Julia set. But then it may happen that supremum is not attained, see [ADU] and [L].

REFERENCES

[ADU] Aaronson J., Denker M., Urbański M., *Ergodic theory for Markov fibred systems and parabolic rational maps*, vol. 32, Göttingen, 1990, Preprint.

[B] Bowen R., *Equilibrium States and the Ergodic Theory of Anosov Diffeomorphisms*, L. N. Math. 470, Springer-Verlag, Berlin–Heidelberg–New York, 1975.

[DU] Denker M., Urbański M., *Ergodic theory of equilibrium states for rational maps*, Nonlinearity 4 (1991), 103–134.

[EL] Eremenko A. E., Levin G. M., *On periodic points of polynomials*, Ukr. Mat. Journal **41.11** (1989), 1467–1471.

[L] Ledrappier F., *Quelques propriétés ergodiques des applications rationelles*, Sér. I Math. **299** (1984), C. R. Acad. Sci., Paris, 37–40.

[P] Przytycki F., *On the Perron–Frobenius–Ruelle operator for rational maps on the Riemann sphere and for Hölder continuous functions*, Bol. Soc. Bras. Mat. **20.2** (1990), 95–125.

[PUZ] Przytycki F., Urbanski M., Zdunik A., *Harmonic, Gibbs and Hausdorff measures for holomorphic maps, Part 1*, Annals of Math. **130** (1989), 1–40; *Part 2*, Studia Math. **97.3** (1991), 189–225.

[PS] Przytycki F., Skrzypczak J., *Convergence and pre-images of limit points for coding trees for iterations of holomorphic maps*, Math. Annalen **290** (1991), 425–440.

[R] Rees M., *Positive measure sets of ergodic rational maps*, Ann. scient. Éc. Norm. Sup. **19** (1986), 383-407.

[Z1] Zdunik A., *Parabolic orbifolds and the dimension of the maximal measure for rational maps*, Inventiones Math. **99** (1990), 627–649.

[Z2] Zdunik A., *Harmonic measure versus Hausdorff measures on repelers for holomorphic maps*, Trans. AMS **326.2** (1991), 633–652.

INSTYTUT MATEMATYCZNY PAN
UL. ŚNIADECKICH 8, S.P. 137
00950 WARSZAWA
POLAND

OPEN QUESTIONS
IN NON-RATIONAL COMPLEX DYNAMICS

ROBERT DEVANEY

The dynamics of complex analytic functions have been studied by many authors during the past decade. Much of this work has been confined to the study of either rational or polynomial maps. The study of other analytic functions is still in its infancy and there are many unsolved problems in this area. In this note we describe a few of these problems.

1. Entire functions. The dynamics of entire functions are quite different from the dynamics of rational maps, mainly because of the essential singularity at infinity. By the Picard theorem, any neighborhood of this singularity is mapped infinitely often over the entire plane missing at most one point. This injects considerable hyperbolicity into the map and often causes the topology of the Julia set of the map to be vastly different from that of a rational map. In addition, the No Wandering Domains Theorem of Sullivan does not hold for this class of maps, so there may be both wandering domains and domains at infinity in the stable sets.

There is one class of entire maps whose dynamics are fairly well understood, namely the entire maps that have finitely many asymptotic and critical values (maps of finite type). With few exceptions (notably examples of Baker [B], Herman [H], and Eremenko and Lyubich [EL], most work has centered around this class of maps. Extending the study to a wider class of maps is an important problem.

PROBLEM: *Find a collection of representative examples of entire maps whose dynamics may be understood.*

As a starting point, one might ask

PROBLEM: *What are the dynamics of maps of the form $\lambda e^z \sin z$ or $\lambda e^z \cos z$?*

2. Entire functions of finite type. Most of the work thus far on the dynamics of entire maps has been concentrated on the class of finite type maps. These are the maps which have only finitely many singular (i.e., critical and asymptotic) values. This class includes λe^z, $\lambda \sin z$, and $\lambda \cos z$. It is known [GK, EL] that the No Wandering Domains theorem holds for this class, and that the Julia sets of these maps often contain Cantor bouquets [DT].

3. The exponential map. Of all entire maps, the exponential family $E_\lambda(z) = \lambda e^z$ has received the most attention. This is natural since E_λ, like the well-studied quadratic family $Q_c(z) = z^2 + c$, has only one singular value, the asymptotic value at 0. Thus this family is a "natural" one parameter family.

The parameter space for E_λ has been studied in [DGK]. However, there remain significant gaps in this picture. It is known that there exists Cantor sets of curves (called hairs) in the parameter plane for which the corresponding exponential maps have Julia sets that are the whole plane.

PROBLEM: *Describe completely the set of λ-values for which the Julia set of E_λ is $\mathbb{C}$.*

PROBLEM: *Many of these λ-values lie on curves or hairs. Are these hairs C^∞? Analytic? Where and how do they terminate?*

There are some interesting topological structures embedded in the dynamics of the exponential that warrant further study. For example, it is known that for $\lambda > 1/e$, $J(E_\lambda) = \mathbb{C}$. However, if $\lambda, \mu > 1/e$, then E_λ and E_μ are not topologically conjugate [DG]. If one looks at the invariant set consisting of $\{z \| 0 \leqslant \operatorname{Im} E_\lambda^n(z) \leqslant \pi \text{ for all } n\}$, it is known that this set is a Knaster-like continuum.

PROBLEM: *Are each of these Knaster-like continua homeomorphic? (for any $\lambda, \mu > \frac{1}{e}$)*

4. The Trigonometric Functions. The parameter spaces for families such as

$$S_\lambda(z) = \lambda \sin z \quad \text{or} \quad C_\lambda(z) = \lambda \cos z$$

also deserve special attention. They also contain curves on which the Julia set is the entire plane. The fundamental difference here is that C_λ and S_λ have no finite asymptotic values (only critical values), whereas the opposite is true for E_λ.

PROBLEM: *Describe the structure of the parameter space for C_λ and S_λ.*

One fundamental difference between the trigonometric and exponential families is the following. Both maps are known to possess Cantor bouquets [DT] in their Julia sets. And any two planar Cantor bouquets are homeomorphic [AO]. Finally, McMullen [Mc] has shown that these Cantor bouquets always have Hausdorff dimension 2. However, the Lebesgue measure of these bouquets is quite different: they always have measure zero in the exponential case, but infinite measure in the trigonometric case.

PROBLEM: *What is the measure and dimension of the hairs i the parameter space for E_λ, S_λ and C_λ.*

5. Other families of non-rational maps. Newton's method applied to non-rational maps offers a fertile area for further investigation. Outside of the work of Haruta [Ha] and van Haesler and Kriete [HK], there is little that is known. So a general problem is:

PROBLEM: *Describe the dynamics of Newton's method applied to general classes of entire functions?*

This, of course, immediately leads to the question of iteration of meromorphic functions. Some work has been done here in case the map has polynomial Schwarzian derivative [DK] or when the map has finitely many singular values [BKY]. But not much else is known.

Finally, there is an intriguing object called the tricorn introduced by Milnor [M] as one of his basic slices of parameter space for higher dimensional maps. This object arises as the analogue of the Mandelbrot set for the anti-holomorphic family $A_c(z) = \bar{z}^2 + c$. It is known [La] that the tricorn is not locally connected, but it also contains smooth arcs in the boundary (with no decorations attached) [W]. As this object arises in slices of the cubic connected locus, it certainly warrants further study. Winters [W] also has introduced a family of fourth degree polynomials whose parameter space is "naturally" $\mathbb{R}^3$ and which contains perpendicular slices given by the Mandelbrot set and the tricorn. Winters suggests that this family can model cubics since there are only two critical orbits.

REFERENCES

[AO]   Aarts J., Oversteegen L., *A Characterization of smooth Cantor bouquets*, Preprint.

[B]   Baker I. N., *Wandering domains in the iteration of entire functions*, Proc. London. Math. Soc. **49** (1984), 563–576.

[BKY]   Baker I. N., Kotus J., Lú Yinian, *Iterates of meromorphic functions*, I, II, and III. Preprints.

[DK]   Devaney R. L., Keen L., *Dynamics of meromorphic maps: maps with polynomial Schwarzian derivative*, Annales Scientifiques de l'Ecole Normale Supérieure. **22** (1989), 55–79.

[DG]   Douady A., Goldberg L., *The nonconjugacy of certain exponential functions*, Holomorphic functions and moduli I., MSRI Publ., Springer Verlag, 1988, pp. 1–8.

[DGK]   Devaney R. L., Goldberg L., Hubbard J., *A dynamical approximation to the exponential map by polynomials*, Preprint.

[DT]   Devaney R. L., Tangerman F., *Dynamics of entire functions near the essential singularity*, Ergodic Thy. Dynamical Syst **6** (1986), 489–503.

[EL]   Eremenko A., Lyubich M. Yu., *Iterates of entire functions*, Dokl. Akad. Nauk SSSR **279** (1984), 25–27 (Russian); English translation in Soviet Math. Dokl. **30** (1984), 592–594.

[EL1]   Eremenko A. and Lyubich M. Yu., *Structural stability in some families of entire functions*, Funk. Anal. i Prilozh. **19** (1985), 86–87. (Russian)

[GK]   Goldberg L. R., Keen L., *A finiteness theorem for a dynamical class of entire functions*, Ergodic Theory and Dynamical Systems **6** (1986), 183–192.

[H]   Herman M., *Exemples de fractions rationelles ayant une orbite dense sur la sphère de Riemann*, Bull. Soc. Math. France **112** (1984), 93–142.

[Ha]   Haruta M., *The dynamics of Newton's method on the exponential in the complex plane*, Dissertation, Boston University, 1992.

[HK]   von Haesler F., Kriete H., *The relaxed Newton's method for rational functions*, Preprint.

[La]   LaVaurs P., *Le lieu de connexité des polynômes du troisième degré n'est pas localement connexe*, Preprint.

[M]   Milnor J., *Remarks on iterated cubic maps*, Preprint.

[Mc]   McMullen C., *Area and Hausdorff dimension of Julia sets of entire functions*, Trans. A.M.S. **300** (1987), 329–342.

[W]   Winters R., *Dissertation*, Boston University, 1990.

DEPARTMENT OF MATHEMATICS
BOSTON UNIVERSITY
111 CUMMINGTON ST.
BOSTON MA 02215
USA

WANDERING DOMAINS FOR HOLOMORPHIC MAPS

A. Eremenko,   M. Lyubich

Let f be a rational or entire function. A connected component D of the complement of the Julia set $J(f)$ is called wandering domain if for all $m > n \geq 0$ we have $f^m D \cap f^n D = \phi$, where f^m stands for m-th iterate of f. One of the most important theorems in holomorphic dynamics due to D. Sullivan states that rational functions have no wandering domains [11]. We ask for possible generalizations of this theorem. All known proofs of Sullivan's theorem use heavily the fact that the space of quasiconformal deformations of a rational function is finitely dimensional (see e.g. [3]).

Here is one situation where a similar result could be proved. We say that an entire function f belongs to the class S if there is a finite set of points $\{a_1, \ldots, a_q\}$ such that

$$f : \mathbb{C} \setminus f^{-1}\{a_1, \ldots, a_q\} \to \mathbb{C} \setminus \{a_1, \ldots, a_q\}$$

is a covering map. The space of quasiconformal deformations of an entire function of the class S has finite dimension and the following result can be proved by extending the Sullivan's method: entire functions of the class S have no wandering domains [6], [8], [2].

On the other hand it is known that wandering domains D may exist for some entire functions f. The examples with the following properties have been constructed:

(1) $f^n D \to \infty$, [1], [2], [5], [9]. The example in [5] has an additional property that the iterates f^n are univalent in D.

(2) The orbit $\{f^n D\}$ has infinitely many limit points, including ∞, [5].

QUESTION 1. *Does there exist an entire function f with a wandering domain D such that the orbit $\{f^n D\}$ is bounded?*

Remark that there are entire functions not in the class S, for which the negative answer can be obtained easily. We say that the function f has order less then one half if

$$\log \log^+ |f(z)| \leq \alpha \log |z|, \quad |z| > r_0$$

for some $\alpha < 1/2$. It follows from a classical theorem by Wiman and Valiron (see, for example, [10]) that such functions have the following property: there exists a sequence $r_k \to \infty$ such that

$$\left| f(r_k e^{i\theta}) \right| > r_k, \quad 0 \leq \theta \leq 2\pi.$$

It follows that there is an increasing sequence of domains G_k, $\cup G_k = \mathbb{C}$ such that the restrictions of f on G_k are polynomial-like maps [4]. So f has no wandering domains with bounded orbit because polynomial-like maps have no wandering domains.

Now we consider a special type of wandering domains whose orbits tend to a finite point z_0. Let φ be a germ of holomorphic function with the point z_0 fixed. Suppose that $\lambda = \varphi'(z_0) = \exp 2\pi i \alpha$, α irrational. It was proved by Fatou [7] that in this situation $\varphi^n(z)$ cannot tend to z_0 in an *invariant* domain. So we have the following

PROBLEM 19.10

QUESTION 2. *Is it possible that $\varphi^n(z) \to z_0$ uniformly in some domain D?*

In the case when φ can be analytically continued to an entire function positive answer would imply the existence of wandering domain whose orbit tends to z_0. It would be also interesting to know the answer to the Question 2 with other additional assumptions on the germ φ, for example, when φ is a germ of an algebraic function.

Finally remark that the answer to the following question is also unknown

QUESTION 3. *Under the assumptions of Question 2 can it happen that there is an orbit tending to z_0?*

REFERENCES

1. Baker I. N., *Multiply connected domains of normality in iteration theory*, Math. Z. **104** (1968), 252–256.
2. Baker I. N., *Wandering domains in the iteration of entire functions*, Proc. London Math. Soc. **49** (1984), 563–576.
3. Carleson L., *Complex Dynamics*, UCLA Course Notes, 1990 Winter.
4. Douady A., Hubbard J. H., *On the dynamics of polynom-al-like mappings*, Ann. Sci. ENS **18** (1985), 287–343.
5. Eremenko A., Lyubich M., *Examples of entire functions with pathological dynamics*, J. London Math. Soc. **36** (1987), 458–468.
6. Eremenko A., Lyubich M., *Dynamical properties of some classes of entire functions*, Preprint, SUNY Inst Math. Sci., 1990/4.
7. Fatou P., *Sur les équations fonctionnelles*, Bull. Soc. Math. France **48** (1920), 33–94, 208–314.
8. Goldberg L., Keen L., *A finiteness theorem for a dynamical class of entire functions*, Erg. Theory and Dynam. Syst. **6** (1986), 183–192.
9. Herman M., *Exemples de fractions rationnelles ayant une orbite dense sur la sphère de Riemann*, Bull. Soc. Math. France **112** (1984), 93–142.
10. Levin B. Ja., *Distribution of zeros of entire functions*, AMS Translations Math. Monographs, 1964, v. 5.
11. Sullivan D., *Quasi conformal homeomorphisms and dynamics I. Solution of Fatou–Julia problem on wandering domains*, Annals Math. **122** (1985), 401–418.

DEPARTMENT OF MATHEMATICS
UNIVERSITY OF KENTUCKY
949 PATTERSON OFFICE TOWER
LEXINGTON, KY 40506
USA

INST. FOR MATH. SCI.
STATE UNIV. AT STONY BROOK
STONY BROOK, NY 11794-3660
USA

BAD POLYNOMIALS FOR NEWTON'S METHOD

Scott Sutherland

Newton's method for solving $f(z) = 0$ corresponds to iteration of $z \mapsto z - f(z)/f'(z)$, which is a degree d rational map of $\overline{\mathbb{C}}$ in the case where f is a polynomial of degree d with distinct roots. Newton's method has long been an important source of examples and theorems in complex dynamical systems (for example, the work of Schröder [Sch, Sch1], Fatou [Fa], and more recently Douady and Hubbard [DH]), as well as being one of the most commonly used numerical schemes for approximating roots. See [HP] and [Sm] for an introduction to the dynamics of Newton's method.

Describing the set of polynomials for which the corresponding Newton's method has periodic sinks which are not roots is an important open problem, (Problem 6 of [Sm]). We shall refer to such polynomials as "bad polynomials". This question is essentially answered for cubic polynomials by the work of Tan Lei [Ta] and Janet Head [He], in which the more comprehensive task of giving a combinatorial description of the parameter space for Newton's method is undertaken. A complete description of the parameter space for higher degrees still seems some way off, however.

In order to answer Smale's question for higher degree polynomials, it may be helpful to consider the relationship between the "relaxed Newton's method"

$$N_{h,f}(z) = z - h\frac{f(z)}{f'(z)}$$

and the "Newton flow" $\mathcal{N}_f$ given by the ordinary differential equation

$$\dot{z} = -\frac{f(z)}{f'(z)}.$$

One sees immediately that the map is an Euler approximation to the flow using step size h. The attractors of $\mathcal{N}_f$ are sinks located at the zeros of $f(z)$, ∞ is the only source, and the other fixed points are at the singularities corresponding to the critical points of f. We can rescale time for $\mathcal{N}_f$ to obtain $\dot{z} = -f(z)\overline{f'(z)}$ (or alternatively $\dot{z} = -\nabla\|f(z)\|^2$), from which we can easily see that these singularities are hyperbolic saddles. Furthermore, solution curves of $\mathcal{N}_f$ are mapped by f to straight lines emanating from the origin. Thus, if f has two critical values with the same argument, then the flow $\mathcal{N}_f$ is degenerate in the sense that there are solution curves which begin at one singularity and terminate at another. Refer to [JJT], [Sa], [STW], [Sm], and [Su] for more details about $\mathcal{N}_f$.

We propose the following conjecture (for which we have some numerical evidence) relating the degenerate flows and bad polynomials. This basically says that one can connect a polynomial which is bad for Newton's method to one which is bad for the flow.

CONJECTURE 1. Let f_1 be a bad polynomial of degree d, that is one for which Newton's method has an attractor which is not a root of f. Then there is a one-parameter family of polynomials $\{f_h\}_{0 < h \leqslant 1}$ which are bad for the relaxed Newton's method N_{h,f_h}. Furthermore, as $h \to 0$, the corresponding flow $\mathcal{N}_{f_h}$ tends to a flow $\mathcal{N}_{f_0}$ which is degenerate.

This conjecture is consistent with the following, as explained below.

CONJECTURE 2. Let f be a polynomial of degree d with all its roots in the unit disk, let α be a root of multiplicity m for f, and let $A_h^*(\alpha)$ be the immediate attractive basin of α for the map $N_{h,f}$. Then the intersection of the set

$$\mathcal{A} = \bigcap_{0 < h \leqslant m} A_h^*(\alpha)$$

with any circle of radius $R \geqslant 3$ contains arcs whose total length is at least $\dfrac{2\pi R}{cd}$, where c is a constant not depending on α, f or d.

This second conjecture says that there is a definite neighborhood of the singular trajectories of $\mathcal{N}_f$ in which the Julia set of $N_{h,f}$ must be contained for all $h \in (0, m]$. Since the periodic orbits for $N_{h,f}$ which are not roots must be contained in the complement of $\bigcup_{f(\alpha)=0} A_h^*(\alpha)$, Conjectures 1 and 2 taken together give some idea of the structure of the parameter space for $N_{h,f}$.

Conjecture 2 has been partially established by Benzinger [Be] (for all h sufficiently near 0), and is a generalization of the main result of [Su], which shows this for $h = 1$. I believe that with slight modifications, the proof in [Su] can be made to work for $0 < h \leqslant m$, which should nearly complete the proof of Conjecture 2.

REFERENCES

[Be]   Benzinger H., *Julia sets and differential equations*, Proc Amer. Math. Soc. (to appear).
[DH]   Douady A., Hubbard J. H., *On the dynamics of polynomial-like mappings*, Ann. Sci. Ecole Norm. Sup., 4e série **t.18** (1985), 287–343.
[He]   Head J., *The combinatorics of Newton's method for cubic polynomials*, Thesis, Cornell University, 1989.
[Fa]   Fatou P., *Sur les équations fonctionnelles*, Bull. Soc. Math. France **47** (1919), 161–271; Bull. Soc. Math. France **48** (1920), 33–94, 208–314.
[HP]   Haessler F. V., Peitgen H.-O., *Newton's method and complex dynamical systems*, Acta Appl. Math. **13** (1988), 3–58.
[JJT]   Jongen H. T. H., Jonker P., Twilt F., *The continuous desingularized Newton's method for meromorphic functions*, Acta Appl. Math. **13** (1988), 81–121.
[Sa]   Saupe D., *Discrete versus continuous Newton's method a case study*, Acta Appl. Math. **13** (1988), 59–80.
[Sch]   Schröder E., *Über unendlich viele Algorithmen zur Auflösung der Gleichungen*, Math. Annalen **2** (1870), 317–365.
[Sch1] Schröder E., *Über iterierte Funktionen*, Math. Annalen **3** (1871), 296–321.
[STW]   Shub M., Tischler D., Williams R., *The Newtonian graph of a complex polynomial*, SIAM J. Math. Anal. **19** (1988), 246–256.
[Sm]   Smale S., *On the efficiency of algorithms of analysis*, Bull. Amer. Math. Soc. **13** (1985), 87–121.

[Su]   Sutherland S., *Finding roots of complex polynomials with Newton's method*, Thesis, Boston University, 1989.

[Ta]   Tan Lei, *Cubic Newton's method of Thurston's type*, Ecole Norm. Sup. Lyon, 1991, Preprint.

INSTITUTE FOR MATH. SCIENCES
SUNY AT STONY BROOK
STONY BROOK, NY 11794
USA

Chapter 20

MISCELLANEOUS

PROBLEMS

CONVEXITY OF THE JOINT NUMERICAL RANGE
OF A k-TUPLE OF OPERATORS

S. L. SHISHKIN,   V. A. YAKUBOVICH

Let F be a family of bounded self-adjoint linear operators on real or complex Hilbert space H and $G = [G_1, G_2, \ldots, G_k]$ be a k-tuple of operators from F, $k \geqslant 2$ (we'll write $G \in F^k$). Define

$$\mathcal{G}_j(x) = (G_j x, x), \qquad j = 1, \ldots, k, \quad x \in H.$$

Recall that joint numerical range $W(G)$ is

$$W(G) = \{\, y = (y_1, \ldots, y_k) \in \mathbb{R}^k \mid \exists x \in H : \quad y_1 = \mathcal{G}_1(x), \ldots, y_k = \mathcal{G}_k(x) \,\}.$$

1. Convexity of the numerical range and S-procedure. The well-known Hausdorff theorem (for complex H) and Dines' theorem (for real H) say that the range $W(G)$ is convex for $k = 2$ and for all $G = [G_1, G_2]$. If $k \geqslant 3$ then the range $W(G)$ can be nonconvex.

PROBLEM 1. *Specify general conditions on the family F that guarantee the convexity of the joint numerical range $W(G)$ (or of its closure $W(G)^-$) for all $G \in F^k$.*

Examples of such families F (for arbitrary k) are given below.

Importance of this problem is based on its relation with so-called "S-procedure" in control theory [1–4]. Let (for simplicity) $G \in F^k$ satisfy the condition

$$(1) \qquad P = \{\, x \in H : \mathcal{G}_j(x) > 0, \quad j = 2, \ldots, k \,\} \neq \varnothing.$$

By S-procedure one means replacement of the condition

$$(2) \qquad \mathcal{G}_1(x) \geqslant 0 \quad \text{if} \quad \mathcal{G}_2(x) \geqslant 0, \ldots, \mathcal{G}_k(x) \geqslant 0, \qquad x \in H$$

including constraints by the condition without constraints

$$(3) \qquad \exists \tau_j \geqslant 0 : S(\tau, z) \overset{\text{def}}{=} \mathcal{G}_1(x) - \sum_{j=1}^{k-1} \tau_j \mathcal{G}_{j+1}(x) \geqslant 0, \quad x \in H.$$

Obviously (3) implies (2). S-procedure is called lossless if (3) is equivalent to (2), in the contrary case it is called lossy. It is simple to verify [2,5] that the convexity of $W(G)$ (or $W(G)^-$) for all $G \in F^k$ implies losslessness of S-procedure for $G_j \in F$. In control theory the forms $\mathcal{G}_j$ depend on certain "constructive" parameters. S-procedure is applied as an intermediate step to obtain stability conditions and other system's qualities. Lossness of S-procedure leads to the "loss" of a certain region in the space of constructive parameters.

Define the set

$$V(G) = \{\, y = (y_1, \ldots, y_k) \mid \exists x \in H : \quad y_1 \geqslant \mathcal{G}_1(x), y_2 \leqslant \mathcal{G}_2(x), \ldots, y_k \leqslant \mathcal{G}_k(x) \,\}.$$

S-procedure is lossless if $V(G)$ (or $V(G)^-$) is convex. Therefore the following problems arise.

PROBLEM 2. *Specify general conditions on the family F that guarantee the convexity of $V(G)$ (or $V(G)^-$) for all $G \in F^k$.*

PROBLEM 3. *Specify general conditions on the family F that guarantee the losslessness of S-procedure for all $G \in F^k$.*

Analogous problems are of practical interest too if $\mathcal{G}_j$ are functions of some other classes instead of being quadratic forms, for example they can be quadratic functionals: $\mathcal{G}_j(x) = (G_j x, x) + (g_j, x) + \gamma_j$. S-procedure is lossless if F is the set of quadratic functionals on the subspace $\mathcal{M}$ of $L_2\{[0, \infty) \to \mathbb{R}^n\}$ such that the corresponding quadratic forms and $\mathcal{M}$ are invariant with respect to shifts, see 6. The well-known A. A. Lyapunov theorem gives an example of a class of nonquadratic functionals for which the joint numerical range is convex and consequently S-procedure is lossless. Some criteria for convexity of joint numerical range are given in [7–11].

S-procedure is closely related to the following extremal problem:

$$(4) \qquad \mathcal{G}_1(x) \to \inf \quad \text{under conditions} \quad \mathcal{G}_2(x) \geqslant 0, \ldots, \mathcal{G}_k(x) \geqslant 0, \quad x \in H.$$

The duality relation

$$\mathrm{Inf}[\mathcal{G}_1(x) \mid x \in H, \, \mathcal{G}_2(x) \geqslant 0, \ldots \mathcal{G}_k(x) \geqslant 0] = \mathop{\mathrm{Sup}}_{\tau_j \geqslant 0} \mathrm{Inf}|S(\tau, x) \mid x \in H]$$

takes place if and only if S-procedure is lossless for k-tuple $\mathcal{G}_1 + \gamma, \mathcal{G}_2, \ldots, \mathcal{G}_k$ with any $\gamma \in \mathbb{R}^1$ (see [5]).

2. Examples of operator classes for which the joint numerical range is convex.

(i) F is a linear operator space, $k = 2$ for real H and $k \leqslant 3$ for complex H; see [3,5].

In the following examples k is any positive number more than 2.

(ii) F is the set of Toeplitz self-adjoint operators on $\mathbb{R}^n$ or $\mathbb{C}^n$; see [9].

(iii) F is the set of Hankel self-adjoint operators on $\mathbb{C}^n$; see [9].

(iv) F is a set of pairwise commuting normal operators [10].

(v) F is the space of operators of the convolution type on $L_2\{(0, \infty), \mathbb{R}^n\}$ or on the $L_2\{(0, \infty), \mathbb{C}^n\}$, so that

$$\mathcal{G}_j[x(\cdot)] = \int_{-\infty}^{+\infty} \tilde{x}(i\omega)^* \Phi_j(i\omega) \tilde{x}(i\omega) \, d\omega,$$

where $\tilde{x}(i\omega)$ is the Fourier transform of a function $x(\cdot) \in L^2$, $\Phi_j(i\omega) = \Phi_j(i\omega)^*$ are Hermitian matrices with L_∞-entries; see [11].

It can be shown that in case $H = \mathbb{R}^n$ ($H = \mathbb{C}^n$) and $\mathcal{G}_j$ are linearly independent the convexity of $W(G)$ is possible only if $k \leqslant n$ ($k \leqslant 2n$).

3. Other applications of S-procedure.

The relationship between duality relation and graph division problems is pointed out in [12]. In particular, Lovas' estimation of Shennon's graph capacity is exact if and only if the duality relation takes place [12].

In [13], necessary conditions of Pareto optimality are obtained using S-procedure. There exists an evident application of S-procedure to optimal approximation by ellipsoid of intersection of a finite number of ellipsoids and polyhedrons. S-procedure is the most effective tool for characterization of the cone of copositive forms (about actuality of this problem see [13]). Some examples of classes F for which S-procedure is lossless and some applications to the optimal control theory may be found in [6].

References

1. Aizerman M. A., Gantmaher F. R., *Absolute Stability of Control Systems*, Ac. Sci. USSR, Moscow, 1963. (Russian)
2. Yakubovich V. A., *The S-procedure in nonlinear control theory*, Vestnik Leningrad Univ., Math. **1** (1971), 62–77 (Russian); English transl. in Vestnik Leningrad Univ., Math. **4** (1977), 73–93.
3. Yakubovich V. A., *Minimization of quadratic functionals under the quadratic constraints and the necessity of a frequency condition in the quadratic criterion for absolute stability of nonlinear control systems*, Dokl. Akad. Nauk SSSR **209** (1973), 1039–1042 (Russian); English transl. in Soviet Math. Dokl. **14** (1973), 593–597.
4. Gelig A. Ch., Leonov G. A., Yakubovich V. A., *The Stability of Nonlinear Systems with a Nonunique Equilibrium State*, Nauka, Moscow, 1978. (Russian)
5. Fradkov A. L., Yakubovich V. A., *S-procedure and the duality relation in nonconvex problems of quadratic programming*, Vestnik Leningrad Univ., Math. **1** (1973), 81–87 (Russian); English transl. in Vestnik Leningrad Univ., Math. **6** (1978).
6. Yakubovich V. A., *One method for solving special global optimization problems*, Vestnik Leningrad Univ., Math., I. **2** (1992), no. 8, 58–62. (Russian)
7. Dines L. L., *On the mapping of n quadratic forms*, Bull. Am. Math. Soc. **48** (1942), no. 6, 467–471.
8. Fan Ming, *Two problems on an n-tuple of operators*, Uppsala University Department of Mathematics, February 1990, Report 1990:2.
9. Abramov Ju. Sh., *Variation Methods in Operator Pencils Theory*, Leningrad University, 1983. (Russian)
10. Dash A. T., *Joint numerical range*, Glasnik Matematički, ser. III **7(27)** (1972), 75–81.
11. Megretsky A., Treil S., *S-procedure and power distribution inequalities: a new method in optimization and robustness of uncertain systems*, Preprint of Mittag-Leffler Institute **1** (1990/91).
12. Shor N. Z., Stecenco S. I., *Quadratic Extremal Problems and Nondifferentiable Optimization*, Naukova Dumka, Kiev, 1989. (Russian)
13. Vershik A. M., *Quadratic forms positive on the cone and quadratic duality*, In: Automorphic functions and number theory II, Zap. Nauch. Semin. LOMI **134** (1984), 59–83. (Russian)

UL. TURKU 3, APT. 22
ST. PETERSBURG, 192212
RUSSIA

DEPT. OF MATHEMATICS AND MECHANICS
ST. PETERSBURG STATE UNIVERSITY
BIBLIOTECHNAYA PL. 2.
STARYI PETERHOF, 198904
RUSSIA

THE DUALITY THEOREM FOR AN OPTIMIZATION PROBLEM

S. G. Krein, G. A. Kukina

Let H be a real Hilbert space with a cone K and a monotone (with respect to the cone) scalar product $\langle \cdot, \cdot \rangle$. Let $U(t)$ be a positive semi-group of bounded operators acting in H. The following problems I, II are considered.

I. $\max_{x(\cdot)} \langle c, x(T) \rangle$,

$U(t)x(s) \leqslant x(s-t),\ x(s) \geqslant 0,\ x(0) = x_0,\ x_0, c \geqslant 0,\ s \in [0,T],\ t \in [0,s],$ (1)

where $x(\cdot)$ is a function (trajectory) with values in H;

II. $\min_{p(\cdot)} \langle x_0, p(0) \rangle$,

$U'(t)p(s) \geqslant p(s+t),\ p(s) \geqslant 0,\ p(T) = c,$ (2)

where $p(s) \in H,\ t \in [0, T-s],\ s,\ x_0,\ c$ are the same as in problem I, the prime denotes the adjoint operator.

Problem I is a generalization of the discrete problem examined in [1, p. 89].

LEMMA. *The inequality*

$$\langle c, x(T) \rangle \leqslant \langle p(s), x(s) \rangle \leqslant \langle x_0, p(0) \rangle$$

holds for arbitrary trajectories $x(\cdot)$, $p(\cdot)$ satisfying (1), (2) respectively.

Proof. The chain of inequalities

$$\langle c, x(T) \rangle = \langle p(T), x(T) \rangle \leqslant \langle U'(T-s)p(s), x(T) \rangle \langle p(s), U(T-s)x(T) \rangle \leqslant$$

$$\langle p(s), x(s) \rangle \leqslant \langle U'(s)p(0), x(s) \rangle = \langle p(0), U(s)x(s) \rangle \leqslant \langle p(0), x_0 \rangle$$

follows from the monotonicity of the scalar product and the relations (1), (2). Considering the underlined terms of the inequalities we get the lemma statement.

THEOREM. *Assume that the trajectory $x^*(\cdot)$ satisfies the relations (1), the trajectory $p^*(\cdot)$ satisfies the relations (2) and $\langle c, x^*(T) \rangle = \langle x_0, p^*(0) \rangle$. Then $x^*(\cdot)$, $p^*(\cdot)$ are the solutions of problems I, II.*

The Theorem follows easily from the Lemma.

PROBLEM. *Let $x^*(\cdot)$ be a solution of problem I. Under what condition does a trajectory $p^*(\cdot)$ exist such that the relations (2) are satisfied and $\langle c, x^*(T) \rangle = \langle x_0, p^*(0) \rangle$, i.e. $p^*(\cdot)$ is the solution of problem II?*

If the operators $U(t)$ have positive inverses for every t (for example $H = \mathbb{R}$), then obviously $x^*(s) = U^{-1}(s)x_0$, $p^*(s) = \left(U'(T-s)\right)^{-1}c$ and our problem is solved.

REFERENCE

1. Ashmanov S. A., *Introduction to mathematical economics*, Nauka, Moscow, 1984. (Russian)

VORONEZH FOREST TECHNICAL INSTITUTE
UL. TIMIRYAZEVA 8, VORONEZH, 394613
RUSSIA

UL. HOLZUNOVA 114, APT. 12
VORONEZH, 394088
RUSSIA

SIMPLE p-CYCLIC PONTRJAGIN SPACE
SELFADJOINT OPERATORS

T. Ya. Azizov

Denote by $\Pi_\varkappa$ a Pontrjagin space [1], [2], [3] and let A be a $\Pi_\varkappa$-selfadjoint operator. The operator A is said to be a simple p-cyclic operator if A is a p-cyclic operator and there is no $\Pi_\varkappa$-orthogonal decompositions $\Pi_\varkappa = \Pi_{\varkappa_1}[+]\Pi_{\varkappa_2}$ that $\Pi_{\varkappa_1}$ and $\Pi_{\varkappa_2}$ are A-invariant subspaces, the operators $A|\Pi_{\varkappa_i}$ are p_i-cyclic, $p_i > 0$ $(i = 1,2)$ and $p_1 + p_2 = p$.

PROBLEM. *Describe simple p-cyclic $\Pi_\varkappa$-selfadjoint operators.*

Remark. It is well known that every simple p-cyclic Hilbert space selfadjoint operator is cyclic. If $\varkappa > 0$ this statement is true only for $\varkappa = 1$. If $\varkappa = 2$ there exists a simple two-cyclic operator. Let H be the Hilbert space

$$H = \text{l.s.}\{e_1\} + \text{l.s.}\{e_2\} + L_2(-1,1) + \text{l.s.}\{e_4\} + \text{l.s.}\{e_5\}$$

with indefinite inner product $[x,y] = (Jx, y)$, where

$$J: \quad J(\alpha_1 e_1 + \alpha_2 e_2 + f(t) + \alpha_4 e_4 + \alpha_5 e_5) =$$
$$= \alpha_4 e_1 + \alpha_5 e_2 + (-f(t)) + \alpha_1 e_4 + \alpha_2 e_5$$

is the Pontrjagin space $\Pi_\varkappa$ with $\varkappa = 2$. Then the $\Pi_\varkappa$-selfadjoint operator A:

$$A(\alpha_1 e_1 + \alpha_2 e_2 + f(t) + \alpha_4 e_4 + \alpha_5 e_5) =$$
$$= \left(\int_{-1}^{1} f(t)\,dt + \alpha_5 \right) e_1 + \alpha_4 e_2 + (tf(t) - \alpha_4)$$

is simple two-cyclic. In fact, since $\text{Ker}\,A = \text{l.s.}\{e_1, e_2\}$ and $\dim \text{Ker}\,A > 1$, the operator A is not cyclic [3], but A is two-cyclic: $\Pi_\varkappa = \text{c.l.s.}\{A^s e_i \mid i = 4, 5\}_{s=0}^{\infty}$. The root lineal $L_0(A) = \text{l.s.}\{e_1, e_2, e_5\}$ corresponding to the eigenvalue $\lambda = 0$ of the operator A has no $\Pi_\varkappa$-orthogonal decompositions into sum of A-invariant subspaces. Hence A is a simple two-cyclic operator.

REFERENCES

1. Azizov T. Ya., Iokhvidov I. S., *Linear Operators in Spaces with Indefinite Metric*, John Wiley and Sons, Chichester–New-York–Brisbane–Toronto–Singapore, 1989.
2. Bognar J., *Indefinite Inner Product Spaces*, Ergebnisse der Mathematik und ihrer Grenzgebiete, vol. 78, Springer–Verlag, Berlin–Heidelberg–New-York, 1974.
3. Iokhvidov I. S., Krein M. G., *Spectral theory of operators in spaces with an indefinite metric*, II, Amer. Math. Soc. Translations 34, no. 2, 283–373.

A/JA 9, VORONEZH-68,
394068,
RUSSIA

NORMS AND EXTREMALS OF CONVOLUTION OPERATORS
ON SPACES OF ENTIRE FUNCTIONS

E. A. Gorin

Given a compact subset $K \subset \mathbb{R}^n$ let $B(K)$ be the Bernstein class of all bounded functions f on (the dual copy) $\mathbb{R}^n$ with Fourier transform $\hat{f}$ supported on K. In fact, every function $f \in B(K)$ can be extended to an entire function of exponential type on $\mathbb{C}^n$. The linear space $B(K)$ with the uniform norm on $\mathbb{R}^n$ is a Banach space (in fact, a dual Banach space).

EXAMPLE. Let K be the unit ball in $\mathbb{R}^n$, i.e.,

$$K = \left\{ \xi \in \mathbb{R}^n : \sum_1^n |\xi_k|^2 \leqslant 1 \right\}.$$

Then $f \in B(K)$ if and only if the function f is a restriction to $\mathbb{R}^n \subset \mathbb{C}^n$ of an entire function on $\mathbb{C}^n$ satisfying

$$|f(z)| \leqslant C \exp |\operatorname{Im} z|, \qquad z \in \mathbb{C}^n,$$

for some constant C.

We shall consider operators $T \colon B(K) \to B(K)$ of the form

$$(T_\mu f)(x) = \int_{\mathbb{R}^n} f(x - t) \, d\mu(t),$$

μ being a complex-valued regular Borel measure of bounded variation on $\mathbb{R}^n$. In other words, $T_\mu f = f * \mu$. The function $\tau = \hat{\mu}|K$ is said to be the *symbol* of $T = T_\mu$. The representation $\mu \to T_\mu$ is not an isomorphism, but nevertheless the symbol τ is uniquely determined by T. The spectrum of T coincides with the range of τ and its norm with the norm of the functional $f \to (Tf)(0)$. If K is a set of spectral synthesis then the symbol τ uniquely determines the corresponding operator $T = \tau(D)$, $D = \frac{1}{i}\partial/\partial x$ in $B(K)$. Moreover, in this case

$$\|T\| = \inf\{ \|\nu\| : \hat{\nu}|K = \tau \}.$$

DEFINITION. *A normal extremal for T is any element $f \in B(K)$ such that $\|f\| = 1$, $(Tf)(0) = \|T\|$.*

It can be easily shown that the normal extremals always exist (and form a convex set). For example, in the case $n = 1$, $K = [-\sigma, \sigma]$, $\sigma > 0$, and $Tf = f'$ the classical result asserts that $\|T\| = \sigma$ and all normal extremals have the following form $a \exp i\sigma x + b \exp(-i\sigma x)$.

A measure ν is called *extremal* for T if $Tf = \nu * f$, $f \in B(K)$ and $\|T\| = \|\nu\|$.
The set of extremal measures may be empty even in case of finite K.

Such problems as calculation of norms and description of extremals go back to the classical papers of S. N. Bernštein. A survey of results obtained in the field up to the middle of 60-ies can be found in [1]. For additional aspects of the topic see [2], which is, unfortunately, flooded by misprints, so be careful.

A compact set K in $\mathbb{R}^n$ is said to be a *star* if with every ξ it contains $\rho\xi$ for each $\rho \in [0,1]$. Every star K is a set of spectral synthesis and $B(K)$ contains sufficiently many functions vanishing at infinity. If $\xi_0 \in K$ and $\sup_{\xi \in K} |\tau(\xi)| = \tau(\xi_0) = 1$ then $\|\tau(D)\| = 1$ (i.e. the norm of $\tau(D)$ coincides with its spectral radius) if and only if the function $\xi \to \tau(\xi + \xi_0)$ admits a positive definite extension to $\mathbb{R}^n$. The operator $\tau(D)$ has extremal measures. If $\|\tau(D)\| > 1$ then every extremal measure is supported on a proper analytic subset of $\mathbb{R}^n$ and the extremal measure is unique provided $n = 1$. For $n > 1$ the uniqueness does not hold (example: K is the unit ball and $\tau(D)$ is the Laplace operator).

Problems in the one-dimensional case. Every polynomial (in one variable) is related to a wide stock of positive definite functions. Suppose that the zeros of a polynomial τ are placed in the half-plane $\operatorname{Re}\zeta \geqslant \frac{1}{2}\sigma > 0$ and that $\tau(0) = 1$. Then the restriction of τ to $[0,\sigma]$ extends to a positive definite function on $\mathbb{R}$. It follows that for all linear τ

$$\|\tau(D)\|_{B_\sigma} = \max\{\, |\tau(\xi)| : \xi \in [-\sigma,\sigma]\,\},$$

where $B_\sigma \overset{\text{def}}{=} B([-\sigma,\sigma])$. In this case all normal extremals can be easily determined and there exists an extremal measure (at least one). At the same time *for polynomials τ of degree 2 these problems still do not have a full solution*. The simplest operator is provided by $y \to -y'' + \lambda y$, $\lambda \in \mathbb{C}$. For $\lambda \in \mathbb{R}$ the problems are solved (see papers of Boas–Shaeffer, Ahiezer and Meiman). For some complex λ (in particular those for which the zeros of the symbol τ satisfy the above mentioned condition) τ admits (after a proper normalization) a positive definite extension, so that the norm of $\tau(D)$ coincides with its spectral radius.

Is it possible to calculate the norm for all $\lambda \in \mathbb{C}$? How do the "Euler equations" look in this case?

Note that according to the Krein theorem the extremal measure is unique provided $K = [-\sigma,\sigma]$ and the spectral radius is less than the norm. *Of course, these problems remain open for polynomials of higher degree.*

The Bernštein inequality for fractional derivatives leads to the following PROBLEM. Consider on $[-1,1]$ the function $\tau(\xi) = (1 - |\xi\,|)^\alpha$, $\alpha > 0$. *The problem is to find*

$$\inf\{\, \|\mu\|_{M(\mathbb{R})} : \mu \geqslant 0,\ \hat\mu\,|[-1,1] = \tau\,\}.$$

If $\alpha \geqslant 1$ then τ is even and is convex on $[0,1]$, and therefore coincides on $[-1,1]$ with a restriction of a positive definite function by the Polya theorem. For $\alpha < 1$ τ becomes concave on $[0,1]$ and moreover τ cannot be extended to a positive definite function on $\mathbb{R}$. Indeed, if φ is positive definite then $-\varphi''$ is a positive definite distribution. At the same time, $-\tau''$ is non-negative, locally integrable on a neighbourhood of zero and non-integrable on a left neighbourhood of the point $\xi = 1$. A positive definite function cannot satisfy this list of properties.

The best known estimate of the norm for $\alpha \in (0,1)$ is $2(1+\alpha)^{-1}$. It is evidently not exact but it is asymptotically exact for $\alpha \to 0$ and $\alpha \to 1$. It should be noticed that in the space of trigonometric polynomials of degree $\leqslant m$ the norm of the operator of fractional differentiation coincides with its spectral radius for $\alpha \geqslant \alpha_0$, where $\alpha_0 = \alpha_0(m) < 1$.

Another example is related to the family $\{\tau_\alpha\}$ of functions $\tau_\alpha(\xi) = 1 - |\xi|^\alpha$, $\alpha > 0$, $-1 \leqslant \xi \leqslant 1$. For $\alpha \leqslant 1$ they are positive definite. Consider the family for $\alpha > 1$. Since every characteristic function φ satisfies the inequality $|\varphi(t)|^2 \leqslant \frac{1}{2}(1+|\varphi(2t)|)$, there are no positive definite extensions for $\alpha \geqslant 2$.

Consider now the case $1 < \alpha < 2$. The following idea has been suggested by A. V. Romanov. Extend $\tau(\xi)$ to $(1,2)$ by the formula

$$\tau(1 + \xi) = -\tau(1 - \xi), \qquad 0 < \xi < 1.$$

Extend now the obtained function on $(0,2)$ to an even periodic function of period 4 keeping the same notation τ for this function. We have

$$\tau(\xi) = \sum_k a_k \cos \frac{\pi k}{2} \xi, \qquad \xi \in \mathbb{R},$$

where the sum is taken over odd positive integers. It is easily verified (integration by parts) that a_k and

$$b_k = \int_0^{\frac{k\pi}{2}} \frac{\cos t}{t^{2-\alpha}} \, dt$$

are of the same sign. Clearly $b_1 > 0$ and $b_k \geqslant b_3$ for $k \geqslant 3$. Hence τ is positive definite if

$$\rho(\alpha) = \int_0^{\frac{3\pi}{2}} \frac{\cos t}{t^{2-\alpha}} \, dt \geqslant 0$$

(cf.[4], Ch.V, Sec.2.29). The function $\rho(\alpha)$ decreases on $(1,2)$ and $\rho(1) = \infty$, $\rho(2) < 0$. Therefore the equation $\rho(\alpha) = 0$ has a unique solution $\alpha_0 \in (1,2)$ (Romanov's number). The function τ is positive definite on $[-1,1]$ if $\alpha \leqslant \alpha_0$. At the same time slightly modifying the arguments from [5], Th.4.5.2 one can easily show that for $0 < \alpha_1 < \alpha_2 < 2$ the function

$$\frac{1 - |\xi|^{\alpha_1}}{1 - |\xi|^{\alpha_2}}$$

is positive definite on $\mathbb{R}$. Hence τ_{α_1} is positive definite on $[-1,1]$ if so is τ_{α_2}. The "separation point" β_0 is clearly $\geqslant \alpha_0$. Is it true that $\beta_0 > \alpha_0$? For $\alpha > \beta_0$ the above problems remain open.

Problems in the multidimensional case. For $n \geqslant 2$, except the case when the norm of $\tau(D)$ coincides with its spectral radius, very few cases of exact calculation of $\|\tau(D)\|$ and description of extremals are known. The GENERAL PROBLEM here is to obtain proper generalization of Boas–Schaeffer's and Ahiezer–Meiman's theorems, i.e. to obtain the "Euler equations" at least for real functionals. Our problems concern concrete particular cases; however it seems that the solution of these problems may throw a light on the problem as a whole.

If $\psi \colon \mathbb{R}^n \to \mathbb{R}$ is a linear form then the operator with the symbol $\psi|K$ is Hermitian on $B(K)$ and hence its norm coincides with the spectral radius. This again will be the case for some operators with the symbol of the form $(p \circ \psi)|K$ where p is a polynomial. The following simple converse statement is true. If τ is a polynomial and

$$\|\tau(D)\|_{B(K)} = \max\{\,|\tau(\xi)| : \xi \in K\,\}$$

for every symmetric convex star K then $\tau = p \circ \psi$ where ψ is a linear form and p is a polynomial.

Does the similar converse statement hold when K ranges over the balls?

Let

$$K_p = \left\{\, \xi \in \mathbb{R}^n : \left(\sum |\xi_k|^p\right)^{1/p} \leqslant 1 \right\},$$

where $p \geqslant 1$ and $\tau(\xi) = -\sum_1^n \xi_k^2$. The operator $\tau(D)$ is obviously the Laplacian Δ. The norm of Δ coincides with the spectral radius in the following cases: $n = 1$; $n = 2, p = 1$; $n \geqslant 2, p = \infty$. The proof is based on the following well-known fact: if μ is a probability measure then $\{\, \xi : |\hat\mu(\xi)| = 1 \,\}$ is a subgroup. The case $p = 2$ turns out to be the most pathological and perhaps the most interesting. We have $\|\Delta\|_{B(K_2)} = n$. In this case extremal measure is not unique and *it would be interesting to describe all extremal measures* (notice that they form a compact convex set). The problem of calculation of the norm can be reduced to the one-dimensional case for operators of the form $p(\Delta)$ in $B(K_2)$. It is possible to calculate the norms by operators with linear symbol in the space $B(K_2)$ explicitly. For example, the norm of Cauchy–Riemann operator equals 2 and its normal extremal is unique. Namely,

$$(x_1 - ix_2)\frac{\sin(x_1^2 + x_2^2)^{1/2}}{(x_1^2 + x_2^2)^{1/2}}.$$

However, for the operators of the second order the things are more complicated. If the symbol $\tau(\xi) = (A\xi, \xi)$ is a positive real quadratic form then $\|\tau(D)\|_{B(K_2)} = \operatorname{tr} A$, the spectral radius of $\tau(D)$ coincides with that of A and normal extremals are of the form

$$\cos|x| + i(a, x)\frac{\sin|x|}{|x|},$$

where $|x|$ is the norm, $a \in \mathbb{R}^n$, $|a| \leqslant 1$.

At the same time *nothing is known about extremals and norm of the operator* $\dfrac{\partial^2}{\partial x_1 \partial x_2}$ in $B(K_2)$ for $n = 2$.

References

1. Akhiezer N. I., *Lectures on Approximation Theory*, "Nauka", Moscow, 1965. (Russian)
2. Gorin E. A., *Bernstein inequalities from the perspective of operator theory*, Vestnik Kharkov. Gos. Un-ta. Ser. Mekh.-Matem., vyp. 45 **205** (1980), 77–105. (Russian)

3. Gorin E. A., Norvidas S. T., *Extremals of some differential operators*, School on theory of operators on functional spaces, Abstracts of lectures, Minsk, July 4-11, 1982, pp. 48–49. (Russian)
4. Zygmund A., *Trigonometric Series*, vol. 1, Cambr. Univ. Press, 1959.
5. Lukacs E., *Characteristic Functions*, 2-nd ed., Griffin, London, 1970.

UL. OSTROVITYANOVA 20, 135
MOSCOW 117321
RUSSIA

COMMENTARY BY THE AUTHOR (ADDENDA TO THE PROBLEM)

1. Let ω be an entire function of exponential type $\sigma > 0$ and such that $|\omega(-z)| \leqslant |\omega(z)|$ for $\operatorname{Im} z > 0$. Consider the linear space E of all entire functions f of exponential type $\leqslant \sigma$ with

$$\|f\| := \sup_{\mathbb{R}} |f/\omega| < +\infty.$$

This is a Banach space (see, e.g., [6]). If $\omega' \in E$, then the differentiation maps E into itself, and its norm is attained on ω. What is its spectrum?

2. Suppose $1 < p < \infty$, $p \neq 2$. The set of all $f \in B_\sigma$ with the finite $L^p(\mathbb{R})$-norm is a Banach space. Are there some feasible criteria for the norm of the operator $f \mapsto \mu * f$ to coincide with its spectral radius? If such a criterion exists, then it "tends" to positive definiteness, as $p \to \infty$, and degenerates, as $p \to 2$.

3. Let $|\cdot|$ be the euclidean norm on $\mathbb{R}^n$. Put

$$k_{\alpha,\beta}(x) = \left(1 - |x|^\alpha\right)/\left(1 - |x|^\beta\right), \qquad \alpha, \beta > 0.$$

There exists a sequence of continuous functions $\tau_n \colon \mathbb{R}_+ \to \mathbb{R}_+$ such that for $k_{\alpha,\beta}$ the positive definiteness on $\mathbb{R}^n$ occurs iff $\alpha \leqslant \tau_n(\beta)$; $\tau_n(\beta) = \beta$ if $\beta \leqslant 2$; $\tau_n(\beta) = 2$, if $2 \leqslant \beta \leqslant 4$. For $\beta > 4$ the number $\tau_n(\beta)$ depends on n. In particular, $\lim_{\beta \to \infty} \tau_n(\beta) = 1$, 0, $1/2$ respectively for $n = 1, 2, 3$, and for $n \geqslant 4$ the function τ_n has a compact support. On the right ends of the support $\beta_n \downarrow \beta_*$, where $4 < \beta_* < 8$. What is this number β_*? An analogous sequence $\tau_n(\beta)$ exists also for the family $\left(1 - |x|^\alpha\right)/\left(1 - |x|^\beta\right)$, but this time we have $\tau_n(\beta) = 0$ if $\beta \leqslant 2n$. Hence, positive definiteness occurs for no choice of positive α's and β's if n is large enough. May be, this phenomenon starts with $n = 4$? The facts above can be proved by means of (not too cumbersome) calculations involving the Cauchy integral and Bessel functions.

4. The numerical range of the operator $y \mapsto -y''$ can be in the space B_σ (in principle) described. This results in a rather explicit formula for the norm of the operator $y \mapsto -y'' + \lambda y$ for all $\lambda \in \mathbb{C}$. But it seems that a long way separates us from a complete complex analogue of the Boas–Scheffer theorem.

5. Consider the Cauchy–Riemann operator $D = \frac{\partial}{\partial x_1} + i \frac{\partial}{\partial x_2}$ in $B(K_2)$, $n = 2$. For any k the operator D^k has a unique normal extremal and a unique extremal measure; they can be found by a reduction to the one-dimensional problem (A. V. Kim). The author noticed that $\|D\| = 2$, $\|D^2\| = 8/\pi$; $\|D^3\|$ can be expressed by complete elliptic integrals and approximately equals 2.97. The following estimate is almost trivial: $c_1 k^{1/3} \leqslant \|D^k\| \leqslant c_2 k^{1/2}$. What is the asymptotics of $\|D^k\|$ for $k \to \infty$?

6. Almost everything is clear with homogeneous operators of the 2-nd order in $B(K_2)$. In particular, the norm of $\partial^2/\partial x_1 \partial x_2$ is $2/\pi$. If $0 < a < 1$, then the norm of $\partial^2/\partial x_1^2 - a\partial^2/\partial x_2^2$ is

$$\max_{0 \leqslant t \leqslant \pi} \left\{ \frac{\sin t}{t} + a\frac{\sin t}{\pi - t} \right\};$$

it is attained on

$$f(x) := -\frac{x_1^2}{x_1^2 + x_2^2} \cos(p^2 + x_1^2 + x_2^2)^{1/2} + \frac{x_2^2}{x_1^2 + x_2^2} \cos(q^2 + x_1^2 + x_2^2)^{1/2},$$

p being the maximum point of the function above, $q = \pi - p$. It is interesting to note that the dependence of the norm on the parameter loses its analyticity by the passage from the elliptic zone to the hyperbolic one. Along with the above family there is one more functional family of real normal extremals. They coincide (as do their gradients) on coordinate axes and have the same Hessean at the origin. Nevertheless, the extremal measure is unique. Thus, the elliptic case is in a sharp contrast with the hyperbolic case.

The real situation (in any dimension) is practically clear. The results can be given an invariant form using operators $\operatorname{tr}(AH_f)$ where H_f is the Hessean. In the complex situation the linear combination of pure second derivatives is completely settled (the main difficulty being caused by the linear algebra), along with some other cases (elliptic real component, a square of a first order operator). But a more general investigation could be useful from the point of view of harmonic analysis.

7. Let f be a continuous complex function on $\mathbb{R}^n$. Suppose that for any compact $Q \subset \mathbb{R}^n$ and for any $x_0 \in Q$ satisfying $|f(x)| \leqslant |f(x_0)| \neq 0$ ($x \in Q$), the function $f(x + x_0)/f(x_0)$ can be extended from $-x_0 + Q$ to a positively definite function on $\mathbb{R}^n$. What can be said about f? If f is bounded, then $f = c\chi$, χ being a character. Moreover, no typically multidimensional polynomial of this kind can exist. This is why the one-dimensional case is especially interesting (there is a lot of examples; see [7] in connection with this problem).

The most part of the results quoted above was published in proceedings of conferences held in various provinces of the late USSR (some more accessible publications are planned).

REFERENCES

6. Brudny Yu. A., Gorin E. A., *Isometric representations and differential inequalities*, Izd. Yarosl. Univ., Yaroslavl', 1981, pp. 1–92. (Russian)
7. Norvidas S. T., *On stability of differential operators in spaces of entire functions*, Dokl. AN SSSR **291** (1986), no. 3, 548–552. (Russian)
8. Norvidas S. T., *On differential inequalities in Banach spaces of entire functions*, Litovskii Mat. Zh. **30**, no. 2, 345–358. (Russian)

HOLOMORPHIC MAPPINGS OF SOME SPACES
CONNECTED WITH ALGEBRAIC FUNCTIONS

V. Ya. Lin

1. For any integer n, $n \geqslant 3$ and any $z = (z_1, \ldots, z_r) \in \mathbb{C}^n$ consider the polynomial $p(\lambda) = \lambda^n + z_1 \lambda^{n-1} + \cdots + z_n$, and let $d_n(z)$ be the discriminant of p. Then d_n is a polynomial in $z_1, \ldots, z_n$ and the sets

$$G_n = \{ z : d_n(z) \neq 0 \},$$
$$G_n^0 = G_n \bigcap \{ z : z_1 = 0 \},$$
$$SG_n = \{ z : z_1 = 0, d_n(z) = 1 \}$$

are non-singular irreducible affine algebraic manifolds, G_n being isomorphic to $G_n^0 \times \mathbb{C}$. The restriction $d_n^0 = d_n | G_n^0 : G_n^0 \to \mathbb{C}^* = \mathbb{C} \setminus \{0\}$ is a locally trivial holomorphic fibering with the fiber SG_n. These three manifolds play an important role in the theories of algebraic functions and of algebraic equations over function algebras. Each of the manifolds is $K(\pi_1, 1)$ for its fundamental group π_1, $\pi_1(G_n)$ and $\pi_1(G_n^0)$ being both isomorphic to the Artin braid group $B(n)$ with n strings and $\pi_1(SG_n)$ being isomorphic to the commutator subgroup of $B(n)$, denoted $B'(n)$ [1], [2]. $\mathbb{Z}$- and $\mathbb{Z}_p$-cohomologies of G_n are known [1], [3], [4]. However, our knowledge of analytic properties of G_n, G_n^0, SG_n essential for some problems of the theory of algebraic functions is less than satisfactory ([5]–[10]). We propose several conjectures concerning holomorphic mappings of G^0 and SG_n. Some of them have arisen (and all have been discussed) on the Seminar of E. A. Gorin and the author on Banach Algebras and Analytic Functions at the Moscow State University.

2. A group homomorphism $H_1 \to H_2$ is called *abelian* (resp. *integer*) if its image is an abelian subgroup of H_2 (resp. a subgroup isomorphic to $\mathbb{Z}$ or $\{0\}$). For complex spaces X and Y, $C(X, Y)$, $\mathrm{Hol}(X, Y)$ and $\mathrm{Hol}^*(X, Y)$ stand for the sets of, respectively, continuous, holomorphic and *non-constant* holomorphic mappings from X to Y. A mapping $f \in C(G_k^0, G_n^0)$ is said to be *splittable* if there is $h \in C(\mathbb{C}^*, G_n^0)$ such that f is homotopic to $h \circ d_k^0$, $d_k^0 : G_k^0 \to \mathbb{C}^*$ being the standard mapping defined above; f is splittable if and only if the induced homomorphism $f_* : B(k) \approx \pi_1(G_k^0) \to \pi_1(G_n^0) \approx B(n)$ is integer. There exists a simple explicit description of splittable elements of $\mathrm{Hol}(G_k^0, G_n^0)$ [6].

CONJECTURE 1. *Let $k > 4$ and $n \neq k$. Then*

(a) *every $f \in \mathrm{Hol}(G_k^0, G_n^0)$ is splittable;*
(b) $\mathrm{Hol}^*(SG_k, SG_n) = \emptyset$.

It is easy to see that (b) implies (a).

Let $\Pi(k)$ be the union of four increasing arithmetic progressions with the same difference $k(k-1)$ and whose first members are k, $(k-1)^2$, $k(k-1)$, $k(k-1)+1$. According to [6], if $k > 4$ and $n \notin \Pi(k)$ then all f in $\mathrm{Hol}(G_k^0 \, G_n^0)$ are splittable. A complete description of all non-splittable f in $\mathrm{Hol}(G_k^0, G_k^0)$ has been also given in [6]. If $k > 4$ and $n < k$, there are only trivial homomorphisms from $\mathrm{B}'(k)$ to $\mathrm{B}(n)$ [11]. Thus for such n and k all elements of $C(G_k^0, G_n^0)$ are splittable and all elements of $C(SG_k, SG_n)$ are contractible. The last assertion implies rather easily that $\mathrm{Hol}^*(SG_k, SG_n) = \emptyset$. It is proved in [10] that for $k \neq 4$ each $f \in \mathrm{Hol}^*(SG_k, SG_k)$ is biholomorphic and has the form $f(z_2, \ldots, z_k) = (\varepsilon^2 z_2, \varepsilon^3 z_3, \ldots, \varepsilon^k z_k)$ with $\varepsilon^{k(k-1)} = 1$.

3. Let $\mathbb{C}^{**} = \mathbb{C} \setminus \{0, 1\}$. A useful technical device in the topic we are discussing is provided by explicit descriptions of all functions $f \in \mathrm{Hol}^*(X, \mathbb{C}^{**})$ for some algebraic manifolds X associated with G_n, G_n^0 and SG_n ([6], [8], [9], [10]). This has led to the questions and results discussed in this section.

Let $\mathcal{A}$ be the class of all connected non-singular affine algebraic manifolds. For every $X \in \mathcal{A}$ the cardinality $q(X)$ of $\mathrm{Hol}^*(X, \mathbb{C}^{**})$ is *finite* (E. A. Gorin). Besides, if $m > \max\{r(X), 1\}$ ($r(X)$ is the rank of the cohomology group $H^1(X, \mathbb{Z})$) then $\mathrm{Hol}^*(X, \mathbb{C} \setminus \{\xi_1, \ldots, \xi_n\}) = \emptyset$ for any *distinct* points $\xi_1, \ldots, \xi_n \in \mathbb{C}$. Using these two assertions, it is not difficult to prove that, given $X \in \mathcal{A}$ and $n \geqslant 3$, the set $\mathrm{Hol}^*(X, SG_n)$ is finite. In particular, for every $k, n \geqslant 3$ the set $\mathrm{Hol}^*(SG_k, SG_n)$ is finite. Let $\mathrm{Top}(X)$ be the class of all Y in $\mathcal{A}$ homeomorphic to X; it is plausible that for any $x \in \mathcal{A}$ the function $q \colon \mathrm{Top}(X) \to \mathbb{Z}_+$ is bounded. I even do not know any example disproving the following stronger

CONJECTURE: *there exists a function* $\nu \colon \mathbb{Z}_+ \to \mathbb{Z}_+$ *such that* $q(X) \leqslant \nu(r(X))$ *for all* $X \in \mathcal{A}$.

A function $\nu_1 \colon \mathbb{Z}_+ \to \mathbb{Z}_+$ with $q(\Gamma) \leqslant \nu_1(r(\Gamma))$ for all *curves* $\Gamma \in \mathcal{A}$ does exist. It has been proved in [14] that there exists a function $\nu_2 \colon \mathbb{Z}_+ \times \mathbb{Z}_+ \to \mathbb{Z}_+$ such that $q(X) \leqslant \nu_2(r_1(X), r_2(X))$ for all manifolds $X \in \mathcal{A}$ of dimension two (here $r_i(X) = \mathrm{rank}\, H^i(X, \mathbb{Z})$, $i = 1, 2$). For $X \in \mathcal{A}$ it is known that

(i) if $r(X) \leqslant 1$ then $q(X) = 0$;
(ii) if $r(X) = 2$ then $q(X)$ is 0 or 6;
(iii) if $r(X) = 3$ then $q(X)$ is 0, 6, 24 or 36 (all cases do occur).

4. SG_k contains a curve $\Gamma_k' = SG_k \cap \{z : z_1 = \cdots = z_{k-2} = 0\}$ isomorphic to $\Gamma_k = \{(x, y) \in \mathbb{C}^2 : x^k + y^{k-1} = 1\}$. It can be proved that if f_1, f_2 in $\mathrm{Hol}(SG_k, SG_n)$ agree on Γ_k' then $f_1 \equiv f_2$. Since $\mathrm{Hol}^*(SG_k, SG_n) = \emptyset$ provided $k > 4$ and $n < k$, the following assertion admittedly implies CONJECTURE 1.

CONJECTURE 2. *If* $n > k > 4$ *then* $\mathrm{Hol}^*(\Gamma_k, SG_n) = \emptyset$.

The curve Γ_k can be obtained from a non-singular projective curve of genus $(k-1)(k-2)/2$ by removing a single point. It seems plausible that $\mathrm{Hol}^*(\Gamma^{(g)}, SG_n) = \emptyset$ for all $n > 4$ and all curves $\Gamma^{(g)} \in \mathcal{A}$ of genus $g < (n-1)(n-2)/2$. In any case the following weaker conjecture is likely to be true (this is really the case if $m \leqslant 1$ or $n \leqslant 4$, E. A. Gorin).

CONJECTURE 3. *Let* $n \geqslant 3$, $m \geqslant 0$ *and* $\zeta_1, \ldots, \zeta_m \in \mathbb{C}$. *Then*

$$\mathrm{Hol}^*\big(\mathbb{C} \setminus \{\zeta_1, \ldots, \zeta_m\}, S(G_n)\big) = \emptyset.$$

5. Even the following weakened variant of CONJECTURE 1 would be useful for applications. Let $X^l \in \mathcal{A}$, $\dim X^l = l$; $k > 4$, $n \geqslant 3$.

CONJECTURE 4.

(a) Let $f_1 \in \mathrm{Hol}(G_k^0, X^l)$, $f_2 \in \mathrm{Hol}(X^l, G_n^0)$. If $l \leqslant k - 2$ then $f_2 \circ f_1$ is splittable.

(b) Let $f_1 \in \mathrm{Hol}(SG_k, X^l)$, $f_2 \in \mathrm{Hol}(X^l, SG_n)$. If $l \leqslant k - 3$ then $f_2 \circ f_1$ is a constant mapping.

It follows from results of [6], [7], [10], [11] that the assertions 4(a) and 4(b) hold if either $k > 4$, $n \leqslant k$ or $k > 4$, $l = 1$ (of course, 4(a) is true for $k > 4$ and $n \notin \Pi(k)$). Maybe even the following sharpenings of 4(a) and 4(b) hold, though they look less probable.

CONJECTURE 5.

(a) If $k > 4$, $l \leqslant k - 2$ and $f \in \mathrm{Hol}(G_k^0, X^l)$ then the induced homomorphism $f_* : \pi_1(G_k^0) \to \pi_1(X^l)$ is abelian.

(b) If $k > 4$, $l \leqslant k - 3$ and $f \in \mathrm{Hol}(SG_k, X^l)$ then the induced homomorphism $f_* : \pi_1(SG_k) \to \pi_1(X^l)$ is trivial.

If $k > 4$ and $l = 1$, f_* really has these properties. It can be proved also that if $k \geqslant 4$ and $l \leqslant k - 3$ then for any *rational* $f \in \mathrm{Hol}(SG_k, X^l)$ the kernel of f_* is non-trivial. CONJECTURE 5 looks a little more realistic in case when X^l is the complement to an algebraic hypersurface in $\mathbb{C}^l$ and f is holomorphic and rational.

6. We formulate here an assertion concerning algebraic functions. To prove this assertion it suffices to verify CONJECTURE 1 for *polynomial* mappings from G_k^0 to G_n^0. Let $\lambda_n = \lambda_n(z)$ be an algebraic function in z ($\in \mathbb{C}^n$) defined by the equation $\lambda^n + z_1 \lambda^{n-1} + \cdots + z_n = 0$ and let Σ_n be the discriminant set of this function, i.e. $\Sigma_n = \{z : d_n(z) = 0\}$.

CONJECTURE 6. *For $n > 4$ there exists no entire algebraic function $F = F(z)$ with the following properties:*

(1) *F is a composition of polynomials and entire algebraic functions in less than $n - 1$ variables;*

(2) *the discriminant set of F coincides with Σ_n;*

(3) *in some domain $U \subset \mathbb{C}^n$ the functions λ_n and F have at least one joint irreducible branch.*

Condition (2) means that F is forbidden to have "extra" branching points (compared with λ_n). It is known that CONJECTURE 6 becomes true if this condition is replaced by that of absence of "extra branches" (which is much stronger) [7, 13].

REFERENCES

1. Arnold V. I., *Certain topological invariants of algebraic functions*, Trudy Moskov. Mat. Obsc. **21** (1970), 27–46. (Russian)

2. Gorin E. A., Lin V. Ya., *Algebraic equations with continuous coefficients and some problems of the algebraic theory of braids*, Mat. Sb. **78 (120)** (1969), no. 4, 579–610 (Russian); English transl. in Math. USSR Sbornik **7** (1969), no. 4, 569–596.

3. Fuks D. B., *Cohomologies of the braid group* mod 2, Funkts. Anal. i Prilozh. **4** (1970), no. 2, 62–73 (Russian); English transl. in Funct. Anal. and Appl. **4** (1970), no. 2, 143–151.

4. Vainshtein F. V., *Cohomologies of braid groups*, Funkts. Anal. i Prilozh. **12** (1978), no. 2, 72–74 (Russian); English transl. in Funct. Anal. and Appl. **12** (1978), no. 2, 135–137.

5. Lin V. Ya., *Algebroid functions and holomorphic elements of homotopy groups of a complex manifold*, Doklady Akad. Nauk SSSR **201** (1971), no. 1, 28–31 (Russian); English transl. in Soviet Math. Dokl. **12** (1971), no. 6, 1608–1612.

Chapter 20. MISCELLANEOUS PROBLEMS

6. Lin V. Ya., *Algebraic functions with universal discriminant manifolds*, Funkts. Anal. i Prilozh. 6 (1972), no. 1, 81–82 (Russian); English transl. in Funct. Anal. and Appl. 6 (1972), no. 1, 73–75.

7. Lin V. Ya, *On superpositions of algebraic functions*, Funkts. Anal. i Prilozh. 6 (1972), no. 3, 77–78 (Russian); English transl. in Funct. Anal. and Appl. 6 (1972), no. 3, 240–241.

8. Kaliman Sh. I., *A holomorphic universal covering of the space of polynomials without multiple roots*, Funkts. Anal. i Prilozh. 9 (1975), no. 1, 71 (Russian); English transl. in Funct. Anal. and Appl. 9 (1975), no. 1, 67–68.

9. Kaliman Sh. I., *A holomorphic universal covering of the space of polynomials without multiple roots*, Teor. Funktsii Funkts. Anal. i Prilozh. Vyp. 28, Kharkov, 1977, pp. 25–35. (Russian)

10. Kaliman Sh. I., *Holomorphic endomorphisms of the variety of complex polynomials with determinant 1*, Uspekhi Mat. Nauk 31 (1976), no. 1, 251–252. (Russian)

11. Lin V. Ya., *Representations of the braid group by permutations*, Uspekhi Mat. Nauk 27 (1972), no. 3, 192. (Russian)

12. Lin V. Ya., *Representations of braids by permutations*, Uspekhi Mat. Nauk 29 (1974), no. 1, 173–174. (Russian)

13. Lin V. Ya., *Superpositions of algebraic functions*, Funkts. Anal. i Prilozh. 10 (1976), no. 1, 37–45 (Russian); English transl. in Funct. Anal. and Appl. 10 (1976), no. 1, 32–38.

14. Bandman T. N., *Holomorphic functions omitting two values on an affine surface*, Vestnik Moskov. Univ. Ser. I, Matem., Mekh. (1980), no. 4, 43–45. (Russian)

15. Lin V. Ya., *Artin braids and related groups and spaces*, Itogi Nauki i Tekhniki., Algebra. Topology. Geometry. Vol. 17, Akad. Nauk SSSR, VINITI, Moscow, pp. 159–227. (Russian)

DEPT. OF MATH., TECHNION,
32000 HAIFA,
ISRAEL

COMMENTARY BY THE AUTHOR

1) About CONJECTURE 2. Now it is known that the condition $n > k > 4$ in this conjecture is essential. Namely, the affine algebraic curve

$$\Gamma_3 = \{\, (x,y) \in \mathbb{C}^2 \, : \, x^3 + y^2 = 1 \,\}$$

has nonconstant polynomial mapping both into SG_6 and SG_9. Moreover, there exists a polynomial mapping $f \colon G_3 \to G_6$, $f(z) = \big(f_1(z), \ldots, f_6(z)\big)$, such that the polynomial with simple roots of degree 6

$$p_6(t, f(z)) = t^6 + f_1(z)t^5 + \cdots + f_6(z),$$

which corresponds to the f-image $f(z)$ of any point $z = (z_1, z_2, z_3) \in G_3$, is **coprime** with the polynomial

$$p_3(t, z) = t^3 + z_1 t^2 - z_2 t + z_3,$$

corresponding to the original point $z = (z_1, z_2, z_3)$.

2) Recently I proved (to appear) that CONJECTURE 6 is true. The proof is based on the following

THEOREM. *Let* $n > k > 3$. *Then there is no holomorphic mappings* $f \colon G_k \to G_n$, $f(z) = \big(f_1(z), \ldots, f_n(z)\big)$ *for each* $z \in G_k$, *having the property that for each point* $z \in G_k$ *the polynomials*

$$p_k(t, z) = t^k + z_1 t^{k-1} + \cdots + z_k$$

and

$$p_n(t, f(z)) = t^n + f_1(z)t^{n-1} + \cdots + f_n(z)$$

are coprime.

ON THE NUMBER OF SINGULAR POINTS
OF A PLANE AFFINE ALGEBRAIC CURVE

V. Ya. Lin,   M. G. Zaidenberg

Let $p(x, y)$ be an irreducible polynomial on $\mathbb{C}^2$. It has been proved in [1] that if the algebraic curve $\Gamma_0 = \{(x, y) \in \mathbb{C}^2 : p(x, y) = 0\}$ is simply connected then there exist a polynomial automorphism α of the space $\mathbb{C}^2$ and positive integers k, l with $(k, l) = 1$ such that $p(\alpha(x, y)) = x^k - y^l$. It follows from this theorem that an irreducible simply connected algebraic curve in $\mathbb{C}^2$ cannot have more than one singular point. (Note that such a curve in $\mathbb{C}^3$ may have as many singularities as you like).

In view of this result the following QUESTION arises:

Does there exist a connection between the topology of an irreducible plane affine algebraic curve and the number of its irreducible singularities? Is it true, for example, that the number of irreducible singularities of such a curve Γ does not exceed $2r + 1$, where $r = \operatorname{rank} H^1(\Gamma, \mathbb{Z})$?

The above assertion on the singularities of the irreducible simply connected curve may be reformulated as follows: let u and v be polynomials in one variable $z \in \mathbb{C}$, such that for any distinct points $z_1, z_2 \in \mathbb{C}$ either $u(z_1) \neq u(z_2)$ or $v(z_1) \neq v(z_2)$; then the system of equations $u'(z) = 0$, $v'(z) = 0$ has at most one solution. It would be very interesting to find a proof of this statement not depending on the above theorem about the normal form of a simply connected curve Γ_0. Maybe such a proof will shed some light onto the following question (which is a slightly weaker form of the question about the irreducible singularities of a plane affine algebraic curve). Let X be an open Riemann surface of finite type (g, n) (g is its genus and $n \geqslant 1$ is the number of punctures), and let u, v be regular functions on X (i.e. rational functions on X with poles at the punctures only). Suppose that the mapping $f \colon X \to \mathbb{C}^2$, $f(x) = (u(x), v(x))$, $x \in X$, is injective.

How many solutions (in X) may have the system of equations $\partial u = 0$, $\partial v = 0$? (Here $\partial = \frac{\partial}{\partial z} \cdot dz$, where z is a holomorphic local coordinate on X.)

REFERENCE

1. Lin V. Ya., Zaidenberg M. G., Doklady Akad. Nauk SSSR **271** (1983), no. 5 1048–1052. (Russian)

DEPT. OF MATH., TECHNION,
32000 HAIFA
ISRAEL

UFR DE MATHÉMATIQUES
UNIVERSITÉ JOSEPH FOURIER
B. P. 74, 38402
SAINT-MARTIN-D'HERES CEDEX
FRANCE

REARRANGEMENT-INVARIANT HULLS OF SETS

A. B. GULISASHVILI

Let (S, Σ, μ) be a non-atomic finite measure space. Denote by Ω_ε ($\varepsilon > 0$) a family consisting of all μ-preserving invertible transformations $\omega : S \to S$ such that $\mu\{x \in S : \omega(x) \neq x\} \leqslant \varepsilon$. Each $\omega \in \Omega_\varepsilon$ generates a linear operator $T_\omega : \mathcal{M} \to \mathcal{M}$ (where $\mathcal{M}$ denotes the space of all measurable functions on S) by the formula $T_\omega f(x) = f(\omega(x))$, $x \in S$, $f \in \mathcal{M}$. The elements of a set $R_\varepsilon \overset{\text{def}}{=} \{T_\omega : \omega \in \Omega_\varepsilon\}$ are called the ε-rearrangements. Each T_ω preserves the distribution of a function, hence the integrability properties of functions are also preserved.

Given a subset A of $\mathcal{M}$ define the *rearrangement-invariant hulls* of A as follows:

$$RH_0(A) \overset{\text{def}}{=} \bigcap_{\varepsilon > 0} R_\varepsilon(A), \qquad RH(A) = \bigcup_{\varepsilon > 0} R_\varepsilon(A).$$

The general problem of characterization of such hulls for a given concrete set A has been posed by O. Tsereteli. We refer to [5] for the contribution of O. Tsereteli to the solution in some concrete cases.

The following results have been obtained in [2] and [3]. Consider for the simplicity the case when S is $[0, 1]$ equipped with the usual Lebesgue measure.

a) Let $\Phi = \{\varphi_n\}$, $n \geqslant 1$ be a family of bounded functions such that for any $f \in L^\infty$

$$\sum |c_n(f; \Phi)|^2 \leqslant \text{const}\|f\|_2^2$$

with a constant independent of f. Here $c_n(f; \Phi) = \int f\varphi_n \, dx$, $n \geqslant 1$. For a non-negative sequence $\{\rho_n\}$, $n \geqslant 1$, such that $\rho_n \to 0$ when $n \to \infty$, define a class

$$A \overset{\text{def}}{=} \left\{ f \in L^1 : \sum [c_n(f; \Phi)^*]^2 \rho_n < \infty \right\},$$

where $c_n(f; \Phi)^*$ denotes the non-increasing rearrangement of $\{|c_n(f; \Phi)|\}$. Then $RH_0(A) = RH(A) = L^1$ [2].

b) Let $\Phi = \{\varphi_n\}$ be a complete orthonormal family of bounded functions in L^2. For given p, $1 \leqslant p < 2$, define a class

$$A \overset{\text{def}}{=} \left\{ f \in L^1 : \|f - (S_n f)\|_p \to 0 \right\},$$

where $S_n f$ denotes the n-th partial sum of the Fourier series of f with respect to Φ. Then $RH_0(A) = RH(A) = L^1$ [3], i.e. any complete orthonormal family of bounded functions is in some sense a basis in L^p, $1 \leqslant p < 2$, modulo rearrangements.

A different effect occurs for the class $A = \{ f \in L^1 : \tilde{f} \in L^1 \}$, where L^1 is taken over the unit circle $\mathbb{T}$ and $\tilde{f}$ denotes the conjugate function of a function f. In that case [4,5]

$$(1) \qquad RH_0(A) = RH(A) = \left\{ f \in L^1 : \int\limits_1^\infty \frac{dt}{t} \left| \int\limits_\mathbb{T} {}^t f(x)\, dx \right| < \infty \right\}.$$

Here

$$ {}^t f(x) = \begin{cases} f(x), & |f(x)| > t \\ 0, & |f(x)| \leqslant t \end{cases}. $$

The class

$$ M \ln^+ M \overset{\text{def}}{=} \left\{ f \in L^1 : \int\limits_1^\infty \frac{dt}{t} \left| \int\limits_\mathbb{T} {}^t f(x)\, dx \right| < \infty \right\} $$

arising in (1) (in [4,5] this class is denoted by Z) coincides, on non-negative functions, with the class $L \ln^+ L$. Moreover, $L \ln^+ L \subset M \ln^+ M$.

In addition to (1) it has been proved in [3] that if $A \overset{\text{def}}{=} \{ f \in L^1 : \|\tilde{f} - \tilde{S}_n(f)\|_1 \to 0 \}$ then

$$ RH_0(A) = RH(A) = M \ln^+ M, $$

i.e. any function from $M \ln^+ M$ can be rearranged on a set of small measure so that the obtained function has L^1-convergent conjugate trigonometrical series.

For any $p, p > 1$, define a class M^p over (S, Σ, μ) as follows

$$ M^p \overset{\text{def}}{=} \left\{ f \in L^1 : \int\limits_0^\infty t^{p-2}\, dt \left| \int\limits_S {}^t f(x)\, dx \right| < \infty \right\}. $$

It is clear that $L^p \subset M^p$ and M^p coincides with L^p on non-negative functions. The class M^p in comparison with L^p takes into account not only the degree of integrability of function but also the degree of cancellation of the positive and negative values of the function. It is known [6] that $M \ln^+ M$ is linear. As for M^p, it has been proved in [3] that 1) $M^p, p > 2$, is non-linear, moreover, there exists $f \in M^p$ such that $f + 1 \notin M^p$; 2) $M^p + L^\infty \subset M^p$, $1 < p \leqslant 2$.

PROBLEM 1. *Is the class M^p linear for $1 < p \leqslant 2$?*

Consider a family $\Phi = \{\varphi_n\}$ such as in b). Denote by S^* the maximal operator for the sequence of partial sum operators $\{S_n\}$ with respect to Φ, i.e. $S^* f = \sup_n |S_n f|$, $f \in L^1$. The problem of finding the rearrangement invariant hulls of a set $A = \{ f \in L^1 : S^* f \in L^1 \}$ is not solved even for classical families Φ. Some partial results have been obtained in [3]. For the trigonometrical system $\{e^{inx} : -\infty < n < +\infty\}$, on $\mathbb{T}$ the following inclusion holds:

$$ M^p \subset RH_0(A), \qquad p > 1. $$

PROBLEM 2. *Find $RH_0(A)$ and $RH(A)$ for the trigonometrical system.*

Very interesting is the case when Φ is the family of Legendre polynomials $\{L_n\}$ on the interval $[-1, +1]$. It is true [3] that

$$
(2) \qquad\qquad M^p \subset RH_0(A), \qquad p > \frac{8}{7}.
$$

PROBLEM 3. *Find $RH_0(A)$ and $RH(A)$ for the family of Legendre polynomials.*

We pose also two easier problems related to Problem 3.

PROBLEM 3'. *Is the inclusion $L^{8/7} \subset RH_0(A)$ true?*

PROBLEM 3''. *Is the inclusion $M^{8/7} \subset RH_0(A)$ true?*

The number $\frac{8}{7}$ in (2) appears from the general theorem proved in [3]. The theorem states that if a sequence of integral operators $\{T_n\}, n \geqslant 1$, has a localization property in L^∞, and the maximal operator $T^*f = \sup |T_n f|$ has a weak type (p, p) with some $p > 1$, then

$$
(3) \qquad\qquad M^{\frac{2p}{p+1}} \subset RH_0(A),
$$

where $A = \{ f \in L^1 : T^*f \in L^1 \}$. It is not known, whether the power $\frac{2p}{p+1}$ in (3) is sharp on the whole class of the operators T^* under consideration. The maximal operator with respect to the Legendre polynomial system has weak type (p, p), $\frac{4}{3} < p < 4$ [1]. The number $\frac{8}{7}$ is the value of $\frac{2p}{p+1}$ at $p = \frac{4}{3}$.

The problems analogous to Problems 3, 3', 3'' can also be formulated for Jacobi polynomial systems.

REFERENCES

1. Badkov B. M., *Convergence in the mean and almost everywhere of Fourier series in polynomials orthogonal on an interval*, Matem. Sborn. **95** (1974), no. 2, 229–262. (Russian)
2. Gulisashvili A. B., *Singularities of summable functions*, Zapiski nauchn. sem. LOMI **113** (1981), 76–96 (Russian); English transl. in J. Soviet Math. **22** (1983), no. 6, 1743–1757.
3. Gulisashvili A. B., *Rearrangements, arrangements of signs, and convergence of sequences of operators*, Zapiski nauchn. sem. LOMI **107** (1981), 46–70 (Russian); English transl. in J. Soviet Math. **36** (1987), no. 3, 326–341.
4. Tsereteli O. D., *Conjugate functions*, Matem. Zametki **22** (1977), no. 5, 771–783. (Russian)
5. Tsereteli O. D., *Conjugate functions*, Doctoral dissertation, Tbilisi, 1976. (Russian)
6. Tsereteli O. D., *A certain case of summability of conjugate functions*, Trudy Tbilisskogo Matem. Inst. AN Gruz.SSR **34** (1968), 156–159. (Russian)

DEPARTMENT OF MATHEMATICS
BOSTON UNIVERSITY
111 CUMMINGTON ST.
BOSTON, MA 02215 USA

SUBJECT INDEX

SUBJECT INDEX

local Fréchet space 12.1
local operator 14.10
local ring S.11.27
local theory of normed spaces 1.1
local Toeplitz operator 7.8
localization S.7.23, 11.4
locally analytic equivalence 4.9
locally compact group 2.3, 2.4, 2.23
Loewner theory 18.5
logarithmic capacity 2.14, 10.16, 12.0, 12.3,
    12.21, 12.23, 13.3, 14.3, 14.5, 18.1, 19.8
logarithmic coefficients 18.7, 18.13
logarithmic convexity S.14.19
logarithmic potential 13.5, 16.10
Lomonosov's theorem 5.2
long-range potential 6.10
Lorentz group 17.8
Lorentz scale 6.2
Lorentz space 1.5, 1.12, 1.13
lower density of zeros 16.5
lower indicator of entire function 16.0, 16.4, 16.5
lower triangular operator 7.13
Lubinsky-Mhaskar-Rahmanov-Saff's number 13.4
Lyapunov curve 8.10, 10.6
Lyapunov theorem 20.1

Macaev ideal (*see* Matsaev ideal)
Malliavin's functional 11.19
Mandelbrot set 12.9, 19.0, 19.1, 19.2, 19.6, 19.7,
    19.9
mapping universal covering 2.25
Marcinkiewicz theorem 14.14
marginal 3.6
Markov operator 3.5
Markov process 3.5
Martin's boundary 10.20
masa 5.3
mating of polynomials 19.1, 19.3
matrix function 4.3, 4.5, 7.20
Matsaev ideal 6.1, 6.2, 7.13, 9.1
Matsaev's conjecture 5.12, 5.13
maximal function 10.5, S.10.25
maximal ideal space 2.5, 2.9, 2.14, 2.16, 2.22,
    2.23, 2.24, 2.26, 2.27, 7.8, S.7.23, 11.16,
    12.22, 17.13
maximum principle 14.0
measure algebra 2.0, 2.21, 2.22
median 1.10
Meixner–Pollaczek polynomials 13.6
Meixner polynomials 13.6
Mellin convolution 13.6
Mellin transform 1.7
Mergelyan set 12.11
Mergelyan's theorem 12.19
meromorphic continuation 11.15

meromorphic function 4.5, 4.8, 10.20, 12.8, 14.1,
    S.16.19, 17.15
Miklyukov theorem 14.15
Milin conjecture 18.7
Milin's constant 18.7
Miller's conjecture 2.22
minimal family S.9.5, 15.3
minimal subharmonic function 16.9
Misiurewicz point 19.5
Möbius transformation 2.9, S.7.22, 10.15, 17.2,
    18 12, 19.4
model space 9.2, 15.3
module 2.1, 2.2, 17.13
modulus of an operator 3.3, 8 9
modulus of continuity S.6.11
modulus of quasitriangularity 8.8
moment problem 7.17, 14.0, 14.1, S.14.20, 15.12
Muckenhoupt condition 7 9, 10.7 10.14
Muckenhoupt weight 15.4
multiplication operator 5.3
multiplicative commutator 5.7
multiplier 2.22, 2.27, 5.12, 7.2, 10.9, 10.23, 15.1,
    15 10
multi-valued function 12.17
Müntz condition 14.11

Naimark theorem 3.3
Nehari problem 4.3
Nehari's theorem 3.3, 7.1, 7.2
von Neumann algebra 2.1, 2.2, 2.4
von Neumann operator 8.0, 8.7
von Neumann's inequality 5.12, 5.13, 7.2
Nevai–Ullman density 13.6
Nevanlinna counting function 18.3
Nevanlinna exceptional values S.16.19, S.16.20
Nevanlinna extremal measure 12.5
Nevanlinna theory S.14.19, 16.0, 16.12
Newtonian capacity 12.15
Newton's method 19.9, 19.11
Noether normalization 11.8
non-dissipative operator 9.2
non-negative invariant subspace 5.9
non-quasianalyticity 11.16
normal extension 8.10
normal family 2 12, 14.0, 14.4
normal matrices 5.1
normal operator 2.13, 3.4, 5.7, 5.15, S.5.18, 8.0,
    8.2, 8.7, 8.10, 8.11, 9.4
normal path inequality 5.1
nuclear algebra 2.1, 2.2
nuclear operator 2.13, 3.2, 4.4, 5.14, 6.4, S.6.11,
    S.6.12, 8.8, 8.11
nuclear space 5 4
numerical range 5.6, 7.11, 20.1, 20.4
Nullstellensatz theorem 11.8, 11.9

SUBJECT INDEX

SUBJECT INDEX

AUTHOR INDEX

AUTHOR INDEX

AUTHOR INDEX

AUTHOR INDEX

504

AUTHOR INDEX

AUTHOR INDEX

STANDARD NOTATION

Symbols $\mathbb{N}$, $\mathbb{Z}$, $\mathbb{R}$, $\mathbb{C}$ denote respectively the set of positive integers, the set of all integers, the real line, and the complex plane:

$$\mathbb{R}_+ = \{\, t \in \mathbb{R} : t \geq 0 \,\};$$
$$\mathbb{Z}_+ = \mathbb{Z} \cap \mathbb{R}_+;$$
$$\mathbb{T} = \{\, z \in \mathbb{C} : |z| = 1 \,\};$$
$$\mathbb{D} = \{\, z \in \mathbb{C} : |z| < 1 \,\};$$

$\widehat{\mathbb{C}}$ stands for the one-point compactification of $\mathbb{C}$

m denotes the normed Lebesgue measure on $\mathbb{T}$ ($m(\mathbb{T}) = 1$);

$f|X$ is the restriction of a mapping (function) f to X;

$\operatorname{clos}(\cdot)$ is the closure of the set $(\cdot)$;

$\vee(\cdot)$ is the closed span of the set $(\cdot)$ in a linear topological space;

$|T|$ denotes the norm of the operator T;

$\hat{f}(\cdot)$ denotes the sequence of Fourier coefficients of f;

$\mathcal{F}f$ denotes the Fourier transform of f;

$\mathfrak{S}_p$ is a class of operators A on a Hilbert space satisfying trace $(A^*A)^{p/2} < +\infty$;

H^p is a Hardy class in $\mathbb{D}$, i.e., the space of all holomorphic functions on $\mathbb{D}$ with

$$\|f\|_p \overset{\text{def}}{=} \sup_{0<r<1} \left(\int_{\mathbb{T}} |f(r\zeta)|^p \, dm(\zeta) \right)^{1/p} < +\infty, \qquad p > 0.$$